# THE BEST TEST PREPARATION FOR THE
# GRE
## GRADUATE RECORD EXAMINATION
# ENGINEERING

**Staff of Research & Education Association**
**Dr. M. Fogiel, Director**

Research and Education Association
61 Ethel Road West
Piscataway, New Jersey 08854

The Best Test Preparation for the
GRADUATE RECORD EXAMINATION
(GRE) IN ENGINEERING

1997 PRINTING

Copyright © 1996, 1988 by Research & Education Association. All rights reserved. No part of this book may be reproduced in any form without permission of the publisher.

Printed in the United States of America

Library of Congress Catalog Card Number 95-68125

International Standard Book Number 0-87891-601-6

Research & Education Association
61 Ethel Road West
Piscataway, New Jersey 08854

REA supports the effort to conserve and protect environmental resources by printing on recycled papers.

# CONTENTS

**STUDY SCHEDULE** ............................................................ xii

**YOU CAN SUCCEED ON THE GRE ENGINEERING** ............................. A-1
    About the Book .................................................... A-3
    About the Test .................................................... A-3
    How to Use the Book ............................................... A-5
    Format of the GRE Engineering ..................................... A-5
    About the Review .................................................. A-6
    About the Practice Tests .......................................... A-6
    Scoring the GRE Engineering Exam .................................. A-7
    Studying for the GRE Engineering .................................. A-9
    The Day of the Test ............................................... A-9

**GRE ENGINEERING REVIEW** .................................................. A-11

*Chapter 1*
**Systems Modeling** ....................................................... A-13
    Network Flow Analysis ............................................. A-15
    PERT/CPM .......................................................... A-15
    Markovian Decision Models ......................................... A-17
    Decision Analysis ................................................. A-18
    Game Theory ....................................................... A-18
    Linear Programming ................................................ A-19
    Non-Linear Programming ............................................ A-19
    Queuing ........................................................... A-20
    Simulation ........................................................ A-21
    Criteria Function ................................................. A-22

*Chapter 2*
**Engineering Economics** .................................................. A-23
    Cost .............................................................. A-23
    Cost Estimation ................................................... A-25
    Life-Cycle Costing ................................................ A-26
    Nominal and Effective Interest Rates .............................. A-28

    Time Value of Money .................................................................. A-28
    Cash Flow ..................................................................................... A-30
    Present Worth Analysis ............................................................... A-30
    Interest Rate-of-Return Analysis ................................................ A-30
    Depreciation ................................................................................ A-31
    Equivalent Uniform Annual Cost Evaluation ......................... A-31
    Capitalized Cost .......................................................................... A-31
    Replacement Analysis ................................................................ A-32
    Selecting an Appropriate Rate of Return ................................. A-32

## *Chapter 3*
## Elementary Computer Operation and Applications ........................ A-33
    Scalar Variables .......................................................................... A-33
    Data Types ................................................................................... A-34
    Binary Notation .......................................................................... A-35
    Encoding Data ............................................................................ A-36
    Iteration vs. Recursion ............................................................... A-39
    Programming .............................................................................. A-40
    Translation .................................................................................. A-40
    Language Description ............................................................... A-40

## *Chapter 4*
## Chemistry .............................................................................................. A-43
    Matter and Its Properties .......................................................... A-43
    Stoichiometry ............................................................................. A-43
    Balancing Equations .................................................................. A-44
    Molecular Weight and Formula Weight ................................. A-44
    Calculations Based on Chemical Equations ........................... A-44
    Theoretical Yield and Percentage Yield .................................. A-45
    Percentage Composition ........................................................... A-45
    Atomic Spectra ........................................................................... A-45
    Quantization of Energy ............................................................. A-45
    Pauli Exclusion Principle and Hund's Rule ............................ A-46
    Isotopes ....................................................................................... A-46
    Periodic Table ............................................................................. A-46
    Properties Related to the Periodic Table ................................. A-47
    Types of Bonds ........................................................................... A-48
    Intermolecular Forces of Attraction ........................................ A-48
    Gas Laws ..................................................................................... A-49
    The Kinetic Molecular Theory .................................................. A-50
    Equilibrium of Liquids, Solids, and Vapors ............................ A-51
    Equilibrium Constants .............................................................. A-51
    Types of Chemical Reactions .................................................... A-51
    Measurements of Reaction Rates ............................................. A-52

    Factors Affecting Reaction Rates ................................................. A-52
    The Arrhenius Equation: Relating Temperature
    and Reaction Rate ........................................................................... A-53

## *Chapter 5*
## Waves and Oscillations ................................................................. A-55
    Nomenclature and Definitions ...................................................... A-55
    Simple Harmonic Oscillators ......................................................... A-55
    Simple Harmonic Motion ............................................................... A-56
    Simple Pendulum ........................................................................... A-57
    Damped Harmonic Motion ............................................................ A-58
    Free Vibration ................................................................................. A-58
    Resonance ....................................................................................... A-60
    Waves ............................................................................................... A-60
    Speed of Sound .............................................................................. A-61
    Intensity of Sound ......................................................................... A-61
    Longitudinal and Transverse Waves ............................................ A-62
    Speed of Light ................................................................................ A-62
    The Poynting Vector ...................................................................... A-62
    Power and Intensity for Electromagnetic Waves ....................... A-63
    Speed and Propagation of Electromagnetic Waves .................. A-63
    Doppler Effect ................................................................................ A-64
    Interference .................................................................................... A-64
    Interference Maxima and Minima ............................................... A-64

## *Chapter 6*
## Electrical Principles and Circuits ............................................... A-67
    Current, Voltage, Power, and Energy ........................................... A-67
    Resistance ....................................................................................... A-69
    Voltage and Current Division ....................................................... A-70
    Ohm's Law ...................................................................................... A-71
    Kirchhoff's Law .............................................................................. A-71
    Capacitance .................................................................................... A-72
    Thevenin's and Norton's Theorems ............................................. A-74
    Alternating Current Circuits ......................................................... A-75
    Magnetic Force on a Differential Current Element ................... A-77

## *Chapter 7*
## Mechanics ........................................................................................ A-79
    Newton's Laws ............................................................................... A-79
    Equilibrium of a Particle ............................................................... A-80
    Moment of a Force About a Point ................................................ A-80
    Equilibrium of Rigid Bodies ......................................................... A-81
    Dry Friction .................................................................................... A-81

Moment of Inertia ........................................................................ A-82
Work of a Force ........................................................................... A-82
Mechanical Efficiency ................................................................. A-83
Motion of a Particle .................................................................... A-83
Uniform Motion .......................................................................... A-83
Relative Motion .......................................................................... A-84
Equations of Motion ................................................................... A-84
Dynamic Equilibrium ................................................................. A-84
Velocity-Dependent Force .......................................................... A-85
Power ........................................................................................... A-86
Work-Energy Principle ............................................................... A-86
Momentum .................................................................................. A-87
Impulse and Momentum ............................................................ A-88

## Chapter 8
**Thermodynamic Systems** ............................................................ A-91
Thermodynamic Systems ........................................................... A-91
Properties of Systems ................................................................. A-91
Processes ..................................................................................... A-92
Thermodynamic Equilibrium ..................................................... A-92
Zeroth Law of Thermodynamics ................................................ A-93
Pressure-Volume Work .............................................................. A-93
Heat ............................................................................................. A-94
Energy of a System ..................................................................... A-94
The First Law of Thermodynamics ........................................... A-95
The First Law for a Change in the State of a System ................ A-95
The Constant-Volume and Constant-Pressure
Specific Heats ............................................................................. A-97
The Heat Engine ......................................................................... A-97
Heat Engine Efficiency ............................................................... A-97
The Second Law of Thermodynamics ...................................... A-98
The Carnot Cycle ........................................................................ A-98
Thermodynamic Temperature Scales ....................................... A-99
Kelvin Temperature Scale .......................................................... A-100
Entropy ....................................................................................... A-100
Perfect Gas (Ideal Gas) ............................................................... A-101

## Chapter 9
**Fluid Mechanics** ............................................................................ A-103
Definition of a Fluid ................................................................... A-103
Basic Laws in Fluid Mechanics .................................................. A-103
Pathlines, Streaklines, Streamlines, and Streamtubes ............. A-103
Newtonian Fluid Viscosity ......................................................... A-104
The Basic Equation of Fluid Statics .......................................... A-105

Fluid Statics .................................................................. A-107
Buoyancy and Stability ................................................ A-109
Forms of Fluid Flow ..................................................... A-110
Laminar and Turbulent Flows ..................................... A-111
Fully-Developed Laminar Flow in a Pipe .................. A-111
Newton's Law of Viscosity .......................................... A-112
Boundary Layer Thickness .......................................... A-113
The Continuity Equation ............................................. A-114
Bernoulli Equation ....................................................... A-114
Flow in Open Channels ............................................... A-115

## Chapter 10
**Heat and Mass Transfer** .................................................. A-117
Basic Mechanism .......................................................... A-117
Fourier's Law (Conduction) ........................................ A-118
Newton's Law of Cooling (Convection) ..................... A-119
Thermal Radiation ........................................................ A-119
Fourier's Law of Heat Conduction ............................. A-120
Thermal Conductivity .................................................. A-121
Shell Energy Balance ................................................... A-122
Steady State Heat Conduction in One Dimension ............... A-122
Thermal Conductance, Thermal Resistance, and
Convective Thermal Resistance .................................. A-123
The Plane Wall with Specified Boundary Temperature ......... A-123
Mass Transfer ................................................................ A-124
Definition of Velocities, Concentrations, and
Mass Fluxes ................................................................... A-126
Convective Mass Transfer ........................................... A-127
Diffusion into a Falling Liquid Film .......................... A-128

## Chapter 11
**Properties of Engineering Materials** ............................. A-131
Crystal Lattice Structure ............................................. A-132
Phase Diagrams ............................................................ A-135
Heat Treatment of Carbon Steels ............................... A-136
Stress .............................................................................. A-136
Property of Ceramic Materials ................................... A-140
Composites .................................................................... A-141
Semi-Conductors .......................................................... A-141

## Chapter 12
**Computational Methods** .................................................. A-143
Properties of Real Functions ...................................... A-143
Bisection Method ......................................................... A-145

Fixed Point Iterative Method ......................................................... A-146
Newton Raphson Method ............................................................ A-146
Fixed Tangent Method .................................................................. A-147
Secant Method ............................................................................... A-148
Regula Falsi ................................................................................... A-149
Newton's and Stirling's Formulas ............................................. A-150
General Applications of Numerical Integration .................... A-151
Trapezoidal, Simpson's Rule ...................................................... A-151
Numerical Quadrature ................................................................. A-152

*Chapter 13*
**Calculus** .............................................................................................. A-153
Exponential and Logarithmic Functions ................................. A-153
Limits ............................................................................................... A-154
Functions ........................................................................................ A-154
Lines and Slopes ........................................................................... A-154
Parametric Equations ................................................................... A-154
Transcendental Functions ........................................................... A-155
The Derivative and the $\Delta$-method ............................................ A-155
Trigonometric Differentiation .................................................... A-157
Exponential and Logarithmic Differentiation ........................ A-157
Maxima and Minima .................................................................... A-158
Velocity and Rate of Change ...................................................... A-158
Definition of Definite Integral .................................................. A-159
Properties of the Definite Integral ........................................... A-160
The Fundamental Theorem of Calculus ................................... A-161
Indefinite Integral ........................................................................ A-162
Area Under A Curve ..................................................................... A-163
Equations of a Line and Plane in Space .................................. A-163

*Chapter 14*
**Numerical Analysis and Differential Equations** ........................... A-165
Infinite Sequence .......................................................................... A-165
Infinite and Geometric Series ................................................... A-166
Tests for Convergence ................................................................. A-167
Alternating Series: Absolute and Conditional
Convergence ................................................................................. A-168
Power Series .................................................................................. A-169
Series Solutions Near an Ordinary Point ................................ A-170
First Order Equations .................................................................. A-171
Second Order Differential Equations ...................................... A-173
Fourier Series Expansion ............................................................ A-175
Parabolic Partial Differential Equation ................................... A-176
Monte Carlo Methods .................................................................. A-176

## Chapter 15
### Linear Algebra .................................................................. A-177
- Types of Matrices .................................................... A-177
- Linear Equations and Matrices ............................... A-178
- Homogenous Systems of Linear Equations ........... A-179
- Matrices ..................................................................... A-180
- Rules of Matrix Arithmetic ....................................... A-182
- Gaussian Elimination ............................................... A-183
- Elementary Matrices ................................................ A-184
- Determinants ............................................................ A-185
- Determinants by Row Reduction ............................ A-185
- Determinant Properties ........................................... A-186
- Cofactor Expansion; Cramer's Rule ........................ A-186
- Linear Independence ............................................... A-188
- Basis and Dimension ............................................... A-188
- Eigenvalues, Eigenvectors ....................................... A-189
- Diagonalization ......................................................... A-189

## Chapter 16
### Probability and Statistics ............................................... A-191
- Sample Spaces ......................................................... A-191
- Events ........................................................................ A-192
- Functions ................................................................... A-192
- Probability Space ...................................................... A-193
- Sampling and Counting ........................................... A-193
- The Fundamental Principle of Counting ................ A-194
- Factorial Notation ..................................................... A-195
- Definition of a Random Variable ............................. A-196
- Probability Distribution Function ............................ A-197
- Counting Procedures Not Involving Order Restrictions (Combinations) ...................................... A-198
- Random Sampling .................................................... A-198
- Conditional Probability ............................................ A-198
- Properties of the Distribution Function ................. A-199
- Probability Density Functions ................................. A-200
- Standard Deviation .................................................. A-200
- Variance ..................................................................... A-200
- Moments .................................................................... A-201
- Coefficients of Skewness-Kurtosis ......................... A-201
- The Normal Distribution .......................................... A-203
- The Poisson Distribution ......................................... A-203

**GRE TEST I** ........................................................................................... 1
   Answer Key .................................................................................. 55
   Detailed Explanations of Answers ................................................ 57

**GRE TEST II** ....................................................................................... 125
   Answer Key ................................................................................ 179
   Detailed Explanations of Answers .............................................. 181

**GRE TEST III** ..................................................................................... 245
   Answer Key ................................................................................ 303
   Detailed Explanations of Answers .............................................. 305

**GRE TEST IV** ..................................................................................... 371
   Answer Key ................................................................................ 429
   Detailed Explanations of Answers .............................................. 431

**GRE TEST V** ...................................................................................... 495
   Answer Key ................................................................................ 551
   Detailed Explanations of Answers .............................................. 553

**APPENDIX** ........................................................................................ 615

# About Research and Education Association

Research and Education Association (REA) is an organization of educators, scientists, and engineers specializing in various academic fields. Founded in 1959 with the purpose of disseminating the most recently developed scientific information to groups in industry, government, and universities, REA has since become a successful and highly respected publisher of study aids, test preps, handbooks, and reference works.

REA's Test Preparation series includes study guides for all academic levels in almost all disciplines. Research and Education Association publishes test preps for students who have not yet completed high school, as well as high school students preparing to enter college. Students from countries around the world seeking to attend college in the United States will find the assistance they need in REA's publications. For college students seeking advanced degrees, REA publishes test preps for many major graduate school admission examinations in a wide variety of disciplines, including engineering, law, and medical schools. Students at every level, in every field, with every ambition can find what they are looking for among REA's publications.

Unlike most Test Preparation books that present only a few practice tests which bear little resemblance to the actual exams, REA's series presents tests which accurately depict the official exams in both degree of difficulty and types of questions. REA's practice tests are always based upon the most recently administered exams, and include every type of question that can be expected on the actual exams.

REA's publications and educational materials are highly regarded and continually receive an unprecedented amount of praise from professionals, instructors, librarians, parents, and students. Our authors are as diverse as the subjects and fields represented in the books we publish. They are well-known in their respective fields and serve on the faculties of prestigious universities throughout the United States.

# Acknowledgments

In addition to our authors, we would like to thank the following:

Dr. Max Fogiel, President, for his overall guidance which has brought this publication to its completion

Stacey A. Daly, Managing Editor, for directing the editorial staff throughout each phase of the project

Dr. Jerry Samples for authoring the review material of this book

# The Graduate Record Examination in ENGINEERING

## Study Schedule

# GRE ENGINEERING STUDY SCHEDULE

The following is a suggested five-week study schedule for the Graduate Records Examination in Engineering. In order for the schedule to benefit you the most, it is necessary that you follow the activities carefully. You may want to condense or expand this schedule depending on how soon you will be taking the actual GRE Engineering. Set aside time each week, and work straight through each activity without rushing. By following a structured schedule, you will be sure to complete an adequate amount of studying, and you will be confident and prepared on the day of the actual exam.

| Week | Activity |
|---|---|
| Week 1 | Acquaint yourself with the GRE Engineering by reading the chapter entitled "You Can Succeed on the GRE." Take Test 1 as a diagnostic test in order to determine your strengths and weaknesses. Make sure to take the test under simulated exam conditions. Carefully check your answers against the explanations, even for the questions you answered correctly. Be sure you worked through the problems properly. Start compiling a list of the topics that could use more work. Score your exam using the Scoring Worksheet provided on page A-7. |
| Week 2 | Begin reading the reviews. Make it your goal to finish five reviews by the end of the week, or one each day. Take notes on the reviews as you work through them, and try to solve the example problems without looking at the solutions. Before you begin a new chapter each day, reread your notes from the day or days before.<br><br>On the last two days of the week, take Test 2 under simulated exam conditions. Try not to allow yourself to be disturbed. Check over your answers and make a note of any sections that need more work. Score your exam using the scoring worksheet. |

| Week | Activity |
|---|---|
| Week 3 | Continue studying the review. Try to finish another five review chapters. Continue reviewing last week's notes as you study; you may even want to create flash cards for the more difficult concepts. Work through the example problems slowly and meticulously. Redo some practice problems from the week before.<br><br>Take Test 3 under simulated exam conditions, and score it with the scoring worksheet. As you check over your answers, continue compiling a list of the topics that need more work. |
| Week 4 | Finish the last six review chapters. Continue to study your notes. Make diagrams and sketches, and create more flash cards to make your study time more interactive.<br><br>Take Test 4 under timed conditions, and score it with the scoring worksheet. Check your answers carefully, and add any topics to your list that still need work. |
| Week 5 | Your goal this week should be to tighten up the areas that give you trouble. Look at the list of weaknesses which you have been compiling, and divide your list into equal portions. Study one section of the list each day. Work through the example problems again, and compare your answers to the detailed solutions. You may even want to redo the practice test questions that you answered incorrectly. |
| Week 6 | Take Test 5 under timed conditions. Do not allow yourself to be disturbed, and try to focus entirely on the exam. Score your test and compare your progress between the exams. Note which areas improved and which could use more work. Take the remainder of the week to study the areas that require more attention. |

Good luck!

# The Graduate Record Examination in
# ENGINEERING

## You Can Succeed on the GRE Engineering

# YOU CAN SUCCEED ON THE GRE ENGINEERING

## About the Book

This book will provide you with an accurate and complete representation of the Graduate Records Examination (GRE) in Engineering. Inside you will find reviews which are designed to provide you with the information you need to help you to excel on the exam, and five practice tests based on the actual GRE Engineering. You are allowed 2 hours and 50 minutes to complete the actual exam. The same amount of time is given to take our practice tests. The practice tests contain every type of question you would expect to appear on the GRE Engineering. Following each test, you will find an answer key with detailed explanations designed to help you more completely understand the test material.

## About the Test

### Who takes the test and what is it used for?

The Graduate Records Examination in Engineering is designed to allow students and graduate schools to make an accurate assessment of students' knowledge and expertise in the field of engineering. With these scores, students and graduate schools will have an understanding of the student's ability to contribute to and learn from a graduate program in Engineering.

Subtest scores provide a further breakdown of the composite score, and serve as an indicator of abilities in specific areas. This information may be used to design curricula or focus a student's education in the various fields of engineering.

### Who administers the test?

The GRE Engineering is developed and administered by the College Board, a division of the Educational Testing Service (ETS). Committees of professors from graduate and undergraduate institutions write and review the test regularly in conjunction with test specialists from ETS.

## When should the GRE Engineering be taken?

The GRE Engineering is taken as part of the admission process for graduate school. Most students take the test many months before their application is due. Seeing your score reports *before* you apply will allow you to determine which schools will be most likely to accept you. You may even find that you will have to take the test again in order to get the scores you need for the graduate institution of your choice. Taking the test early will allow you to do so.

Regardless of the timing, be sure to take the test early enough to allow your scores to arrive at the prospective schools before the application deadline. Testing dates are usually scheduled in October, December, February, and April with other special administrations. Actual testing dates as well as the frequency of specialty administrations vary from year to year. Contact ETS at (609) 921-9000 to determine which dates the GRE Engineering is given in your state. Be sure to ask for specialty administration dates if the official dates are unsuitable.

## When and where is the test given?

Test locations and dates are listed in the *Information and Registration Bulletin*, which is provided upon request by ETS. For more information, or to get a bulletin, see your college advisor, or contact ETS at the following address:

> Graduate Records Examinations
> Educational Testing Service
> PO Box 6000
> Princeton, NJ 08541-6000
> (609) 921-9000

Students with special needs can consult the bulletin in order to arrange convenient testing accommodations.

## Is there a registration fee?

To take the GRE Engineering, you will be asked to pay a registration fee. Financial assistance may be granted in some situations. To find out the registration fee or to determine if you are eligible for a fee waiver, contact ETS at the address and phone number listed above.

# How to Use the Book

### What do I study first?

Before you do anything else, take one of the practice tests in this book to help you to determine which areas may cause you the most difficulty. Carefully reviewing the detailed explanations of answers will help you to determine what you are doing wrong. After you have taken a practice test, you can begin studying the reviews that cover your problem areas. The reviews will help you to refresh your memory of the basic topics covered on each subsection of the exam.

Once you have studied the reviews that cover your problem areas, continue to study the remaining reviews; they will allow you to brush up on your skills. As you take the remaining practice tests, try to simulate actual testing conditions as closely as possible. These sample tests will help you to gauge your possible performance on the actual exam. As you check your answers, note which areas still need more work. Review those areas again before taking the next practice test.

To get the most out of your studying time, we recommend that you follow the Study Schedule appearing before this introduction. It suggests a plan to help budget your study time to your best advantage.

### When should I start studying?

It is never too early to start studying for the GRE Engineering. The earlier you begin, the more time you have to sharpen your skills. Do not procrastinate! Last-minute cramming is not helpful for this test. By studying a little bit each day, you can familiarize yourself with the test format and time limits, which will allow you to be confident and calm so that you can score high on the GRE Engineering.

# Format of the GRE Engineering

### What types of questions can I expect?

The questions for the test are composed by a committee of specialists who are recommended by the American Society for Engineering Education, and selected from various graduate and undergraduate facilities. The tests consists of approximately 140 multiple-choice questions. Some questions are grouped together based on graphs, tables of data, or descriptive paragraphs. Emphasis is placed on material common to the various subdivisions of engineering, usually studied in the first three years of college. The test questions are usually distributed as follows:

I. Science and Engineering Subjects (approximately 100 questions)

| | |
|---|---|
| Mechanics | Dynamics |
| Statics | Thermodynamics |
| Kinematics | Heat |
| Fluid Mechanics and Hydraulics | Mass |
| Transfer and Rate Mechanisms | Electricity |
| Momentum | Chemistry |
| Nature and Properties of Matter, including Particulate | Light and Sound |
| Computer Fundamentals | Engineering Economy |
| Properties of Engineering Materials | Engineering Judgment |

II. Mathematics (approximately 40 questions)

| | |
|---|---|
| Differential Equations | Numerical Analysis |
| Linear Algebra | Probability and Statistics |

The Mathematics section includes factual recall and intuitive calculus questions. Recall questions involve mathematical facts that should be common to all candidates. Intuitive questions require the student to choose the approach that would be most suitable to the problem.

## About the Review

The reviews are designed to help you to prepare for the GRE Engineering test by providing a review of the key concepts that are covered in most undergraduate engineering curricula. Each chapter discusses a subject tested on the GRE Engineering by outlining basic ideas, formulae, and providing diagrams and charts.

## About the Practice Tests

Each practice test represents a complete GRE Engineering test. The practice tests are designed to be taken in the same time period as the actual exam and have the same amount of questions. The questions are modeled after actual GRE Engineering questions, and appear in the same format and at the same level. The design and content of the practice exams will help you to prepare for the exam by familiarizing you with the format and presenting the content under timed exam conditions.

# Scoring the GRE Engineering Exam

### How do I score my Practice Tests?

Use this scoring worksheet to track your score improvement from Practice Test 1 to Practice Test 5. The following procedure describes how to score the exam. Be aware that this worksheet is *not* intended to predict your score on the actual GRE Engineering; it merely serves to measure your progress on the practice tests, and give you an idea of how you may perform on the actual exam.

When receiving your actual scores from ETS, you will also see two subscores for Math and Engineering questions. These subscores are based on a breakdown of the multiple-choice questions under the two headings. You will receive more information regarding the subtest scores when you receive your GRE Score Report.

## Scoring Worksheet

| | Correct Responses | Incorrect Responses (do not count blanks) | | Raw Score | Scaled Score (see conversion table) |
|---|---|---|---|---|---|
| Test 1 | _____ | − ( _____ | ÷ 4) = | _____ | _____ |
| Test 2 | _____ | − ( _____ | ÷ 4) = | _____ | _____ |
| Test 3 | _____ | − ( _____ | ÷ 4) = | _____ | _____ |
| Test 4 | _____ | − ( _____ | ÷ 4) = | _____ | _____ |
| Test 5 | _____ | − ( _____ | ÷ 4) = | _____ | _____ |

Now compare your raw scores to the following chart to find your approximate scaled score. Remember; the scaled score will just give you an idea of your progress on the practice tests. It is not meant to predict your actual scores.

## What do my scores mean?

The actual GRE Engineering is scored in much the same way as you scored your practice tests. The scale that converts raw scores to scaled scores will vary depending on the difficulty of the test, the ability of the test-takers, and the similarity of this test to previous GRE Engineering tests.

Since there is no established "passing" or "failing" grade, you can only evaluate your own scores in terms of the requirements of your prospective schools. Contact the admissions departments of the schools that interest you in order to determine the necessary scores for that program.

| Total Scores | | | |
|---|---|---|---|
| Raw Score | Scaled Score | Raw Score | Scaled Score |
| | 990 | 86-88 | 690 |
| | 980 | 84-85 | 680 |
| | 970 | 82-83 | 670 |
| | 960 | 79-81 | 660 |
| | 950 | 77-78 | 650 |
| | 940 | 74-76 | 640 |
| | 930 | 72-73 | 630 |
| 143-144 | 920 | 69-71 | 620 |
| 141-142 | 910 | 67-68 | 610 |
| 138-140 | 900 | 64-66 | 600 |
| 136-137 | 890 | 62-63 | 590 |
| 133-135 | 880 | 59-61 | 580 |
| 131-132 | 870 | 57-58 | 570 |
| 129-130 | 860 | 54-56 | 560 |
| 126-128 | 850 | 52-53 | 550 |
| 124-125 | 840 | 49-51 | 540 |
| 121-123 | 830 | 47-48 | 530 |
| 119-120 | 820 | 44-46 | 520 |
| 116-118 | 810 | 42-43 | 510 |
| 114-115 | 800 | 39-41 | 500 |
| 111-113 | 790 | 37-38 | 490 |
| 109-110 | 780 | 34-36 | 480 |
| 106-108 | 770 | 32-33 | 470 |
| 104-105 | 760 | 30-31 | 460 |
| 101-103 | 750 | 27-29 | 450 |
| 99-100 | 740 | 25-26 | 440 |
| 96-98 | 730 | 22-24 | 430 |
| 94-95 | 720 | 20-21 | 420 |
| 91-93 | 710 | 17-19 | 410 |
| 89-90 | 700 | 15-16 | 400 |

| Total Scores | | | |
|---|---|---|---|
| Raw Score | Scaled Score | Raw Score | Scaled Score |
| 12-14 | 390 | | 290 |
| 10-11 | 380 | | 280 |
| 7-9 | 370 | | 270 |
| 5-6 | 360 | | 260 |
| 2-4 | 350 | | 250 |
| 0-1 | 340 | | 240 |
| | 330 | | 230 |
| | 320 | | 220 |
| | 310 | | 210 |
| | 300 | | |

## Studying for the GRE Engineering

It is very important to chose the time and place for studying that works best for you. Only you can determine when and where your study time will be most effective, but be consistent and use your time wisely. Work out a study routine and stick to it!

## The Day of the Test

### Before the Test

On the day of the test, try to be as well-rested and comfortable as possible. Get a good night's sleep, and wear comfortable clothes so you are not distracted by the temperature. Wearing layers will allow you to adjust to the temperature.

Make sure to wake up early enough to allow yourself plenty of time to get to the test center. This will not only eliminate the anxiety associated with being late, it will also provide you with enough time to eat a good breakfast.

Try to get to the test center a little early. Any stress added by rushing or getting lost will probably not help your score. Be sure to bring your examination admission ticket and two forms of I.D. Each form of I.D. should have your signature on it, and at least one should have your photo (i.e., drivers license, student I.D., etc.). You should also bring two passport-sized shots of yourself for your "photo-record." Understand that you will not be admitted into the test center without proper I.D. You should also bring several sharpened #2 pencils.

You will not be permitted to bring calculators, calculator watches, books, papers, slide rules, beepers, compasses, rulers, or dictionaries into the test.

## During the Test

When you arrive at the test center, try to find a seat where you will be comfortable. If you are left-handed, you may make arrangements in advance for a left-handed desk. A copy of the supplied reference book will be given to you before the test, and must be returned with your test materials at the end of the testing session.

Mark your circles in the appropriate spaces on the answer sheet. Fill in the oval that corresponds to your answer choice as neatly as possible, and make sure it is dark. You can change your answer, but remember to completely erase your old answer. Only one answer choice should be marked. This is very important, as your answer sheet is scored by machine, and stray lines may cause the machine to score your answers incorrectly.

## After the Test

When you have finished the test, you will be asked to turn in your materials and then you will be dismissed from the testing center. Then, go home and relax! Your score report will arrive in approximately five weeks.

# The Graduate Record Examination in ENGINEERING

## GRE Engineering Review

# CHAPTER 1

# SYSTEMS MODELING

Systems is the management of technology in the development of complex systems. These complex systems are broken down, or decomposed, into manageable subsystems for design purposes and then reassembled to build the complete system from the subsystem parts. Careful consideration of the life cycle (concept to retirement) of the design through activities such as formulation, analysis, and interpretation assists in the understanding of the systems behavior.

Trial and error solutions are suitable for simple designs which are user based. More complex designs spawn the realization that a design must be more efficient and effective when human interaction is involved. Thus a major objective in the development of systems engineering is to improve the effectiveness (availability, reliability, maintainability, quality, and trustworthiness) while reducing costs. An established framework for the application of **systems engineering** includes the following steps:

1. Requirements and specifications identification
2. Preliminary conceptual design
3. Logical design and systems specifications
4. Detailed design, production, and testing
5. Operational implementation
6. Evaluation and modification
7. Operational deployment including retirement

The life cycle can be divided into a **three-phase, four-phase,** or an **n-phase** model depending on the desired complexity. The life cycle divided into a three-phase model might include:

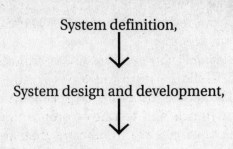

The steps on the previous page (1 thru 7) are put into the phases as appropriate. Morphological diagrams are produced to discover possible alternatives, through study of structure or form using a definite behavioral approach and a specific methodology. Thus any problem is divided into a structure which outlines a plausible solution, one step at a time.

An example morphological diagram is shown below:

| | |
|---|---|
| **System definition** | Requirements and specification identification<br>Preliminary conceptual design |
| **System design and development** | Logical design and system specification<br>Detailed design, production, and testing |
| **System operation and maintenance** | Operational implementation<br>Evaluation and modification<br>Operational deployment including retirement |

Not all decisions are direct, and most have multiple solutions of differing value to the user. Time to project completion, probability that the predicted time will be met, economic analysis, decision criteria, and finally a decision, are parts of the process that make a good decision. **Decision analysis** is a generic term which means that the problem type requires some sort of optimization to select the best alternative.

The major problem types are **deterministic** and **stochastic.** Deterministic problems are those with data which is known with certainty. Some of these methods include: **network flow analysis, linear, nonlinear, integer,** and **dynamic programming.** Stochastic problems are those where the data is not certain and may vary significantly. Use of probability theory is necessary to estimate the future trends in this case. The most often used methods include **Markovian decision models, decision analysis, game theory, queuing theory, simulation, stochastic programming** and **sto-**

chastic dynamic programming. The critical path method (CPM) and the program evaluation and review technique (PERT) are used in inventory theory and project management.

The two methods, decision analysis, and computer simulation, are deterministic or stochastic in nature and depend on the data used. Quantitative empirical data lead to deterministic evaluation through experience description, inference, and testing of propositions. Implied probability data introduces variance and indicates that stochastic solutions follow. These solution methods employ classical mathematical statistics methods which are discussed in a later section entitled, **Probability and Statistics.**

## Network Flow Analysis

Network flow analysis was originally developed in electrical engineering; since then the theory has been applied to transportation, information, and communication systems. A network is a set of points, or **nodes,** which are connected by a set of **branches.** The most common network flow problems are **shortest route** and **maximal flow.** In the shortest route, the network is designed with non-negative cost of distance values for each branch. The solution finds the shortest route from the beginning node, the **source** or **origin,** to the final node, **sink** or **destination.** In the maximal flow problem, the total flow from the source to the sink is maximized. The flows throughout the entire network, dependent upon the capacity for each branch, are determined.

## PERT/CPM

The **Gantt chart,** which is a simple bar chart plotting against time, was the precusor of the PERT chart. Both PERT and CPM were designed to estimate total project time, identify the bottlenecks, evaluate the consequences of deviations from a predetermined schedule, and determine the effects on total project cost after changing the allocations of project resources. When stochastic PERT is used, it can determine the probability of meeting deadlines. When updates are introduced throughout the project executions, PERT/CPM is a dynamic control method.

The PERT method assumes that the length of time to perform a future task is known within a degree of uncertainty. This uncertainty is measured by estimating three parameters: optimistic time, most-likely time, and pessimistic time. In contrast, CPM assumes that the length of time required for a task is known with certainty, and thus CPM is a deterministic procedure. CPM finds the lowest total project cost by employing linear and integer programming techniques.

The PERT chart below shows the activities, duration in days, the critical path, the activity (arrow), the event (circle), and a dummy (dotted line). The critical path is the line where there is no slack time; everything must move smoothly along the critical path.

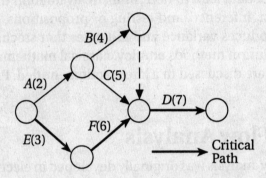

**Figure 1. PERT chart**

The activity arrows are labelled A–F. The number in parenthesis, i.e., (5), is the time unit of the activity. These units can be hours, days, etc.

Further analysis can be accomplished to ensure activities remain on schedule. The following parameters are used: the earliest start time of an activity $i$ ($ES_i$), the earliest finish time of an activity $i$ ($EF_i = ES_i + t_i$), where $t_i$ is the amount of time the activity takes, the latest finish an activity can have without delaying the project ($LF_i$), and the latest start for an activity without delaying the project ($LS_i = LF_i - t_i$). Along the critical path, corresponding start and finish values will be equal as shown below.

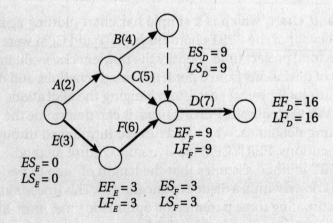

**Figure 2. Critical path**

Stochastic PERT is applicable in most situations, as the time values are somewhat random. For each activity, there is an optimistic time $a_i$, a most likely time $m_i$, and a pessimistic time $b_i$. These times come from a beta probability distribution. The expected activity time $t_i$ is then:

$$t_i = \frac{a_i + 4m_i + b_i}{6}$$

and by the standard deviation $\sigma_i$ of the beta distribution can be given by

$$\sigma_i = \frac{b_i - a_i}{6}.$$

Then to find the probability, $Z$, of finishing at a specified time, the variance along the expected critical path is summed. The square root of this sum is added and subtracted from the mean. Placing all this on a normal distribution we can find the probability of completing the project on time by normalizing the probability between 0 – 1.

$$Z = \frac{(x - \mu)}{\sigma}$$

where
- $x$ = Nonstandard normal variate
- $\mu$ = Mean of the nonstandard normal
- $\sigma$ = Standard deviation of the nonstandard normal

The value, $Z$, is then compared with the areas of a standard normal distribution table to obtain final probability.

PERT charts can be used for cost comparisons, resource constrained problems, earliest time and latest time analysis, and for critical path considerations. Some related methods are the **Precedence Diagramming Method (PDM)**, the **Line of Balance (LOB)**, and **Precedence Networking (PN)**.

## Markovian Decision Models

In a schedule, the time that elapses between any two tasks may be constant or it may vary with some characteristics of the project. Since the definition of a **stochastic process** is an indexed family of random variables, these times may be considered as a stochastic process. In addition, the set of time points may be defined as a **Markovian process** is the probability of any future events is independent upon the previous history of events. Several Markov processes can be combined in a **Markov Chain**.

Markov chain models are used to predict future changes in dynamic systems.

## Decision Analysis

Problems with some degree of uncertainty can be analyzed by a group of several systematic procedures. This is known as **decision analysis.** The result of the decision is dependent upon external uncertain circumstances. However, the possible set of circumstances, or **states of nature,** and their probability distributions are known. Several criteria are used: first a decision may be selected to maximize an expected gain (or minimize an expected loss). This is an **a priori criterion.** Secondly, a decision may be chosen to maximize the possible gain, or an **optimistic criterion.** Finally, if the decision maker chooses to minimize the maximum possible loss, it is known as **pessimistic criterion.**

This analysis is performed with decision trees; each is oriented to represent a given decision process. The trees are composed of nodes and branches. The nodes are further classified as either decision nodes or chance nodes. At the decision nodes, the person must choose a path from several alternatives, while at the chance nodes, the person has no control. The chance node represents a state of nature. Nodes may also be used to represent the end of the decision process. To obtain the optimal decision, the decision maker starts from the end nodes and moves backward to the origin. The expected gain at each point is calculated, and the optimal decision is that which gives the maximum expected gain. The **utility scale** can be used, then the optimal decision is that which provides the maximum expected utility.

## Game Theory

Decisions in a competitive setting with two or more opponents are best modeled by game theory. Here, each opponent wishes to optimize his own objectives, at the expense of the other players. The final outcome, for any player, depends upon both his own actions and those of his opponents. The player must choose his actions or strategy with uncertainty, as he cannot predict the actions of the others. The games are defined by a set of rules, or payoffs, which outline the possible moves. Different players may be restricted to different rules. The simplest game which may be played is the two-person-zero-sum game. In this case, only two persons play, and the one player wins while the other loses.

# Linear Programming

Linear Programming (LP) is a management tool which allocates scarce resources. It is a mathematical technique that will maximize or minimize a linear function subject to a system of linear constraints. The linear function is the object function, and is usually being maximized. A form of an LP model follows:

$$\text{Maximize} \quad c_1x_1 + c_2x_2 + \ldots + c_nx_n$$
$$\text{subject to restrictions} \quad a_{11}x_1 + a_{12}x_2 + \ldots + a_{1n}x_n \leq b_1$$
$$a_{m1}x_1 + a_{m2}x_2 + \ldots + a_{mn}x_n \leq b_m$$
$$\text{and} \quad x_1 \geq 0, x_2 \geq 0, \ldots, x_n \geq 0$$

The constants $a_{11}$, $b_n$, $c_1$, are all parameters of the problem. This basic model will fit most LP problems and with minor changes one can look at minima or constraints which have equalities. The solution lies at the intersection of constraints. The **simplex algorithm** evaluates the objective function at a sequence of intersections until a solution is found.

# Non-Linear Programming

Many problems cannot be described by a set of linear functions. Instead, non-linear equations and/or constraints are used. As in linear programming, the goal is to optimize an objective function. In many cases integer valued variables and constraints, such as differential equations, are omitted from non-linear programming, as their solutions tend to be problem-specific. The non-linear problems are solved in two phases: First the search direction is selected, and second, the objective function is minimized or maximized (to some degree) in the search.

There are several methods for optimization when one variable is involved, or unidirectional optimization. The best known technique for minimization of a single variable is **Newton's method**, which requires that the function be twice differentiable. For a quadratic function, the procedure will converge within one iteration. For functions which approximate quadratics, several iterations are necessary, but convergence is possible. The **quasi-Newton methods** have the advantage of not requiring the function to have analytical derivatives. Instead, the central differences are used to approximate the values of the derivatives. **Secant methods** approximate the first derivative as a straight line. These methods require two initial points, in contrast to the single point used by Newton's method. The secant methods are rather crude, but they usually work well in practice. Finally, alternatives to unidimensional search methods are quadratic and cubic-curve-fitting techniques. These are grouped into the category of **polynomial approximation methods.**

For functions of multiple variables, variations of the Newton, quasi-Newton, and secant methods can be used in problems without constraints. In addition, the **conjugate gradient search,** which uses a quadratic objective function, is another alternative. When constraints are present, the problem becomes substantially more difficult. The classic solution method is that of the **Lagrange multiplier.** Extensions of this technique include augmented Lagrange and the **penalty function methods.** The generalized reduced-gradient method and the sequential (recursive) quadratic programming may also be employed.

# Queuing

Queuing theory is the last example and is used less than 60% of the time by the companies polled. It applies any time some population is waiting to be serviced. Examples include aircraft taking off and landing, cars at gasoline stations, and people at fast food restaurants. The whole process of flying is an excellent example of queuing; waiting in line. The aim of the queuing theory is to reduce time in lines thus improving customer satisfaction.

In most cases, the facility can serve the customers at a rate that is faster than their arrival rate. If the arrival rate and/or service rate are random, then the demand for service may exceed the facility's current capacity, and thus the customers form a waiting line, or a **queue.** The **queue discipline** is the order in which the waiting customers are selected. Sometimes the customers are selected on a *first-come-first-served* basis. Other schemes may include service dependent upon a priority rating or the customers may be randomly served.

A service unit is one that can only serve one customer at a time. A multichannel system has many units. The analysis of the system depends upon: the customer arrival pattern, the queue discipline, the service time distribution, and the number of channels in the system.

Queuing systems are modeled as single-channel, single-phase systems to multichannel, multiphase systems. A **simplified single-server (phase) model** assuming Poisson arrival, exponential service times, first come, first served, and infinite source and queue produces the following:

Probability of 0 input (empty system)  $P(0) = 1 - \left(\dfrac{\lambda}{\mu}\right)$

| | |
|---|---|
| Probability of $n$ inputs in the system | $P(n) = P(0)\dfrac{\lambda}{\mu}$ |
| Amount of time server is busy | $\delta = \dfrac{\lambda}{\mu}$ |
| Expected number of inputs | $I_s = \dfrac{\lambda}{(\mu-\lambda)}$ |
| Expected number of inputs in queue | $I_q = \dfrac{\lambda^2}{\mu(\mu-\lambda)}$ |
| Expected time in system | $T_s = \dfrac{1}{(\mu-\lambda)}$ |
| Expected time waiting | $T_q = \dfrac{\lambda}{\mu(\mu-\lambda)}$ |
| Mean arrival time | $\lambda$ |
| Mean arrival rate | $\mu$ |

In many service facilities, accurate prediction of the customers' arrival and the amount of time required to be served is nearly impossible to predict. Containing too many service units results in these units being idle, which could be very costly. However, inadequate service facilities could discourage potential customers, resulting in a loss of business. The problem in some queuing systems is to optimize the number of channels so as to minimize the sum of the cost caused by waiting costumers or idle service units. Some examples include the numbers of spaces in a parking lot or the number of check-out counters in a store.

Queuing systems have been mathematically analyzed using queuing networks and computer simulation techniques. One technique is known as the **regenerative method.** In series queuing systems, the departure process from one service station forms the arrival process at the next station.

Queuing models are easy to develop and are less expensive than other simulations, however they sometimes lack robustness and flexibility needed in more complex applications.

# Simulation

Since a simulation of a stochastic system produces an estimate of the true system characteristics, a combination of several simulation runs are needed. One drawback is the high cost of the simulations. However, simulations do have many advantages. Many real systems are too complex to be described by mathematical models and simulations is the only practical method. Once the simulation model is developed and verified, the decision maker can experiment with the model to study the impact of various decisions.

In **discrete systems**, the values of the controllable variables and the outcome are only interesting at specific points in time. In these systems, the objects are defined as **entities**, and a change in the state of the entity is called an **event**. The simulation of the discrete system includes: generating the sequence of events, recording the statistics on all processes, periodically updating the records, and finally, analyzing the records.

# Criteria Function

An additional simple method of evaluating the alternatives in any project is the use of the **criteria function**. Not listed by the Fortune 500 companies, it is used by many to select the "best" alternative. For any decision the engineer must decide the criteria ($i$) which will guide selection of some option. Then the relative merit ($x_i$) of the option being considered (how well this option fulfills the criteria) must be established. Finally the criteria are weighted ($A_i$) to show relative importance of the criteria. There may be ten criteria, with a relative merit from 1–10 or any desired scale, and a weighting factor for each criteria which again may be arbitrary. Care must be taken to ensure that decisions are not made by careless selection of the relative merit and weighting factors.

The **criteria factor** (CF) is the sum of the products of the relative merit and the weighting factor. In equation form:

$$CF = \sum_{i=1}^{n} A_i x_i$$

Most decisions include economic considerations as a primary criteria to be weighed. This topic, which is reviewed in the next section, is certainly an important aspect of systems analysis.

# CHAPTER 2

# ENGINEERING ECONOMICS

Engineering economics involves understanding interest rates and how their application impacts on the feasibility of the alternatives. Most engineering decisions involves choices among several alternatives. Economic considerations must be addressed to develop feasible alternatives from which the solution is selected. It is important to consider the value of an investment throughout its life cycle and to understand the time value of money.

Economic analysis and decision making is an important facet of engineering, especially in the field of design. Several aspects of analysis are involved, and many of these require comparing costs of equipment, services, and contracts.

## Cost

The cost of a project can be divided into many areas. **First cost, or investment cost,** is the cost of initiating the project. First cost is usually limited to one-time expenses, such as the procurement of equipment, shipping and installation costs, and personnel training costs. If any item is not standard, then design and development costs, along with construction, may be included in the first cost. The investment cost measures whether a project should be undertaken. While a project may seem profitable in the long-term, if the first costs are high, a firm or company may not be able to meet the level of investment. Second are the **operation and maintenance (O&M) costs.** These include the labor, fuel and power, materials and supplies, spare parts, insurance, taxes, an allocated amount of indirect cost known as overhead, and with the new environmental concerns, disposal costs. These operation and maintenance costs usually increase with time, while the investment costs decrease. **Fixed costs** are those that remain relatively constant over time. Some examples include depreciation, taxes, insurance, interest on invested capital, sales, and perhaps administrative expenses. **Variable costs** are a function of the production level. As production increases, some costs such as labor, materials, and utilities, also increase. The material costs do sometimes vary with the amount purchased, thus larger quantities may have discounted prices. The labor costs can vary due to the efficiency of the production line, the use of overtime, and requirements for additional personnel.

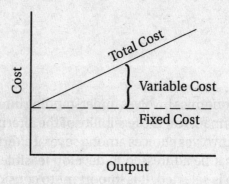

Figure 3. Fixed cost

The increases in cost are measured by the **incremental** or **marginal cost**. Marginal cost describes the additional costs acquired to produce one more unit of output. As production increases, the marginal cost usually decrease to a certain minimum. Past the minimum, the marginal cost increases as production increases. Thus many analyses attempt to minimize the marginal cost.

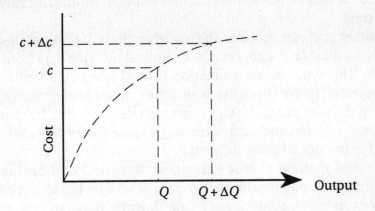

**Figure 4. Marginal cost**

All of the cost can also be grouped into one of two categories: **direct** or **indirect costs**. Direct costs describe material, labor, and any subcontracts which are directly related to manufacturing the finished product. The indirect costs include supervision and management staff, support operations, and all the materials used in the production operation which are not found in the final product. Other indirect costs include utilities, building rentals, depreciation, taxes, insurance, and maintenance.

Another classification of costs may be **recurring** and **nonrecurring**. Recurring costs continue over the life of the project. Manufacturing costs, engineering support during production, customer support over the product life, and ongoing project management are examples of recurring costs. Nonrecurring costs occur only once during the project. They include product applied research, design and development, testing (other than quality control), new construction, and manufacturing tools and equipment. The overall cost of a project is a combination of the recurring and the nonrecurring costs. At the beginning, the design and development costs are mainly nonrecurring. As the project nears production, the costs are mainly recurring.

# Cost Estimation

Cost can be estimated by several methods. The **engineering estimate** totals the cost of all the individual elements of the design and fabrication process. These include all the production tools, pipes, pumps, types of materials, and everything else that is found in an engineering drawing. This method does have two major disadvantages. When the cost estimate is composed of all the individual elements, the overall perspective may be lost in the details. In addition, the cost estimate requires lengthy preparation time.

Another cost estimation is the use of **analogy.** If little detailed information is available, the cost can be estimated by using data from similar projects. This type of cost estimation is used when companies initiate new projects. The cost from the analogous systems are adjusted to correspond to the new program. This cost estimation can also be used at the lower levels of detail, such as determining the number of direct labor hours needed to manufacture a similar part.

The final method of cost estimation, **statistical cost estimating,** is based upon the relationship between cost and its influencing factors. These factors may include power rating, flowrate, mass, volume, and production quantities. This technique is used for long range planning.

In the beginning of a project, little information is available and an analogous cost estimation method may be used. As the project progresses, cost estimates can be made using statistical methods. The statistical methods can also be used to provide information on unforeseen events and costs. Finally as the detailed plans become available, an engineering estimation can be employed.

# Life-Cycle Costing

The cost estimation of all phases of the project, or the entire life of the project, is defined as the **life-cycle cost.** This total cost includes research and development costs, production costs, operation and support costs, and retirement and disposal costs. The company, in the early stages of the design of the project, sets forth cost targets. One method which is used to successfully manage the costs is the design-to-cost approach. The cost is seen as a project constraint. For each stage of the project, several alternatives may be evaluated, in trade-off studies, to find ways to accomplish the goal with reduced cost. However, the reduced cost should not lead to reduced quality of the final product.

A method of managing cost in large projects is a **cost breakdown structure.** The project activities are related to the available resources, and the total cost is divided into logical categories. The level of detail of the cost breakdown structure is dependent upon how the cost is managed. Cost breakdown structures may have the following characteristics:

1. All cost of the project are included.
2. Cost categories are clearly defined, and the project personnel understand these categories.
3. The structure is defined to a level which management can iden-

tify the high cost areas and the cause and effect relationships.
4. The structure incorporates a coding system so that data can be gathered about specific areas, such as disposal costs as a function of manufacturing.
5. The coding scheme separates the producer, supplier, and consumer costs.
6. The cost breakdown structure is compatible with planning documentation and other management structures.

The life cost can be calculated by the following equation:

$$\text{ALCC} = \frac{P}{n} + O + (n-1)\frac{M}{2}$$

where
- ALCC = Average annual life cost
- $O$ = Constant annual operating cost
- $M$ = Annual increase in operating cost
- $n$ = Life of the project, in years
- $P$ = First cost of the project

Note that this formula does not account for the time value of money, which will be discussed later. The minimum cost life, $n^*$, is found by:

$$n^* = \sqrt{2\frac{P}{M}}$$

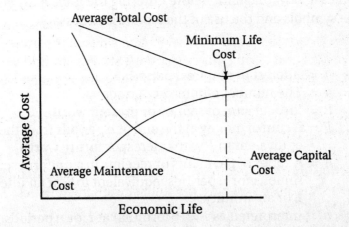

Figure 5. Average Life Cost

# Nominal and Effective Interest Rates

The **nominal interest rate**, $r$, is the "advertised" rate and is usually based on a period of one year. The interest rate per interest period, $i$, is the quotient of the nominal interest rate and the number of compounding periods, $n$, per year.

$$i = \frac{r}{n}$$

From this definition the **effective interest rate**, $i_e$, becomes,

$$i_e = (1+i)^n - 1$$

The more frequent the compounding, the larger the variance between the effective interest rate and the nominal interest rate.

Simple interest can be calculated by:

$$I = Pni$$

where

$I$ = Interest earned, in dollars
$P$ = Principal or amount of money loaned

# Time Value of Money

The **time value of money** is expressed using six common equations to relate **present, future,** and **periodic sums.** The table below contains the algebraic formula, common symbols, the name of the factor, and the associated cash flow diagram. Before entering the table a brief explanation of the symbols and the use of the factors is necessary.

## Symbols

$i$ = Interest per interest period
$n$ = The number of interest periods
$P$ = Present sum of money or **present worth**
$F$ = A sum of money at the end of $n$ periods from the present equivalent to $P$ with interest or **future worth**
$A$ = A series of payments or receipts in a uniform series continuous for $n$ periods, equivalent to $P$ at an interest rate $i$, or an **annuity**.

($F/P$, $i\%$, $n$) is interpreted as $F = P$ (formula) at $i$ for $n$ periods.

## Factors

**Compound Amount Factor** = Compound interest

**Present Worth Factor** = Amount of money needed now to ensure that a specified amount of money is available in the future

**Uniform Series Compound Amount Factor** = Future worth when equal payments are made periodically

**Sinking Fund Factor** = Amount which must be paid periodically to ensure that a specified amount of money is available in the future

**Capital Recovery Factor** = Amount a fund can pay in periodic payments for some number of periods. Can be used to pay bills that are long term.

**Uniform Series Present Worth Factor** = Amount needed now to provide equal payments for some number of periods in the future

### Table 1. Summary of Interest Factors Relating Present, Future, and Periodic Sums

| Formula | Symbol | Factor |
|---|---|---|
| $(1+i)^n$ | (F/P, i%, n) | Compound Amount Factor |
| $\dfrac{1}{(1+i)^n}$ | (P/F, i%, n) | Present Worth Factor |
| $\dfrac{(1+i)^n - 1}{i}$ | (F/A, i%, n) | Uniform Series Compound Amount Factor |
| $\dfrac{i}{(1+i)^n - 1}$ | (A/F, i%, n) | Sinking Fund Factor |
| $\dfrac{i + (1+i)^n}{(1+i)^n - 1}$ | (A/P, i%, n) | Capital Recovery Factor |
| $\dfrac{(1+i)^n - 1}{i(1+i)^n}$ | (P/A, i%, n) | Uniform Series Present Worth Factor |

# Cash Flow

A **cash flow diagram** can provide a pictorial description of the receipts and disbursements over time. These diagrams are used to visualize the activity of the funds over time. An increase in cash is shown by an upward arrow, and a decrease by a downward arrow. If during an interest period, both negative and positive cash flows occurred, the net value is shown as an arrow which is the sum of all the cash flows. The direction of the arrow is dependent upon the sign of the sum, negative-downwards, and positive-upwards.

# Present Worth Analysis

The **present worth** of an amount over a period of $n$ years at an interest rate, $i$, is:

$$PW(i) = \sum_{t=0}^{n} F_t(1+i)^{-t}$$

where
$\quad PW(i)$ = Present worth at interest rate $i$
$\quad F_t$ = Future amount after $t$ years

The present worth includes the time value of money.

# Interest Rate-of-Return Analysis

Another way to evaluate investment opportunities is the **rate-of-return analysis.** Rate-of-return is defined as the interest rate that causes the future value of an investment to be equal to the present value. The rate of return analysis determines the interest rate that reduces the present, annual, and future worth of payments to zero. After the rate of return has been determined, the payback period is calculated. The **payback period** is the length of time necessary for the investment to pay for itself. The equation:

$$\sum_{t=0}^{n} F_t \geq 0$$

is used to calculate the minimum number of periods required to recover the initial investment. However, the formula does not account for the time value of money. The discounted payback period method incorporates the time value of money. The discounted payback period is the smallest value of $n^*$ that satisfies the condition:

$$\sum_{t=0}^{n^*} F_t(1+i)^{-1} \geq 0$$

where

$F_t$ = Future worth
$i$ = Interest, as before.

## Depreciation

**Depreciation** is defined as the reduction in the value of an asset over time. An asset is depreciated over time to compensate for reduced ability to perform. The main cause of depreciation is normal wear and tear, although new technology may also devalue older equipment. If the asset is used to produce income, the depreciation can be used for tax purposes. The depreciation may be calculated by one of two methods: the **straight-line method**, or the **Modified Accelerated Cost Recovery System (MACRS) method** provided by the Internal Revenue Service. The straight-line method assumes the final salvage value of an asset to be zero. The IRS requires that an asset be depreciated over a life of ten years.

Both **inflation** and **deflation** describe the change in price with time. The **price index** is the ratio of the price of a commodity at one time relative to the price at some other point in time. The effect of inflation on the rate of return on an investment depends on how the future returns respond, that is on the inflation or deflation rates.

## Equivalent Uniform Annual Cost Evaluation

To compare nonuniform series of money distributions each series must be annualized reducing them to an **equivalent uniform annual series** of payments. Approximations are often easier than actual determination of the cost factor in the following equation.

EUAC =
(initial investment) (uniform annual cost factor) + annual cost

## Capitalized Cost

The **capitalized cost** represents the present total cost of financing and maintaining a given alternative for an indefinite period of time.

Capitalized Cost = EUAC / interest rate

# Replacement Analysis

When determining whether or not to maintain the present equipment or to replace it with new equipment, the present value of the two alternatives must be compared. The alternative with the smallest present value is the financially correct choice.

# Selecting an Appropriate Rate of Return

A company compares alternatives to ensure that the maximum return is obtained. For a comparison, a **minimum attractive rate of return** (MARR) is established that is equal to the greatest of either the cost of borrowing money, or the cost of capital, or the rate of return the company can earn from other investments. Another alternative, known as the **do-nothing alternative,** is not to invest in any new projects but to buy surplus funds in a business unrelated to the companies interests. Since many events cannot be predicted, an element of risk or uncertainty exists. The expected value of the future outcomes can be estimated with **risk analysis.** Reasonable probabilities are established for each outcome, and then the expected value can be calculated by:

$$\text{Expected value} = \text{outcome } A \times P(A) + \text{outcome } B \times P(B) + \ldots$$

where

$A, B$ = Values of the outcomes
$P(A), P(B)$ = Probabilities of the outcomes.

# CHAPTER 3

# ELEMENTARY COMPUTER OPERATION AND APPLICATIONS

Fundamental understanding of computer operations is necessary to maximize the benefits of high powered software and for the creation of complex computer codes developed by many engineers. Computational manipulations transparent to the user, along with some common definitions, are reviewed in this chapter.

## Scalar Variables

### Computer Memory

Computer memory can be envisioned as a huge collection of locations that can store information or data, similar to the banks of post office boxes in a post office. Each individual memory location consists of a number of two-valued (i.e., binary) information storage units. Each of these two-valued storage units is usually called a **bit** (for "*b*inary dig*it*"), and stores a value of 0 or 1 (or "off" and "on"). Each memory location has a unique address so information can be stored and retrieved easily, and the addresses are usually numbered sequentially (often starting at 0). Thus if a small computer has 256,000 memory locations, they are sequentially numbered from 0 to 255,999.

A standard memory location on a mainframe computer is traditionally called a **word** and typically consists of 8, 16, 32, 36, 40, or 60 bits. An (addressable) subsection of a word is called a **byte** and is commonly used to represent an encoded character. A byte usually consists of 8 bits, even though only 7 may be used to represent a character in code. On occasion, a half of a byte is called a **nibble.**

Larger computers (i.e., "mainframes") usually have a longer word size, and these words can sometimes be subdivided. In most personal computers, memory is usually arranged in bytes, which are joined together if needed for larger data.

## Data Types

In most contemporary programming languages, there are at least four standard **types:**

| | |
|---|---|
| **Integer** | whole numbers such as 2, 34, −234 |
| **Real** | numbers that can contain a decimal point |
| **Character** | letters, symbols, and numbers stored as characters |
| **Boolean** | values related to two-valued logic, sometimes called **Logical.** |

Any unit of information that is used in a program must be classified according to one of the allowable types, and in most languages this classification cannot be changed during the course of the program's execution. Information stored in memory is also classified as to whether it remains constant throughout the program (such as $\pi = 3.1415926$) or whether the contents of that memory location are allowed to be changed. Memory locations that contain unchangeable data are called **constants.** Memory locations that contain changeable data are called **variables.**

Since computer memory can only store binary information, all information, numeric or non-numeric, has to be translated into some sort of binary code before storage. The code must be unique as to **type** and easy to use in operations. In addition, there should be some way of determining what type of information is stored in which memory location, so that the information can be interpreted correctly.

To aid the computer in determining what type of information is stored where, when a program is compiled, a **symbol table** is created in which each variable is listed along with its **type.** Normally, other information is also stored in a symbol location, especially the variable's memory location, and any initial value.

# Binary Notation

Binary notation, like decimal notation, is positional. Only two digits (bits) are possible in each place, 0 or 1. The place to the left of the binary point is called the **ones** place (as in decimal notation), but in binary notation it can also be indicated as the $2^0$ place. The next position to the left is the **twos** (or $2^1$) place. Next comes the **fours** place, then the **eights**, then the **sixteens**, etc., all labeled in successive powers of two.

Similarly, the places to the right of the binary point are labeled as negative powers of 2. The first place to the right of the binary point is the **halves** place, followed by the **quarters** place, followed by the **eighths** place, etc.

To translate from binary into decimal, a binary number should be expanded according to the appropriate (positional) power of two, and rewritten in decimal notation. As an example, take 11010.101.

$$
\begin{aligned}
11010.101_2 &= 1\times 2^4 + 1\times 2^3 + 0\times 2^2 + 1\times 2^1 + 0\times 2^0 + 1\times 2^{-1} + 0\times 2^{-2} + 1\times 2^{-3} \\
&= 1\times 16 + 1\times 8 + 0\times 4 + 1\times 2 + 0\times 1 + 1\times\left(\frac{1}{2}\right) + 0\times\left(\frac{1}{4}\right) + 1\times\left(\frac{1}{8}\right) \\
&= 16 + 8 + 2 + 0.5 + 0.125 \\
&= 25.625_{10}
\end{aligned}
$$

To translate an **integer** from decimal into binary, the number should be repeatedly **divided** by two 2 and the bits that form the **remainders** saved. This procedure stops when zero is reached as a quotient. These remainder bits, when read from last to first, form the binary equivalent. As examples, take 5 and 4.

| Number | Remainder        | Number | Remainder       |
|--------|------------------|--------|-----------------|
| 2: 5   | (none initially) | 2: 4   | (non initially) |
| 2: 2   | 1                | 2: 2   | 0               |
| 2: 1   | 0                | 2: 1   | 0               |
| 0      | 1                | 0      | 1               |

**Bits read last to first:**
$5_{10} = 101_2$

**Bits read last to first:**
$4_{10} = 100_2$

To translate a **fraction** from decimal to binary, the number should be

A–35

repeatedly **multiplied** by 2 and the bits that form the **overflow** (into the integer section left of the decimal point) saved. Once saved, the overflow bit is removed from the number as far as further calculations are concerned. This procedure stops when zero is reached as a product, or when it is obvious that the number is a repeating binary fraction, or when enough bits accurately have been achieved. These overflow bits, when read from first to last, form the binary equivalent. As examples, take 0.25 and 0.75.

| Number | Overflow | Number | Overflow |
|---|---|---|---|
| $2 \times 0.25$ | (none initially) | $2 \times 0.75$ | (non initially) |
| 0.50 | 0 | 1.50 | 1 (now remove) |
| $2 \times 0.50$ | | $2 \times 0.50$ | |
| 1.00 | 1 (now remove) | 1.00 | 1 (now remove) |
| 0.00 | | 0.00 | |

**Bits read last to first:**     **Bits read first to last:**
$0.25_{10} = 0.01_2$             $0.75_{10} = 0.11_2$

# Encoding Data

### Integers

In the binary representation of integers, the left-most bit is interpreted as a **sign-bit,** which is 0 for positive numbers and 1 for negative numbers. The other bits store the **magnitude** of the number (sometimes called the **mantissa**). This magnitude is interpreted in different ways depending on whether the number is positive or negative and depending on which method is used by the computer for representing signed integers.

There are three common schemes used to store signed integers. The actual method employed depends on the computer being used and each computer employs only one scheme.

Positive integers are encoded in direct binary notation no matter which of the three schemes is used, e.g.,

$$2_{10} = 00\ldots00010_2$$
$$9_{10} = 00\ldots01001_2$$

Negative integers are encoded differently according to the rules of the scheme being used.

**Sign magnitude**—The first bit indicates the **sign,** and the other bits indicate the number in standard (i.e., positive = "magnitude") form. E.g.,

$$2_{10} = 00\ldots00010_2$$
$$-2_{10} = 10\ldots00010_2$$

**Note:** In this scheme, there is only one representation for 0, and the arithmetic is fairly easy.

Technically, the three schemes of sign-magnitude, one's complement and two's complement are applicable to all signed integers, both positive and negative. However, there is no difference in the resulting coded number for positive numbers. Only when encoding and decoding negative numbers must the scheme be known in order to perform the coding correctly.

## Real Numbers

Real numbers are stored in two sections in one word using a format related to "scientific notation." A real number expressed in scientific notation is written with a section containing the decimal point (usually called the **mantissa** or the **significant digits**), multiplied by 10 raised to some power (called the **exponent**). For example, one million (1,000,000) can be written as $1.0 \times 10^6$ or as $100.0 \times 10^4$. When real numbers are stored in a computer, the mantissa is **normalized** (i.e., usually there are no digits to the left of the decimal point and no leading zeros to the right of the decimal point). E.g.,

$$0.0025 \Rightarrow .25 \times 10^{-2} \quad \Rightarrow \text{often written } .25E-2$$
$$3000.0 \Rightarrow .30 \times 10^4 \quad \Rightarrow \text{often written } .30E+4$$

Whether the "binary" point is assumed before or after the digits of the mantissa varies with the system. The point itself is never stored.

Thus, for any real number, a total of four units of information must be stored in a word: the binary version of the mantissa, the sign of the mantissa, the binary version of the exponent, and the sign of the exponent. Note that as seen in the example above, the sign of the exponent can be negative while the sign of the mantissa can be positive.

For purpose of example, assume that a computer has a 40 bit word. One possible way in which the bits of a word are used for storing a real number might be the following:

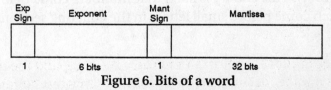

**Figure 6. Bits of a word**

It should be noted that real number arithmetic is more difficult than integer arithmetic. A simple arithmetic example will illustrate the problem and sketch the steps a computer takes:

How are the following numbers added: $0.25E-2$ and $0.30E+4$?

To solve, first: Shift the decimal (or binary) point of one number (adjusting both the mantissa and exponent) until the exponents of both numbers are equal. E.g.,

$$.25E-2 \Rightarrow .00000025E+4.$$

Second: Add the mantissas only. Note that on computers, the limited machine accuracy means that one number may not changed the other number, i.e., the sum may actually equal one of the two addends. In our example, the sum would be $0.30000025E+4$.

Third: Normalize the computed sum (if necessary). On a computer, after normalization, the number from the computational register is stored in memory, truncating low order bits if necessary. If only six decimal digits can be stored, the stored sum would be the same as one of the two original numbers, i.e., $0.300000E+4$.

**Characters**

Characters are stored via a coding scheme. Each character, whether it is a letter of the alphabet (upper case or lower case), a digit, or a special symbol (printable or non-printing), is assigned a number in the coding scheme, often called the **collating sequence** (especially when the characters are listed in the numerical order of the code numbers). There are two major schemes in use.

EBCDIC (pronounced "eb-see-dick") is a scheme produced by IBM. It is an acronym for Extended Binary Coded Decimal Information Code, and is still used in some IBM mainframes. This coding is such that small letters come before the capital letters, which come before the numbers in the collating sequence.

ASCII (pronounced "as-key") is an acronym for American Standard Code for Information Interchange. This is a national standard, in use on most mainframes other than IBM and on most personal computers (including IBM). This coding is such that numbers come before capital letters, which come before small letters in the collating sequence.

# Iteration vs. Recursion

In the vast majority of programs, some action or computation is usually repeated a number of times. There are two basic approaches to repetitive programming: (1) iteration, and (2) recursion. An **iterative** program is one in which a loop (i.e., repetitive code) is explicit. A **recursive** program is one in which there may not be any explicit loop, but in which a concept is defined (and computed) by calling itself.

Some concepts naturally are defined recursively, others can be expressed recursively, while others cannot be (easily) expressed recursively. The basic mathematical operations can either be expressed recursively or by expansion in terms of a simpler operation (which can be considered as one form of iteration).

$$\text{Factorial} \quad n! = (1)(2)(3)\ldots(n-1)(n)$$
$$= n(n-1)!$$

(Here $n!$ is expressed in terms of a simpler version of itself, $(n-1)!$.)

$$\text{Multiplication} \quad ab = a(b-1) + a = \underbrace{a + a + a + a + \ldots a}_{b}$$

(Here $ab$ is expressed in terms of $a(b-1)$.)

$$\text{Exponentiation} \quad a^b = a^{(b-1)} a = \underbrace{a\, a\, a\, a \ldots a}_{b}$$

(Here $a^b$ is expressed in terms of $a^{b-1}$.)

Every recursive definition must have some alternate definition for a fundamental case that does not involve itself. In the case of the three arithmetic operations just given, the following are the foundational values.

| | |
|---|---|
| Factorial | $1! = 0! = 1$ |
| Multiplication | $a1 = a$ |
| Exponentiation | $a^1 = a$ |

Many routines can be written in either an iterative version or a recursive version. Sometimes the iterative version is the one that has been developed as the fundamental algorithm and it is converted into a recursive version. Many times, however, the recursive version is the first one described, and then it is converted into an iterative form.

# Programming

Armed with the understanding of computer operations, a problem which is most suitable to computer solution, and knowledge of a **programming language** such as **FORTRAN, Pascal, C,** or **Basic**, the engineer can write a program to solve the problem. This program, sometimes referred to as a **code**, is not in binary notation and must be interpreted by the computer, solved, and should produce a result which is easily understood by the programmer. The computer must translate from the programming language to a language understood by the machine.

# Translation

There are two approaches to language translation: **compilation** and **interpretation**. A **compiler** will translate the entire program from, for example, Pascal, into the machine language. After it translates the entire program, it starts to execute it. On the other hand, an **interpreter** translates the program one statement at a time. The interpreter will translate the first statement, execute it, translate the second statement, execute it, and so on. There are advantages to both approaches. With interpretation, if there is an error in one of the statements the computer can pinpoint which statement caused it, because it is executing each statement as it comes across it. Compilation is useful when there is a task that is repeated several times. The translated version of the task is saved and can be executed directly whenever it is needed without having to translate the same statement over and over again. The compiler is usually more advantageous than an interpreter, and is the approach that is most widely used in translation.

# Language Description

A programming language can be described by its **syntax** and **semantics**. **Syntax** are the rules that tell us how to put statements together. Semantics are descriptions of how a program is actually written. For example, the syntax of Pascal tell us that

$$a := b;$$

is a correct assignment statement. We use a **grammar** to describe the syntax. The grammar is a set of rules for defining all valid constructs of a language. An example of a grammar is **Backus-Nawr Form (BNF)**. An expression in BNF grammar would be described as:

expression :: = value operator expression.

Another type of grammar are **syntax diagrams.** Here the expression is described as:

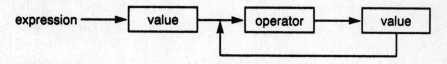

Figure 7. Syntax Diagram

# CHAPTER 4

# CHEMISTRY

Chemistry is the foundation for combustion studies, energy conversion work, materials development, and is a natural link to **thermodynamics**. Chemistry is concerned with the properties and composition of matter and the changes that matter undergoes. Engineers are concerned with the design and operation of processes and things made of matter: the link is obvious.

## Matter and Its Properties

Matter occupies space and possesses mass. It can occur in three states, or **phases: solid, liquid,** or **vapor**. The composition may be divided into two categories: distinct substances (**elements** or **compounds**) and **mixtures**.

The **law of conservation of matter** states that matter can neither be created nor destroyed, but only changed from one form to another. This law requires that a "material balance" be maintained in chemical equations. Likewise, the **law of conservation of energy** states that energy can neither be created nor destroyed, but only changed.

## Stoichiometry

One **mole** of any substance is the amount which contains $6.02 \times 10^{23}$ particles. This number is known as **Avogadro's number**. The gram-atomic weight of any element is defined as the mass in grams, which contains 1 mole.

## Balancing Equations

When balancing chemical equations, one must make sure that there are the same number of atoms of each element on both the left and the right side of the arrow. For example:

$$2\,NaOH + H_2SO_4 \rightarrow Na_2SO_4 + 2H_2O$$

Na: 2 atoms
O: 6 atoms
H: 4 atoms
S: 1 atoms

## Molecular Weight and Formula Weight

The formula weight (**molecular weight**) of a molecule or compound is determined by the addition of its component atomic weights. For example:

$$\text{F.W. of } CaCO_3 = 1(40) + 1(12) + 3(16) = 100\,\frac{g}{mole}$$

Molecular weight =

$$\underbrace{\underbrace{density \times volume\ per\ molecule}_{\text{mass of one molecule}} \times Avogadro's\ number}_{\text{mass of one mole of molecules}}$$

## Calculations Based on Chemical Equations

The **coefficients** in a chemical equation provide the ratio in which moles of one substance react with moles of another. For example:

$$C_2H_4 + 3O_2 \rightarrow 2CO_2 + 2H_2O \quad \text{represents}$$

1 mole of $C_2H_4$ + 3 moles of $O_2$ → 2 moles $CO_2$ + 2 moles $H_2O$

In this equation, the number of moles of $O_2$ consumed is always equal to three times the number of moles of $C_2H_4$ that react.

# Theoretical Yield and Percentage Yield

The **theoretical yield** of a given product is the maximum yield that can be obtained from a given reaction if the reaction goes to completion (rather than to equilibrium).

The **percentage yield** is a measure of the efficiency of the reaction. It is defined

$$\text{percentage yield} = \frac{\text{actual yield}}{\text{theoretical yield}} \times 100\%$$

# Percentage Composition

The percentage composition of a compound is the percentage of the total mass contributed by each element:

$$\%\text{ composition} = \frac{\text{mass of element in compound}}{\text{mass of compound}} \times 100\%$$

# Atomic Spectra

The **ground state** is the lowest energy state available to the atom.

The **excited state** is any state of energy higher than that of the ground state.

The formula for changes in energy ($\Delta E$) is

$$\Delta E_{electron} = E_{final} - E_{initial}$$

When the electron moves from the ground state to an excited state, it absorbs energy.

When it moves from an excited state to the ground state, it emits energy.

This exchange of energy is the basis for atomic spectra.

# Quantization of Energy

Light behaves as if it were composed of tiny packets, or **quanta**, of energy (now called "photons").

$$E_{photon} = h\nu$$

where $h$ is **Plank's constant** and $\nu$ is the **frequency** of light.

$$E = \frac{hc}{\lambda}$$

where $c$ is the speed of light and $\lambda$ is the wavelength of light.

The electron is restricted to specific energy levels in the atom. Specifically,

$$E = -\frac{A}{n^2}$$

where $A$ is $2.18 \times 10^{-11}$ erg, and $n$ is the quantum number.

## Pauli Exclusion Principle and Hund's Rule

The **Pauli exclusion principle** states that no two electrons within the same atom may have the same four quantum numbers.

**Hund's Rule** states that for a set of equal-energy orbitals, each orbital is occupied by one electron before any orbital has two. Therefore, the first electrons to occupy orbitals within a sublevels have parallel spins.

## Isotopes

If atoms of the same element (i.e., having identical atomic numbers) have different masses, they are called **isotopes.**

The relative abundance of the isotopes is equal to their fraction in the element.

The average atomic weight, $A$, is equal to

$$M_{avg} = X_1 M_1 + X_2 M_2 + \ldots + X_N M_N$$

where $M_i$ is the atomic mass of isotope $i$ and $X_i$ is the corresponding probability of occurrence.

There are slight differences in chemical behavior of the isotopes of an element. Usually, these differences, called isotope effects, influence the rate of reaction rather than the kind of reaction.

## Periodic Table

**Periodic law** states that chemical and physical properties of the elements are periodic functions of their atomic numbers.

Vertical columns are called **groups**, each containing a family of elements possessing similar chemical properties. The **horizontal** rows in the periodic table are called **periods.** The elements lying in two rows just below the main part of the table are called the **inner transition elements.** In the first of these rows are elements 58 through 71, called the **lanthanides** or **rare earths.** The second row consists of elements 90 through 103, the **actinides.** Group IA elements are called the **alkali metals.** Group IIA elements are called the **alkaline earth metals.** Group VIIA elements are called the **halogens,** and Group O elements, **noble gases.**

The metals in the first two groups are the light metals, and those toward the center are the heavy metals. The elements found along the dark line in the chart are called the **metalloids.** They have characteristics of both metals and nonmetals. Some examples of metalloids are boron and silicon.

## Properties Related to the Periodic Table

The most active metals are found in the lower left corner. The most active nonmetals are found in the upper right corner.

Metallic properties include high electrical conductivity, luster, generally high melting points, ductility (ability to be drawn into wires), and malleability (ability to be hammered into thin sheets). Nonmetals are uniformly very poor conductors of electricity, do not possess luster of metals and form brittle solids. Metalloids have properties intermediate between those of metals and nonmetals.

### Atomic Radii

The **atomic radius** generally decreases across a period from left to right. The atomic radius increases down a group.

### Electronegativity

The **electronegativity** of an element is a number that measures the relative strength with which the atoms of the element attract valence electrons in a chemical bond. This electronegativity number is based on an arbitrary scale from 0 to 4. Metals have electronegativities less than 2. Electronegativity increases from left to right in a period and decreases as you go down a group.

### Ionization Energy

**Ionization energy** is defined as the energy required to remove an electron from an isolated atom in its ground state. As we proceed down a group, a decrease in ionization energy occurs. Proceeding across a period from left to right, the ionization energy increases. As we proceed to the right, base-forming properties decrease and acid-forming properties increase.

# Types of Bonds

An **ionic bond** occurs when one or more electrons are transferred from the valence shell of one atom to the valence shell of another. The atom that loses electrons becomes a positive ion (cation), while the atom that acquires electrons becomes a negatively-charged ion (anion). The ionic bond results from the coulomb attraction between the oppositely-charged ions.

The **octet rule** states that atoms tend to gain or lose electrons until there are eight electrons in their valence shell.

A **covalent bond** results from the sharing of a pair of electrons between atoms. In a **nonpolar covalent bond**, the electrons are shared equally. Nonpolar covalent bonds are characteristic of homonuclear diatomic molecules. For example, the fluorine molecule:

$$\cdot \ddot{\underset{\cdot\cdot}{F}}: \quad \cdot \ddot{\underset{\cdot\cdot}{F}}: \quad \longrightarrow \quad :\ddot{\underset{\cdot\cdot}{F}}:\ddot{\underset{\cdot\cdot}{F}}:$$

Fluorine atoms ⟶ Fluorine molecule

When there is an unequal sharing of electrons between the atoms involved, the bond is called a **polar covalent bond**. An example:

H ˟ Cl:      ˟ hydrogen electron
                      • chlorine electrons

H ˟ O:      ˟ hydrogen electron
    H         • chlorine electrons

Because of the unequal sharing, the bonds shown are said to be polar bonds (dipoles). The more electronegative element in the bond is the negative end of the bond dipole. In each of the molecules shown here, there is also a non-zero molecular dipole moment, given by the vector sum of the bond dipoles.

A pure crystal of elemental metal consists of roughly $6.023 \times 10^{23}$ (Avogadro's number) atoms held together by metallic bonds.

# Intermolecular Forces of Attraction

A **dipole** consists of a positive and negative charge separated by a distance. A dipole is described by its **dipole moment**, which is equal to the charge times the distance between the positive and negative charges:

$$\text{net dipole moment} = \text{charge} \times \text{distance}$$

In polar molecular substances, the positive pole of one molecule attracts the negative pole of another. The force of attraction between polar molecules is called a dipolar force.

When a hydrogen atom is bonded to a highly electronegative atom, it will become partially positively-charged, and will be attracted to neighboring electron pairs. This creates a **hydrogen bond.** The more polar the molecule, the more effective the hydrogen bond is in binding the molecules into a larger unit.

The relatively weak attractive forces between molecules are called **Van der Waals forces.** These forces become apparent only when the molecules approach one another closely (usually at low temperatures and high pressure). They are due to the way the positive charges of one molecule attract the negative charges of another molecule. Compounds of the solid state that are bound mainly by this type of attraction have soft crystals, are easily deformed, and vaporize easily.

# Gas Laws

### Boyle's Law

Boyle's law states that, at a constant temperature, the volume of a gas inversely proportional to the pressure:

$$V \propto \frac{1}{P} \text{ or } V = \text{constant} \times \frac{1}{P} \text{ or } PV = \text{constant}.$$

$$P_i V_i = P_f V_f$$

### Charles' Law

Charles' law states that at constant pressure, the volume of a given quantity of a gas varies directly with the temperature:

$$\frac{V_1}{T_1} = \frac{V_2}{T_2} \text{ or } \frac{V_1}{V_2} = \frac{T_1}{T_2}$$

### Dalton's Law of Partial Pressures

The pressure exerted by each gas in a mixture is called its **partial pressure.** The total pressure exerted by a mixture of gases is equal to the sum of the partial pressures of the gases in the mixture.

$$P_T = P_a + P_b + P_c + \ldots$$

**Law of Guy-Lussac**

The law of Guy-Lussac states that at constant volume, the pressure exerted by a given mass of gas varies directly with the absolute temperature:

$$\frac{P_1}{T_1} = \frac{P_2}{T_2}$$

**Ideal Gas Law**

$$V \propto \frac{1}{P}, \quad V \propto T, \quad V \propto n$$

then

$$V \propto \frac{nT}{P}$$

$$PV = nRT$$

Other forms are: $PV = mRT$ and $Pv = RT$.

**Avogadro's Law (The Mole Concept)**

Avogadro's law states that under conditions of constant temperature and pressure, equal volumes of different gases contain equal numbers of molecules.

$$\frac{V_f}{V_i} = \frac{N_f}{N_i}$$

# The Kinetic Molecular Theory

The kinetic molecular theory is summarized as follows:

1. Gases are composed of tiny, invisible molecules that are widely separated from one another in otherwise empty space.
2. The molecules are in constant, continuous, random, and straight-line motion.
3. The molecules collide with one another, but the collisions are perfectly elastic (that is, they result in no net loss of energy).
4. The pressure of a gas is the result of collisions between the gas molecules and the walls of the container.

5. The average kinetic energy of all the molecules collectively is directly proportional to the absolute temperature of the gas. The average kinetic energy of equal numbers of molecules of any gas is the same at the same temperature.

## Equilibrium of Liquids, Solids, and Vapors

In a closed system, when the rates of evaporation and condensation are equal, the system is in **phase equilibrium**.

In a closed system, when opposing changes are taking place at equal rates, the system is said to be in **dynamic equilibrium**.

## Equilibrium Constants

The rate of an elementary chemical reaction is proportional to the concentrations of the reactants raised to powers equal to their coefficients in the balanced equation.

For $aA + bB \leftrightarrow eF + fF$,

$$\text{rate}_f = k_f[A]^a[B]^b,$$

$$\text{rate}_r = k_r[E]^e[F]^f$$

and

$$\frac{k_f}{k_r} = \frac{[E]^e[F]^f}{[A]^a[B]^b} = K_c$$

where $k_f$ and $k_r$ are **rate constants** for the forward and reverse reactions, respectively.

## Types of Chemical Reactions

The four basic kinds of chemical reactions are: **combination, decomposition, single replacement,** and **double replacement**. ("Replacement" is sometimes called "metathesis.")

Combination can also be called **synthesis**. This refers to the formation of a compound from the union of its elements. For example:

$$Zn + S \rightarrow ZnS$$

Decomposition, or **analysis**, refers to the breakdown of a compound into its individual elements and/or compounds. For example:

$$C_{12}H_{22}O_{11} \rightarrow 12\ C + 11\ H_2O$$

The third type of reaction is called **single replacement** or **single displacement**. This type can best be shown by some examples where one substance is displacing another. For example:

$$Fe + CuSO_4 \rightarrow FeSO_4 + Cu$$

The last type of reaction is called **double replacement** or **double displacement**, because there is an actual exchange of "partners" to form new compounds. For example:

$$AgNO_3 + NaCl \rightarrow AgCl + NaNO_3$$

# Measurements of Reaction Rates

The measurement of **reaction rate** is based on the rate of appearance of a product or disappearance of a reactant. It is usually expressed in terms of change in concentration of one of the participants per unit time:

$$\text{rate of chemical reaction} = \frac{\text{change in concentration}}{\text{time}}$$

$$= \frac{\text{moles/liter}}{\text{sec}}$$

# Factors Affecting Reaction Rates

There are five important factors that control the rate of a chemical reaction. These are summarized below:

1. The nature of the reactants and products, i.e., the nature of the transition state formed. Some elements and compounds, because of the bonds broken or formed, react more rapidly with each other than do others.
2. The surface area exposed. Since most reactions depend on the reactants coming into contact, increasing the surface area exposed, proportionally increases the rate of the reaction.
3. The concentrations. The reaction rate usually increases with increasing concentrations of the reactants.
4. The temperature. A temperature increase of 10°C above room temperature usually causes the reaction rate to double.
5. The catalyst. **Catalysts** speed up the rate of a reaction but do not change the equilibrium constant (i.e., it simply speeds up the rate of approach to equilibrium).

# The Arrhenius Equation: Relating Temperature and Reaction Rate

The following is the Arrhenius equation:

$$k = Ae^{-E_a/RT}$$

where $k$ is the rate constant, $A$ = the Arrhenius constant, $E_a$ = activation energy, $R$ = universal gas constant, and $T$ = temperature in Kelvin. $k$ is small when the activation energy is very large or when the temperature of the reaction mixture is low.

# CHAPTER 5

# WAVES AND OSCILLATIONS

From physics, wave and oscillation theory forms the basis for many engineering applications including electromagnetics, vibrations, radiation heat transfer, acoustics, optics, and electricity.

## Nomenclature and Definitions

**Mechanical vibration:** The oscillation of a body about an equilibrium position.
**Period:** Time taken by the system to complete one cycle.
**Frequency:** The number of cycles per unit time.
**Amplitude:** Maximum displacement from the equilibrium position.
**Free vibration:** Motion is sustained by restoring forces only.
**Forced vibration:** Caused by periodic forces applied external to the system.
**Undamped vibration:** Friction forces are neglected.
**Damped vibration:** Internal and external frictional forces and other damping forces are included.

## Simple Harmonic Oscillators

The movement of an object going through a periodic motion can be broken down into simple harmonic motions having frequencies which are multiples of the frequency of the periodic motion. Simple harmonic motion can be described by the equation:

$$m\frac{d^2x}{dt^2} = -fx$$

where $m$ is the mass, $x$ is the displacement, and $f$ is the force-per-unit displacement. If we let $w^2 = f/m$, the equation above becomes:

$$\frac{d^2x}{dt^2} + w^2x = 0$$

where $w$ is the angular frequency of vibration.

## Simple Harmonic Motion

**Simple Harmonic Motion:** Linear motion of a body where the acceleration is proportional to the displacement from a fixed origin and is always directed towards the origin. The direction of acceleration is always opposite to that of the displacement.

Equation of Motion

$$m\ddot{x} + kx = 0 \quad \text{or}$$
$$\ddot{x} + p^2x = 0$$

where $p^2 = \dfrac{k}{m}$

General Solution of Equation

$$x = c_1 \sin pt + c_2 \cos pt$$

where $c_1$ and $c_2$ may be obtained from initial conditions.

An alternate form of equation

$$x = x_m \sin(pt + \phi)$$

where $x_m$ = the amplitude, and $\phi$ is the phase angle.

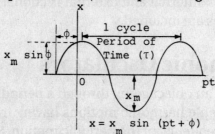

Figure 8. Period $= t = \dfrac{2\pi}{P}$    Frequency $= \dfrac{1}{t} = \dfrac{P}{2\pi}$

Maximum Values:

$$V_m = X_m p$$
$$a_m = X_m p^2$$

# Simple Pendulum

The simple pendulum consists of a weightless cord fixed at one end and has a particle at the other end that oscillates in an arc of a circle. The forces involved are gravity and tension.

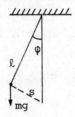

**Figure 9. Pendulum**

For small angles of vibration

$$\phi = \frac{s}{\ell}$$

Equation of Motion in Terms of $\phi$:

$$\ddot{\phi} + \frac{g}{\ell} \sin \phi = 0$$

If $\phi$ (in radians) is small, the $\sin \phi \approx \phi$. The equation then becomes

$$\ddot{\phi} + \frac{g}{\ell} \phi = 0.$$

The solution is given by

$$\phi = \phi_0 \cos(\omega_0 t + \alpha_0)$$

where    $\omega_0 = \sqrt{\frac{g}{\ell}};$

$\phi_0$ = Max. amplitude of oscillation and
$\alpha_0$ = Phase factor.

The period of oscillation is given by

$$\tau_0 = \frac{2\pi}{\omega_0} = 2\pi\sqrt{\frac{\ell}{g}}$$

## Damped Harmonic Motion

All vibrating bodies lose energy to frictional forces and, thereby, radiate sound energy into the surrounding medium.

Including friction, the equation of simple harmonic motion becomes:

$$\frac{d^2x}{dt^2} + 2K\frac{dx}{dt} + w^2x = 0$$

This is the equation of damped harmonic motion. The displacement can be solved for:

$$x = Ac^{-kt}\cos(w't - \theta)$$

where

$$w' = \sqrt{w^2 - k^2}.$$

## Free Vibration

Consider the spring-mass system in Figure 10 in which the mass, after being displaced, vibrates freely and indefinitely, neglecting all resistances. Such a condition is known as free vibration.

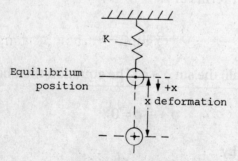

Figure 10. Spring system

Using Newton's second law, the equation of motion becomes

$$\frac{d^2x}{dt^2} = -w^2x,$$

where

$$\omega = \sqrt{\frac{Kg}{W}}.$$

General solution of equation
$$x = A\sin\omega t + B\cos\omega t$$
Constants $A$ and $B$ are determined from initial conditions.

Period:
$$\tau = 2\pi\sqrt{\frac{W}{Kg}}$$

$W$ = Weight of body

Note: $\tau$ depends only on the weight of the body and the spring constant $K$.

Frequency:
$$f = \frac{1}{2\pi}\sqrt{\frac{Kg}{W}}$$

Specifying initial conditions as $x_0$ and $v_0$ at $t = 0$,

$$x = \frac{v_0}{\omega}\sin\omega t + x_0 \cos\omega t$$
$$= C\cos(\omega t - \alpha),$$

where

$$C = \sqrt{x_0^2 + \left(\frac{v_0}{\omega}\right)^2}, \text{ and } \alpha \text{ is the phase angle, equal to}$$

$$\tan^{-1}\left(\frac{v_0}{\omega x_0}\right)$$

The equation may be graphically interpreted as below:

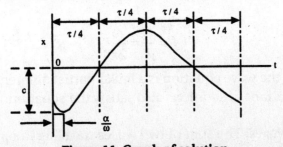

Figure 11. Graph of solution

# Resonance

At resonance, the amplitude of the displacement is at its maximum. The frequency at which resonance occurs is given by:

$$\omega_r = \sqrt{\omega^2 - 2k^2}$$

Note: This frequency differs from the frequency of the damped harmonic motion.

# Waves

The typical wave may be written as

$$\frac{\partial^2 q}{\partial t^2} = v^2 \frac{\partial^2 q}{\partial y^2}$$

where

$v$ is the wave speed equal to $\sqrt{\dfrac{Ka^2}{m}}$ or $\sqrt{\dfrac{s}{\rho}}$

where  $K = \dfrac{s}{a}$,

$s = $ The tension of the string,
$a = $ The distance between particles at equilibrium and
$\rho = $ The density of the string.

The general solution of the equation may be expressed as

$$q = \alpha(y+vt) + \phi(y-vt)$$

**Sinusoidal Waves:** Given that $q = q(y, t)$ then

$$q = c_1 g\left(\frac{2\pi}{\lambda}(y+vt)\right) + c_2 g\left(\frac{2\pi}{\lambda}(y-vt)\right)$$

where $g$ represents either sin or cos and $\lambda$ is the wavelength.

Frequency:

$$f = \frac{v}{\lambda} = \frac{\omega}{2\pi}$$

Note: Because the wave equation is a linear partial differential equation, linear combinations of solutions also satisfy the equation.

**Standing waves:** The sum of two waves traveling in opposing directions with equal amplitudes results in standing waves which are represented by the following equation:

$$q = c \sin\left(\frac{2\pi y}{\lambda}\right) \cos \omega t$$

where

$$\omega = \frac{2\pi \upsilon}{\lambda}.$$

# Speed of Sound

Wave Speed in Fluid

$$\upsilon = \sqrt{\frac{B}{\rho_0}} \quad \rightarrow \quad \text{units}: \frac{\text{meters}}{\text{seconds}}$$

$\upsilon$ = Speed of sound in a fluid
$B$ = Modulus of elasticity
$\rho_0$ = Density of medium

Note: Gas is also a fluid, yet a more precise equation may be given, as seen below.

Wave Speed in a Gas

$$\upsilon_g = \sqrt{\frac{\gamma p_0}{\rho_0}} \quad \rightarrow \quad \text{units}: \frac{\text{meters}}{\text{seconds}}$$

(Ideal gas)

$\upsilon_g$ = Speed of sound
$\gamma$ = Ratio of specific heats for a gas
$p_0$ = Undisturbed pressure
$\rho_0$ = Density of medium

# Intensity of Sound

Average Intensity

$$I = \frac{1}{2} \frac{P_m^2}{\sqrt{B\rho_0}}$$

$I$ = Average intensity
$P_m$ = Pressure amplitude
$B$ = Bulk modulus of elasticity
$\rho_0$ = Density of medium

# Longitudinal and Transverse Waves

### Longitudinal Waves

Longitudinal waves are so named because the displacement from equilibrium is along the direction of propagation.

### Transverse Waves

The displacement in transverse waves is perpendicular to the direction of propagation.

# Speed of Light

$$c = \frac{1}{\sqrt{\varepsilon_0 \mu_0}} \quad \text{units:} \quad \frac{\text{meters}}{\text{second}}$$

$c$ = Speed of light (electromagnetic radiation) in free space
$\varepsilon_0$ = Electric permittivity constant
$\mu_0$ = Magnetic permeability constant

# The Poynting Vector

Definition: The rate of energy transport per unit area in an electromagnetic wave may be described as the **Poynting vector S.**

### Vector Equation

$$\mathbf{S} = \frac{1}{\mu_0} \mathbf{E} \times \mathbf{B} \quad \text{units:} \quad \frac{\text{watts}}{\text{meter}^2} \quad \text{(mks system)}$$

$\mathbf{S}$ = Rate of energy
$\mu_0$ = Permeability constant
$\mathbf{E}$ = Electric field
$\mathbf{B}$ = Magnetic field

# Power and Intensity for Electromagnetic Waves

The **intensity** of an electromagnetic wave is the average power per unit area and is described by the equation:

$$I = \tfrac{1}{2} c \varepsilon_0 E_0^2$$

where

    $c$ = Speed of light
    $\varepsilon_0$ = Permittivity of free space
    $E_0$ = Amplitude of the electric field

# Speed and Propagation of Electromagnetic Waves

### Speed of an Electromagnetic Wave

The speed of an electromagnetic wave in any medium is given by:

$$V = \frac{1}{\sqrt{\varepsilon \mu}}$$

where

    $\varepsilon$ = Permittivity of the medium
    $\mu$ = Permeability of the medium

The **index of refraction** is the measure of how the speed of an electromagnetic wave is reduced in a medium other than a vacuum.

The index of refraction is:

$$n = \sqrt{\frac{\varepsilon \mu}{\varepsilon_0 \mu_0}}$$

where

    $\varepsilon_0$ = Permittivity of free space
    $\mu_0$ = Permeability of free space

Therefore, the speed of an electromagnetic wave in any medium is:

$$V = \frac{c}{n}$$

## Propagation of an Electromagnetic Wave

The direction of propagation is indicated by the propagation vector, K. The magnitude of the propagation vector is the wave number $K$, which relates the speed of a wave to its frequency.

$$V = \frac{2\pi f}{K}$$

# Doppler Effect

The frequency of an electromagnetic wave depends on the motion of the source, the receiver, and the medium. Accounting for the Doppler effect, the frequency becomes:

$$f_D = f\left(\frac{v_s + v_{med} - v_{rec}}{v_s + v_{med} - v_{source}}\right)$$

where $v_s$ is the speed of wave, $v_{med}$ is the speed of the medium, $v_{rec}$ is the speed of the receiver, and $v_{source}$ is the speed of the source.

The direction of $v_{rec}$ and $v_{source}$ determine whether the frequency increases or decreases. Doppler's principle states that when the receiver and source are moving toward one another the frequency increases and when they are moving apart the frequency decreases.

# Interference

Interference is the combination of two or more waves. Since any wave can be expressed as a linear combination of sinusoidal waves with different frequencies, the interference of two or more waves can itself be expressed as a singular linear combination of sinusoidal waves.

# Interference Maxima and Minima

**Maxima**

$$d \sin \theta = m\lambda$$

where $m = 0, 1, 2, \ldots$ (maxima)

**Minima**

$$d\sin\theta = \left(m - \tfrac{1}{2}\right)\lambda$$

where

$m$ = 1, 2, ... (minima)
$d$ = Distance between slits
$\theta$ = Angle between the midpoint of the two slits and a point on the screen
$\lambda$ = Wavelength

# CHAPTER 6

# ELECTRICAL PRINCIPLES AND CIRCUITS

Knowledge of the fundamentals of electrical circuits facilitates the understanding of those branches of engineering, other than electrical engineering, where appropriate laws apply. Electrical circuit analogies are used in mechanical engineering in the study of fluid dynamics and heat transfer. Armed with a few fundamentals the engineer can solve complex interdisciplinary problems when cast in circuit theory.

## Current, Voltage, Power, and Energy

### Current

The **current** is the measurement of the rate of the number of charges moving through a given reference point in a circuit in 1 second. For steady current,

$$i = \frac{q}{t}$$

where $q$ is the net charge passing through the point in $t$ seconds.

Unit of current = ampere (A) = 1 coulomb of charge moving past a point in 1 second.

Instantaneous current ($i$) = time rate of change of charge = $\dfrac{dq}{dt}$.

**Current flow:** The current flow in a wire is opposite to the motion of the electrons by convention. (See Figure below.)

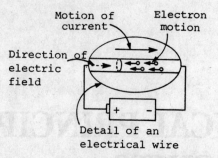

Figure 12. Electric current

Note: Current flow in the opposite direction of the Figure is given a negative sign.

### Voltage

The **voltage** (V), or the potential difference between two points, is the measure of the work required to move a unit charge from one point to another.

$$\text{Unit of voltage} = \text{volt} = 1 \frac{\text{joule}}{\text{coulomb}}$$

**Voltage Sign Convention:** Assume a positive current supplied by an external source is entering terminal 1. Then,

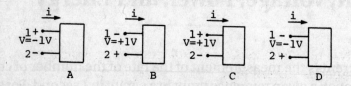

Figure 13. Voltage convention

Terminal 1 is 1 volt positive with respect to terminal 2 and terminal 2 is 1 volt positive with respect to terminal 1.

### Power and Energy

($p$) power [watts] = $v$ [volts] × $i$ [amperes]

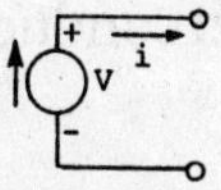

**Figure 14. Voltage diagram**

**Energy:** Since power ($p$) is the time rate of energy transfer $p = \dfrac{dW}{dt}$

$$W = \int_{t_1}^{t_2} p\,dt$$

(the energy transferred during a given time interval $x$) or $W$ (energy in watt-seconds or joules) = $p$ (power in watts) × $t$ (time in seconds).

## Resistance

Resistance is the measure of the tendency of a material to impede the flow of electric charges through it.

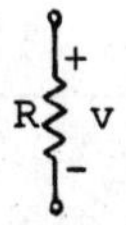

**Figure 15. Circuit symbol**

$R$ = resistance of the resistor having units [volts/ampere] or ohm ($\Omega$).

### Resistivity

Resistivity is the characteristic of a material which indicates how much a material impedes current flow

$$R = \frac{\rho \ell}{A} \quad \text{(at constant temperature)}$$

$R$ = Resistance in ohms
$\ell$ = Length [m]
$A$ = Cross-sectional area [m²]
$\rho$ = Resistivity [$\Omega$ – m]

**Figure 16. Current flow**

Note: Resistivity is low in a good conductor but high in a poor conductor (insulator).

## Resistor Combinations

For **series** combination of $N$ resistors:
$$R_{eq} = R_1 + R_2 + \ldots + R_N$$

For **parallel** combination of $N$ resistors:
$$\frac{1}{R_{eq}} = \frac{1}{R_1} + \frac{1}{R_2} + \ldots + \frac{1}{R_N}$$

# Voltage and Current Division

## Voltage Division

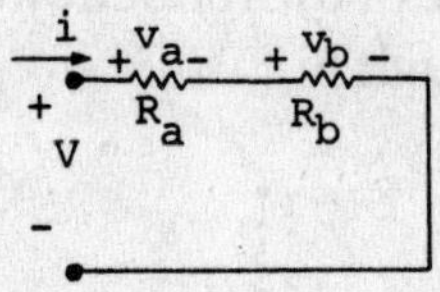

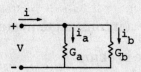

Figure 17. Circuit with voltage division

The formula for voltage division is:
$$V_b = \frac{R_b}{R_a + R_b} V$$

## Current Division

Figure 18. Circuit with current division

The formula for a circuit with a current division is:
$$i_b = \frac{R_a}{R_a + R_b} i$$

# Ohm's Law

Ohm's Law states the voltage across a conducting material is directly proportional to the current through the material, i.e., $v = Ri$, where $R$ (resistance) is the proportionality constant.

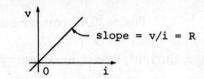

**Figure 19. Ohm's law**

Hence, absorbed power in a resistor is given by

$$p = Vi = i^2 R = \frac{V^2}{R}.$$

Note: This power is in the form of heat because a resistor is a passive element; it neither delivers power nor stores energy.

**Short circuit:** Circuit as a zero ohms resistance, i.e., voltage across a short circuit = 0.

**Open circuit:** Circuit as an infinite resistance, i.e., current across an open circuit = 0.

# Kirchhoff's Law

### Kirchhoff's Current Law

Kirchhoff's current law states the algebraic sum of all currents entering a node equals the algebraic sum of all currents leaving it. For a given node, $\Sigma$ currents entering = $\Sigma$ currents leaving or

$$\sum_{n=1}^{N} i_n = 0.$$

### Kirchhoff's Voltage Law

Kirchhoff's voltage law states the algebraic sum of all voltages around a closed loop (or path) is zero. For a closed loop, $\Sigma$ potential rises = $\Sigma$ potential drops.

$$\sum_{n=1}^{N} v_n = 0$$

# Capacitance

**Circuit Symbol:**

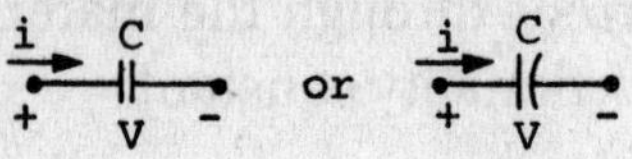

$$\overset{i}{\underset{+\quad v\quad -}{\bullet\!\!-\!\!|\!|\!-\!\!\bullet}} \quad \text{or} \quad \overset{i}{\underset{+\quad v\quad -}{\bullet\!\!-\!\!|(\!-\!\!\bullet}}$$

**Figure 20. Circuit symbol**

**Capacitor Voltage, Current, Power, and Energy:**

$$v(t) = \frac{1}{C} \int_{-\infty}^{t} i(\tau)d\tau$$

$$i(t) = \frac{dq(t)}{dt} = C\frac{dv(t)}{dt}$$

$$p = Cv\left(\frac{dv}{dt}\right)$$

and

$$W = \frac{1}{2}Cv^2 = \text{stored energy [joules]}$$

or

$$W = \frac{Q^2}{2C}$$

where $Q$ is the charge and is defined as $Q = C_v$.

**Characteristic:** A capacitor acts as an open circuit to $dc$.

# Inductor

**Concept of Self-Inductance:**

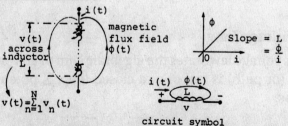

**Figure 21. Self-inductance**

**Inductance** $(L) = \dfrac{N d\phi(i)}{di}$ [henry ($H$) or volt-second/ampere].

Voltage $v(t) = N$ (no. of turns of coil) $\times \dfrac{d\phi(t)}{dt}$

(rate of change of $\phi$ with respect to time), or

$$v(t) = \dfrac{N d\phi(i)}{di} \dfrac{di}{dt} = L \dfrac{di(t)}{dt}$$

$$i(t) = \dfrac{1}{L} \int_{-\infty}^{t} v\, dt$$

$$W = \dfrac{1}{2} L i^2$$

$$p = vi = Li\dfrac{di}{dt}\,[W]$$

An inductor acts as a short circuit to $dc$.

## Iron-Core Transformer

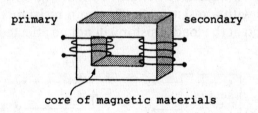

**Figure 22. Transformer**

**Turns ratio:** The determination of how much a transformer steps up or steps down a voltage.

$$\text{Turns ratio} = \dfrac{\text{no. of turns on the primary}\,(N_p)}{\text{no. of turns on the secondary}\,(N_s)}$$

or $\dfrac{V_p}{V_s} = \dfrac{N_p}{N_s}$ and $\dfrac{I_p}{I_s} = \dfrac{N_s}{N_p}$

**Voltage step-up transformer $N_p < N_s$:**

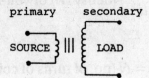

Figure 23. Voltage step-up transformer

**Voltage step-down transformer $N_p > N_s$:**

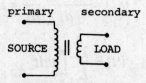

Figure 24. Voltage step-down transformer

# Thevenin's and Norton's Theorems

### Thevenin's Theorem

In any linear network, it is possible to replace everything except the load resistor by an equivalent circuit containing only a single voltage source in *series* with a resistor ($R_{th}$ Thevenin resistance), where the response measured at the load resistor will not be affected.

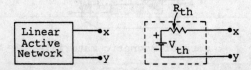

Figure 25. Voltage source in series with a resistor

### Procedures to find Thevenin equivalent:
1) Solve for the open circuit voltage $v_{oc}$ across the output terminals.
$$V_{oc} = V_{th}$$
2) Place this voltage $v_{oc}$ in series with the Thevenin resistance which is the resistance across the terminals found by setting all independent voltage and current sources to zero. (i.e., short circuits and open circuits, respectively.)

## Norton's Theorem

Given any linear circuit, the passive and active components can be converted into an equivalent two-terminal network consisting of a single current source in *parallel* with a resistor ($R_N$ – Norton resistance).

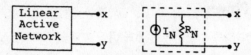

Figure 26. Voltage source in parallel with a resistor

**Procedures to find Norton equivalent:**
1) Set all sources to zero (i.e., voltage sources → short circuits, and current sources → open circuits). Then find resulting resistance $R_N$ between the output terminals.

2) $I_N$ is the current through a short circuit applied to the two terminals of the given network.

The Norton's equivalent is obtained by connecting current source $I_N$ and $R_N$ in parallel.

# Alternating Current Circuits

## Sinusoidal Forcing Function – General Form

$$v(t) = V_m \sin(\omega t + \theta)$$

where

$V_m$ = Maximum value

$f$ = Frequency = $\dfrac{1}{T}$ = $\dfrac{\text{cycles}}{\text{sec}}$ = hertz (Hz)

$T$ = Period (time duration of 1 cycle = sec)

$\omega$ = Angular frequency = $2\pi f$ = $\dfrac{2\pi}{T}$ = radians/sec

$\theta$ = The lead or lag of the current depending on the expression for the current.

## Phasor Notation

In general, the phasor form of a sinusoidal voltage or current is

$$V = V_m \angle \theta \quad \text{and} \quad I = I_m \angle \theta.$$

Thus, for the voltage source $v(t) = V_m \cos \omega t$, the corresponding phasor form is $V_m \angle 0°$. For the current response $i(t) = I_m \cos(\omega t + \theta)$, the corresponding phasor form is $I_m \angle \theta°$.

**Instantaneous Power**

$$p = vi$$

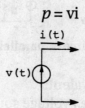

Figure 27. Passive element

Note: $p$ = positive energy transfer from source to network.
$p$ = negative transfer from network to source.

In a resistive circuit, $p = i^2 R = \dfrac{v^2}{R}$.

In a inductive circuit, $p = Li\dfrac{di}{dt} = \dfrac{1}{L} v \int_{-\infty}^{t} v\, dt$.

In a capacitive circuit, $p = Cv\dfrac{dv}{dt} = \dfrac{1}{C} i \int_{-\infty}^{t} i\, dt$.

**Average Power**

Average power $(P) = \dfrac{1}{2} V_m I_m \cos\theta$

$$= \underbrace{V_{rms} I_{rms} \cos\theta}_{\text{apparent power}}$$

where

$$V_{rms} = \dfrac{V_m}{\sqrt{2}}$$

$$I_{rms} = \dfrac{I_m}{\sqrt{2}}$$

Note: $rms$ = effective values

# Magnetic Force on a Differential Current Element

$$\bar{F} = Q\bar{U} \times \bar{B} \quad \text{and} \quad d\bar{F} = dQ\bar{U} \times \bar{B}$$

$$d\bar{F} = \bar{J} \times \bar{B} dv$$
$$d\bar{F} = \bar{K} \times \bar{B} ds$$
$$d\bar{F} = I d\bar{L} \times \bar{B}$$

where

$$\bar{J} = \rho \bar{U}$$
$$\bar{J} dv = \bar{K} ds = I d\bar{L}$$

$$\bar{F} = \int_{vol} \bar{J} \times \bar{B} dv$$
$$\bar{F} = \int_s \bar{K} \times \bar{B} ds$$
$$\bar{F} = \oint I d\bar{L} \times \bar{B} = -I \oint \bar{B} \times d\bar{L}$$
$$\bar{F} = I\bar{L} \times \bar{B}$$

The current density is a vector and it is defined in terms of the charge density and the drift density as

$$\bar{J} = \rho \bar{U} \frac{A}{m^2}$$

# CHAPTER 7

# MECHANICS

Mechanics is the science which describes and predicts the conditions of rest or motion of bodies under the action of forces. It is a physical science, since it deals with the study of physical phenomena. Mechanics is the foundation to most engineering sciences and is an indispensable prerequisite to their study.

## Newton's Laws

**First Law**

Every body remains in its state of rest or uniform linear motion, unless a force is applied to change that state.

A. This law describes a common property of matter, called **inertia.**
B. It defines a reference system called the Newtonian or inertial reference system. Rotating or accelerating systems are not inertial.
C. The quantitative measure of inertia is called mass.

**Second Law**

If the vector sum of the forces $F$ acting on a particle of mass $m$ is different from zero, then the particle will have an **acceleration, $a$,** directly proportional to, and in the same direction as $F$, but inversely, proportional to mass $m$. Symbolically:

$$F = ma$$

### Third Law

For every action, there exists a corresponding equal and opposing reaction, or the mutual actions of two bodies are always equal and opposing.

$$F_A = -F_B$$

## Equilibrium of a Particle

If the resultant force acting on a particle is zero, then the particle is in equilibrium.

Equation of equilibrium:

$$\Sigma F = 0$$

$$\Sigma F_x i + \Sigma F_y j + \Sigma F_z k = 0$$

$\Sigma F$ is the vector sum of all the forces acting on the particle, and these forces are resolved into their respective $i, j,$ and $k$ directions.

Component equations:

$$\Sigma F_x = 0$$

$$\Sigma F_y = 0$$

$$\Sigma F_z = 0$$

These three equations can solve for no more than three unknowns in a problem.

## Moment of a Force About a Point

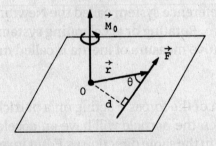

Figure 28. Moment diagrams

Movement of $F$ about point 0:

$$M_0 = r \times F$$

# Equilibrium of Rigid Bodies

The necessary and sufficient conditions for equilibrium of a rigid body are that the vector sum of all the forces and couple moments (about any point in space) equal zero:

$$\Sigma F = 0$$

$$\Sigma M_0 = \Sigma(r \times F) = 0$$

# Dry Friction

Dry Friction (also called coulomb friction) is friction between bodies in the absence of lubrication.

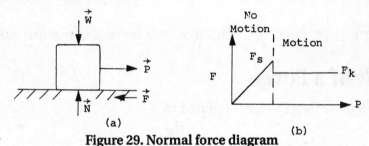

Figure 29. Normal force diagram

$N$ = Normal Force
$F$ = Static Friction Force

As $P$ increases (refer to Figure 1), the body remains in equilibrium until $F$ reaches a maximum value $F_s$. If $P$ continues to increase, then $F$ drops to a lower value, $F_k$, called kinetic-friction force.
$F_s$ becomes, $F_s$ called the maximum static friction force.

Coefficient of **static friction**

$$\mu_s = \frac{F_s}{N}$$

Coefficient of **kinetic friction**

$$\mu_k = \frac{F_k}{N}$$

A) During equilibrium phase:

$$F_s < \mu_s N$$

$\mu_s$ = Coefficient of Static Friction
$N$ = Normal Force

B) At the point of impeding motion:
$$F_s = \mu_s N$$

C) When motion begins:
$$F_k < \mu_k N$$
$\mu_k$ = Coefficient of Kinetic Friction

# Moment of Inertia

### Area
General Formula:

$$I = \int_A s^2 dA$$

where $s$ = perpendicular distance from the axis to the area element.

# Work of a Force

Work done by a force $F$ is defined as
$$dw = F \times ds$$

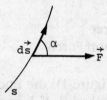

$$dw = Fds \cos\alpha$$

Work is a scalar quantity. Its unit is force-length.
Special cases:

A) $\xrightarrow{\;\;\;ds\;\;\;} F$     $dw = Fds$

B) $\xleftarrow{\;\;\;ds\;\;\;} F$     $dw = -Fds$

C) $\Big\uparrow^{ds} \rightarrow F$     $dw = 0$

The total work of the internal forces holding particles of a body together, is zero.

Work of a Moment, $M$:
$$dw = M d\theta$$
$\theta$ = Angle the body has rotated

## Mechanical Efficiency

**Ideal machines:** Input work equals output work
**Real machines:** Output work is always less then input work because of losses due to friction.
**Mechanical efficiency** is defined as:
$$\eta = \frac{\text{output work}}{\text{input work}}$$

For an ideal machine: $\eta = 1$.

## Motion of a Particle

The motion of a particle is known if its position is known for all values of time $t$.

If acceleration is given in terms of $t$,
$$a = f(t)$$
Velocity: $dv = a\,dt$
$$v - v_0 = \int_0^t f(t)\,dt$$

## Uniform Motion

A. For constant velocity:
$$x = x_0 + vt$$

B. For constant acceleration:
$$v = v_0 + at$$
$$x = x_0 + v_0 t + \frac{1}{2} a t^2$$
$$v^2 = v_0^2 + 2a(x - x_0)$$

$x_0$, $v_0$ are initial values for $x$ and $v$.

# Relative Motion

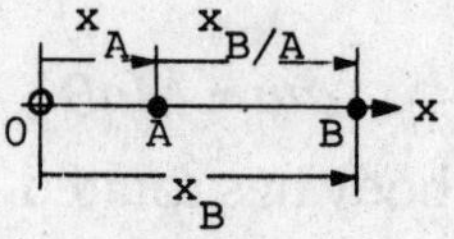

**Figure 31. Relative motion**

Relative position: $x_B = x_A + x_{B/A}$

Relative velocity: $v_B = v_A + v_{B/A}$

Relative acceleration: $a_B = a_A + a_{B/A}$

# Equations of Motion

### Rectangular Components

$$\Sigma(F_x i + F_y j + F_z k) = m(a_x i + a_y j + a_z k)$$

or

$$\Sigma F_x = m\ddot{x}$$
$$\Sigma F_y = m\ddot{y}$$
$$\Sigma F_z = m\ddot{z}$$

# Dynamic Equilibrium

A particle is in **dynamic equilibrium** if the sum of all forces acting on the particle, including inertial forces, is equal to zero:

$$\Sigma F - ma = 0$$

The term $-ma$ is called the **inertia vector**.

The first term $\Sigma F$ is sometimes called effective forces, i.e., it is the resultant of all external forces.

When expressed in tangential and normal form:
1. The tangential component of inertia measures the resistance of the particle to change in speed.
2. The normal component of inertia, also called centrifugal

force, measures the resistance of the particle to leave its curved path.

# Velocity-Dependent Force

The force acting on a particle can be a function of time. Some examples of such forces are:

A. Viscous resistance
B. Air resistance

General equation:

$$F(v) = m\frac{dv}{dt}$$

The equation can be integrated to solve for time:

$$t_2 - t_1 = \int_1^2 \frac{m}{F(v)} d(v)$$

A second integration yields position:

$$x_2 - x_1 = \int_1^2 v(t) dt$$

A. **Linear Case:** Assume the fluid resistance is proportional to the first power of velocity. The differential equation of motion:

$$-(mg + cv) = m\frac{dv}{dt}$$

where $g$ = gravitational constant and $c$ is a function of viscosity and geometry of the object. If $c$ is constant then,

$$v = -\frac{mg}{c} + \left(\frac{mg}{c} + v_0\right)e^{-ct/m}$$

where $v_0$ = initial velocity.

As the $t \to \infty$, the equation becomes

$$v_t = -\frac{mg}{c}$$

$v_t$ is the terminal velocity.

A second integration of the equation yields:

$$x - x_0 = \int_0^t v(t)dt = -\frac{mg}{c}t + \left(\frac{m^2 g}{c^2} + \frac{mv_0}{c}\right)\left(1 - e^{-ct/m}\right)$$

Characteristic time is defined as:

$$\tau = \frac{m}{c}$$

$$v = -v_t + (v_t + v_0)e^{-t/\tau}$$

B. **Quadratic Case:** If the fluid resistance is proportional to the square of velocity.

Differential equation of motion:

$$-mg \pm cv^2 = -m\frac{dv}{dt}$$

# Power

**Power** is defined as the time-rate of change of work and is denoted by $dw/dt$,

$$\text{Power} = \frac{dw}{dt} = F \times v$$

# Work-Energy Principle

**Kinetic energy** for a particle of mass $m$ and velocity $v$ is defined as

$$\text{K.E.} = \frac{1}{2}mv^2$$

Kinetic energy is the energy possessed by a particle by virtue of its motion.

**Principle of Work and Energy:** Given that a particle undergoes a displacement under the influence of a force $F$, the work done by $F$ equals the change in kinetic energy of the particle.

$$w_{1-2} = (KE)_2 - (KE)_1$$

## Potential Energy

The stored energy of a body or particle in a force field associated with its position from a reference frame.

$$PE = mgz$$

where $g$ is the gravitional constant and $z$ is the height of the reference point.

If potential energy is denoted by $PE$, then

$$w_{1-2} = (PE)_1 - (PE)_2$$

where $PE$ = weight $(x)$. A negative value of $w_{1-2}$ is associated with an increase in potential energy.

## Conservation of Energy

For a particle under the action of conservative forces:

$$(KE)_1 + (PE)_1 = (KE)_2 + (PE)_2 = E$$

The sum of kinetic and potential energy at a given point is constant. The equation above can also be written as:

$$E = \tfrac{1}{2} mv^2 + (PE)$$

# Momentum

## Linear Momentum

Linear momentum is defined as

$$P = mv$$

It is directed in the same direction as the vector $v$.

Results:
A. Newton's first law can be written as:
   If the resultant force acting on a particle is zero, then the linear momentum $P$ of the particle is constant. This is the Law of Conservation of Linear Momentum.

B. Newton's second law can be expressed as

$$\Sigma F = \frac{d}{dt} mv = \frac{dP}{dt}$$

i.e., the resultant force is equal to the rate of change of linear momentum.

C. Newton's third law:
For two particles A and B,
$$P_A + P_B = \text{constant}$$

**Angular Momentum**

Angular momentum is defined as the moment about the origin 0 of the angular momentum vector $mv$. It is denoted by $H_0$:
$$H_0 = r \times mv$$

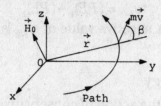

Figure 32. Path vector

Note: The angular momentum vector acts in a direction perpendicular to the plane containing the position and the linear momentum vectors.

# Impulse and Momentum

An alternate method to solve problems in which forces are expressed as functions of time. It is applicable to situations when forces act over a small interval of time.

**Linear Impulse-Momentum Equation:**

$$\int_1^2 F dt = \text{impulse} = mv_2 - mv_1$$

Figure 6 expresses the idea that the vector sum of the initial momentum and impulse equals the final momentum of the particle.

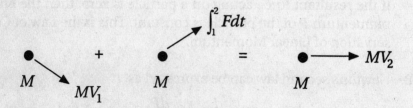

Figure 33. Impulse

Impulse is a vector quantity and acts in the direction of the force if the force remains constant.

$$\int_1^2 F\,dt = i\int_1^2 F_x\,dt + j\int_1^2 F_y\,dt + k\int_1^2 F_z\,dt$$

To obtain solutions, it is necessary to replace the equation with its component equations.

When several forces are involved, the impulse of each must be considered:

$$mv_1 + \Sigma\int_1^2 F\,dt = mv_2$$

When a problem involves more than one particle, each particle must be considered separately and then added:

$$\Sigma mv_1 + \Sigma\int_1^2 F\,dt = \Sigma mv_2$$

If there are no external forces, conservation of total momentum results:

$$\Sigma mv_1 = \Sigma mv_2$$

Note: There are cases in which an impulse is exerted by a force which does no work on the particle. Such force should be considered when applying the principle of impulse and momentum in solving problems.

# CHAPTER 8

# THERMODYNAMIC SYSTEMS

Thermodynamics is the study of heat and work. Thermodynamic principles are fundamental to the study of fluid dynamics and heat transfer. Since it is a science, thermodynamics is experimentally based with laws which have evolved from observations. Some of these experiments are easily reproduced and are instructive to those learning the subject.

## Thermodynamic Systems

The term **system** as used in thermodynamics refers to a definite quantity of matter bounded by some closed surface which is impervious to the flow of matter. This surface is called the **boundary** of the system. Everything outside the boundary of a system constitutes its **surroundings.**

Depending on the nature of the boundary involved, we can classify a thermodynamic system in one of the following three categories:

1. An **isolated system** allows neither matter nor energy transfer across the boundary.
2. An **open system** allows exchange of both matter and energy.
3. A **closed system** allows only exchange of energy.

## Properties of Systems

The state of a system is its condition as identified by coordinates which can usually be observed quantitatively, such as volume, density, temperature, etc. These coordinates are called properties.

All properties of a system can be divided into two types:

1. An **intensive** property is independent of the mass (pressure, density, temperature). Two independent intensive properties are required to fix the state of a pure simple compressible substance.
2. An **extensive** property has a value which varies directly with the mass (volume, energy, entropy). Extensive properties divided by mass are called **specific properties** and are treated as intensive.

# Processes

When a thermodynamic system changes from one state to another, it is said to execute a process, which is described in terms of the end states. A cycle is a series of processes in which the initial and end states are identical.

Processes are classified according to the following categories:

1. An **isothermal** process is a constant-temperature process.
2. An **isobaric process** is a constant-pressure process.
3. An **isometric process** is a constant-volume process.
4. An **adiabatic process** is a process in which heat does not cross the system boundary.
5. A **quasistatic process** consists of a succession of equilibrium states, such that at every instant the system involved departs infinitesimally from the equilibrium state.
6. In a **reversible process**, the initial state of the system involved can be restored with no observable effects in the system and its surroundings. This is called an Ideal Process.
7. In an **irreversible process**, the initial state of the system involved cannot be restored without observable effects in the system and its surroundings.

# Thermodynamic Equilibrium

When a system is not subject to interactions and a change of state cannot occur, then the system is in a state of equilibrium. There are four kinds of equilibrium: stable, neutral, unstable, and metastable. Of these, a stable equilibrium is one of the most encountered in thermodynamics. A system is in a state of stable equilibrium if a finite change of state of the system cannot occur without leaving a corresponding, finite alteration in the state of the environment.

A system is in thermal equilibrium when its temperature is uniform throughout, and equal to the temperature of the surroundings.

A system is in mechanical equilibrium when it has no unbalanced force within it and when the force it exerts on its boundary is balanced by external forces.

A system is in chemical equilibrium when the chemical composition of the system remains unchanged.

## Zeroth Law of Thermodynamics

When two systems are each in thermal equilibrium with a third system, they are also in thermal equilibrium with each other.

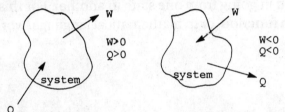

Figure 34. Signs of work and heat related with a system

Note: This sign convention is not a universal convention.

## Pressure-Volume Work

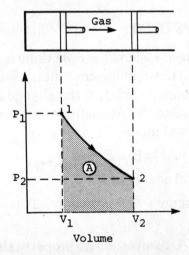

Figure 35. Gas in a cylinder

Consider as a system the gas contained in a cylinder as shown in

Figure 35.
For any small expansion in which the gas increases in volume by $dv$, the work done by the gas is

$$W = \int_1^2 p\,dv$$

where $p$ is the pressure exerted on the piston.

The integral $\int_1^2 p\,dv$ is the area under the curve on the P-V diagram. Since we can go from state 1 to state 2 along many different paths, it is evident that the amount of work involved in each case is a function not only of the end states of the process, but it is also dependent on the path that is followed in going from one state to another. For this reason, work is called a **path function**, or in mathematical language, work is an inexact differential.

## Heat

Heat is defined as the form of energy that is transferred across the boundary of a system at a given temperature to another system at a lower temperature by virtue of the temperature difference between the two systems.

Positive heat transfer is heat addition to a system and negative heat transfer is heat rejection by the system.

## Energy of a System

Since for a given closed system, the work done is the same in all adiabatic processes between the equilibrium states, it follows that a property of the system can be defined such that the change between two equilibrium end states is equal to the adiabatic work. We define this property as the **energy**, $E$, of the system or

$$\Delta E = E_2 - E_1 = W_{adiabatic}$$

For a system the total energy is,

$E$ = Internal energy + Kinetic energy + Potential energy

where:
1. **Internal energy** $(U)$ is an extensive property, since it depends upon the mass of the system, it represents energy modes on the microscopic level, such as energy associated with nuclear spin, molecular binding, magnetic dipole moment, and so on.

2. **Kinetic energy** *(K.E.)* is the kind of energy that a body has during motion. The kinetic energy of a system having a mass *m* with a velocity *v* is given by

$$KE = \frac{1}{2}mv^2$$

3. **Potential energy** *(P.E.)* is the kind of energy that a body has because of its position in a potential field. The potential energy of a system having mass *m*, and an elevation *z*, above a defined plane in a gravitational field with constant *g* is given by

$$PE = mgz$$

# The First Law of Thermodynamics

The principle of conservation of energy, the first law of thermodynamics, may simply be stated as:

change in stored energy = energy input − energy output

# The First Law for a Change in the State of a System

For a system undergoing a cycle, changing from state 1 to state 2, we have:

$$\delta Q - \delta W = dE \qquad (A)$$

where

$\delta Q$ = Heat transfer occurring during the process
$\delta W$ = Work transfer done during the process
$dE$ = Change in total energy of the system.

The inexact differentials $\delta$ are used since heat and work are path functions and are not defined at state points.

The integrated form of equation (A) with *g* constant becomes:

$$_1Q_2 - _1W_2 = U_2 - U_1 + \frac{m(v_2^2 - v_1^2)}{2} + mg(z_2 - z_1) \qquad (B)$$

where

$_1Q_2$ = Heat transferred during the process from 1-2
$U_2 - U_1$ = Change in internal energy

$$\frac{m\left(v_2^2 - v_1^2\right)}{2} = \text{Change in kinetic energy}$$

$mg(z_2 - z_1) = $ Change in potential energy

$_1W_2 = $ Work transfer done by or in the system during the process from 1-2

A rate expression for open systems is

$$\dot{Q}_{cv} - \dot{W}_{cv} = \dot{m}_2 h_2 - \dot{m}_1 h_1 + \dot{m}_2 \frac{v_2^2}{2} - \dot{m}_1 \frac{v_1^2}{2} + \dot{m}_2 gz_2 - \dot{m}_1 gz_1$$

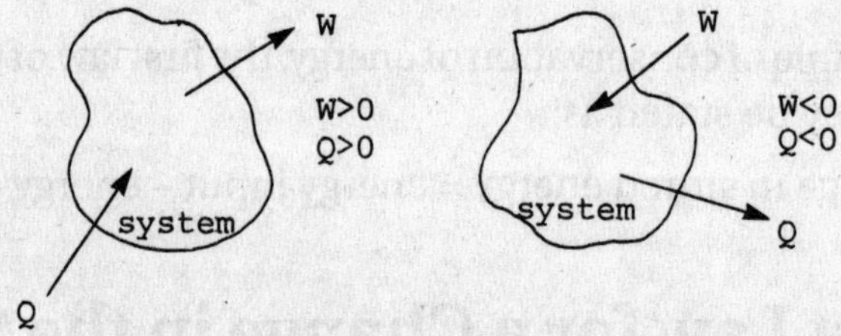

**Figure 36. Signs of work and heat related with a system**

Note: This sign convention is not a universal convention.

1. *E* depends only on the initial and final states and not on the path followed between the two states; therefore, *E* is a point function and is considered the differential of a property of the systems.
2. The net change of the energy of the system is always equal to the net transfer of energy (heat and work) across the system boundary.
3. Equation (B) gives only changes in internal energy, kinetic energy, and potential energy. We cannot learn absolute value of these quantities from the equation.
4. Equation (A) is a consequence of the first law, and not the first law itself. The first law includes the additional information that energy *E* is a property.
5. The first step for applying equation (B) to the solution of any problem must be the description of a closed system and its boundaries.

# The Constant-Volume and Constant-Pressure Specific Heats

A. $C_v = \left(\dfrac{\partial u}{\partial T}\right)_v$    Constant-volume specific heat

B. $C_p = \left(\dfrac{\partial h}{\partial T}\right)_p$    Constant-pressure specific heat

# The Heat Engine

A **heat engine** is a system which operates in a cycle while only heat and work cross its boundaries.

A steam power plant is a heat engine which receives heat from a high-temperature system at the boiler, rejects heat to a lower-temperature system at the condenser, and delivers useful work.

# Heat Engine Efficiency

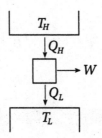

Figure 37. Heat engine

We define the **efficiency** of a work producing heat engine as the ratio of the net work delivered to the environment to the heat received from the high temperature source.

$$\eta = \dfrac{W}{Q_H} = \dfrac{Q_H - Q_L}{Q_H} = 1 - \dfrac{Q_L}{Q_H}$$

where

$Q_H$ = Amount of heat added to an engine
$Q_L$ = Amount of heat rejected by an engine
$W$ = Net amount of work produced by an engine

# The Second Law of Thermodynamics

There are two classical statements of the second law:

## Kelvin-Plank Statement

It is impossible to construct a device that will operate in a cycle and produce no effect other than the raising of a weight and the exchange of heat with a single reservoir.

## Clausius Statement

It is impossible to construct a device that operates in a cycle and produces no effect other than the transfer of heat from a cooler body to a hotter body.

# The Carnot Cycle

The Carnot Cycle is the reversible approximation of the ideal heat engine or refrigerator. It consists of reversible processes for a reversible cycle. Since the cycle is reversible, a Carnot heat engine can be reversed to operate as a Carnot refrigerator/heat pump as seen in the Figure below.

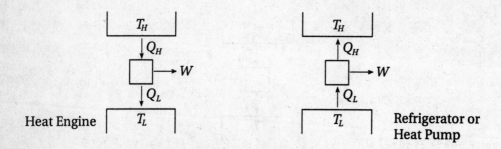

**Figure 38. Carnot cycle**

When the cycle operates between two temperature reservoirs, high and low, the cycle consists of four processes depicted in the Figure below. These four processes are the same for any Carnot cycle.

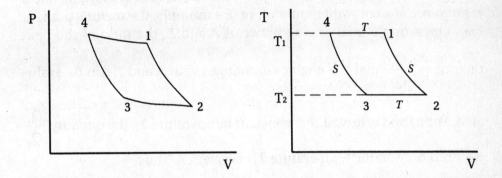

Figure 39. PV and TS diagrams for Carnot cycle

Process 1–2
   A reversible adiabatic process in which the temperature of the working fluid decreases from the high temperature to the low temperature.
Process 2–3
   A reversible isothermal process in which heat is transferred to or from the low-temperature reservoir.
Process 3–4
   A reversible adiabatic process in which the temperature of the working fluid increases from the low temperature to the high temperature.
Process 4–1
   A reversible isothermal process in which heat is transferred to or from the high-temperature reservoir.

# Thermodynamic Temperature Scales

The second law permits the definition of a temperature scale. Such a scale can be defined as follows:

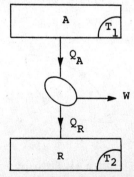

Figure 40. Temperature scales

Let us consider a system $A$ (Figure 40) in a state of equilibrium, and a reservoir $R$, at a constant temperature (for example, the melting point of ice). Operating a heat engine between $A$ and $R$, the ratio of the heat quantities $\left(\dfrac{Q_A}{Q_R}\right)$ that the engines exchanges with $A$ and $R$ can be evaluated. For a fixed reservoir $R$ at constant temperature $T_2$, the quantity $\dfrac{Q_A}{Q_R}$ depends only on the temperature $T_1$ of system $A$. Thus

$$\frac{Q_A}{Q_R} = f(T) .$$

This can also be expressed in the form,

$$T = F_R\left(\frac{Q_A}{Q_R}\right)$$

The function $F_R$ defines a temperature scale which is independent of the nature of any particular thermometric substance and can be determined experimentally.

## Kelvin Temperature Scale

The Kelvin Scale defined by the relation:

$$T = F_R\left(\frac{Q_A}{Q_R}\right) = -T_R\left(\frac{Q_A}{Q_R}\right)$$

where $T_R$ denotes the temperature in degrees Kelvin of a heat reservoir at the triple state of water, $T_R = 273.16°K$.

## Entropy

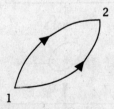

Figure 41. Entropy

It can be shown that for two reversible processes operating on a cycle that the endpoints alone define the quantity:

$$dS = \left(\frac{\delta Q}{T}\right)_{rev}$$

where

$\delta Q$ = Heat supplied to the system
$T$ = Absolute temperature of the system

The quantity $dS$ represents the change in the value of the property called **entropy**.

Notes:
1. The equation above is valid for any reversible process.
2. Entropy is an extensive property, and it is a function of the end points only (a point function).
3. The change in the entropy of a system may be found by integrating the equation above.

$$S_1 - S_2 = \int_1^2 \left(\frac{dQ}{T}\right)_{rev}$$

**Entropy Change of a System During an Irreversible Process**

$$dS \geq \frac{dQ}{T} \quad \text{or} \quad S_2 - S_1 \geq \int \frac{dQ}{T}$$

# Perfect Gas (Ideal Gas)

The perfect gas is a special case of a semiperfect gas obeying the equation,

$$\frac{Pv}{RT} = 1$$

This equation may be written in the following alternative forms:

$$PV = n\overline{R}T$$
$$PV = mRT$$

where $m$ is the mass, $n$ is the number of moles, and $v$ is specific volume.

## Properties of a Perfect Gas

### P–V–T Relation

$$Pv = RT$$

### Internal Energy

$$u = u(T)$$

$$u - u_0 = \int_{T_0}^{T} C_v dt = C_v(T - T_0)$$

### Enthalpy Change

$$h = h(T)$$

$$h - h_0 = \int_{T_0}^{T} C_p dT = C_p(T - T_0)$$

### Entropy Change

$$s - s_0 = \int_{T_0}^{T} \frac{C_v dT}{T} + R \ln \frac{v}{v_0} = C_v \ln \frac{T}{T_0} - R \ln \frac{v}{v_0}$$

$$s - s_0 = \int_{T_0}^{T} \frac{C_p dT}{T} + R \ln \frac{p}{p_0} = C_p \ln \frac{T}{T_0} - R \ln \frac{p}{p_0}$$

### Specific Heats

$$k = k(T) \qquad C_v = C_v(T) \qquad C_p = C_p(T)$$

$$C_p - C_v = R$$

$$C_p = \frac{Rk}{k-1}$$

$$C_v = \frac{R}{k-1}$$

$$k = \frac{C_p}{C_v}$$

# CHAPTER 9

# FLUID MECHANICS

Fluid mechanics is the study of fluids at rest or in motion, and the effects of the fluid on the boundaries which contain the fluid or contact it. The common boundaries are solids in static cases, solid surfaces over which the fluids flow, and other fluids. Fluids are used to perform work, to move heat, to provide life to aircraft, but are also destructive and must be contained to prevent erosion and flooding.

## Definition of a Fluid

A **fluid** is a substance that changes its shape continuously under application of a shear stress no matter how slight the shear stress.

## Basic Laws in Fluid Mechanics

Five basic laws are used for the solution of any problem in fluid mechanics. These basic laws are:

1. Conservation of mass
2. Newton's second law of motion
3. Conservation of momentum
4. The first law of thermodynamics
5. The second law of thermodynamics

## Pathlines, Streaklines, Streamlines, and Streamtubes

**Pathline** is a line traced out by a moving particle in time.

**Streakline** is a line traced out by all fluid particles which at some time passed through one fixed location in space.

**Streamlines** are imaginary lines drawn in the flow whose tangents are parallel to the velocity vector at a given instant of time.

**Streamtube** is an elementary region of the flow whose walls are made up of streamlines (Figure 42).

In the case of steady flow $\left(\frac{\partial v}{\partial t} = 0\right)$, pathlines, streaklines, and streamlines all coincide.

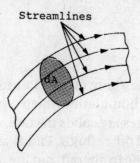

Figure 42. Streamtube

# Newtonian Fluid Viscosity

Fluids may be classified according to the relation between the applied shear stress and the rate of deformation of the fluid. Fluids in which the shear stress is proportional to the rate of deformation are called **Newtonian fluids** (Figure 44) or

$$\tau_{yx} \propto \frac{du}{dy}$$

where
- $u$ = The velocity of the upper plate
- $\frac{du}{dy}$ = Rate of deformation
- $\tau_{yx}$ = The shear stress

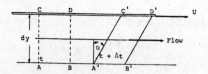

Figure 43. Fluid movement

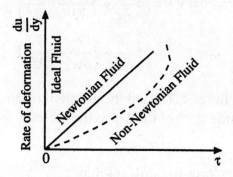

Figure 44. Fluid deformation

The constant of proportionality is the coefficient of viscosity $\mu$. Thus, the equation may be written as

$$\tau_{yx} = \mu \frac{du}{dy}$$

which is the Newton's law of viscosity.

The ratio of the absolute viscosity, $\mu$, to the density, $\rho$, of a fluid is called the kinematic viscosity and is represented by the symbol $\nu$, therefore,

$$\nu = \frac{\mu}{\rho}.$$

# The Basic Equation of Fluid Statics

To determine the pressure field within the fluid a differential element shown in Figure 45 is used.

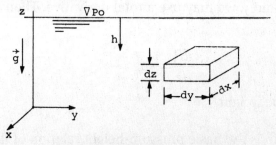

Figure 45. Differential element

For this differential fluid element the following equation applies:
$$-\text{grad } P + \rho g = 0$$
where:

$$\text{grad } p \equiv \nabla P \equiv \left(\hat{i}\frac{\partial P}{\partial x} + \hat{j}\frac{\partial P}{\partial y} + \hat{k}\frac{\partial P}{\partial z}\right)$$

$$\equiv \left(\hat{i}\frac{\partial}{\partial x} + \hat{j}\frac{\partial}{\partial y} + \hat{k}\frac{\partial}{\partial z}\right)P$$

The term in parenthesis is called the **gradient of the pressure.**
The physical significance of the $\nabla P$ is the pressure force per unit volume at a point.

$\rho g$ = Body force per unit volume
$\rho$ = Density
$g$ = The local gravity vector

By expanding into components, we find

$$\text{x-direction} \quad -\frac{\partial P}{\partial x} + \rho g_x = 0$$

$$\text{y-direction} \quad -\frac{\partial P}{\partial y} + \rho g_y = 0$$

$$\text{z-direction} \quad -\frac{\partial P}{\partial z} + \rho g_z = 0$$

If the coordinate system is chosen such that the z-axis is directed vertically, then

$$g_x = 0, \quad g_y = 0, \quad g_z = -g$$

Because the pressure $P$ varies only in the z direction and because it is not a function of $x$ and $y$, we may use a total derivative. Then the equation becomes

$$\frac{dP}{dz} = -\rho g = -\gamma$$

where $\gamma$ = specific weight.

This equation is the basic pressure-height relation of fluid statics. It holds under the following restrictions:

1. Static fluid
2. Gravity is the only body force
3. The z-axis is vertical

**Incompressible Fluid**

In an equation for an incompressible fluid, $\rho$ = constant. Therefore, for constant gravity, the equation becomes

$$P = P_0 + \rho g h$$

where

$P_0$ = The pressure $P$ at the reference level
$h = z - z_0$ ($h$ measured positive downward)
$\rho$ = Density

**Compressible Fluid**

If we know the manner in which the specific weight, $\gamma$, varies, the pressure variation in a compressible fluid can be evaluated.

# Fluid Statics

**Pressure in a Static Fluid**

The pressure at any point of a static fluid is the result of the force exerted by its liquid mass per unit of area. Consider for instance, a vessel of constant cross-sectional area containing a fluid as shown in Figure 46.

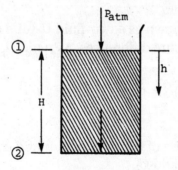

Figure 46. Static fluid

The pressure exerted by the fluid at any height $h$ is given by

$$P = \rho g h$$

where
$\rho$ = Density
$g$ = Gravity
$h$ = Height

From the equation above, the one can conclude that the pressure will be constant at any cross-section, but will vary with height. Thus, the pressure at the bottom of the vessel will be

$$P = \rho g H$$

If the total pressure at the bottom is desired, one must add the pressure at section (1) ($P_{atm}$ in this case). Therefore,

$$P_1 = \rho g H + P_{atm}$$

This leads to the general pressure equation which can be used when gauge pressures are known, with a manometer, or in a fluid statics sense as above. The pressure equation is

$$P_{absolute} = P_{gauge} + P_{ref}$$

where
$P_{absolute}$ = The total pressure: the desired result
$P_{gauge}$ = The pressure from a standard gauge or the result of the static pressure $\rho g h$. In the case of the gauge pressure, the value is positive unless the gauge is reading a vacuum at which time the pressure enters the equation as a negative value.
$P_{ref}$ = Is normally atmospheric pressure unless otherwise specified.

Differential Manometer:
This device, also known as a two-fluid U-tube manometer, is helpful to measure small pressure differences, $\Delta P$ which is given by

$$\Delta P = P_1 - P_2 = H(\rho_C - \rho_B)\frac{g}{g_C}$$

where
$H$ = Reading of the manometer and $\rho_B < \rho_C$ (fluid $B$ is lighter than fluid $C$).

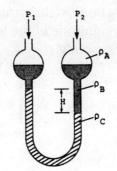

**Figure 47. Differential manometer**

# Buoyancy and Stability

## Buoyancy of a Body

The resultant vertical force exerted on a body by a static fluid in which it is submerged or floating, is called the buoyant force.

The magnitude of this force is

$$F_z = \int \rho g(z_2 - z_1) dA = \rho g V$$

where

$\rho$ = Density of fluid
$g$ = Gravitational constant
$V$ = Volume of the body

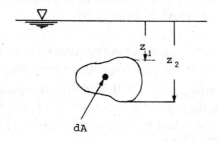

**Figure 48. Forces on a submerged body**

The buoyant force of an incompressible fluid goes through the centroid of the volume displaced by the body.

## Stability

For a completely submerged body, the condition for stability (stable equilibrium) is that the center of gravity, $G$, must be directly below the center of buoyancy.

The vertical alignment of $B$ and $G$ is important for stability. If $G$ and $B$ coincide, neutral equilibrium is obtained.

In a floating body, stable equilibrium can be achieved even when *G* is above *B*. The magnitude of the length *GA* serves as a measure of the stability of a floating body.

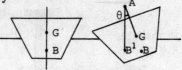

Figure 49. Stability

# Forms of Fluid Flow

### Reynolds number

Reynolds number is used to determine the type of flow regime present in a flowfield. The Reynolds number is defined by

$$\mathrm{Re}_x = \frac{\rho v x}{\mu}$$

where
- $\rho$ = The fluid density
- $v$ = The fluid velocity
- $x$ = The distance from the leading edge of the flow or the diameter of a cylinder in pipe flow
- $\mu$ = The viscosity of the fluid

The critical Reynolds number is the point where the flow transitions from smooth to random. For flow over flat plates and surfaces the critical Reynolds number is $\mathrm{Re}_{crit} = 500{,}000$. For flow inside of pipes, the critical Reynolds number is $\mathrm{Re}_{crit} = 2{,}300$.

### Laminar or Viscous Flow

This is a well ordered flow pattern at Reynolds numbers below the critical value in which the fluid layers appear to slide over one another.

### Turbulent Flow

This flow pattern is present at Reynolds number above the critical value and is characterized by the random motion of the fluid particles in all directions.

### Critical Velocity

This is the velocity at which the flow changes from laminar to turbulent or from turbulent to laminar.

# Laminar and Turbulent Flows

**Laminar flow** is one in which the fluid flows in a well-ordered pattern whereby fluid layers are assumed to slide over one another. If a transition is taking place from a previously well-ordered flow to an unstable flow, the flow is called a turbulent flow.

The difference between these two flows is shown in Figure 9. The figure shows the traces of velocity at a fixed position for steady laminar, steady turbulent, and unsteady turbulent flows.

In turbulent flow the velocity, $V'$, indicates fluctuations with respect to the mean velocity $\overline{V}$. The instantaneous velocity, $V$, may be written as

$$V = \overline{V} + V'$$

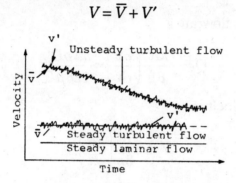

**Figure 50. Flow v. velocity**

# Fully-Developed Laminar Flow in a Pipe

The entrance length, $L$, which is the distance from the entrance to the position in the pipe, for fully-developed laminar flow has been determined to be where $D$ is the pipe diameter.

$$L = 0.058 \text{ Re } D$$

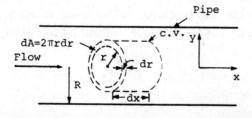

**Figure 51. Control volume for analysis of laminar flow in a pipe**

Boundary condition:
$$u = 0 \quad \text{at} \quad r = R$$

The velocity profile is

$$u = \frac{1}{4\mu}\left(\frac{\partial p}{\partial x}\right)(r^2 - R^2) = -\frac{R^2}{4\mu}\left(\frac{\partial p}{\partial x}\right)\left[1 - \left(\frac{r}{R}\right)^2\right].$$

The shear distribution will be

$$\tau_{rx} = \mu\frac{du}{dr} = \frac{r}{2}\left(\frac{\partial p}{\partial x}\right).$$

The volumetric flowrate is

$$Q = -\frac{\pi R^4}{8\mu}\left(\frac{\partial p}{\partial x}\right) = \frac{\pi \Delta p D^4}{128 \mu L}.$$

The average velocity is

$$V = -\frac{R^4}{8\mu}\left(\frac{\partial p}{\partial x}\right).$$

The point of maximum velocity is at $r = 0$ when $\frac{du}{dr} = 0$.

Thus,

$$\text{at } r = 0, \quad u = u_{max} = -\frac{R^2}{4\mu}\left(\frac{\partial p}{\partial x}\right) = 2\overline{V}.$$

## Newton's Law of Viscosity

According to Newton's Law of Viscosity the shear force per unit is directly proportional to the negative of the local velocity gradient. The fluids which behave in this fashion are termed **Newtonian Fluids**.

$$\tau_{yx} = -\mu\frac{\partial v_x}{\partial y}$$

where

$$\mu = \text{Viscosity}\left(\frac{\text{Kg}}{\text{m sec}} \text{ or } \frac{\text{lb}_m}{\text{ft sec}}\right)$$

$\dfrac{\partial V_x}{\partial y}$ = Velocity gradient

$\tau_{yx}$ = Shear stress applied perpendicular to the y-surface in the x-direction.

# Boundary Layer Thickness

The boundary layer is a thin region adjacent to a solid boundary which is particularly sensitive to the effect of viscosity. A laminar region begins at the leading edge and grows in thickness as the flow moves along the surface. Then, a transition zone is reached where the flow becomes turbulent and the boundary layer is thicker. Figure 52 shows the boundary layer variation along a flat plate.

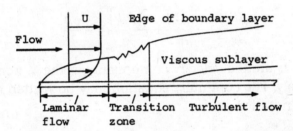

Figure 52. Boundary layer details on a flat plate

The thin layer near the solid boundary in the turbulent region is the viscous sublayer where the viscous effect is predominant.

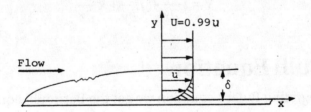

Figure 53. Boundary layer thickness

The boundary layer thickness ($\delta$) is defined as the distance from the surface where the fluid velocity is ninety-nine percent (99%) of the freestream velocity (Figure 53).

The boundary layer displacement thickness ($\delta^*$) is defined as the distance by which the solid boundary would have to be displaced to keep the same mass flow in a hypothetical frictionless flow (see Figure 54).

**Figure 54. Displacement thickness**

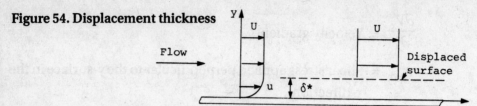

For incompressible flow we have:

$$\delta^* = \int_0^\infty \left(1 - \frac{u}{U}\right) dy$$

# The Continuity Equation

The differential form of the continuity equation is given by

$$\frac{\partial(\rho u)}{\partial x} + \frac{\partial(\rho v)}{\partial y} + \frac{\partial(\rho w)}{\partial z} = -\frac{\partial \rho}{\partial t}$$

where $u$, $v$, and $w$ are the velocity components in the $x$, $y$, and $z$ directions, respectively. The continuity equation may be written more compactly in vector notation as

$$\nabla \bullet \rho v = -\frac{\partial \rho}{\partial t}$$

where

$$\nabla = \hat{i}\frac{\partial}{\partial x} + \hat{j}\frac{\partial}{\partial y} + \hat{k}\frac{\partial}{\partial z}$$

# Bernoulli Equation

By integrating Euler's equation we get the Bernoulli equation:

$$\frac{p}{\rho} + gz + \frac{v^2}{2} = \text{constant}$$

(along a streamline)

Applied between any two points on a streamline the Bernoulli equation becomes

$$\frac{p_1}{\rho} + \frac{v_1^2}{2} + gz_1 = \frac{p_2}{\rho} + \frac{v_2^2}{2} + gz_2$$

or

$$\frac{p_1}{\gamma}+\frac{v_1^2}{2g}+z_1=\frac{p_2}{\gamma}+\frac{v_2^2}{2g}+z_2$$

$\gamma = \rho g$ or specific weight

## Flow in Open Channels

An open channel is a conduit in which the liquid flows with a free surface subjected to certain prescribed conditions of pressure (usually atmospheric pressure). Streams, rivers, artificial channels, and irrigation ditches are some examples of open channel flows. However, a pipe not completely full of fluid will behave as an open channel as well.

A complete solution of open channel flow problems is more complicated because there are a large number of variables to be taken into account. Drastic variations of cross sections and boundary surfaces make the choice of an appropriate friction factor difficult. Besides that, the free surface allows unpredictable phenomena to occur which can change the behavior of the fluids.

# CHAPTER 10

# HEAT AND MASS TRANSFER

In thermodynamics, the amount of energy transfer in the form of heat could be determined based on the end states of the medium. There was however no concern for the mechanism for the transfer of this heat. **Heat transfer** describes the three mechanisms; conduction, convection, and radiation which cause the transfer of heat. Fluid mechanics describes the flow of bulk fluids but did not consider the mixing of fluids due to concentration gradients. Mass transfer deals with the mixing of fluids and, as with conduction heat transfer, mass transfer is a molecular transport mechanism.

## Basic Mechanism

There are three basic mechanisms by which heat transport takes place: **conduction, convection,** and **radiation**.

### Conduction

Conduction is the transport of heat from one part of a body to another part of a body under the influence of a temperature gradient, without appreciable movement of molecules. Example: the heating of a metal spoon in boiling water.

### Convection

Convection is the transport of heat energy which takes place by bulk movement and mixing of molecules of warmer portions with cooler portions of the same material.

There are two types of convection:
1. **Forced convection:** In this type of convection the fluid is forced to flow past a solid small surface by a pump, fan or by mechanical means.
2. **Free convection:** In this type of convection, the heat is transferred due to the density difference resulting from the temperature gradient in the fluid. This is classified as free convection. Examples are:
   (a) Loss of heat from circuit boards in still air.
   (b) Heat rising from a hot roadway surface.

## Radiation

It is the transfer of heat energy through space or vacuum by means of electromagnetic waves.

In general:

$$\text{Heat transfer rate} = \frac{\text{Driving force}}{\text{Resistance}}$$

Driving force is a temperature difference, and the resistance is the barrier across or through which heat must flow.

# Fourier's Law (Conduction)

The rate of heat flux is directly proportional to the temperature gradient.

Mathematically,

$$\frac{q_x}{A} = -k\frac{dT}{dx}$$

$q_x$ = Heat transfer rate in the direction of the x-coordinate (watts)
$A$ = Cross-sectional area [$M^2$], normal to the direction of flow
$T$ = Temperature in K°
$k$ = Proportionality constant usually termed as thermal conductivity in $\frac{W}{M \times K}$

$$\frac{dT}{dx} = \text{temperature gradient}$$

The minus (–) sign is required because heat flows from a higher temperature region to a lower temperature region.

## Newton's Law of Cooling (Convection)

The heat flux is directly proportional to the overall temperature difference or

$$q = hA(T_s - T_\infty)$$

where
- $q$ = Heat flux
- $A$ = Heat transfer area
- $T_\infty$ = Ambient temperature of the cooling medium, i.e., air, water, etc.
- $T_s$ = Surface temperature of solid
- $h$ = Heat transfer coefficient (convection) in $\dfrac{\text{BTU}}{\text{hr-ft}^2 \, F^\circ}$

## Thermal Radiation

The basic equation for heat transfer by radiation from a black body with an emissivity $\varepsilon = 1.0$

$$dq_r = dA \times \sigma \times T^4$$

where
- $dq_r$ = Rate of heat transfer by radiation from one side of the black element of area $dA$
- $\sigma$ = Stefan-Boltzmann's dimensional constant which is approximately equal to $5.676 \times 10^{-8} \text{W/m}^2\text{-K}^4$
- $T$ = Temperature of black body in $K^\circ$

With a gray surface (which has emissivity less than a black body) at a lower temperature than the enclosure temperature, $T_1$, the net adsorption of energy due to radiation between the two bodies can be estimated by the equation

$$q_{\text{ner}} = A_1 \varepsilon \sigma \left( T_2^4 - T_2^4 \right)$$

where
- $\varepsilon$ = Emissivity of the small body at $T_2$

σ = Stefan-Boltzmann's constant
$A_1$ = Area of the gray body (m²)

# Fourier's Law of Heat Conduction

This can be explained using the transport phenomena theory by considering heat transfer through a slab between two large parallel plates which are at a distance Y apart as shown in Figure 55. It is noticeable from the Figure that as time proceeds the temperature profile in the slab changes, and ultimately a linear steady temperature distribution is attained.

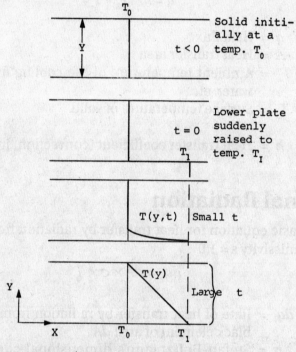

Figure 55. Temperature profile

When this steady state condition is reached, a constant rate of heat flow Q through the slab is required to maintain the temperature difference, $T_1 - T_0 = \Delta T$.

It is found that

$$\frac{Q}{A} = k \frac{\Delta T}{Y}$$

That is, heat flow per unit area is proportional to the temperature drop in the distance Y, where k is the constant of proportionality which is usually termed the thermal conductivity of the slab.

The heat flow per unit area in the positive directions $x$, $y$, and $z$ respectively:

$$q_x = -k\frac{\partial T}{\partial x}, \quad q_y = -k\frac{\partial T}{\partial y}, \quad q_z = -k\frac{\partial T}{\partial z}$$

These three relations are the components of the single vector equation

$$q = -k\nabla T$$

which is the three-dimensional form of Fourier's law.

# Thermal Conductivity

The thermal conductivity of monoatomic and polyatomic gases and liquids can be obtained by using the following formulae:

**Chapman-Enskog**

$$k = 1.9891 \times 10^{-4} \frac{\left[\frac{T}{M}\right]}{\sigma^2 \Omega_K} \quad \text{(Monoatomic gas)}$$

where $\sigma$ is angstroms (A°) and $k$ is in Cal cm$^{-1}$sec$^{-1}$k$^{-1}$.

**Bridgman's Equation**

$$k = 3\left(\frac{\tilde{N}}{\hat{V}}\right)^{\frac{2}{3}} K V_s \quad \text{(Monoatomic liquid)}$$

**Euken's Formula**

$$k = \left(C_p + \frac{5}{4}\frac{R}{M}\right)\mu \quad \text{(Polyatomic gas)}$$

$$k = 2.80\left(\frac{\tilde{N}}{\hat{V}}\right)^{\frac{2}{3}} K V_s \quad \text{(Polyatomic liquid)}$$

where

$$\frac{\hat{V}}{\tilde{N}} = \text{volume per molecule}$$

$$V_s = \text{sonic velocity} = \left[\frac{C_p}{C_v}\left(\frac{\partial p}{\partial \rho}\right)_T\right]^{\frac{1}{2}}$$

$\dfrac{C_p}{C_v} = 1$ for liquids (except at critical point)

$\left(\dfrac{\partial p}{\partial \rho}\right)_T =$ isothermal compressibility correlation

$K =$ molecular heat capacity

# Shell Energy Balance

Energy transport may occur by:

1. Overall fluid motion: convective transport.
2. Dissipation of electric energy.
3. Nuclear fission.
4. Viscous dissipation of mechanical energy.
5. Conversion of chemical energy to heat.

For steady state conditions a shell energy balance is written below:

$$\begin{pmatrix}\text{Rate of}\\ \text{energy}\\ \text{input}\end{pmatrix} - \begin{pmatrix}\text{Rate of}\\ \text{energy}\\ \text{output}\end{pmatrix} + \begin{pmatrix}\text{Rate of thermal}\\ \text{energy}\\ \text{production}\end{pmatrix} = 0$$

The boundary conditions employed are as follows:

1. The temperature and heat flux at a surface to be specified.
2. Newton's law of cooling is defined at the solid-fluid interface.
3. The continuity of heat flux and temperature are to be specified at the solid-state interface.

# Steady State Heat Conduction in One Dimension

Governing differential equation:
For the steady state one-dimensional conduction with no internal heat

generation, the general heat conduction reduces to:

Cartesian Coordinates: $\dfrac{\partial^2 T}{\partial x^2} = 0$

Cylindrical Coordinates: $\dfrac{\partial^2 T}{\partial r^2} + \dfrac{1}{r}\dfrac{\partial^2 T}{\partial r} = 0$

# Thermal Conductance, Thermal Resistance, and Convective Thermal Resistance

Thermal conductance $C$: $C = \dfrac{k}{\Delta x}$

Thermal resistance $R_k$: $R_k = \dfrac{1}{CA} = \dfrac{\Delta x}{kA}$

Convective thermal resistance $R_c$: $R_c = \dfrac{1}{hA}$

Overall heat transfer coefficient $U_o$: $U_0 = \dfrac{1}{AR_{total}}$

$k$ = Thermal conductivity
$h$ = Heat transfer coefficient
$\Delta x$ = Thickness of the material

# The Plane Wall with Specified Boundary Temperature

Conditions:

1. Steady state
2. One-dimensional conduction
3. Each face is maintained at a uniform temperature
4. No internal heat generation
5. Constant conductivity

**Figure 56. Temperature profile**

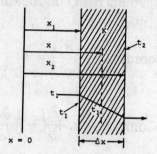

Governing Equation:

$$\frac{q}{A} = -k\frac{dT}{dx} \quad \text{(Fourier's Law)}$$

Boundary Conditions:

$$T(x = x_1) = T_1$$
$$T(x = x_2) = T_2$$

Solution:

$$\frac{q}{A} = k_m \frac{T_1 - T_2}{\Delta x}$$

# Mass Transfer

### Molecular Diffusion in Fluids

This is the transfer of molecules (mass aggregates) from one distinct phase to another or through a single phase due to a concentration gradient by means of the random motion of the molecules.

**Ficks Law:** The mass transport of an *i*th species per unit area is directly proportional to the concentration gradient in the direction of flow

$$j_i^* = \frac{m}{A}\rho\nabla C$$

or

$$j_{Ax}^* = -D_{AB}\frac{dC_A}{dx}$$

(for binary mixture of species *A* and *B*)

where

$j^*_{Ax}$ = mass flux $\left(\dfrac{\text{kg}}{\text{m}^2\text{hr}}\right)$

$\dfrac{dC_A}{dx}$ = concentration gradient

$D_{AB}$ = diffusion coefficient $\left(\dfrac{\text{m}^2}{\text{hr}}\right)$

**Molecular Diffusion in Gases**

$$-dP_A = \dfrac{RT}{D_{AB}P}(N_A P - N_A P_A - N_B P_A)dz$$

where

$$D_{AB} = \dfrac{R^2 T^2}{\beta P} = \text{diffusivity of } A \text{ to } B$$

**Steady State Equimolal Counter Diffusion:**
For equimolal counter diffusion $N_A$ and $N_B$ are constant.

$$\therefore N_A = -N_B \quad \text{and} \quad Z = (Z_2 - Z_1)$$

$$\therefore N_A = \dfrac{D_{AB}}{RTZ}(P_{A_1} - P_{A_2})$$

**Steady-State Diffusion of $A$ Through Stagnant (non-diffusing) $B$:**

$$N_B = 0 \quad N_A = \text{constant}$$

$$N_A = \dfrac{D_{AB}P}{RTZP_{Bm}}(P_{A_1} - P_{A_2})$$

where

$$P_{Bm} = \dfrac{P_{B_2} - P_{B_1}}{\ln\left(\dfrac{P_{B_2}}{P_{B_1}}\right)}$$

**Steady-State Diffusion of *A* Through a Stagnate Multicomponent Mixture:**

$$N_A = \frac{D'_A P}{RTZP_{Bm}}(P_{A_1} - P_{A_2})$$

where

$$D'_A = \frac{1}{\dfrac{Y_B}{D_{AB}} + \dfrac{Y_C}{D_{AC}} + \dfrac{Y_D}{D_{AD}}}$$

$Y$ = mole fraction component on an *A*-free basis

# Definition of Velocities, Concentrations, and Mass Fluxes

1. Mass Concentration $\rho_i = \dfrac{\text{Mass of species } i}{\text{Volume of solution}}$

2. Molar Concentration $c_i = \dfrac{\text{No. of moles of species } i}{\text{Volume of solution}} = \dfrac{\rho_i}{M_i}$
   where $M_i$ = molar weight of species $i$.

3. Mass fraction $\omega_i = \dfrac{\text{Mass concentration of species } i}{\text{Total mass density of solution}} = \dfrac{\rho_i}{\rho}$

4. Mole fraction $x_i = \dfrac{\text{Molar concentration of species } i}{\text{Total molar density of solution}} = \dfrac{c_i}{c}$

5. Local mass average velocity $v = \dfrac{\sum\limits_{i=1}^{n} \rho_i v_i}{\sum\limits_{i=1}^{n} \rho_i}$

   where
   $v_i$ = Velocity of the $i$th species with respect to stationary coordinate axis.

6. Local molar average velocity $v^* = \dfrac{\sum_{i=1}^{n} c_i v_i}{\sum_{i=1}^{n} c_i}$

7. Mass flux relative to stationary coordinate axes
$$n_i = \rho_i v_i$$

8. Molar flux relative to stationary coordinate axes
$$N_i = c_i v_i$$

9. Mass flux relative to the mass average velocity $v$
$$j_i = \rho_i (v_i - v)$$

10. Molar flux relative to the mass average velocity $v$
$$J_i = c_i (v_i - v)$$

11. Mass flux relative to the molar average velocity $v^*$
$$j_i^* = \rho_i (v_i - v^*)$$

12. Molar flux relative to the molar average velocity $v^*$
$$J_i^* = c_i (v_i - v^*)$$

# Convective Mass Transfer

Definition:

$$J_{A_1}^* = K_C' (C_{A_1} - C_{A_2})$$

$J_{A_1}^*$ = Flux of $A$ from the surface $A_1$ relative to the whole bulk phase

$K_C'$ = Convective mass-transfer coefficient

$K_C' = \dfrac{D_{AB} + \bar{\varepsilon}_M}{Z_2 - Z_1}$ = experimental mass transfer coefficient

where

$\bar{\varepsilon}_M$ = Mass eddy diffusion in $\dfrac{m^2}{s}$.

$C_{A_1}$ and $C_{A_2}$ are the concentration of points (1) and (2).

## Mass Transfer Coefficient for Equimolar Counter Diffusion

$$N_A = -N_B \quad N_A = K_{C'}(C_{A_1} - C_{A_2})$$

where

$$K_{C'} = \frac{D_{AB} + \bar{\varepsilon}_M}{Z_2 - Z_1}$$

Let $Y_A$ = Mole fraction in gas phase
$X_A$ = Mole fraction in liquid phase

Then for equimolar counter diffusion:

Gases: $N_A = K'_C(C_{A_1} - C_{A_2}) = K'_G(P_{A_1} - P_{A_2})$
$= K_Y(Y_{A_1} - Y_{A_2})$

Liquids: $N_A = K'_C(C_{A_1} - C_{A_2}) = K'_L(C_{A_1} - C_{A_2})$
$= K_X(X_{A_1} - X_{A_2})$

Relationship between the Mass-transfer Coefficients:

Gases: $K'_C C = K'_C \dfrac{P}{RT} = K'y = K_C \dfrac{P_B M}{RT}$

Liquids: $K'_C C = K'_L C = K_L \dfrac{\rho}{M} = K'_X$

$\rho$ = Density of the liquid
$M$ = Molecular weight of the liquid

## Diffusion into a Falling Liquid Film

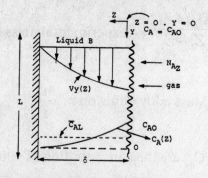

Figure 57. Diffusion gradient

Assumptions:
1. Mass transfer of $A$ into a falling liquid film.
2. The diffusion is negligible in the y-direction.
3. The inlet concentration $C_A = 0$, at steady state.
4. The gas $A$ is slightly soluble in $B$ (liquid).
5. The depth of penetration is small compared to the thickness of the film $\delta$.

General Equation:

$$V_{max}\frac{\partial C_A}{\partial y} = D_{AB}\frac{\partial C_A}{\partial z}$$

Boundary Conditions:
(1) at $y = 0$    $C_A = 0$
(2) at $z = 0$    $C_A = C_{A_0}$
(3) at $Z = \infty$  $C_A = C_0$

Solution:
(1) Concentration profile

$$\frac{C_A}{C_{A_0}} = erfc\frac{Z}{\sqrt{4D_{AB}^Y/V_{max}}}$$

(2) Local mass influx at $Z = 0$

$$N_{AZ}(y)\big|_{Z=0} = C_{A_0}\sqrt{\frac{D_{AB}V_{max}}{\pi y}} = -D_{AB}\frac{\partial C_A}{\partial z}\bigg|_{Z=0}$$

(3) Molar flow rate of $A$ across the plane $Z = L$ is

$$N_A = W_A = WLC_{A_0}\sqrt{\frac{4D_{AB}V_{max}}{\pi L}}$$

# CHAPTER 11

# PROPERTIES OF ENGINEERING MATERIALS

Every engineer is involved with the selection of materials to fit the needs of a design. Weight, strength, toughness, durability, fatigue, and other properties are considered when selecting a material. Properties of materials are determined by the structure at the microscopic level. Materials engineering looks at that structure to determine properties and to develop materials which satisfy often conflicting needs, such as lightweight and strong.

Most materials may be categorized into three classes: **metals, ceramics,** and **polymers.** Metals are composed of elements that easily give up electrons to form the "electron cloud" of a metallic bond. These include iron, copper, aluminum, nickel, and alloys or combinations of metals. Metals are characterized by high thermal and electrical conductivity, high strength and ductility. These properties will be further explained later in the text. Iron is the most commonly used metal, iron oxides, or pig iron are extracted from the iron ore in a blast furnace. The pig iron, which contains carbon and other impurities, is further processed to reduce the carbon content and obtain various grades of steel. Examples include **Bessemer** and **oxygen processes,** also known as the L-D or **Linz-Donawitz processes.** Carbon steels are the simple grades and have carbon as the major non-ferrous element. Nickel, copper, manganese and other metals are alloyed with steel to obtain desirable properties. For example, adding chromium improves steel by reducing corrosion, thus giving stainless steel.

# Crystal Lattice Structure

A crystal lattice structure is an arrangement of atoms or ions in a solid which is repeated in three dimensions. These lattice structures are identified by a unit cell which is repeated throughout the structure.

## Body Centered Cubic

In this structure, there are atoms at each of the corners of a cube and one at its center. This is the least closely packed structure of the three principal structures.

## Face Centered Cubic

In this structure, there are atoms at each of the corners of the cubic and at the center of each face. This structure and the Hexagonal Close Packed are the most closely packed.

## Hexagonal Close Packed

This structure consists of three layers of atoms. The top and bottom layers have atoms at each of the vertices of a hexagon as well as one at the center. The middle layer has three atoms arranged so the three layers are as closely packed as possible.

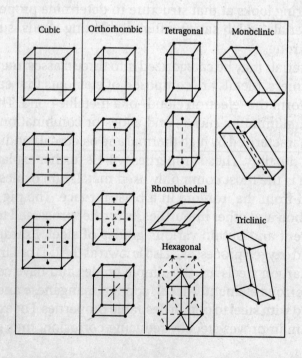

Figure 58. Conventional unit cells of the 14 Bravais space lattices

## Crystallographic Planes

The planes and directions within a crystal can be identified with specific notation. Suppose a plane which intercepts the coordinates $(x, y, z)$ at distances $u$, $v$, and $w$, is represented by the equation:

$$\frac{x}{u} + \frac{y}{v} + \frac{z}{w} = 1$$

The **Miller indices** ($h$, $k$, and $l$) can be used to represent any crystal face. They are defined as:

$$h = \frac{1}{u}$$

$$k = \frac{1}{v}$$

and $$l = \frac{1}{w}$$

Thus, $hx + ky + lz = 1$.

The Miller indices can be evaluated by the following steps.
1. Examine the intercepts of the plane with respect to the crystallographic axes, and be certain that the plane does not intercept with the origin. If it does, choose another plane.
2. Determine the reciprocals of the intercepts.
3. Multiply each of the reciprocals by the least common denominator to obtain a set of whole numbers.

The Miller indices are written as $[hkl]$. Negative indices are marked with a bar atop the index, i.e., negative $k$ is written as $\bar{k}$.

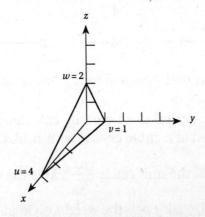

Figure 59. Plane indices for different lengths of cuts

A–133

Most metals in the solid phase assume crystal structures. The most convenient method for determining the structure is **x-ray diffraction**. Diffraction occurs because the wavelength of the x-ray is on the same order of magnitude as the distance between atoms in the crystal, about 0.1 to 0.2 nm. Consider three layers (or planes) of atoms. The difference between the paths DEF and ABC is GEH, and GEH = 2 GE. From trigonometry, we have GE = EH = d sin$\theta$. Bragg discovered that GE is an integer if

$$n\lambda = 2d\sin\theta$$

with $n = 1, 2, 3 \ldots$ (an integer) and $\lambda$ = wavelength. This relationship can be combined with the Miller indices [$hkl$] to give:

$$\frac{\lambda n}{2\sin\theta} = \frac{a}{\sqrt{(h^2 + k^2 + l^2)}}$$

where $a$ is the lattice constant of the crystal.

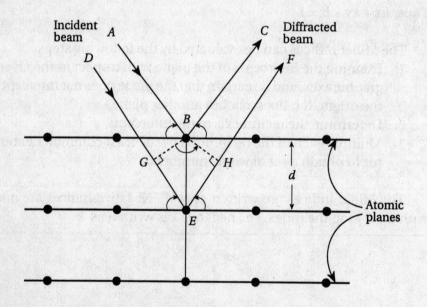

**Figure 60. Diffraction from planes of atoms**

The **density** of a crystal can be easily calculated if the structure is known. For a unit cell of a cubic crystal with side $a$, and $n$ atoms in the pattern, the weight of the unit cell is $\frac{nx}{N_A}$. $M$ is the atomic weight, $N_A$ is Avogadro's number. Density, $\rho$, is the weight divided by the volume, $a^3$.

$$\rho = \frac{nM}{a^3 N_A} = \frac{nM}{V N_A}$$

where $V$ is the volume of the unit cell. Since for a face centered cubic (fcc), the lattice constant, $a$ is related to the atomic radius, $\rho$, by

$$a = \frac{4\rho}{\sqrt{2}},$$

then $V = 32 \dfrac{r^3}{\sqrt{2}}$.

# Phase Diagrams

### Phase

A phase is a region in the microstructure of a material that differs in structure and/or composition from the regions around it.

### Phase Diagrams

Phase diagrams show what phases are present in a material at different temperatures, pressures, and compositions. For metals, phase diagrams show the composition of an alloy that is slow cooled at different temperatures and pressures.

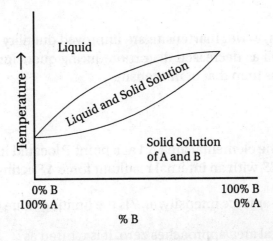

Figure 61. Phase diagram

# Heat Treatment of Carbon Steels

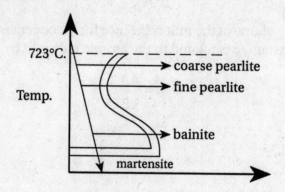

Figure 62. Phase diagram of steel

Figure 62 shows the quenching of carbon steels and isothermal transformation.

### Martensite

The hardness of **martensite** increases with its carbon content whereas the ductility decreases with increases carbon content.

### Pearlite

The hardness of **pearlite** increases as the steel transformation goes from coarse to fine pearlite.

### Bainite

Its advantage over martensite are improved ductility and impact resistance as well as decreased distortion during quenching. However its hardness is less than that of martensite.

# Stress

Consider the elemental area $\Delta A$ at a point $P$ located in the cross-sectional plane, $RS$, with an internal resultant force $\Delta F$ acting on the area.

The stress, or force intensity, at $P$ is the limiting value of the ratio $\dfrac{\Delta F}{\Delta A}$ as the elemental area approaches zero. It is written as

$$\text{Stress at point } P = \left(\frac{\Delta F}{\Delta A}\right)$$

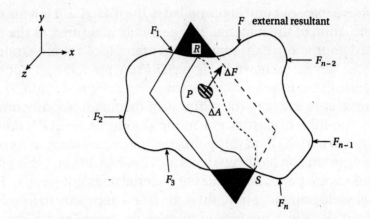

**Figure 63. Arbitrary solid being acted on by external forces**

Consider a small rectangular element *ABCD* in a homogeneous, isotropic body, with sides $dx$ and $dy$ in the plane before deformation. External forces are applied to the body, and the element is displaced to a final position $A'B'C'D'$ (See Figure 63).

The displacement consists of two basic geometric deformations of the element:

1. A change in length of an initial straight line in a certain direction
2. A change in the value of the given angle

These deformations are classified, respectively, as the **longitudinal strain** and the **shear strain**.

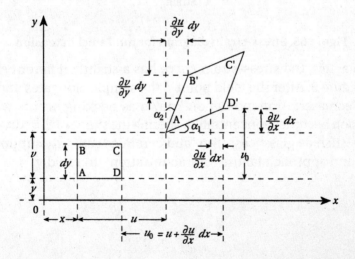

**Figure 64. Rectangular element before and after deformation**

A–137

**Stress** is a force per unit area applied to the material. It is a measure of the deformation of the material. It is generally measured as the relative length of deformation of the material. An example of a stress-strain curve for a ductile sample is shown in Figure 65. From point O to P, Hooke's law applies and a linear relationship exists between stress and strain. The slope of the curve is an index of the **stiffness** of the material. Sometimes it is called the **modulus of elasticity** or **Young's modulus**. Point E is the **elastic limit** of the material. Up to this point, the material returns to its original state after the tension has been removed. However, beyond this point, all deformations are permanent and the material does not recover. Point Y, the **upper yield point** (or strength) is the stress necessary to free dislocations. The dislocations are moved through the lattice at the **lower yield point** L. The **ultimate strength** of the material is shown at point U. Point R represents the **breaking point** or **fracture strength**.

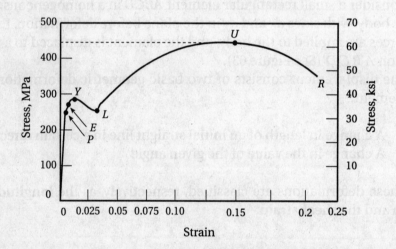

Figure 65. Stress-strain diagram for mild steel in tension

For **plastics**, the stress-strain curve has a slightly different shape as seen in Figure 9. After the yield point Y, the sample elongates and the diameter decreases. A phenomenon known as **necking** occurs when the deformation becomes concentrated, altering the shape of the curve. Brittle materials such as glass, cast iron, and ceramics can only support small stresses, and approach failure (or fracture strength) rapidly.

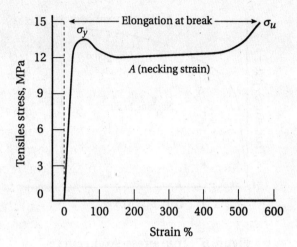

**Figure 66. Stress-strain curve for polyethylene**

A stress-strain curve predicts a fracture stress that is lower than the ultimate strength of the material. This error is due to the fact that the stress is calculated on the basis of the original cross-sectional area. The **true stress-strain curve** calculates the instantaneous deformation. A differential strain element $d\varepsilon$ is the ratio of the instantaneous length, $dl$ to the original length, $l$.

$$d\varepsilon = \frac{dl}{l}$$

The total strain during deformation is then the integral

$$\varepsilon = \int_{l_0}^{l} \frac{dl}{l} = \ln\left(\frac{l}{l_0}\right)$$

Figure 10 compares the true stress-strain curve (corrected) to the nominal curve (uncorrected). In addition, the part of the true stress-strain curve from Y to R can be approximated by the **Power law equation:**

$$\sigma = K\varepsilon^n$$

where $\sigma$ is the true stress, $K$ is the strength coefficient of the material, and $n$ is the strain hardening coefficient.

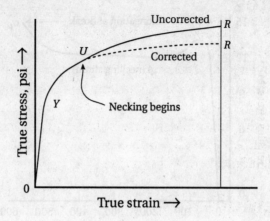

**Figure 67. True stress-strain curve**

# Property of Ceramic Materials

Ceramics are materials containing compounds of both metals and nonmetals that are bound by both ionic and covalent bonds. Ceramics usually have high thermal and chemical resistances, and they are generally poor conductors of heat and electricity. Some simple examples include: magnesium oxide, beryllium oxide, silicon carbide, and silicon nitride. The structure may be a combination of highly ordered crystals and glassy regions. Ceramics may be roughly classified into four groups: **clays, refractories, cements,** and **glasses.** When wet, the clays can be easily blended and molded. Upon drying and **firing,** materials such as bricks, tiles, porcelain, and stoneware can be manufactured. The type of drying process alters the porosity and permeability of the substance. Refractories are designed to withstand high temperatures in industrial operations: gas turbines, ramjet engines, and nuclear reactors. Some examples include alumina-silica compositions, carbides, nitrides, carbon, and graphite. Cements are characterized by their ability to set and harden after being mixed with water, a common type is Portland cement. Finally glasses, mainly made of silica, are materials that have been cooled to a rigid condition but have not been crystallized. For economic reasons, metal oxides are usually added to the glass mixture to reduce the melting temperature. Commercial products include soda lime or lime glass, lead glasses, borosilicate glasses, and high-silica glasses.

# Composites

Composites are produced by joining two materials together to produce properties not exhibited in either material. Composites may be created to be lightweight and strong while resisting corrosion. Composites are formed in three ways: as a particulate, as a fiber, or as a laminate. Cement mixed with sand and gravel could be considered a rough particulate composite. Graphite fiber reinforcements in racquets is an example of the fiber based composite. Plywood is a laminate which roughly fits the definition of a composite.

# Semi-Conductors

There are two types of semiconductors: **intrinsic** and **extrinsic.**

### Intrinsic semiconductors

Intrinsic semiconductors, like silicon and germanium, are neither conductors nor insulators. They require a larger amount of energy than do conductors to excite electrons into the conductor band.

### Extrinsic semiconductors

Extrinsic semiconductors are pure semiconductors doped with an impurity containing one more or one less valence electron. The impurity atoms enter the existing crystal lattice replacing the host atoms. The two types of extrinsic semiconductors are **n-type** and **p-type,** respectively.

In the n-type extrinsic semiconductor, the excess electron is free to conduct. In the p-type, the missing electron of the impurity, known as a "hole," acts as a positive charge carrier.

# CHAPTER 12

# COMPUTATIONAL METHODS

Computational methods are used to solve problems where the complexity of the solution lends itself to an algorithm which can be solved easily on a computer. The steps listed with several of these methods constitute the algorithm mentioned above. Convergence criteria allows definition of the point when the solution is considered "accurate."

## Properties of Real Functions

Let $f$ be a function defined on a set of real numbers $X$. Then:

$$\lim_{x \to x_0} f(x) = L \text{ if } \forall \varepsilon > 0, \, \exists \delta > 0 \, \ni |f(x) - L| < \varepsilon$$

whenever $|x - x_0| < \delta$.

$f$ is continuous at $x_0$, if $f(x_0)$ is defined and

$$\forall \varepsilon > 0, \, \exists \delta > 0 \, \ni |x - x_0| < \delta$$
$$\Rightarrow |f(x) - f(x_0)| < \varepsilon$$

then $\lim_{x \to x_0} f(x) = f(x_0)$.

$f$ is continuous on $X$, if $f$ is continuous at every $x_0 \in X$.
$f$ is differentiable at $x_0$, if

$$\lim_{x \to x_0} \frac{f(x) - f(x_0)}{x - x_0} \text{ exists.}$$

This limit is denoted by $f'(x_0)$.

If $f$ is differentiable at $x_0$ then $f$ is continuous at $x_0$. The converse needs not be true.

### Rolles Theorem

If $f$ is continuous on $[a,b]$, differentiable on $(a,b)$, with $f(a) = f(b) = 0$, then $f'(c) = 0$ for some $c$ in $(a,b)$; $c$ may not be unique.

### Mean Value Theorem

If $f$ is continuous on $[a,b]$, differentiable on $(a,b)$, then there exists

$$c \in (a,b) \ni f'(c) = \frac{f(b) - f(a)}{b - a}$$

If $f$ is continuous on $[a,b]$, and $f^n$ exists on $(a,b)$, then there exists a point

$$c \in (a,b) \ni f''(c) = 0.$$

### Extreme Value Theorem

If $f$ is continuous on $[a,b]$, and there exists $c_1$ and $c_2 \in [a,b] \ni$:

$$f(c_1) \le f(x) \le f(c_2) \quad \forall x \varepsilon [a,b]$$

If $f$ is differentiable on $(a,b)$, then $c_i = a$ or $c_i = b$ or $f'(c_i) = 0$.

### Weighted Mean Value Theorem for Integrals

If $f$ is continuous on $[a,b]$, with $g$ integrable on $[a,b] \ni g(x) > 0$, then $\exists$ a point $c \in (a,b) \ni$

$$\int_a^b f(x) g(x) dx = f(c) \int_a^b g(x) dx$$

if $g(x) \equiv 1$ we get the average value of $f(x)$ on $[a,b]$.

### Intermediate Value Theorem

If $f$ is continuous on $[a,b]$, with $d$ any number $\ni : f(a) < d < f(b)$, and $f(a) \ne f(b)$, then there exists a $c \in (a,b) \ni f(c) = d$. If $f(a) < 0$ and $f(b) > 0$, or vice versa, then there is a $c \varepsilon (a,b) \ni f(c) = 0$. A function that is continuous on $[a,b]$ takes on every value $\in (f(a), f(b))$.

## Continuity and Uniformness

A function $f$ is uniformly continuous on $[a,b]$, if

$$\forall \varepsilon > 0, \; \exists \delta > 0 \; \ni |x - x_0| < \delta \Rightarrow |f(x) - f(x_0)| < E$$

$\forall x_0 \in [a,b]$, $\delta$ independent of $x_0$.

Any function continuous on $[a, b]$, is uniformly continuous on $[a, b]$. Note: the converse need not be true.

## Lipschitz Condition

A function $f$ is defined on $[a,b]$, $f$ satisfies a Lipschitz condition on $[a,b]$, if

$$\exists L \ni |f(x_1) - f(x_2)| \le L|x_1 - x_2| \forall x_1, x_2 \in [a,b].$$

$L$ is called a Lipschitz constant. If $f$ has a Lipschitz constant $L$ on $[a,b]$, then $f \in C[a,b]$. If $f'$ is bounded on $[a,b]$ by $L$, the $f$ satisfies a Lipschitz condition with Lipschitz constant $L$ on $[a,b]$.

$C$ will be used as a constant throughout this text.

# Bisection Method

## Basis for Bisection Method

If $f$ is continuous on $[a,b]$, with $f(a)$ and $f(b)$ having opposite signs, the intermediate value theorem states that there is $p \in (a,b) \ni f(p) = 0$. By repeatedly halving the subintervals, keeping the half containing $p$, we solve $f(x) = 0$.

Step 1. Find endpoints $a, b$.

Step 2. Set $p = a + \dfrac{(b-a)}{2}$

Step 3. If $f(p) = 0$ we finish. The accuracy is a matter of choice.)

Step 4. If $f(p) \ne 0$, $f(a)f(p) > 0$. Let $a = p$. Otherwise set $b = p$.

Step 5. Repeat procedure from Step 2.

The sequence generated while approximating $p$ is one such that

$$|p_n - p| \le \frac{b-a}{2^n} \quad n \ge 1$$

If we choose an error bound of $\varepsilon$, the minimum, number of iterations $N$ needed to ensure accuracy within $\varepsilon$ is

$$N \geq \frac{\log_{10}(b-a) - \log_{10}\varepsilon}{\log_{10} 2}$$

# Fixed Point Iterative Method

A fixed point of a continuous function $f$ is a number $x$ such that $f(x) = x$.
Basis for fixed point: Let $h(x) = f(x) - x$.

If $h$ is defined on $[a,b]$ and $h(p) = 0$ for some $p \in [a,b]$, then $h$ has a fixed point $p \in [a,b]$.

If $h \in C[a,b]$, and $h(x) \in [a,b]\ \forall\ x \in [a,b]$, $h$ has a fixed point in $[a,b]$. If $h'(x)$ exists on $[a,b]$, with $|h'(x)| < 1\ \forall\ x \in [a,b]$, then $h$ has a unique point in $[a,b]$.

We rewrite $f(x) = 0$ in the form $h(x) = x$ after we have tested $h$ for a fixed point. This test is important since it ensures convergence.

Step 1. Find initial approximation $p_0$
Step 2. Set $p = h(p_0)$
Step 3. Check $|p - p_0|$ to decide termination. Otherwise
Step 4. Set $p_0 = p$
Step 5. Repeat procedure from Step 2.

If the function $h(x) - x = 0$ satisfies a **Lipschitz condition** on $[a,b]$ with $h < 1\ \forall\ x \in [a,b]$, then the sequence generated by $p_n = h(p_{N-1})$, $N \geq 1$, converges to the unique fixed point in $[a,b]$.

# Newton Raphson Method

If a function $h \in C^2[a,b]$ with $h(p) = 0$, and $x$ an approximation to $p$ such that $h'(x) \neq 0$, $|x - p|$ small, the first degree Taylor polynomial for $h(x)$ about $x$ is

$$h(x) = h(\hat{x}) + (x - \hat{x})h'(\hat{x}) + \frac{(x-\hat{x})^2 h''(\zeta(x))}{2}$$

$\zeta(x)$ between $x$ and $\hat{x}$. With $x = p$

$$0 = h(\hat{x}) + (p - \hat{x})h'(\hat{x}) + \frac{(p-\hat{x})^2}{2} h''(\zeta(p))$$

Dropping the last term, and solving for $p$

$$p \approx (\hat{x}) - \frac{h(\hat{x})}{h'(\hat{x})}$$

We can generate the sequence $\{p_n\}$ until we are as close as we wish, where

$$p_n = p_{n-1} - \frac{h(p_{n-1})}{h'(p_{n-1})} \qquad n \geq 1$$

To find the solution to $f(x) = 0$:

Step 1.     Set $p_1 = p_0 - \frac{f(p_0)}{f'(p_0)}$.

Step 2.     Check $|p - p_0|$ to decide termination. Otherwise
Step 3.     Set $p_0 = p_1$
Step 4.     Repeat procedure from Step 1.

It is important to choose an initial approximation that is close to the root. Otherwise convergence may be slow.

The Newton Raphson method has quadratic convergence. If $e_n$ is the $n$th absolute error, then

$$e_{n+1} = \frac{\frac{1}{2} e_n^2 f''(p)}{f'(p)} \qquad f'(p) \neq 0$$

as

$$\hat{x}_n \to p \qquad \frac{e_{n+1}}{e_n} \to \frac{f''(p)}{2 f'(p)} \qquad f'(p) \neq 0$$

The number of correct decimal places should double with each iteration.

## Fixed Tangent Method

This method is based upon Newton Raphson method. The slope of the first tangent is used for all iterations.

**Figure 68.** Tangent method

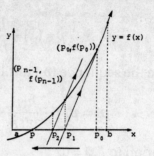

The equation of the tangent at $f(p_0)$ is

$$y = f(p_{n-1}) + f'(p_0)(x - p_{n-1})$$

at $x = p_n$ we have:

$$p_n = p_{n-1} - \frac{f(p_{n-1})}{f'(p_0)} \quad n \geq 1$$

then iterate until $p_n$ is satisfactory.

# Secant Method

This method is a variation of Newton Raphson Method, applied to functions whose derivatives are cumbersome to evaluate.

We know

$$f'(p_{n-1}) = \lim_{x \to p_{n-1}} \frac{f(x) - f(p_{n-1})}{x - p_{n-1}}$$

at $x = p_{n-2}$

$$f'(p_{n-2}) \approx \frac{f(p_{n-2}) - f(p_{n-1})}{p_{n-2} - p_{n-1}}$$

$$= \frac{f(p_{n-1}) - f(p_{n-2})}{p_{n-1} - p_{n-2}}$$

If we use this approximation of $f'(p_{n-1})$ in Newton's formula

$$p_n = p_{n-1} - \frac{f(p_{n-1})}{f'(p_{n-1})}$$

we have

$$p_n = p_{n-1} - \frac{f(p_{n-1})(p_{n-1} - p_{n-2})}{f(p_{n-1}) - f(p_{n-2})} \quad n \geq 2$$

This method requires two initial approximations.

Step 1: Set $r_0 = f(p_0)$, $r_1 = f(p_1)$, ($p_0$, $p_1$ initial approximation)
Step 2: Find

$$p = \frac{p_1 - r_1(p_1 - p_0)}{r_1 - r_0}$$

$$p_2 = p_1 - \frac{r_1(p_1 - p_0)}{r_1 - r_0}$$

Step 3: If $|p - p_1|$ is satisfactory, terminate. Otherwise
Step 4: Set $p_0 = p_1$
$r_0 = r_1$
$p_1 = p_2$
$r_1 = f(p_2)$
Step 5: Continue from Step 2.

The secant method converges to the root $p$ with

$$\left|\frac{e_{n+1}}{e_n e_{n-1}}\right| \to \frac{1}{2}\left|\frac{f''(p)}{f'(p)}\right|$$

providing $f$ is twice continuously differentiable, with $f'(p) \neq 0$. The order of convergence is bounded between linear and quadratic.

## Regula Falsi

In the Regula Falsi method: to find $x \ni f(x) = 0$, $x \in [a,b]$, the intervals $[a_i,b_i]$ are chosen in the same way as the bisection method, new approximate intervals are generated in a similar fashion to the secant method.

Let $f$ be continuous on $[a,b]$, with $f(a)f(b) < 0$.

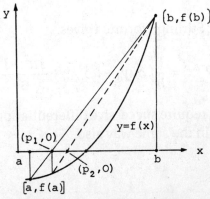

**Figure 69. Regula falsi method**

The secant through $(a, f(a))$, and $(b, f(b))$ has slope

$$\frac{f(b) - f(a)}{b - a}.$$

The equation of the secant is

$$y - f(a) = \frac{f(b) - f(a)}{b - a}(x - a)$$

using point $(p_1, 0)$ gives

$$p_1 = a - \frac{f(a)(b - a)}{f(b) - f(a)}.$$

For the iterative procedure we get

$$p_n = a_n - \frac{f(a_n)(b_n - a_n)}{f(b_n) - f(a_n)} \quad n \geq 1$$

For each new $p$, check whether $f(p_n)f(a) < 0$. This allows us to determine the interval containing the root.

## Newton's and Stirling's Formulas

If we used interpolating polynomials to approximate a function $f$, a reasonable approach to approximate differentiation is to differentiate these polynomials.

Differentiating Newton's forward polynomial gives

$$f'(x) \approx P'(x) = \frac{1}{h}\left(\Delta y_0 + (u - \tfrac{1}{2})\Delta^2 y_0 + \left(\frac{3u^2 - 6u + 2}{6}\right)\Delta^3 y_0 + \ldots\right)$$

For approximation to higher derivatives, we can find $f''$, ..., $f^n$. We use $x = x_0 + uh$ to get to the derivatives relative to $x$; $u$ can also be replaced by $i$.

Differentiating Stirling's formula gives

$$f'(x) \approx P'(x) = \frac{1}{h}\left(\delta u y_0 + u\delta^2 y_0 + \frac{3u^2 - 1}{6}\delta^3 u y_0 + \ldots\right)$$

Higher derivatives require successive differentiations. The other polynomials can be used in the same way also.

# General Applications of Numerical Integration

Sometimes we encounter the problem of evaluating a definite integral of a function whose antiderivative cannot be explicitly represented with elementary functions, then approximation is the only alternative.

One can integrate approximating polynomials such as Newton's, Stirling's, and a variety of others, not forgetting Taylor polynomials. Composite formulas are obtained when we use other formulas over a subdivided interval which otherwise would have been too long.

Some of these are presented in the following section.

## Trapezoidal, Simpson's Rule

### Trapezoidal Rule

To find $\int_a^b f(x)dx$, choose a step size $h = \dfrac{b-a}{n}$, $n$ is a positive integer.

Let $x_i = x_0 + ih$, $i = 1, 2, \ldots, n$. The method sums the areas $A_i$ of the trapezoids constructed, whose corners are $x_i, x_{i+1}, f_i, f_{i+1}$, so the smaller $h$ is, the better the approximation. The integral becomes

$$\int_a^b f(x)dx \approx \sum_{i=1}^n A_i \approx \frac{h}{2}\left[f_0 + 2f_1 + 2f_2 \ldots + 2f_{n-1} + f_n\right]$$

with approximation error

$$|E_n| = \left|\frac{-(x_n - x_0)h^2 f^{(2)}(\zeta)}{12}\right|$$

$$x_0 < \zeta < x_n$$

If $x_0 = a$, $x_1 = x_n = b$ then

$$\int_a^b f(x)dx = \frac{h}{2}\left[f_0 + f_1\right]$$

### Simpson's Rule

Simpson's rule uses the points

$$(x_i, x_{i+1}), (x_{i+1}, x_{i+2}), \quad i = 0, 1, \ldots, n-2$$

and replaces the curve between $(x_i, f_i)$, $(x_{i+1}, f_{i+1})$, $(x_{i+2}, f_{i+2})$ with a quadratic, then sums the areas $A_i$ of all the segments.

To find $\int_a^b f(x)dx$, let $h = \dfrac{b-a}{n}$, $n$ an even integer, $x_i = x_0 + ih$, $i = 0, 1, \ldots, n$. The sum is

$$\int_a^b f(x)dx \approx \sum A_i \approx \frac{h}{3}\left[f_0 + 4f_1 + 2f_2 + 4f_3 + \ldots + 4f_{n-1} + f_n\right]$$

With approximation error

$$|E_n| = \left|\frac{-(x_n - x_0)h^4 f^{(4)}(\zeta)}{180}\right|$$

$$x_0 < \zeta < x_n$$

Note: The trapezoidal and Simpson's Rule can be obtained by integrating the Lagrange interpolating polynomial over the interval $[x_0, x_n]$.

## Numerical Quadrature

Numerical Quadrature methods involve approximating $\int_a^b f(x)dx$ using sums of the form $\Sigma c_i f(x_i)$ which comes from integrating the Lagrange polynomial $P_n$ over $[a, b]$, giving

$$\int_a^b \sum_{i=0}^n f(x_i) L_i(x) dx = \sum_{i=0}^n c_i f(x_i)$$

$$c_i = \int_a^b L_i(x)dx \quad \text{for each } i = 0, 1, \ldots, n.$$

# CHAPTER 13

# CALCULUS

Fundamental to the understanding of engineering, calculus is the methodology of describing infinitesimally small changes. Calculus allows us to compute areas and volumes, and also define rates of change. The importance of a thorough knowledge of calculus cannot be overstressed.

## Exponential and Logarithmic Functions

If $f$ is a nonconstant function that is continuous and satisfies the functional equation $f(x+y) = f(x) \times f(y)$, then $f(x) = a^x$ for some constant $a$. That is, $f$ is an exponential function.

Consider the exponential function $a^x$, $a > 0$ and the logarithmic function $\log_a x$, $a > 0$. Then $a^x$ is defined for all $x \in R$, and $\log_a x$ is defined for only positive $x \in R$.

These functions are inverses of each other,

$$a^{\log_a x} = x \,; \quad \log_a(a^y) = y$$

### The Natural Logarithmic Function

1. To every real number $y$ there corresponds a unique positive real number $x$ such that the natural logarithm, ln, of $x$ is equal to $y$. That is $\ln x = y$.
2. The natural exponential function, denoted by exp, is defined by

   $$\exp x = y \quad \text{if and only if} \quad \ln y = x$$

   for all $x$, where $y > 0$.
3. The natural log and natural exponential are inverse functions. $\ln(\exp x) = x$ and $\exp(\ln y) = y$.

4. Properties of the natural logarithmic function
$$\ln(ab) = \ln a + \ln b$$
$$\ln \frac{a}{B} = \ln a - \ln b$$
$$\ln(a^b) = b \ln a$$

# Limits

Let $f$ be a function that is defined on an open interval containing $b$, but possibly not defined at $b$ itself. Let $L$ be a real number. The statement

$$\lim_{x \to b} f(x) = L$$

defines the limit of the function $f(x)$ at the point $b$. Very simply, $L$ is the value that the function has as the point $b$ is approached.

# Functions

A **function** is a correspondence between two sets, the **domain** and the **range**, such that for each value in the domain there corresponds exactly one value in the range.

# Lines and Slopes

Each straight line in a coordinate plane has an equation of the form $Ax + By + C = 0$, where $A$ and $B$ are not zero.

The equation for a line can be conveniently written as

$$y = mx + b$$

where $m = \text{slope} = \dfrac{y_1 - y_0}{x_1 - x_0}$, and $b$ = y-intercept, or the value of $y$ when $x = 0$.

# Parametric Equations

A **parameter** is a quantity whose value determines the value of other quantities. Given the equation $y = t^2 + t + 3$ and $x = t^2 + 2t$, where the value of $t$ determines the value of $x$ and $y$; $t$ is considered a parameter.

Let $x$ and $y$ represent two differentiable functions on the interval $[a,b]$ which are continuous at the endpoints and $(x(t), y(t))$ represents a point in the plane. If $t$ is in the interval $[a,b]$, then the curve traced out by the point $(x(t), y(t))$ is given parametrically as

$$x = x(t) \text{ and } y = y(t)$$

# Transcendental Functions

## Trigonometric Functions

The trigonometric functions are defined in terms of a point $P$ which moves in a circular track of unit radius.

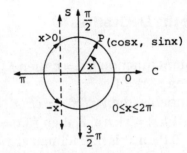

Figure 70. Polar coordinates

## Continuity

A function $f$ is continuous at a point $b$ if

$$\lim_{x \to b} f(x) = f(b).$$

This implies that three conditions are satisfied:

1. $f(b)$ exists, that is, $f$ is defined at $b$
2. $\lim_{x \to b} f(x)$ exists, and
3. the two numbers are equal

To test continuity at a point $x = b$ we test whether

$$\lim_{x \to b^+} M(x) = \lim_{x \to b^-} M(x) = Mb$$

# The Derivative and $\Delta$-Method

The **derivative** of a function expresses its rate of change with respect to an independent variable. The derivative is also the slope of the tangent line to the curve.

We denote the derivative of the function $f$ to be $f'$. So we have

$$f'(x) = \lim_{x \to 0} \frac{f(x + \Delta x) - f(x)}{\Delta x}$$

## The Derivative at a Point

If $f$ is defined on an open interval containing $b$ then

$$f'(b) = \lim_{x \to b} \frac{f(x) - f(b)}{x - b}$$

provided the limit exists.

## Rules for Finding the Derivatives

### General rule:
1. If $f$ is a constant function $f(x) = c$, then $f'(x) = 0$.
2. If $f(x) = x$, then $f'(x) = 1$.
3. If $f$ is differentiable, then $(cf(x))' = cf'(x)$
4. Power Rule: If $f(x) = x^n$, $n \in Z$, then $f'(x) = nx^{n-1}$; if $n < 0$ then $x^n$ is not defined at $x = 0$.

   When the exponent is a quotient: $f(x) = x^{\frac{m}{n}}$, then $f'(x) = \frac{m}{n} x^{\frac{m}{n}-1}$

   where $m, n \in Z$, and $n \neq 0$.

5. If $f$ and $g$ are differentiable on the interval $(a, b)$ then:
   
   a) Summation Rule
   
   $(f + g)'(x) = f'(x) + g'(x)$
   
   b) Product Rule
   
   $(fg)'(x) = f(x)g'(x) + g(x)f'(x)$
   
   Example: Find $f'(x)$ if $f(x) = (x^3 + 1)(2x^2 + 8x - 5)$
   $f'(x) = (x^3 + 1)(4x + 8) + (2x^2 + 8x - 5)(3x^2)$
   $= 4x^4 + 8x^3 + 4x + 8 + 6x^4 + 24x^3 - 15x^2$
   $= 10x^4 + 32x^3 - 15x^2 + 4x + 8$
   
   c) Quotient Rule
   
   $$\left(\frac{f}{g}\right)'(x) = \frac{g(x)f'(x) - f(x)g'(x)}{[g(x)]^2}$$
   
   Example: Find $f'(x)$ if $f(x) = \dfrac{3x^2 - x + 2}{4x^2 + 5}$
   
   $$f'(x) = \frac{(4x^2 + 5)(6x - 1) - (3x^2 - x + 2)(8x)}{(4x^2 + 5)^2}$$
   
   $$= \frac{(24x^3 - 4x^2 + 30x - 5) - (24x^3 - 8x^2 + 16x)}{(4x^2 + 5)^2}$$

$$= \frac{(4x^2 + 14x - 5)}{(4x^2 + 5)^2}$$

6. Polynomials:
   If $f(x) = (a_0 + a_1 x + a_2 x^2 + \ldots + a_n x^n)$
   then $f'(x) = a_1 + 2a_2 x + 3a_3 x^2 + \ldots + na_n x^{n-1}$. This employs the power rule and rules concerning constants.
7. Chain Rule:
   Let $f(u)$ be a composite function, where $u = g(x)$.
   Then $f'(u) = f'(u)g'(x)$ or if $y = f(u)$ and $u = g(x)$ then $D_x y = (D_u y)(D_x u)$
   $= f'(u)g'(x)$

# Trigonometric Differentiation

$$D_x \sin u = \cos u \, D_x u$$

$$D_x \cos u = -\sin u \, D_x u$$

$$D_x \tan u = \sec^2 u \, D_x u$$

$$D_x \sec u = \tan u \sec u \, D_x u$$

$$D_x \cot u = -\csc^2 u \, D_x u$$

$$D_x \csc u = -\csc u \cot u \, D_x u$$

# Exponential and Logarithmic Differentiation

The exponential function $e^x$ has the simplest of all derivatives. It's derivative is itself.

$$\frac{d}{dx} e^x = e^x$$

$$\text{and } \frac{d}{dx} e^u = e^u \frac{du}{dx}$$

Since the natural logarithmic function is the inverse of $y = e^x$ and $\ln e = 1$, it follows that

$$\frac{d}{dx} \ln y = \frac{1}{y} \frac{dy}{dx}$$

and $\dfrac{d}{dx}\ln u = \dfrac{1}{u}\dfrac{du}{dx}$

## Maxima and Minima

Suppose that $c$ is a critical value of a function, $f$, in an interval $(a,b)$, then if $f$ is continuous and differentiable we can say that,

1. If $f'(x) > 0$ for all $a < x < c$ and $f'(x) < 0$ for all $c < x < b$, then $f(c)$ is a local maximum.
2. If $f'(x) < 0$ for all $a < x < c$ and $f'(x) > 0$ for all $c < x < b$, then $f(c)$ is a local minimum.
3. If $f'(x) > 0$ or if $f'(x) < 0$ for all $x \in (a,b)$ then $f(c)$ is not a local extrema.

## Velocity and Rate of Change

### Velocity

Instantaneous velocity at time $t$ is defined as

$$v = Ds(t) = \lim_{\Delta h \to 0} \dfrac{f(t+\Delta h) - f(t)}{\Delta h}$$

where $s(t)$ is the function which describes the location, $s$, at any time $t$.

We usually write

$$v(t) = \dfrac{ds}{dt}.$$

Acceleration, the rate of change of velocity with respect to time is

$$a(t) = \dfrac{dv}{dt}.$$

### Rate of Change

In general, we can speak about the rate of change of any function with respect to an arbitrary parameter (such as time in the previous section).

For linear functions $f(x) = mx + b$ (the equation of a line), the rate of change is simply the slope $m$.

For non-linear functions we define the average rate of change between points $c$ and $d$ to be

$$\frac{f(d) - f(c)}{d - c}$$

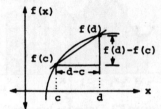

Figure 71. Rate of change

and the instantaneous rate of change of $f$ at the point $x$ to be

$$f'(x) = \lim_{\Delta h \to 0} \frac{f(x + \Delta h) - f(x)}{\Delta h}$$

# Definition of Definite Integral

A **partition** $P$ of a closed interval $[a,b]$ is any decomposition of $[a,b]$ into subintervals of the form,

$$[x_0, x_1], [x_1, x_2], [x_2, x_3], \ldots, [x_{n-1}, x_n],$$

where $n$ is a positive integer and $x_i$ are numbers, such that

$$a = x_0 < x_1 < x_2 < \ldots < x_{n-1} < x_n = b.$$

The length of the subinterval is $\Delta x_i = x_i - x_{i-1}$. The largest of the numbers $\Delta x_1, \Delta x_2, \ldots, \Delta x_n$ is called the **norm** of the partition $P$ and denoted by $\|P\|$.

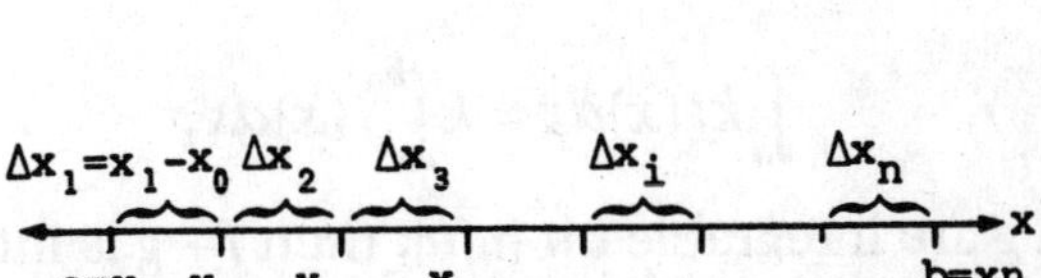

Figure 72. An interval

Let $f$ be a function that is defined on a closed interval $[a,b]$ and let $P$ be a partition of $[a,b]$. A **Riemann Sum** of $f$ for $P$ is any expression $R_p$ of the form,

$$R_p = \sum_{i=1}^{n} f(w_i)\Delta x_i$$

where $w_i$ is some number in $[x_{i-1}, x_i]$ for $i = 1, 2, ..., n$.

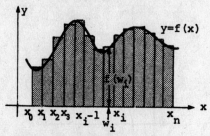

Figure 73. Definite integral

Let $f$ be a function that is defined on a closed interval $[a,b]$. The **definite integral** of $f$ from $a$ to $b$, denoted by $\int_a^b f(x)dx$ is given by

$$\int_a^b f(x)dx = \lim_{\|P\| \to 0} \sum_i f(w_i)\Delta x_i$$

provided the limit exists.

If $f$ is continuous on $[a,b]$, then $f$ is integralable on $[a,b]$.

If $f(a)$ exists, then $\int_a^b f(x)dx = 0$.

# Properties of the Definite Integral

1. If $f$ is integrable on $[a,b]$, and $k$ is any real number, then $kf$ is integrable on $[a,b]$, and

$$\int_a^b kf(x)dx = k\int_a^b f(x)dx.$$

2. If $f$ and $g$ are integrable on $[a,b]$, then $f + g$ is integrable on $[a,b]$, and

$$\int_a^b [f(x) + g(x)]dx = \int_a^b f(x)dx + \int_a^b g(x)dx.$$

3. If $a < c < b$ and $f$ is integrable on both $[a,c]$, and $[c,b]$, then $f$ is inte-

grable on [a,b], and

$$\int_a^b f(x)dx = \int_a^c f(x)dx + \int_c^b f(x)dx.$$

4. If $f$ is integrable on a closed interval and if $a$, $b$, and $c$, are any three numbers in the interval, then

$$\int_a^b f(x)dx = \int_a^c f(x)dx + \int_c^b f(x)dx.$$

5. If $f$ is integrable on $[a,b]$, and if $f(x) \geq 0$ for all $x$ in $[a,b]$, then

$$\int_a^b f(x)dx \geq 0.$$

# The Fundamental Theorem of Calculus

The fundamental theorem of calculus establishes the relationship between the indefinite integrals and differentiation by use of the mean value theorem.

### Mean Value Theorem for Integrals

If $f$ is continuous on a closed interval $[a,b]$, then there is some number $P$ in the open interval $(a,b)$ such that

$$\int_a^b f(x)dx = f(P)(b-a).$$

To obtain $f(P)$ we divide both sides of the equation by $(b-a)$ obtaining

$$f(P) = \frac{1}{(b-a)} \int_a^b f(x)dx.$$

### Definition of the Fundamental Theorem

Suppose $f$ is continuous on a closed interval $[a,b]$, then
1. If the function $G$ is defined by

$$G(x) = \int_a^x f(t)dt$$

for all $x$ in $[a,b]$, then $G$ is an antiderivative of $f$ on $[a,b]$.
2. If $F$ is any antiderivative of $f$, then

$$\int_a^b f(x)dx = F(b) - F(a).$$

# Indefinite Integral

The indefinite integral of $f(x)$, denoted by $\int f(x)dx$ is the most general integral of $f(x)$, that is

$$\int f(x)dx = F(x) + C.$$

$F(x)$ is any function such that $F'(x) = f(x)$. $C$ is an arbitrary constant.

## Integration Formulas

1. $\int x^n dx = \dfrac{x^{n+1}}{n+1} + C, \quad n \neq -1$

2. $\int \dfrac{dx}{x} = \ln|x| + C$

3. $\int \dfrac{dx}{x-a} = \ln|x-a| + C$

4. $\int \dfrac{dx}{x^2 + a^2} = \dfrac{1}{a}\tan^{-1}\dfrac{x}{a} + C$

5. $\int \dfrac{xdx}{x^2 + a^2} = \dfrac{1}{2}\ln|x^2 + a^2| + C$

6. $\int \dfrac{dx}{(a^2 - x^2)^{\frac{1}{2}}} = \sin^{-1}\dfrac{x}{a} + C$

7. $\int \sin ax\,dx = -\dfrac{1}{a}\cos ax + C$

8. $\int \cos ax\,dx = \dfrac{1}{a}\sin ax + C$

9. $\int \sec^2 x\,dx = \tan x + C$

10. $\int e^{ax} dx = \dfrac{e^{ax}}{a} + C$

11. $\int \sinh ax\,dx = \dfrac{1}{a}\cosh ax + C$

12. $\int \cosh ax\,dx = \dfrac{1}{a}\sinh ax + C$

# Area Under A Curve

If $f$ and $g$ are two continuous functions on the closed interval $[a,b]$, then the area of the region bounded by the graphs of these two functions and the ordinates $x = a$ and $x = b$ is

$$A = \int_a^b \left[ f(x) - g(x) \right] dx$$

where

$$f(x) \geq 0 \text{ and } f(x) \geq g(x)$$
$$a \leq x \leq b$$

The area below $f(x)$ and above the x-axis is represented by $\int_a^b f(x)$.

# Equations of a Line and Plane in Space

## Line

The equations of a line joining two points $(x_0, y_0, z_0)$ and $(x_1, y_1, z_1)$ are expressed parametrically as

$$x = x_0 + (x_1 - x_0)t$$
$$y = y_0 + (y_1 - y_0)t$$
$$z = z_0 + (z_1 - z_0)t$$

## Plane

In the three-dimensional system, the equation of a plane is represented by the equation,

$$Ax + By + Cz + D = 0.$$

A plane is determined by three points, all of which are not present on the same straight line.

# CHAPTER 14

# NUMERICAL ANALYSIS AND DIFFERENTIAL EQUATIONS

Numerical analysis and differential equations are an extension of the calculus and enable the engineer to solve more complex problems. Solution convergence and series solutions are emphasized because of their importance in engineering. Ordinary differential equations are an essential element in the study and solution of complex problems.

## Infinite Sequence

An **infinite sequence** is a function whose domain is the set of positive integers.

A sequence $\{a_n\}$ has the limit $\ell$ (denoted by $\lim_{n \to \infty} a_n = \ell$) if for every $c > 0$, there exists a positive number $N$ such that if $n > N$, then $|a_n - \ell| < c$.

If the limit ($\lim_{n \to \infty} a_n$) does not exist, then the sequence $\{a_n\}$ has no limit.

The statement $\lim_{n \to \infty} a_n = \infty$ means that for every positive real number $s$, there exists a number $N$ such that if $n > N$, then $a_n > s$.

1. $\lim_{n \to \infty} r^n = 0$, if $|r| < 1$.

2. $\lim_{n \to \infty} r^n = \infty$, if $|r| > 1$.

**Properties**

Suppose that $\lim_{n \to \infty} a_n = \ell$ and $\lim_{n \to \infty} b_n = m$, where $m$ and $\ell$ are real numbers, then it can be proven that:

1. $\lim_{n \to \infty} (a_n + b_n) = \ell + m$

2. $\lim_{n \to \infty} (a_n - b_n) = \ell - m$

3. $\lim_{n \to \infty} (a_n b_n) = \ell \times m$

4. $\lim_{n \to \infty} \dfrac{a_n}{b_n} = \dfrac{\ell}{m}$

Suppose that $\{a_n\}$ represents a sequence and that $\lim_{n \to \infty} |a_n| = 0$ then $\lim_{n \to \infty} a_n = 0$.

A **monotonic sequence** is a sequence with successive terms that are non-decreasing

$$(a_1 \leq a_2 \leq \ldots \leq a_n)$$

or non-increasing

$$(a_1 \geq a_2 \geq \ldots \geq a_n).$$

A sequence is said to be **bounded** if there is a positive real number $\alpha$ such that $|a_m| \leq \alpha$ for all $m$.

An infinite sequence that is bounded and monotonic is said to have a limit.

# Infinite and Geometric Series

An infinite series $\sum_{k=0}^{\infty} a_k$, sometimes written $a_0 + a_1 + a_2 + \ldots$, is the sequence $\{s_0, s_1, s_2, s_3, \ldots\}$ of partial sums.

An infinite series converges if $\lim_{n \to \infty} s_n = s$ for some real number $s$. The series diverges if the sequence of the partial sums diverges (the limit does not exist).

Geometric series have the form $a + ar + ar^2 + \ldots + ar^{n-1}$, where $a$ and $r$

are real numbers and $a \neq 0$.

The geometric series $a + ar + ar^2 + \ldots + ar^{n-1}$, with $a \neq 0$:

1. Converges and has the sum $\dfrac{a}{1-r}$ if $|r| < 1$.
2. Diverges if $|r| \geq 1$.

If an infinite series $\Sigma a_n$ is convergent, then $\lim\limits_{n \to \infty} a_n = 0$. The infinite series $\Sigma a_n$ is divergent if $\lim\limits_{n \to \infty} a_n \neq 0$. If $\lim\limits_{n \to \infty} a_n = 0$, this does not mean that the series is convergent.

For every $c > 0$, if there exists an integer $N$ such that $|s_k - s_\ell| < c$ whenever $k, \ell > N$, then the infinite series $\Sigma a_n$ is convergent.

If $\Sigma a_n$ and $\Sigma b_n$ are convergent series with the sums $A$ and $B$ respectively, then:

1. $\Sigma(a_n + b_n)$ converges and has the sum $A + B$.
2. If $c$ is a real number, $\Sigma c a_n$ converges and has the sum $cA$.
3. $\Sigma(a_n - b_n)$ converges and has the sum $A - B$.

## Tests for Convergence

If $\Sigma a_n$ is a positive term series and if there exists a number $m$ such that $s_n < m$ for every $n$, then the series converges and has a sum $s \leq m$. If no such $m$ exists, the series diverges.

### The Integral Test

If a function $f$ is positive, continuous and decreasing on the interval $[1, \infty)$, then the infinite series:

1. Converges if $\int_1^\infty f(x)\,dx$ converges.
2. Diverges if $\int_1^\infty f(x)\,dx$ diverges.

### The P-Series

The p-series expressed as $\sum_{n=1}^{\infty} \frac{1}{n^p}$, converges if $p > 1$ and diverges if $p \leq 1$.

### Comparison Test

Let $\Sigma a_n$ and $\Sigma b_n$ represent positive term series.

1. If $a_n \leq b_n$ and the series $\Sigma b_n$ converges, then $\Sigma a_n$ converges.
2. If $a_n \geq b_n$ and the series $\Sigma b_n$ diverges, then $\Sigma a_n$ also diverges.

### The Limit Comparison Test

Let $\Sigma a_n$ and $\Sigma b_n$ represent positive term series.

If $\lim_{n \to \infty} \frac{a_n}{b_n} = \ell$, where $\ell$ is some positive number, then either both series converge or both diverge.

## Alternating Series: Absolute and Conditional Convergence

An **alternating series** is an infinite series in which successive terms have opposite signs. An alternating series is usually expressed as

$$a_1 - a_2 + a_3 - a_4 + \ldots + (-1)^{n-1} a_n + \ldots,$$

or

$$-a_1 + a_2 - a_3 + a_4 - \ldots + (-1)^n a_n + \ldots,$$

where each $a_i > 0$.

### Alternating Series Test

Let $\{a_k\}$ represent a decreasing series of positive terms. If $a_k > 0$, then $\sum_{k=1}^{\infty} (-1)^k a_k$ converges.

If the conditions $a_k > a_{k+1} > 0$ for every positive integer $k$ hold for the alternating series $\Sigma (-1)^{k-1} a_k$ and $\lim_{n \to \infty} a_k = 0$, then the error in approximating the sum $S$ of the series by the $k + n$ partial sum, $S_k$, is numerically less than $a_{k+1}$.

## Absolute Convergence

The series $\Sigma a_k$ is absolutely convergent if the series $\Sigma |a_k| = |a_1| + |a_2| + \ldots + |a_n|$, obtained taking the absolute value of each term, is convergent.

## Conditional Convergence

The series $\Sigma a_k$ is conditionally convergent if $\Sigma |a_k|$ diverges while $\Sigma a_k$ converges.

## Ratio Test

Suppose that in the series $\Sigma |a_k|$, every $a_k \neq 0$,

$$\lim_{k \to \infty} \left| \frac{a_{k+1}}{a_k} \right| = \rho \quad \text{or} \quad \lim_{k \to \infty} \left| \frac{a_{k+1}}{a_k} \right| = +\infty.$$

Then:

1. If $\rho < 1$, the series $\Sigma a_k$ converges absolutely.

2. If $\rho > 1$, or if $\lim_{k \to \infty} \frac{|a_{k+1}|}{|a_k|} = +\infty$, the series diverges.

3. If $\rho = 1$, the test gives no information.

## Root Test

Let $\Sigma a_k$ represent an infinite series.

1. If $\lim_{k \to \infty} \sqrt[k]{|a_k|} = \ell < 1$, the series is absolutely convergent.

2. If $\lim_{k \to \infty} \sqrt[k]{|a_k|} = \ell > 1$ or $\lim_{k \to \infty} \sqrt[k]{|a_k|} = \infty$, the series is divergent.

3. If $\lim_{k \to \infty} \sqrt[k]{|a_k|} = 1$, the series may be absolutely convergent, conditionally convergent, or divergent.

# Power Series

A **power series** is a series of the form $c_0 + c_1(x-a) + c_2(x-a)^2 + \ldots + c_n(x-a)^n$ in which $a$ and $c_i$, $i = 1, 2, 3$, etc. are constants.

The notations $\sum_{n=0}^{\infty} c_n(x-a)^n$ and $\sum_{n=0}^{\infty} c_n x^n$ are used to describe power series.

A power series $\Sigma c_n x^n$ is said to converge:

1. at $x_1$ if and only if $\Sigma c_n x^n$ converges.
2. on the set $S$ if and only if $\Sigma c_n x^n$ converges for each $x \in S$.

If $\Sigma c_n x^n$ converges at $x_1 \neq 0$, then it converges absolutely whenever $|x| < |x_1|$. If $\Sigma c_n x^n$ diverges at $x_1$, then it diverges for $|x| > |x_1|$.

**The Differentiation of Power Series**

If $f(x) = \sum_{n=0}^{\infty} c_n x^n$ for all $x$ in $(-a,a)$, then $f$ is differentiable on $(-a,a)$ and

$f'(x) = \sum_{n=1}^{\infty} n c_n x^{n-1}$ for all $x$ in $(-a,a)$.

A power series defines an indefinite differentiable function in the interior of its interval of convergence.

The derivatives of this function may be obtained by differentiating term by term.

# Series Solutions Near an Ordinary Point

Consider the equation
$$P(x)y'' + Q(x)y' + R(x)y = 0. \quad (A)$$

**Procedure:**
1. Assume that the solution is of the form
$$y = \sum_{n=0}^{\infty} a_n (x - x_0)^n. \quad (B)$$
2. Substitute (A) and its derivatives into (B).
3. Collect all terms with the same powers of $x$.
4. Combine the equations to obtain the recurrence formula, a relation between the coefficients $a_k$ and $a_{k+1}$.
5. Find the first few terms in the series.
6. Rewrite the general solution as

$$y = \sum_{n=0}^{\infty} a_n (x - x_0)^n = a_0 y_1(x) + a_1 y_2(x),$$

where $a_0$ and $a_1$ are arbitrary, and $y_1$ and $y_2$ are linearly independent series solutions.

# First Order Equations

## Linear Equations

Given a first-order, ordinary linear equation of the general form

$$\frac{dy}{dx} + P(x)y = Q(x) \qquad (C)$$

multiply both sides by an integrating factor

$$u(x) = \int_e^x P(x)dx$$

to get

$$\frac{d}{dx}(u(x)y) = u(x)Q(x).$$

The general solution for (A) is then

$$y = \frac{1}{u(x)}\int u(x)Q(x)dx + c$$

## Separable Equations

If an equation of the form

$$M(x, y) + N(x, y)\frac{dy}{dx} = 0$$

can be written

$$M(x)M(y)dx + N(x)N(y)dy = 0. \qquad (D)$$

The general solution for (D) is

$$\int^x \frac{M(x)}{N(x)}dx + \int^y \frac{N(y)}{M(y)}dy = c$$

## Exact Equations

An exact equation has the form

$$M(x, y) + N(x, y)\frac{dy}{dx} = 0$$

A–171

where
$$\frac{\partial M}{\partial y} = \frac{\partial N}{\partial x}.$$

1. Find $\mu(x,y)$:
$$\mu(x,y) = \int M(x,y)\,dx + \phi(y) \qquad (E)$$
(Holding $y$ as a constant.)

2. Set $\dfrac{\partial \mu}{\partial y} = N(x,y)$, holding $x$ as a constant.

3. Solve for $\phi'(y)$:
$$\phi'(y) = N(x,y) - \int \frac{\partial M(x,y)\,dx}{\partial y} \qquad (F)$$

4. Substituting back into (E) gives the general solution
$$\mu(x,y) = \int M(x,y)\,dx + \int \phi'(y)\,dx$$
where $\phi'(y)$ is given by (F).

**Integrating Factors**

If the equation
$$M(x,y)\,dx + N(x,y)\,dy = 0 \qquad (G)$$
is not exact, then an integrating factor is sought to make it exact.

1. If $\dfrac{\dfrac{\partial M}{\partial y} - \dfrac{\partial N}{\partial x}}{N}$ is a function of $x$ alone, then

$$\mu(x) = \exp\left[\int \frac{\dfrac{\partial M}{\partial y} - \dfrac{\partial N}{\partial x}}{N}\,dy\right]$$

is the integrating factor of (G).

2. If $\dfrac{\dfrac{\partial M}{\partial y} - \dfrac{\partial N}{\partial x}}{-M}$ is a function of $y$ alone, then

$$\mu(y) = \exp\left[\int \dfrac{\dfrac{\partial M}{\partial y} - \dfrac{\partial N}{\partial x}}{-M}\, dx\right]$$

is the integrating factor of (G).

3. If $\dfrac{\partial M}{\partial x} + \dfrac{\partial N}{\partial y} \neq 0$, then

$$\mu(x, y) = \dfrac{1}{\dfrac{\partial M}{\partial x} + \dfrac{\partial N}{\partial y}}$$

is the integrating factor of (G).

### The Initial Value Problem

The initial value problem is to find a function $y(x)$ satisfying the first order differential equation $y' = f(x,y)$ with the initial value $y(x_0) = y_0 = \alpha$.

## Second Order Differential Equations

### Homogenous Equations with Constant Coefficients

Consider the equation

$$ay'' + by' + cy = 0$$

The characteristic or auxiliary equation of this is

$$ar^2 + br + c = 0 \qquad \text{(H)}$$

The roots of (H) are

$$r_1 = \dfrac{-b + (b^2 - 4ac)^{\frac{1}{2}}}{2a}$$

and

$$r_2 = \frac{-b-(b^2-4ac)^{\frac{1}{2}}}{2a}$$

If the roots are real and equal ($b^2 - 4ac = 0$),

$$y = c_1 e^{r_1 x} + c_2 x e^{r_1 x}$$

## Boundary Value Problems

The boundary value problem is described by an equation of the form

$$y'' = f(x,y,y') \qquad x \in [a,b],$$

with boundary conditions

$$y(a) = \alpha \qquad y(b) = \beta$$

The boundary conditions may be given in different ways. Here are some existence theorems that help us to determine whether unique solutions to such problems exist or not.

Given the boundary value problem

$$y'' = f(x,y,y'), \quad x \in [a,b], \quad y(a) = \alpha, \quad y(b) = \beta$$

If $f$ is continuous on the set

$$S = \{(x,y,y') \mid x \in [a,b], y, y' \in (-\infty, \infty)\}$$

with $\dfrac{\partial f}{\partial x}, \dfrac{\partial f}{\partial y'}$ also continuous on $S$, then the boundary value problem has a unique solution if:

1. $\dfrac{\partial f}{\partial x}(x,y,y') > 0 \quad \forall \, (x,y,y') \in S$.

2. $\exists$ a constant, $k \ni \left| \dfrac{\partial f}{\partial y}, (x,y,y') \right| \leq k \,\forall\, (x,y,y') \in S$.

If $f(x,y,y')$ can be written as

$$f(x,y,y') = p(x)y' + q(x)y + r(x)$$

then the equation $y'' = f(x,y,y')$ is linear. We restate the theorem to accommodate this type.

Given the boundary value problem

$$y'' = p(x)y' + q(x)y + r(x), \quad x \in [a,b], y(a) = \alpha, y(b) = \beta$$

has a unique solution if

1. $p$, $q$, and $r$ are continuous on $[a,b]$.
2. $q > 0$ on $[a,b]$.

**Shooting Method**

If the boundary value problem is replaced by two initial value problems, that is, given

$$f(x,y,y') = p(x)y' + q(x)y + r(x),$$

we work the pair:

(A) $y'' = p(x)y' + q(x)y + r(x)$, $x \in [a,b]$, $y(a) = \alpha$, $y(b) = \beta$.
(B) $y'' = p(x)y' + q(x)y$, $x \in [a,b]$, $y(a) = 0$, $y'(a) = 1$.

Assume the solution to (A) is $y_0(x)$, and the solution to (B) is $y_1(x)$. So

$$y(x) = y_0(x) + \frac{(\beta - y_0(b))}{y_1(b)} y_1(x), \quad y_1(b) \neq 0$$

We can use any method available that can be applied to solving initial value problems to generate approximate solutions, $y_0(x)$, $y_1(x)$ for (A) and (B), then use the equation:

$$y(x) = y_0(x) + \frac{(\beta - y_0(b))}{y_1(b)} y_1(x), \quad y_1(b) \neq 0$$

to approximate the unique solution $y(x)$ to the boundary value problem. So we finally should have a table of function values at mesh points in $[a,b]$. The accuracy of the solutions will depend on the method chosen. Runge-Kutta methods of order $\geq 4$ are commonly used.

# Fourier Series Expansion

Sometimes, the need arises to have a function $f$ expressed in a series of the form

$$\sum_{k=1}^{\infty} c_k \psi_k$$

where $\{\psi_k\}$ is a set, chosen so that its members form an orthonormal system with respect to some weight function $w$ on some interval $[a,b]$. With the assumption that such a compound representation for $f$ exists, with suitable convergence, the coefficients $c_k$ would be found to be

$$c_k = \int_a^b w(x)f(x)\psi_k(x)dx \quad k=1,2,\ldots$$

One type of the above representation of a function is known as the Fourier Series, of which there are several classes.

## Parabolic Partial Differential Equation

One example of a parabolic partial differential equation is the diffusion equation

$$\frac{\partial}{\partial t}V(x,t) = \frac{a^2 \partial^2}{\partial x^2}V(x,t)$$

$x \in (a,b)$, $t > 0$,
with conditions

$$V(0,t) = 0, \quad V(b,t) = 0, \quad t > 0$$
$$V(x,0) = f(x), \quad x \in [0,b]$$

$a$ is a constant.

The numerical method uses finite differences.

## Monte Carlo Methods

Monte Carlo methods are used to solve certain kinds of problems by using random numbers. Since these methods have a very slow rate of convergence with fair accuracy, they provide initial approximations for better methods.

When a Monte Carlo method is used to imitate a random process it is called a simulation. Some examples are:
1. A neutron's motion in a reaction chamber.
2. Arrival frequencies for passengers at a train station within certain times.

When properties of a large set of elements are deduced by studying a small but random subset of the same elements, this is called sampling, for example:

> Estimating the average value of a function $f$ over an interval by estimating the average value over a finite random subset of points in the interval.

# CHAPTER 15

# LINEAR ALGEBRA

Solution of simultaneous equations, whether for a single solution or as part of an interactive computational algorithm, lends itself to matrix operations. The associated eigenvectors and eigenvalues give the complete solution to the problem with many unknowns. The basis of calculus of vector functions, or linear algebra, is important to the engineer in practice and while studying advanced principles.

## Types of Matrices

### Identity Matrix

An identity matrix (I) is a square matrix with unity on the main diagonal and zeros everywhere else.

### Elementary Matrix

An $N \times N$ matrix which can be derived from the $N \times N$ identity matrix by performing a single elementary row operation is called an elementary matrix (E).

### Coefficient Matrix

A matrix which contains entries corresponding to the coefficients of a system of linear equations, but excludes the constants of that system is called a coefficient matrix.

### Square Matrix

A square matrix with $N$ rows and $N$ columns is called a square matrix of order $N$.

## Triangular Matrix

A square matrix is in upper triangular form if it has all zero entries below the main diagonal; it is in lower triangular form if it has all zero entries above the main diagonal; it is in triangular form if it is in either upper or lower triangular form.

## Orthogonal Matrix

The square matrix A with the property $A^{-1} = A^t$ is called an orthogonal matrix.

## Diagonal Matrix

A square matrix with non-zero entries on the main diagonal and zeros everywhere else.

# Linear Equations and Matrices

A **linear equation** is an equation of the form $A_1X_1 + A_2X_2 + \ldots + A_NX_N = b$, where $A_1, \ldots, A_N$ and $b$ are real constants.

**Examples:**
a) $2x + 6y = 9$
b) $x_1 + 3x_2 + 7x_3 = 5$
c) $\alpha - 2 = 0$

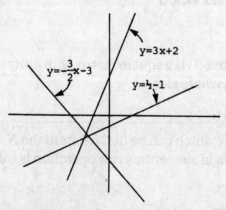

Figure 74. Linear equations are always straight lines

A **system** of linear equations is a finite set of linear equations, all of which use the same set of variables.

The **solution** of a system of linear equations is that set of real numbers which, when substituted into the set of variables, satisfies each equation in the system. The set of all solutions is called the solution set $S$ of the system.

a) $x_1 + x_2 = 5$    $S = \{(5,0), (0,5), (4,1), (1,4), (3,2), (2,3)\}$
b) $y + z = 9$    $S = \{5, 4\}$
    $z = 4$

A **consistent** system of linear equations has at least one solution, while an **inconsistent** system has no solutions.

Every system of linear equations has either one solution, no solution, or infinitely many solutions.

The **augmented matrix** for a system of linear equations is the matrix of the form:

$$\begin{bmatrix} a_{11} & a_{12} & \cdots & a_{1N} & b_1 \\ a_{21} & a_{22} & \cdots & a_{2N} & b_2 \\ \vdots & & & & \\ a_{M1} & a_{M2} & \cdots & a_{MN} & b_M \end{bmatrix}$$

where $a_{ij}$ represents each coefficient in the system and $b_i$ represents each constant in the system.

Elementary row operations are operations on the rows of an augmented matrix, which are used to reduce that matrix to a more solvable form. These operations are the following:

1. Multiply a row by a non-zero constant.
2. Interchange two rows.
3. Add a multiple of one row to another row.

# Homogeneous Systems of Linear Equations

A **homogenous** system of linear equations is a system in which all of the constant terms (those which are not multiplied by any variables) are zero.

For example:
$x_1 + 3x_2 = 0$   is a homogenous system.
$4x_1 + x_2 + 7x_3 = 0$
$2x_2 + 2x_3 = 0$

Every homogenous system of $N$ linear equations has at least one solution, called the **trivial solution**, in which all of the variables $x_1, x_2, \ldots, x_N$ are equal to zero. All other solutions to the system are called **non-trivial solutions**.

Every homogenous system of linear equations has either (a) only the trivial solution, or (b) the trivial solution and an infinite number of non-trivial solutions. If there are more unknowns than equations, then the system has non-trivial solutions.

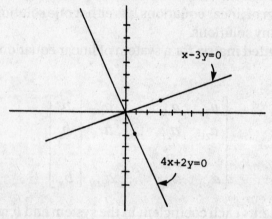

Figure 75. In a homogenous system, all lines pass through an origin

## Matrices

A **matrix** is a rectangular array of numbers, called **entries**.

(a) $\begin{bmatrix} 6 & 2 \\ 3 & 1 \\ 0 & 0 \end{bmatrix}$

(b) $\begin{bmatrix} 3 \\ 1 \end{bmatrix}$

(c) $\begin{bmatrix} 1 & 7 & 2 & 1 \end{bmatrix}$

A matrix with $N$ rows and $N$ columns is called a **square matrix** of order $N$.

$$\begin{bmatrix} a_{11} & a_{12} & \cdots & a_{1N} & b_1 \\ a_{21} & a_{22} & \cdots & a_{2N} & b_2 \\ \vdots & & & & \\ a_{M1} & a_{M2} & \cdots & a_{MN} & b_M \end{bmatrix} \qquad \begin{bmatrix} 2 & 10 & 1 \\ 6 & 2 & 9 \\ 3 & 3 & 7 \end{bmatrix}$$ is a square matrix of order 3.

Two matrices are equal if they have the same size and the same entries.

Entries starting at the top left and proceeding to the bottom right of a square matrix are said to be on the **main diagonal** of the matrix.

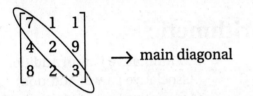 → main diagonal

The sum **B + D** is the matrix obtained when two matrices, **B** and **D**, are added together; they must both be of the same size. **B − D** is obtained by subtracting the entries of **D** from the corresponding entries of **B**.

(a) $\begin{bmatrix} 1 & 2 \\ 2 & 6 \end{bmatrix} + \begin{bmatrix} -4 & 7 \\ 1 & 1 \end{bmatrix} = \begin{bmatrix} -3 & 9 \\ 3 & 7 \end{bmatrix}$

(b) $\overset{B}{\begin{bmatrix} 1 & 2 \\ 2 & 6 \end{bmatrix}} - \overset{D}{\begin{bmatrix} -4 & 7 \\ 1 & 1 \end{bmatrix}} = \begin{bmatrix} 5 & -5 \\ 1 & 5 \end{bmatrix}$

The product of a matrix **A** by a scalar $k$ is obtained by multiplying each entry of **A** by $k$.

If $A = \begin{bmatrix} 4 & 7 \\ -1 & 2 \end{bmatrix}$ and $k = 3$, then $Ak = \begin{bmatrix} 12 & 21 \\ -3 & 6 \end{bmatrix}$.

When multiplying two matrices **A** and **B**, the matrices (the number of columns of **A** must equal the number of rows of **B**) must be of the sizes $M \times N$ and $N \times P$; to obtain the ($ij$) entry of **AB**, multiply the entries in row $i$ of **A** by the corresponding entries in column $j$ of **B**. Add up the resulting products; this sum is the ($ij$) entry of **AB**.

If AB = C, then $c_{ij} = \sum_{k=1}^{M} a_{ik}b_{kj}$.

**Example:**

If $A = \begin{bmatrix} 2 & 3 \\ 4 & 5 \end{bmatrix}$ and $B = \begin{bmatrix} 3 & 3 \\ 7 & 2 \end{bmatrix}$, then $AB = \begin{bmatrix} 27 & 12 \\ 47 & 22 \end{bmatrix}$.

A matrix which contains entries corresponding to the coefficients of a system of linear equations, but excludes the constants of that system is called a **coefficient matrix**.

## Rules of Matrix Arithmetic

1. A + B = B + A  (Commutative Law of Addition)
2. A + (B + C) = (A + B) + C  (Associative Law of Addition)
3. A(BC) = (AB)C  (Associative Law of Multiplication)
4. A(B ± C) = AB ± AC  (Distributive Law)
5. $a$(B + C) = $a$B + $a$C
6. ($a$ ± $b$)C = $a$C ± $b$C
7. ($ab$)C = $a$($b$C)
8. $a$(BC) = ($a$B)C = B($a$C)

A matrix whose entries are all zero is called a **zero matrix, 0**.

If the size of the matrices are such that the indicated operations can be performed, the following rules of matrix arithmetic are valid:

1. A + 0 = 0 + A = A
2. A − A = 0
3. 0 − A = −A
4. A0 = 0

An **identity matrix (I)** is a square matrix with ones on the main diagonal and zeros everywhere else.

If $A$ is a square matrix and a matrix B exists such that AB = BA = I, then A is **invertible** and B is the **inverse** of A, ($A^{-1}$). An invertible matrix has one and only one inverse.

If A and B are invertible matrices of the same size, then:

1. **AB** is invertible
2. $(AB)^{-1} = (A^{-1})(B^{-1})$

Formula for inverting a 2 × 2 matrix:

If $A = \begin{bmatrix} a & b \\ c & d \end{bmatrix}$, then $A^{-1} = \dfrac{1}{ad-bc}\begin{bmatrix} d & -b \\ -c & a \end{bmatrix}$.

If **A** is an invertible matrix, then:

1. $A^{-1}$ is invertible; $(A^{-1})^{-1} = A$
2. $kA$ is invertible (where $k$ is a non-zero scalar); $(kA)^{-1} = \dfrac{1}{k}A^{-1}$
3. $A^N$ is invertible; $(A^N)^{-1} = (A^{-1})^N$

If **A** is a square matrix and $x$ and $y$ are integers, then:

1. $A^x A^y = A^{x+y}$
2. $(A^x)^y = A^{xy}$

# Gaussian Elimination

### Reduced Row-Echelon Matrix

A matrix is in reduced row-echelon form, if it has the following properties:

a. Either the first non-zero entry in a row is 1, or the row consists entirely of zeros.
b. All rows consisting entirely of zeros are grouped together at the bottom of the matrix.
c. If two successive rows do not consist entirely of zeros, then the leading 1 in the lower row occurs farther to the right than the leading 1 in the higher row.
d. Every column with a leading 1 has zeros in every other entry.

The variables corresponding to the leading ones in a reduced row-echelon matrix are called leading variables. Gauss-Jordan elimination is a process using elementary row operations by which any matrix can be brought into reduced row-echelon form. Once this is done the system of linear equations corresponding to that matrix is easily solvable. The process is as follows:

1. Add appropriate multiples of the first row to the rows below so that all entries below the leading one of the column found in (a) become zeros.
2. Cover the first row and, using the remaining submatrix, begin again at step (a). Continue (a) through (d) until the matrix is in row-echelon form.
3. Starting with the last row which does not consist entirely of zeros, add appropriate multiples of each row to the rows above so that each column containing a one has zeros everywhere else. The matrix will now be in reduced row-echelon form.

## Elementary Matrices

An $N \times N$ matrix which can be derived from the $N \times N$ identity matrix by performing a single elementary row operation is called an elementary matrix (E).

If E is an $M \times M$ elementary matrix and B is an $M \times M$ matrix, then the product of EB is equivalent to performing the row operation of E on B.

Every elementary matrix is invertible, and its inverse is also invertible.

Row-equivalent matrices are matrices which can be derived from each other by a finite sequence of row operations.

If A is an $N \times N$ matrix, then the following statements are equivalent (either all true or all false):

1. A is invertible.
2. The system of linear equations represented by $Ax = 0$ has only the trivial solution.
3. A is row-equivalent to the $N \times N$ matrix.

To invert an $N \times N$ matrix, A, where $N > 2$, you must perform the elementary row operations on the $N \times N$ identity matrix which would reduce A to I. That derivation of the $N \times N$ identity matrix will be $A^{-1}$.

If A is an invertible $N \times N$ matrix, then for the system of equations,

$Ax = B$, where $B$ is any $N \times 1$ matrix, there is only one solution; namely $x = A^{-1}B$.

If $A$ is an $N \times N$ matrix, then the following statements are equivalent:

1. A is invertible.
2. $Ax = 0$ has only the trivial solution.
3. A is row-equivalent to the $N \times N$ identity matrix.
4. $Ax = B$ is consistent for every $N \times N$ matrix B.

In a homogenous system, all lines pass through the origin.

## Determinants

The determinant of a square matrix A (det(A)) is the sum of all elementary products of A.

A signed elementary product from the matrix A is an elementary product of multiplies by –1 or +1. We use the + sign if the permutation of the set is even and – if odd.

An **elementary product** from an $N \times N$ matrix A is a product of $N$ entries from A, with no two entries from the same row or column.

A **permutation** of a set of integers is some arrangement of those integers without any repetitions or omissions.

An even permutation has an even number of inversions; an odd permutation has an odd number of inversions.

An inversion in a permutation occurs when a larger integer appears before a smaller one.

## Determinants by Row Reduction

A square matrix containing a row of zeros has a determinant of zero.

A square matrix is in upper triangular form if it has all zero entries below the main diagonal; it is in lower triangular form if it has all zero entries above the main diagonal; it is in triangular form if it is in either upper or lower triangular form.

If A is an $N \times N$ triangular matrix, then the det(A) is the product of the main diagonal entries.

If a square matrix has two proportional rows, then its determinant is zero.

Given A is any $N \times N$ matrix,

1. If A* is the result of multiplying one row of A by a constant $k$, then

$\det(A^*) = k\det(A)$.
2. If $A^*$ is the result of switching two rows of A, then $\det(A^*) = -\det(A)$.
3. If $A^*$ is the result of adding a multiple of a row of A to another row of A, then $\det(A^*) = \det(A)$.

## Determinant Properties

If A is an $M \times N$ matrix, then the transpose of A, denoted by $(A^t)$ is defined as the $N \times M$ matrix, where the rows and columns of A are switched.

Properties of the transpose operation
1. $(A^t)^t = A$
2. $(A + B)^t = A^t + B^t$
3. $(kA)^t = kA^t$ (where $k$ is a scalar)
4. $(AB)^t = A^tB^t$

If A is a square matrix, then $\det(A) = \det(A^T)$. (Because of this theorem, all determinant theorems concerning the rows of a matrix also apply to the columns of a matrix.)

If A and B are square matrices of the same size, and $k$ is a scalar, then:

1. $\det(kA) = k^N\det(A)$      ($N$ is the number of rows of A)
2. $\det(AB) = \det(A)\det(B)$

A square matrix A is invertible if and only if $\det(A) \neq 0$.
If A is invertible, then $\det(A^{-1}) = 1/\det(A)$.

## Cofactor Expansion; Cramer's Rule

If A is a square matrix, then the minor entry of $a_{ij}$, denoted $(M_{ij})$, is defined to be the determinant of the submatrix remaining after the $i^{th}$ row and $j^{th}$ column of A are removed.

If A is a square matrix, then the cofactor of entry $a_{ij}$, denoted $c_{ij}$, is defined to be the scalar $(-1)^{i+j}M_{ij}$.

$c_{ij} = \pm M_{ij}$, depending on the position of the entry in relation to the matrix

$$\begin{bmatrix} + & - & + & - & \cdots \\ - & + & - & + & \cdots \\ + & - & + & - & \cdots \\ - & + & - & + & \cdots \\ \vdots & \vdots & \vdots & \vdots & \end{bmatrix}$$

If A is a square matrix, det(A) can be found by cofactor expansion along the *i*th row or the *j*th column of A. This is done by multiplying the entries in the *i*th row of the *j*th column of A by their cofactors and summing the resulting products. Thus,

$$\det(A) = a_{i1}C_{i1} + a_{i2}C_{i2} + \cdots$$

or

$$\det(A) = a_{1j}C_{2j} + a_{2j}C_{2j} + \cdots$$

If A is a square matrix and $c_{ij}$ is the cofactor of $a_{ij}$, then the matrix cofactors from A is

$$\begin{bmatrix} C_{11} & C_{12} & \cdots \\ C_{21} & C_{22} & \cdots \\ \vdots & \vdots & \end{bmatrix}$$

The transpose of this matrix is called the **adjoint of A (adj(A))**.

If A is an invertible matrix, then

$$A^{-1} = \frac{1}{\det(A)} \operatorname{adj}(A)$$

## Cramer's Rule

If $Ax = B$ is a system of $N$ linear equations having unknowns $x_1, x_2, \ldots x_N$, then the unique solution of the system is:

$$x_1 = \frac{\det(A_1)}{\det(A)}, \quad x_2 = \frac{\det(A_2)}{\det(A)}, \quad \ldots, \quad x_N = \frac{\det(A_N)}{\det(A)}$$

where $A_N$ is the matrix obtained by replacing the *j*th column of A with the column of constants of the system,

$$B = \begin{bmatrix} b_1 \\ b_2 \\ \vdots \\ b_N \end{bmatrix}$$

## Linear Independence

Let V be a vector space, and let $v_1, \ldots, v_n$ be elements of V. Vectors $v_1, \ldots, v_n$ are linearly dependent if there exists numbers $a_1, \ldots, a_n$ not all equal to zero such that,

$$a_1 v_1 + \ldots + a_n v_n = 0.$$

If such numbers do not exist, then $v_1, \ldots, v_n$ are linearly independent.

## Basis and Dimension

If V is any vector space, and $x = \{v_1, v_2, v_3, \ldots, v_R\}$ is a finite set of vectors in V, then $s$ is called a basis for V.

Also:

1. $x$ is linearly independent
2. $x$ spans V

$V = \{(1,0,0), (0,1,0), (0,0,1)\}$ is the basis for $R^3$

The set of vectors $\{(1,0,\ldots,0), (0,1,0,\ldots,0), (0,0,\ldots,1)\}$ in $R^n$ is called the standard basis for $R^n$.

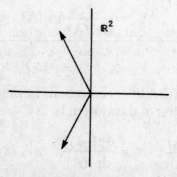

Figure 76. Any two non-colinear vectors in $R^2$ span $R^3$

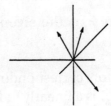

Figure 77. Any three non-coplanar vectors in $R^3$ span $R^3$

If a non-zero vector space contains a basis consisting of a finite set of vectors, then it is called finite dimensional. Otherwise, it is infinite dimensional. The zero vector space is defined to be finite dimensional.

Any two bases for a finite dimensional vector space must have the same number of vectors.

The numbers of vectors in a basis of a finite dimensional vector space is called the dimension of that vector space. The zero vector space is said to have dimension zero.

If $s = \{v_1, v_2, \ldots, v_R\}$ is a linearly dependent set of vectors in an $N$-dimensional space $x$, and $R < n$, then $s$ can be enlarged to form a basis for V.

# Eigenvalues, Eigenvectors

If A is an $N \times N$ matrix, then the non-zero vector $x \in R^N$ is called an **eigenvector** of A if $Ax = \lambda x$, where $\lambda$ is a scalar called the **eigenvalue of A**.

If A is a square matrix, then the characteristic equation of A is defined to be $\det(\lambda I - A) = 0$. When expanded, the $\det(\lambda I - A)$ is called the **characteristic polynomial of A**.

# Diagonalization

If A is a square matrix and there exists a matrix B such that $B^{-1}AB$ is diagonal, then A is diagonalizable and B diagonalizes A.

If A is an $N \times N$ matrix, and A is diagonalizable, then A has $N$ linearly dependent eigenvectors.

### Diagonalizing Procedure

The following is the procedure for diagonalizing an $N \times N$ matrix A:

1. Find the set of $N$ linearly independent eigenvectors of A, $\{v_1, v_2, \ldots, v_n\}$
2. Form the matrix P having $v_1, v_2, \ldots, v_N$ as its column vectors.
3. The matrix $P^{-1}AP$ will then be diagonal, with $\lambda_1, \lambda_2, \ldots, \lambda_N$ as its

diagonal entries where $\lambda_i$ is the eigenvalue corresponding $v_i$, $i = 1, 2, \ldots, n$.

If $v_1, v_2, \ldots, v_N$ are eigenvectors corresponding to distinct eigenvalues $\lambda_1, \lambda_2, \ldots, \lambda_N$, then $\{v_1, v_2, \ldots, v_N\}$ is a linearly independent set.

If an $N \times N$ matrix has $N$ distinct eigenvalues, then it is diagonalizable.

# CHAPTER 16

# PROBABILITY AND STATISTICS

The probability of the occurrence of an event and the uncertainty of the event happening are important "facts" to the engineer. As engineering enters the realm of uncertain events, stochastic modeling becomes more important. A firm understanding of probability and statistics is the essence in communicating the results of studies and the certainty of an event's occurrence.

## Sample Spaces

### Experiments: Deterministic or Random

The word **experiment** is used to describe any act that can be repeated under given conditions. In the chemistry laboratory, the students determine the boiling point of water as 100°C. If the experimental conditions remain the same, then the determination will be the same.

In contrast, there are experiments in which the results vary in spite of all efforts to keep the experimental conditions the same. For example, coin tossing or the birth date can be thought of as experiments. In these experiments, the results are unpredictable.

If the results of the repeated experiments are exactly the same, we say the experiments are deterministic; otherwise, they are said to be random or stochastic.

Probability theory is used to explain and predict, to some degree, the results of random experiments.

# Events

## Outcomes

An **outcome** is a particular result of an experiment.

For example, if two coins, one penny and one nickel, are tossed simultaneously and the outcome for each coin is recorded using ordered pair notation: (penny, nickel), then the sample space is given by

$$S = \{(H,H), (H,T), (T,H), (T,T)\}$$

Note that each of: $(H,H)$, $(H,T)$, $(T,H)$, and $(T,T)$ is a possible outcome of the experiment.

## Definition of Events

An **event** is a subset of the **sample space**. An elementary event is a set consisting of a single element of the sample space.

To say that an event $E$ has occurred is to say that the actual outcome of the random experiment is an outcome in the set of outcomes associated with the event $E$. Thus, if in the draw of a chip one of the six numbers in the set $\{3, 6, 9, 12, 15, 18\}$ occurs, then we say that the event "the number is a multiple of 3" has occurred; otherwise the event does not occur.

# Functions

## Definition of a Function

A rule or a correspondence that assigns to every element of a set a unique element of a set $K$ is called a **function** $f$ from $H$ to $K$. The set $H$ is called the **domain** of the function, and the set $K$ is called to **codomain** of the function. The set of all elements of $K$ that are related to the elements of $H$ by the correspondence is called the **range** of the function $f$.

Let $x$ represent an arbitrary, unspecified element of the set $H$; then it is customary to write $f(x)$ for the element of $K$ that corresponds to $x$. The element $f(x)$ of $K$ is called the value of the function $f$ for the element $x$ and it is read "$f$ of $x$" or "$f$ at $x$," or the image of $x$.

## Set Functions and Real-Valued Functions

A function, $f$, whose domain is the set of real numbers or a subset of the real numbers is called a function of a real variable; and, if the range of $f$ is also the real numbers or a subset of the real numbers, then $f$ is called a real-valued function.

A function whose domain consists of sets of elements, such as the subsets of a sample space is called a **set function**.

## Probability Space

For each random experiment, there is a sample space, for each sample space there are events, and for each event there is the question of the occurrence of the event.

Probability offers a measure of the occurrence of the event. Since we assign a unique real number as the probability of an event in a sample space, probability is a function which has its own distinguishing properties. The following definition lists the properties that the set of events and the probability function must satisfy.

A **probability space** is a sample space $S$ of outcomes, a set $E$ of subsets of $S$ called events, and a set of function $P$ whose domain is the set $E$, and whose codomain is the set of real numbers such that the following two sets of axioms are satisfied:

A. Axioms for the Domain $E$
1. The sample space $S$ is in $E$. That is, $S \in E$.
2. The empty set $\phi$ (the impossible event) is in $E$.
3. If a finite or countable number of events $A_1, A_2, A_3, \ldots$ belong to $E$, then their union belongs to $E$.
4. If an event $A$ belongs to $E$, then the complement $A'$ belongs to $E$.
5. If a finite or countable number of events $A_1, A_2, A_3, \ldots$ belong to $E$, then their intersection belongs to $E$.

B. Axioms for the Set Function $P$ (for the codomain $R$ of real numbers)
1. For every event $A$ of $E$, $0 \leq P(A) \leq 1$
2. $P(S) = 1$, and $P(\phi) = 0$
3. The probability of the union of a finite or countable number of pairwise exclusive events is the sum of the probabilities of the events; that is,
$$P(A_1 \cup A_2 \cup A_3 \cup \ldots) = P(A_1) + P(A_2) + P(A_3) + \ldots$$
where $A_i \cap A_j \neq \phi$ for all $i$ and $j$, $i \neq j$

## Sampling and Counting

There are many instances in the application of probability theory where it is desirable and necessary to count the outcomes in the sample space and the outcomes in an event. For example, in the special instance of a **uniform probability function**, the probability of an event is known when the number of outcomes that comprises the event is known; that is, as soon as the number of outcomes in the subset that defines the event is known.

If a sample space $S = \{e_1, e_2, \ldots, e_n\}$ contains $n$ simple events, $E_i = \{e_i\}$, $i = 1, 2, \ldots, n$, then using a uniform probability model, we assign probability $1/n$ for each point in $S$; that is, $P(E_i) = 1/n$. To determine the probability of an event $A$, we need,

1. The number of possible outcomes in $S$.
2. The number of outcomes in the event $A$. Then,

$$P(A) = \frac{\text{number of outcomes corresponding to } A}{\text{number of possible outcomes in } S}$$

Frequently, it may be possible to enumerate fully all the sample space points in $S$ and then count how many of these correspond to the event $A$. For example, if a class consists of just three students, and the instructor always calls on each student once and only once during each class, then if we label the students 1, 2, and 3, we can easily enumerate the points in $S$ as

$$S = \{(1, 2, 3), (1, 3, 2), (2, 1, 3), (2, 3, 1), (3, 1, 2), (3, 2, 1)\}$$

Assume that the instructor chooses a student at random, it would seem reasonable to adopt a uniform probability model and assign probability $\frac{1}{6}$ to each point in $S$. If $A$ is the event that John is selected last, then

$$\text{John} = B$$
$$A = \{(1, 2, 3), (2, 1, 3)\}$$
$$P(A) = \frac{2}{6} = \frac{1}{3}$$

# The Fundamental Principle of Counting

Suppose a man has four ways to travel from New York to Chicago, three ways to travel from Chicago to Denver, and six ways to travel from Denver to San Francisco, in how many ways can he go from New York to San Francisco via Chicago and Denver?

If we let $A_1$ be the event "going from New York to Chicago," $A_2$ be the event "going from Chicago to Denver," and $A_3$ be the event "going from Denver to San Francisco," then because there are four ways to accomplish $A_1$, three ways to accomplish $A_2$, and six ways to accomplish $A_3$, the number of routes the man can follow is

$$(4) \times (3) \times (6) = 72$$

We can now generalize these results and state them formally as the fundamental principle or multiplication rule of counting:

If an operation consists of a sequence of $k$ separate steps of which the first can be performed in $n_1$ ways, followed by the second in $n_2$ ways, and so on until the $k^{th}$ can be performed in $n_k$ ways, then the operation consisting of $k$ steps can be performed in

$$n_1 \times n_2 \times n_3 \ldots n_k$$

ways.

## Factorial Notation

Consider how many ways the owner of an ice cream parlor can display ten ice cream flavors in a row along the front of the display case. The first position can be filled in ten ways, the second position in nine ways, and the third position in eight ways, and so on. By the fundamental counting principle, there are

$$(10) \times (9) \times (8) \times (7) \times \ldots \times (2) \times (1)$$

or 3,628,800 ways to display the flavor. If there are 16 flavors, there would be $(16) \times (15) \times (14) \times \ldots \times (3) \times (2) \times (1)$ ways to arrange them. In general, if $n$ is a natural number, then the product from 1 to $n$ inclusive is denoted by the symbol $n!$ (read as "$n$ factorial" or as "factorial $n$" and is defined as

$$n! = n(n-1)(n-2) \ldots (3)(2)(1)$$

where $n$ is a positive natural number.

There are two fundamental properties of factorials:
1. By definition, $0! = 1$
2. $n(n-1)! = n!$

### Counting Procedures Involving Order Restrictions (Permutations)

A **permutation** of a number of object is any arrangement of these objects in a definite order.

If we have $n$ items with $r$ objects alike, then the number of distinct permutations taking all $n$ at a time is

$$\frac{n!}{r!}$$

In general, in a set of $n$ elements having $r_1$ elements of one type, $r_2$ elements of a second type and so on to $r_k$ elements of a $k^{th}$ type, then the number of distinct permutations of the $n$ elements, taken all together, is given by

$$nPn = \frac{n!}{r_1! r_2! r_3! \ldots r_k!}$$

where

$$\sum_{i=1}^{k} r_i = n$$

The number of permutations of a set of $n$ distinct objects, taken all together, is $n!$.

An arrangement of $r$ distinct objects taken from a set of $n$ distinct objects, $r \leq n$, is called a permutation of $n$ objects taken $r$ at a time. The total number of such orderings is denoted by $nPr$, and defined as

$$nPr = \frac{n!}{(n-r)!}$$

# Definition of a Random Variable

Many random experiments have a natural numerical description of the outcomes, such as the number of letters in a word or the number of dots on the uppermost face of a die. If a random experiment does not have a numerical description, we can give an assignment of a set of real numbers to represent the outcomes. Thus, to each elementary event in a sample space, we somehow assign a real number, thereby defining a function.

A **random variable** is a function whose domain is a finite or countably infinite sample space $S$ and whose codomain is the set of real numbers.

The usual notation is to let a capital letter, such as $X$ or $Y$, represent the functional correspondence, $X(e_i)$ denote the numerical value of the random variable for the elementary event $e_i$, and let $S_x$ denote the actual range set of the function $X$ in the codomain of real numbers. For example, flip a coin three times in succession. This experiment is a sequence of three Bernoulli trials and the basic sample space $S$ is the eight three-tuples of $H$ and $T$. That is,

$$S = \{(H,H,H), (H,H,T), (H,T,H), (T,H,H), (H,T,T), (T,H,T), (T,T,H), (T,T,T)\}$$
$$e_1 = (H,H,H),\ e_2 = (H,H,T),\ e_3 = (H,T,H),\ e_4 = (T,H,H),\ e_5 = (H,T,T),$$
$$e_6 = (T,H,T),\ e_7 = (T,T,H),\ e_8 = (T,T,T)$$

define the random variable $X$ on $S$ to be the number of heads in any outcome. Then

$$S_X = \{0, 1, 2, 3\},$$

and the correspondence for the sample space $S$ is given by $X(e_1) = 3$, $X(e_2) = 2$, $X(e_3) = 2$, $X(e_4) = 2$, $X(e_5) = 1$, $X(e_6) = 1$, $X(e_7) = 1$, $X(e_8) = 0$.

In general, if $X$ is a random variable whose domain is the sample space $S$ with probability function $P$, and if $S_x$ is the range set of real numbers for $X$ with

$$S_X = \{x_1, x_2, x_3, \ldots, x_n\}$$

then the function $P_x$ defined for any elementary event $\{x_i\}$ by

$$P_X(x_i) = P(\{e_j \mid e_j \in S, X(e_j) = x_i\})$$

is a probability function for $S_X$.

If $A$ is an arbitrary event of $S_X$ then

$$A = \{x_i \mid x_i \in S_X\ a \le x_i \le b\} \quad a \text{ and } b \text{ are real numbers}$$

and

$$P_X(A) = \sum_{x_i \in A} P_X(x_i)$$

# Probability Distribution Function

Sometimes it is convenient to extend the definition of the probability point function $Q$ defined above to the set of all real numbers by defining $Q(x) = 0$ if $x \notin S_x$. Thus, if

$$S_X = \{x_1, x_2, x_3, \ldots, x_n\}$$

then

$$Q(x) = \{x_i \text{ if } x = x_i, \quad i = 1, 2, 3, \ldots, n$$
$$0 \text{ if } x_i \notin S_x$$

# Counting Procedures Not Involving Order Restrictions (Combinations)

## Combination

A subset of $r$ objects selected without regard to order from a set of $n$ different objects, $r \leq n$, is called a **combination** of $n$ objects taken $r$ at a time. The total number of combinations of $n$ things taken $r$ at a time is denoted by $nCr$ or $\binom{n}{r}$ and is defined as

$$nCr = \binom{n}{r} = \frac{n!}{r_1!(n-r)!}$$

# Random Sampling

Essentially, **random sampling** is an application of the uniform probability model. Using the concepts of the uniform model, we define a random sample as follows.

A sample is said to be a **random sample** if all possible samples, of a particular size, chosen under some specified selection schemes, have the same probability of being chosen.

The following property is referred to as the **equivalence law of ordered sampling**:

If a random sample of size $k$ is drawn from a population of size $N$, then on any particular one of the $k$ draws each of the $N$ items has the same probability $1/N$ of being selected. The probability that a specified item is included in a sample of size $k$ taken from a population of size $N$ is $k/N$.

# Conditional Probability

Frequently we are interested in probabilities concerning part, rather than all, of a sample space. For example, the probability that a person chosen at random from a population has blue eyes is different from the probability of blue eyes in a subpopulation of people with blonde hair. Probabilities associated with these subpopulations are called **conditional probabilities**.

## Conditional Probability for Two Events

Let $A$ and $B$ be two events of a probability space $S$ with probability function $P$, and $P(B) > 0$, then the conditional probability of an event $A$, given that the event $B$ has occurred, is denoted by $P(A/B)$ and is defined

$$P\left(\frac{A}{B}\right) = \frac{P(A \cap B)}{P(B)}$$

for $P(B) \neq 0$. If $P(B) = 0$, then $P(A/B)$ is not defined.

The conditional probability for two events in a sample space $S$ is defined in terms of the probability function $P$ for $S$. Hence, all axioms and theorems for general probability functions hold for the conditional probability function.

**Properties of the Conditional Probability Function:**
1. $0 \leq P(E/F) \leq 1$
2. $P(S/F) = 1$
3. $P[(A \cup B)/F)] = P(A/F) + P(B/F)$, if $A \cap B = \phi$
4. $P[(A \cup B)/F)] = P(A/F) + P(B/F) - P[(A \cap B)/F)]$
5. $P(E'/F) = 1 - P(E/F)$

If $A$ is an event of $S_x$, then

$$P_X(A) = \sum_{x_i \in A} Q(x_i)$$

where the sum is taken over all points in $A$ such that $Q(x_i) > 0$.

## Properties of the Distribution Function

Distribution functions have three important properties:
1. $F(x)$ is a nondecreasing function of $x$. That is, if $u$ and $v$ are real numbers such that $u < v$, then $F(u) \leq F(v)$. If the sample space is $S_x = \{x_1, x_2, x_3, \ldots, x_n\}$, where $x_1 < x_2 < x_3 < \ldots < x_n$, then
2. $F(x) = 0$ for $x < x_1$, and
3. $F(x) = 1$ for $x \geq x_1$.

For example, if $S_x = \{0, 1, 2, 3\}$, and $Q$ is defined as

$$Q(x) = \begin{cases} \binom{3}{x} \times \left(\frac{1}{2}\right)^x \times \left(\frac{1}{2}\right)^{3-x} & \text{for } x = 0, 1, 2, 3. \\ 0 \end{cases}$$

Otherwise then, for $x = 0$, $Q(x) = \frac{1}{8}$; for $x = 1$, $Q(x) = 3(\frac{1}{2})(\frac{1}{2})^2 = \frac{3}{8}$; and so forth.

# Probability Density Functions

Let $X$ be a random variable with probability distribution function $F_x$. A **probability density function** is a function $f_x$ of $X$ whose domain is the set $R$ and whose range is also the set $R$; that is $f_x$ takes $R$ into $R$ such that

$$F_X(x) = \int_{-\infty}^{x} f_x(u)du$$

The probability density function $f_x$ has the following properties:
1. $f_x(x) \geq 0$ for all $x$
2. $f_x$ has at most a finite number of discontinuities in every finite interval of the real line.
3. $\int_{-\infty}^{\infty} f_x(u)du = 1$
4. For every interval $[a,b]$,

$$P_X(a \leq X \leq b) = \int_{a}^{b} f_X(x)dx$$

Thus, if $X$ is a random variable defined on a noncountable infinite sample space $S$, then the range of $X$, together with the probability density function $f_x$ form a continuous probability space.

# Standard Deviation

The **standard deviation** of a set $x_1, x_2, \ldots, x_n$ of $n$ numbers is defined by

$$s = \sqrt{\frac{\sum_{i=1}^{n}(x_1 - \bar{x})^2}{n}} = \sqrt{\overline{(x - \bar{x})^2}}$$

The sample standard deviation is denoted by $s$, while the corresponding population standard deviation is denoted by $\sigma$.

# Variance

The **variance** of a set of measurements is defined as the square of the standard deviation. Thus

$$s^2 = \frac{\sum_{i=1}^{n}(x_i - \bar{x})^2}{n}$$

or

$$s^2 = \frac{\sum_{i=1}^{n} f_i(x_i - \bar{x})^2}{\sum_{i=1}^{n} f_i}$$

Usually the variance of the sample is denoted by $s^2$ and the corresponding population variance is denoted by $\sigma^2$.

## Moments

The set of numbers $x_1, x_2, \ldots, x_n$ is given. Their $s^{th}$ **moment** is defined by

$$\overline{x^s} = \frac{x_1^s + x_2^s + \ldots + x_n^s}{n} = \frac{\sum_{i=1}^{n} x_i^s}{n}$$

For $s = 1$ the first moment is the arithmetic mean $\bar{x}$.

The $s^{th}$ moment about the mean $\bar{x}$ is defined as

$$m_s = \frac{\sum_{i=1}^{n}(x_i - \bar{x})^s}{n} = \overline{(x - \bar{x})^2}$$

## Coefficients of Skewness-Kurtosis

Distributions can be symmetric or asymmetric. For example, the normal distribution is symmetric. Among the asymmetric distributions, some can be "more" asymmetric than others. The degree of asymmetry is measured by **skewness**. Consider for example a distribution skewed to the right.

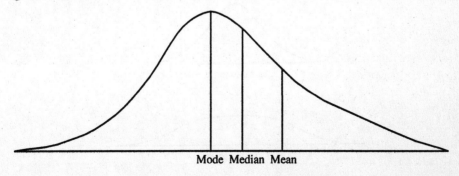

Figure 78. Moments of a distribution

The distribution skewed to the right has positive skewness. The asymmetry is measured by the difference.

$$\text{Skewness} = \frac{\text{Mean} - \text{Mode}}{\text{Standard Deviation}} = \frac{\bar{x} - \text{mode}}{s} \quad (*)$$

For a distribution skewed to the right, mean > mode and skewness is positive. The skewness of a distribution skewed to the left is negative. Using the empirical relation

$$\text{Mean} - \text{Mode} = 3\,(\text{Mean} - \text{Median})$$

we can write

$$\text{Skewness} = \frac{3\,(\text{Mean} - \text{Median})}{\text{Standard Deviation}} \quad (**)$$

**Skewness measures** the degree of asymmetry. Kurtosis measures the shape of the peak of a distribution. Usually this is measured relative to a normal distribution. A distribution can have one of three kinds of peaks.

1. **Leptokurtic,** where a distribution has a relatively high peak.
2. **Mesokurtic,** where the peak is neither very high nor very low; the peak of the normal distribution, for example.
3. **Platykurtic,** where the distribution is flat and the peak is low and not sharply outlined.

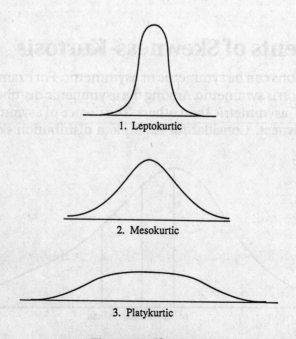

**Figure 79. Skewness**

# The Normal Distribution

The **normal distribution** is one of the most important examples of a continuous probability distribution. The equation

$$y = \frac{1}{\sigma\sqrt{2\pi}} e^{-\frac{(x-\mu)^2}{2\sigma^2}}$$

is called a normal curve or Gaussian distribution. In the equation $\mu$ denotes mean and $\sigma$ is the standard deviation. The total area bounded by the equation and the x-axis is one. Thus the area bounded by the curve, the x-axis, and $x = a$ and $x = b$, where $a < b$, represents the probability that $x \in [a,b]$, denoted by

$$p(a < x < b)$$

A new variable $z$ can be introduced by

$$z = \frac{x - \mu}{\sigma}$$

Then the equation becomes

$$y = \frac{1}{\sigma\sqrt{2\pi}} e^{-\frac{1}{2}z^2}$$

where $\sigma = 1$. The equation is called the standard form of a normal distribution. Hence, $z$ is normally distributed with a mean of zero and a variance of one.

# The Poisson Distribution

The **Poisson distribution** is the discrete probability distribution defined by

$$p(x) = \frac{\lambda^x e^{-\lambda}}{x!}$$

where $x = 0, 1, 2, \ldots$
$e$ and $\lambda$ are constants.

Here are some properties of the Poisson distribution.

| | |
|---|---|
| Mean | $\mu = \lambda$ |
| Standard deviation | $\sigma = \sqrt{\lambda}$ |
| Variance | $\sigma^2 = \lambda$ |

# The Graduate Record Examination in
# ENGINEERING

## Test 1

# GRE ENGINEERING TEST 1
## ANSWER SHEET

1. Ⓐ Ⓑ Ⓒ Ⓓ Ⓔ
2. Ⓐ Ⓑ Ⓒ Ⓓ Ⓔ
3. Ⓐ Ⓑ Ⓒ Ⓓ Ⓔ
4. Ⓐ Ⓑ Ⓒ Ⓓ Ⓔ
5. Ⓐ Ⓑ Ⓒ Ⓓ Ⓔ
6. Ⓐ Ⓑ Ⓒ Ⓓ Ⓔ
7. Ⓐ Ⓑ Ⓒ Ⓓ Ⓔ
8. Ⓐ Ⓑ Ⓒ Ⓓ Ⓔ
9. Ⓐ Ⓑ Ⓒ Ⓓ Ⓔ
10. Ⓐ Ⓑ Ⓒ Ⓓ Ⓔ
11. Ⓐ Ⓑ Ⓒ Ⓓ Ⓔ
12. Ⓐ Ⓑ Ⓒ Ⓓ Ⓔ
13. Ⓐ Ⓑ Ⓒ Ⓓ Ⓔ
14. Ⓐ Ⓑ Ⓒ Ⓓ Ⓔ
15. Ⓐ Ⓑ Ⓒ Ⓓ Ⓔ
16. Ⓐ Ⓑ Ⓒ Ⓓ Ⓔ
17. Ⓐ Ⓑ Ⓒ Ⓓ Ⓔ
18. Ⓐ Ⓑ Ⓒ Ⓓ Ⓔ
19. Ⓐ Ⓑ Ⓒ Ⓓ Ⓔ
20. Ⓐ Ⓑ Ⓒ Ⓓ Ⓔ

21. Ⓐ Ⓑ Ⓒ Ⓓ Ⓔ
22. Ⓐ Ⓑ Ⓒ Ⓓ Ⓔ
23. Ⓐ Ⓑ Ⓒ Ⓓ Ⓔ
24. Ⓐ Ⓑ Ⓒ Ⓓ Ⓔ
25. Ⓐ Ⓑ Ⓒ Ⓓ Ⓔ
26. Ⓐ Ⓑ Ⓒ Ⓓ Ⓔ
27. Ⓐ Ⓑ Ⓒ Ⓓ Ⓔ
28. Ⓐ Ⓑ Ⓒ Ⓓ Ⓔ
29. Ⓐ Ⓑ Ⓒ Ⓓ Ⓔ
30. Ⓐ Ⓑ Ⓒ Ⓓ Ⓔ
31. Ⓐ Ⓑ Ⓒ Ⓓ Ⓔ
32. Ⓐ Ⓑ Ⓒ Ⓓ Ⓔ
33. Ⓐ Ⓑ Ⓒ Ⓓ Ⓔ
34. Ⓐ Ⓑ Ⓒ Ⓓ Ⓔ
35. Ⓐ Ⓑ Ⓒ Ⓓ Ⓔ
36. Ⓐ Ⓑ Ⓒ Ⓓ Ⓔ
37. Ⓐ Ⓑ Ⓒ Ⓓ Ⓔ
38. Ⓐ Ⓑ Ⓒ Ⓓ Ⓔ
39. Ⓐ Ⓑ Ⓒ Ⓓ Ⓔ
40. Ⓐ Ⓑ Ⓒ Ⓓ Ⓔ

41. Ⓐ Ⓑ Ⓒ Ⓓ Ⓔ
42. Ⓐ Ⓑ Ⓒ Ⓓ Ⓔ
43. Ⓐ Ⓑ Ⓒ Ⓓ Ⓔ
44. Ⓐ Ⓑ Ⓒ Ⓓ Ⓔ
45. Ⓐ Ⓑ Ⓒ Ⓓ Ⓔ
46. Ⓐ Ⓑ Ⓒ Ⓓ Ⓔ
47. Ⓐ Ⓑ Ⓒ Ⓓ Ⓔ
48. Ⓐ Ⓑ Ⓒ Ⓓ Ⓔ
49. Ⓐ Ⓑ Ⓒ Ⓓ Ⓔ
50. Ⓐ Ⓑ Ⓒ Ⓓ Ⓔ
51. Ⓐ Ⓑ Ⓒ Ⓓ Ⓔ
52. Ⓐ Ⓑ Ⓒ Ⓓ Ⓔ
53. Ⓐ Ⓑ Ⓒ Ⓓ Ⓔ
54. Ⓐ Ⓑ Ⓒ Ⓓ Ⓔ
55. Ⓐ Ⓑ Ⓒ Ⓓ Ⓔ
56. Ⓐ Ⓑ Ⓒ Ⓓ Ⓔ
57. Ⓐ Ⓑ Ⓒ Ⓓ Ⓔ
58. Ⓐ Ⓑ Ⓒ Ⓓ Ⓔ
59. Ⓐ Ⓑ Ⓒ Ⓓ Ⓔ
60. Ⓐ Ⓑ Ⓒ Ⓓ Ⓔ

61. (A) (B) (C) (D) (E)
62. (A) (B) (C) (D) (E)
63. (A) (B) (C) (D) (E)
64. (A) (B) (C) (D) (E)
65. (A) (B) (C) (D) (E)
66. (A) (B) (C) (D) (E)
67. (A) (B) (C) (D) (E)
68. (A) (B) (C) (D) (E)
69. (A) (B) (C) (D) (E)
70. (A) (B) (C) (D) (E)
71. (A) (B) (C) (D) (E)
72. (A) (B) (C) (D) (E)
73. (A) (B) (C) (D) (E)
74. (A) (B) (C) (D) (E)
75. (A) (B) (C) (D) (E)
76. (A) (B) (C) (D) (E)
77. (A) (B) (C) (D) (E)
78. (A) (B) (C) (D) (E)
79. (A) (B) (C) (D) (E)
80. (A) (B) (C) (D) (E)
81. (A) (B) (C) (D) (E)
82. (A) (B) (C) (D) (E)

83. (A) (B) (C) (D) (E)
84. (A) (B) (C) (D) (E)
85. (A) (B) (C) (D) (E)
86. (A) (B) (C) (D) (E)
87. (A) (B) (C) (D) (E)
88. (A) (B) (C) (D) (E)
89. (A) (B) (C) (D) (E)
90. (A) (B) (C) (D) (E)
91. (A) (B) (C) (D) (E)
92. (A) (B) (C) (D) (E)
93. (A) (B) (C) (D) (E)
94. (A) (B) (C) (D) (E)
95. (A) (B) (C) (D) (E)
96. (A) (B) (C) (D) (E)
97. (A) (B) (C) (D) (E)
98. (A) (B) (C) (D) (E)
99. (A) (B) (C) (D) (E)
100. (A) (B) (C) (D) (E)
101. (A) (B) (C) (D) (E)
102. (A) (B) (C) (D) (E)
103. (A) (B) (C) (D) (E)
104. (A) (B) (C) (D) (E)

105. (A) (B) (C) (D) (E)
106. (A) (B) (C) (D) (E)
107. (A) (B) (C) (D) (E)
108. (A) (B) (C) (D) (E)
109. (A) (B) (C) (D) (E)
110. (A) (B) (C) (D) (E)
111. (A) (B) (C) (D) (E)
112. (A) (B) (C) (D) (E)
113. (A) (B) (C) (D) (E)
114. (A) (B) (C) (D) (E)
115. (A) (B) (C) (D) (E)
116. (A) (B) (C) (D) (E)
117. (A) (B) (C) (D) (E)
118. (A) (B) (C) (D) (E)
119. (A) (B) (C) (D) (E)
120. (A) (B) (C) (D) (E)
121. (A) (B) (C) (D) (E)
122. (A) (B) (C) (D) (E)
123. (A) (B) (C) (D) (E)
124. (A) (B) (C) (D) (E)

125. (A) (B) (C) (D) (E)
126. (A) (B) (C) (D) (E)
127. (A) (B) (C) (D) (E)
128. (A) (B) (C) (D) (E)
129. (A) (B) (C) (D) (E)
130. (A) (B) (C) (D) (E)
131. (A) (B) (C) (D) (E)
132. (A) (B) (C) (D) (E)
133. (A) (B) (C) (D) (E)
134. (A) (B) (C) (D) (E)
135. (A) (B) (C) (D) (E)
136. (A) (B) (C) (D) (E)
137. (A) (B) (C) (D) (E)
138. (A) (B) (C) (D) (E)
139. (A) (B) (C) (D) (E)
140. (A) (B) (C) (D) (E)
141. (A) (B) (C) (D) (E)
142. (A) (B) (C) (D) (E)
143. (A) (B) (C) (D) (E)
144. (A) (B) (C) (D) (E)

# THE GRADUATE RECORD EXAMINATION ENGINEERING TEST

## MODEL TEST I

Time: 170 Minutes
      144 Questions

Directions: Choose the best answer for each question and mark the letter of your selection on the corresponding answer sheet.

1. Which of the following forms an isothermal process?

   (A) The pressure must decrease slowly and the volume must increase slowly while no heat is allowed to enter the system.

   (B) The volume is kept constant while heat is allowed to enter the system.

   (C) The pressure must decrease slowly and the volume must increase slowly while heat is allowed to enter the system.

   (D) Heat is allowed to enter the system while the pressure is increased quickly.

   (E) The pressure must increase slowly and the volume must decrease slowly while heat is allowed to enter the system.

2. A change in voltage produces a change in the instantaneous resistance of a coil. This is due to

   (A) the nonlinear resistivity of the coil

(B) the internal inductivity of the coil

(C) the linear, time variant resistivity of the coil

(D) the temperature coefficient of the resistance

(E) the capacitive behavior of the coil

3. In order to produce the greatest increase in the flow rate of a viscous liquid such as glycerine through a cylindrical tube

(A) the length of the tube must be increased

(B) the radius of the tube must be decreased

(C) the temperature must be increased

(D) the radius of the tube must be increased and the temperature decreased

(E) the length of the tube must be decreased and the temperature increased

4. In Figure 1, point E is the center of square ABCD and EF $\perp$ CD. If the section BEFCB is removed, the centroid moves to point

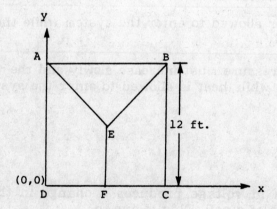

Figure 1

(A) (3,5)  (D) (3,7.2)

(B) (5,8.5)  (E) (3,8)

(C) (4,6.8)

5. A man can run at a constant speed from point O to point Q along the straight line segment OQ in 20 seconds. At the same speed, he can run from O to P along OP in 10 seconds. If the outer edges of Tracks 1 and 2 are circles with center O, and the geometry is as shown below, what is the area of Track 2?

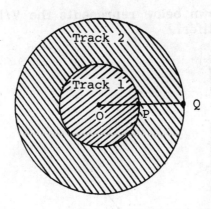

(A) $4\pi(OP)^2$  (D) $\pi(OP)^2$

(B) $3\pi(OP)^2$  (E) $(OQ)^2$

(C) $\pi(PQ)^2$

6. For a glass that is slowly being cooled from its molten state to the glass transition temperature

(A) crystallization will occur upon reaching the glass transition temperature

(B) more thermal stresses will arise as a result of the slow cooling rate

(C) the specific volume of the glass will decrease sharply when the glass transition temperature is reached

(D) the viscosity increases until it becomes rigid and brittle upon reaching the glass transition temperature

(E) the viscosity will keep increasing and undergo an abrupt increase in specific volume at the glass transition temperature

7. Which of the following chemical equations for the burning of propane gas, $C_3H_8$, in oxygen is balanced?

(A) $C_3H_8 + 4O_2 \rightarrow 3C + 4H_2O$

(B) $C_3H_8 + 3O_2 \rightarrow 3CHO + 3O$

(C) $C_3H_8 + 5O \rightarrow CO + 4H_2O$

(D) $C_3H_8 + 5O_2 \rightarrow 3CO_2 + 4H_2O$

(E) $C_3H_8 + 3O_2 \rightarrow 3CO_2 + 3H_2O$

8. The graph shown below represents the V/I characteristics of a thermionic rectifer.

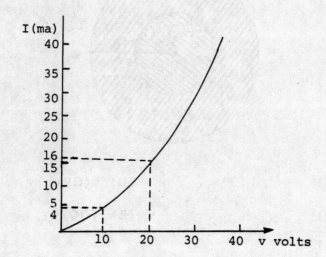

Which of the following equations does the curve represent?

(A) $I = 3V^2$

(B) $\dfrac{I}{100} = \dfrac{3V}{V + 100}$

(C) $10I = \left(\dfrac{V}{5}\right)^3$

(D) $I^2 = 2V^2$

(E) $I = \left(\dfrac{V}{5}\right)^2$

9. A boy traveling in a train moving at a constant velocity on straight track throws a ball vertically upward. The ball comes down after 3 seconds. The ball will land

(A) in his hands

(B) just ahead of him

(C) just behind him

(D) on top of the train

(E) at an unpredictable position

10. In the diagram given below, the value of the unknown current i at time t = 0.5 sec. is

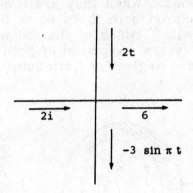

(A) 0
(B) -1
(C) 1
(D) 0.5
(E) 2

11. In the given circuit, the value of current (I) in the 100Ω resistor is

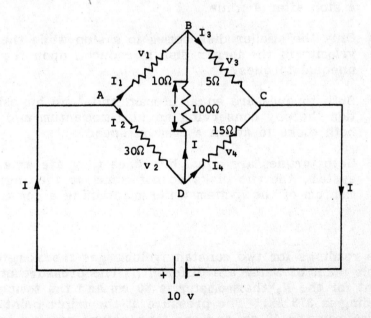

(A) 1.5 A  (D) 0 A

(B) 2 A  (E) 1 A

(C) 0.5 A

12. In the figure below, two disks of different radii are rotating at different speeds. When they are brought together, an impulsive torque acts on both disks so as to change their individual angular moments. Which of the following is true about the state of the system comprised of both disks after they are engaged? Note: Neglect any frictional effects.

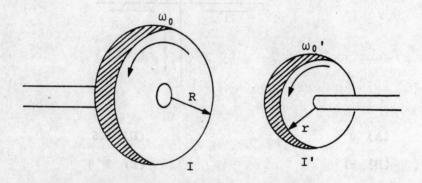

(A) The larger disk exerts a greater torque on the smaller disk causing it to spin faster.

(B) The impulsive torque on both disks is the same in magnitude but opposite in direction and causes them to come to a stop after a while.

(C) Only the smaller disk comes to a stop while the angular velocity of the larger disk is reduced upon the action of unequal torques.

(D) Both torques are equal in magnitude and opposite in direction thereby conserving angular momentum and allowing both disks to attain a common speed.

(E) Both torques are equal, but since they are external to the system, the two disks do not conserve the angular momentum of the system while maintaining a common speed.

13. The readings for two constant volume gas thermometers at the triple point of water are different. The pressure at the triple point for the $N_2$ thermometer is 80 cm and the temperature reading is 373.32. The pressure at the triple point for the $O_2$ thermometer is 40 cm and the temperature reading is 373.43. In order to reduce the variation in temperature readings

(A) the pressure in the $O_2$ thermometer is increased to the pressure of the $N_2$ thermometer

(B) the pressure in the $O_2$ thermometer is reduced below 40 cm

(C) the pressure in the $N_2$ thermometer is decreased to the pressure of the $O_2$ thermometer

(D) both the $O_2$ thermometer and the $N_2$ thermometer are reduced below 40 cm

(E) both (B) and (D)

14. Molecules of an ideal gas exert a force on the container walls due to

(A) surface tension of the gas

(B) gravitational repulsion at the wall

(C) cohesion between molecules of the gas and its container wall

(D) coulomb attraction at the wall

(E) collisions between molecules of the gas and its container wall

15. In a common household, hot water heated by a basement furnace forces cold water from the radiator to the hot side of the furnace. The transfer of heat from the hot side to the cold side is through

(A) conduction

(B) radiation

(C) free convection

(D) forced convection

(E) both (A) and (D)

16. In order to maintain an equilibrium system of ice and water with decreasing temperature

(A) the pressure must be increased and the volume held constant

(B) the pressure must be decreased and the volume held constant

(C) the pressure must be increased but the volume can be changed

(D) the pressure must be held constant but the volume can be changed

(E) both (A) and (C)

17. A sound source is located in the bottom of a lake. What would the speed of the resulting propagating shock waves be (in the lake)?

   (A) Mach number (M) = 0
   (B) M < 1
   (C) M > 1
   (D) M = 1
   (E) None of the above

18. If 15 grams of nitrogen monoxide undergoes combustion, what mass of nitrogen dioxide will appear (atomic weights: N = 14, O = 16)?

   (A) 14g
   (B) 23g
   (C) 32g
   (D) 40g
   (E) 46g

19. The nuclei of isotopes of an element have

   (A) same atomic number and different masses
   (B) same masses and different atomic numbers
   (C) same binding energy
   (D) same half lives
   (E) None of the above characteristics

20. In which of the given processes is the thermal energy transferred to a gas completely converted to internal energy, resulting in an increase in the gas temperature?

   (A) Isochoric process
   (B) Adiabatic process
   (C) Isothermal process
   (D) Free expansion

(E) Throttling process

21. Which of the following is the dimensional formula for the quantity of water flowing in unit time?

(A) MT

(B) $L^3T$

(C) $M^3T^{-1}$

(D) $L^3T^{-1}$

(E) $L^2T^{-1}$

22. Which of the following equations is a better approximation for the behavior of real gases than the ideal gas law (PV = nRT)?

(A) $(P + a)(V + b) = nRT$

(B) $\left[\dfrac{P}{a}\right](V + b) = nRT$

(C) $\left[P + \dfrac{a}{V^2}\right](V - b) = nRT$

(D) $\left[P + \dfrac{V}{a}\right](V - b) = nRT$

(E) None of the above

23. What is the magnitude, T, of the torque which must be exerted on the drum for starting motion? (Coefficient of static friction, $\mu_{S1} = 0.1$)

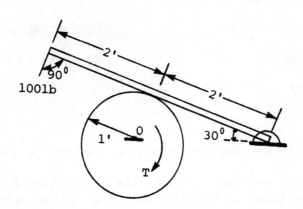

(A) 20 ft. lb

(B) 30 ft. lb

(C) 40 ft. lb

(D) 50 ft. lb

(E) 60 ft. lb

24. A refrigerator operating at full capacity is kept in the center of an adiabatically sealed room. The door of the refrigerator is kept open and the room doors are closed. After 24 hours it is found that

   (A) the temperature of the room was lower than the original temperature

   (B) the temperature of the room remained the same

   (C) the temperature in the room went up

   (D) the temperature in the room went up or down depending on the initial temperature

   (E) None of the above

25. An ac RLC series circuit is said to be in a state of resonance when

   (A) $X_L = R$         (D) $X_L > X_c$

   (B) $|X_L| = |X_c|$   (E) $X_L < X_c$

   (C) $X_c = R$

26. For a gas that is allowed to expand reversibly and adiabatically there is no change in

   (A) internal energy    (D) enthalpy

   (B) temperature        (E) both (A) and (B)

   (C) entropy

27. Cloudy nights are warmer than clear nights because

   (A) the water molecules forming the clouds have higher kinetic energy which is transferred to the surrounding environment

   (B) the interaction between negatively charged cloud ions and positively charged ground ions warms the air.

   (C) the clouds shield the ground, therefore radiation loss is reduced

(D) chemical reactions between pollutants and water molecules heat up the clouds which in turn warm the surroundings

(E) None of the above

28. An inclined plane is used sometimes to assist in lifting heavy objects. This is done since the

(A) force is reduced
(B) work is reduced
(C) potential energy is increased
(D) weight is increased
(E) distance is reduced

29. The graph below represents the speed-time variation for a ball

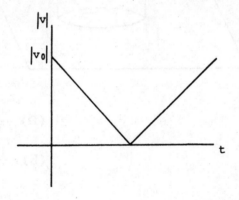

(A) that is dropped from an initial height and bounces elastically off the ground back to that initial height

(B) that is thrown straight up in the air from an initial height and falls back to that height

(C) that is thrown from an initial height in a parabolic path until it returns to that initial height

(D) that is thrown from an initial height horizontally and bounces elastically off the ground back to that initial height

(E) None of the above

30. The earth receives about 2655 $\frac{W}{m^2}$ of radiant energy from the

sun. If the energy is in the form of a plane polarized monochromatic wave, the magnitude of the electric field for normal incidence is

Note: $\begin{pmatrix} \varepsilon_0 = 8.85 \times 10^{-8} C^2 N^{-1} m^{-2} \\ \mu_0 = 4\pi \times 10^{-7} WbA^{-1} m^{-1} \end{pmatrix}$

(A) $\dfrac{100}{\sqrt{3}} \dfrac{V}{m}$

(B) $10\sqrt{3} \dfrac{V}{m}$

(C) $100 \dfrac{V}{m}$

(D) $10 \dfrac{V}{m}$

(E) $\dfrac{10}{\sqrt{3}} \dfrac{V}{m}$

31. A rotating cylinder of mass M and radius R is brought to rest on a flat surface with a coefficient of kinetic friction, $\mu$. Evaluate the magnitude of its angular deceleration if it has a pure rotation about its central axis.

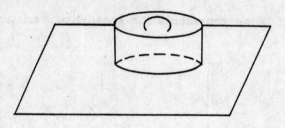

(A) $\dfrac{2g\mu}{R^2}$

(B) $\dfrac{2g\mu m}{R}$

(C) $\dfrac{2g\mu m}{R^2}$

(D) $\dfrac{4\mu g}{3R}$

(E) $\dfrac{4\mu gM}{3R}$

32. If an oscillating system which experiences a small damping force is driven by a periodic force at its natural undamped frequency, then

(A) the amplitude will tend to infinity

(B) resonance will occur but the system will have damped oscillations

(C) resonance will occur and the resonant amplitude will be $\dfrac{F_m}{b\omega}$ in which $\omega$ is the frequency of the natural undamped oscillation

(D) we will have an undamped oscillation with an amplitude not equal to its resonant value

(E) we will have a damped oscillation with a frequency equal to the natural undamped frequency and an amplitude equal to $\frac{F_m}{b\omega}$ in which $F_m$ is the amplitude of the driving force

33. In the circuit given below, the energy absorbed in one hour by the 1 mA source is

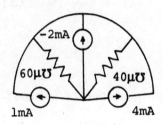

(A) −108 J

(B) 108 J

(C) −100 J

(D) 100 J

(E) 98 J

34. The greater nuclear stability of an alpha particle $^4_2He_2$ as opposed to an isotope of lithium $^5_3Li_2$ is accounted for by

(A) the greater mass of the lithium isotope

(B) the unequal pairing of neutrons and electrons in lithium

(C) the increase in the nuclear density of the lithium isotope

(D) the repulsive energy from the greater number of electrons

(E) the greater rest energy of the lithium isotope as opposed to the rest energy of its constituents

35. A concrete block with a volume of $0.1 m^3$ is submerged in a liquid with a constant density of $1 \times 10^3$ kg/m$^3$. The block hangs from a cord that is attached to a beam immersed in the liquid whose area is $0.2 m^2$. Find the mass density of the block if the system is in mechanical equilibrium.
(Note: gravity (g) = 9.8 m/sec² )

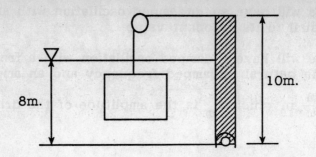

(A)  2000 Kg/m³

(D)  3133.3̄ Kg/m³

(B)  5000 Kg/m³

(E)  4600 Kg/m³

(C)  2550 Kg/m³

36. Which of the following elements is most likely to form a compound with H (Hydrogen)?

(A) Cl

(D) He

(B) Li

(E) Na

(C) Cu

37. An open cylinder contains liquid. A pressure gauge is suspended within the liquid by attaching the gauge to the bottom through a string. When stationary, the gauge reads 20 lb/in². What is the reading when the container is allowed to fall freely? Neglect air friction.

(A)  0 lb/in²

(D)  20 lb/in²

(B)  10 lb/in²

(E)  25 lb/in²

(C)  15 lb/in²

38. A block of mass m kg is attached to a string which can withstand a force of 50m Newtons. The other end of the string is attached to the roof of an elevator. What must the minimum acceleration of the elevator be for the string to break? (Assume the acceleration of gravity is 9.8 m/s² )

(A)  49 m/s² upwards

(C)  40.2 m/s² upwards

(B)  49 m/s² downwards

(D)  40.2 m/s² downwards

(E) 39.2 m/s² upwards

39. A man touches two metals, metal A and metal B. Metal A and metal B have both been kept together for a long time at room temperature. Why would one feel colder than the other?

(A) One's temperature is less than the other.

(B) One has a higher heat transfer coefficient.

(C) One is larger in volume than the other.

(D) One has a higher thermal conductivity.

(E) None of the above

40. Which of the curves given below shows the value of the terminal voltage for different values of load current (i.e. the external characteristics), for a separately excited d.c. generator?

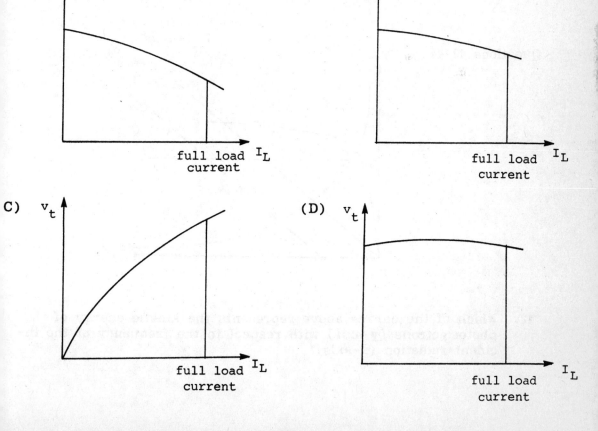

(E)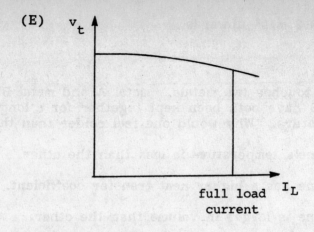

41. For a double slit system the slit width is not negligible in comparison with the wavelength of light that is incident upon it. What happens if one of the slits is covered up?

    (A) The diffraction pattern broadens.

    (B) The number of interference fringes decreases.

    (C) The interference fringes disappear.

    (D) The diffraction pattern becomes narrower.

    (E) Both (B) and (D)

Questions 42-44

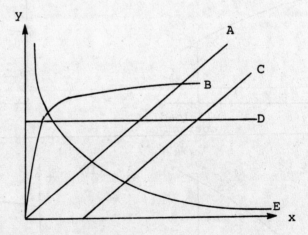

42. Which of the curves above represents the kinetic energy of photoelectrons (y-axis) with respect to the frequency of the incident radiation (x-axis)?

43. Which of the curves above represents the fraction of initial amount of mass remaining (y-axis) as a function of its fraction of time in units of its half time (x-axis)?

44. Which of the curves above represents the capacitive reactance (y-axis) as a function of frequency (x-axis)?

45. A man deposits $800 in a bank for 5 years at a nominal annual interest rate of 10%. If the man withdraws $80 at the end of each year for five consecutive years, the amount left in the bank after five years is

   (A) $800

   (B) $400(1 + .10)$^5$

   (C) $720(1 + .10)$^5$

   (D) $\dfrac{\$720}{(1 + .10)^5}$

   (E) 800(1 + .10)$^5$

Directions: Choose the best answer for each question and mark the letter of your selection on the corresponding answer sheet.

Note: The natural logarithm of x will be denoted by ℓnx.

46. $\displaystyle\int_{-1}^{+1} (x^2 - 4)\,dx =$

   (A) $\dfrac{20}{3}$

   (B) 24

   (C) $\dfrac{-22}{3}$

   (D) $\dfrac{-16}{3}$

   (E) -23

47. What is the probability of getting a 5 on each of two successive rolls of a balanced die?

   (A) $\dfrac{1}{6}$

   (B) $\dfrac{5}{7}$

   (C) $\dfrac{1}{36}$

   (D) $\dfrac{5}{6}$

(E) $\frac{3}{44}$

48. The probability of getting at least two heads when a fair coin is tossed four times is

(A) $\frac{1}{4}$

(B) $\frac{11}{16}$

(C) $\frac{1}{8}$

(D) $\frac{6}{16}$

(E) $\frac{13}{16}$

49. $\int_0^\pi (1 + \sin x)\,dx =$

(A) $\pi - 3$

(B) $\pi + 2$

(C) $\pi + 1$

(D) $\pi$

(E) $\pi - 2$

50. Which of the following matrices has an inverse?

(A) $\begin{pmatrix} 3 & 1 \\ 6 & 2 \end{pmatrix}$

(B) $\begin{pmatrix} 4 & 2 \\ 2 & 1 \end{pmatrix}$

(C) $\begin{pmatrix} 6 & 2 \\ 9 & 3 \end{pmatrix}$

(D) $\begin{pmatrix} 8 & 2 \\ 4 & 1 \end{pmatrix}$

(E) $\begin{pmatrix} 5 & 2 \\ 2 & 1 \end{pmatrix}$

51. $\dfrac{e^{i(\cos^{-1}x)} + e^{-i(\cos^{-1}x)}}{2} =$

(A) $\cos^{-1} x$

(B) $i \cos x$

(C) $x$

(D) $\sin^{-1} x$

(E) None of the above

52. A sample of 9 patients in the intensive care unit of Mt. Sinai Hospital were questioned. The number of heart attacks that they had suffered numbered 2, 3, 4, 5, 5, 6, 7, 7, 10. The statistical mode is

   (A) 7
   (B) 5
   (C) 7 and 5
   (D) 10 and 2
   (E) 6

53. $\dfrac{d^2}{dx^2}(\ln x^2) =$

   (A) $2x$
   (B) $\dfrac{1}{x^2}$
   (C) $\dfrac{-2}{x^2}$
   (D) $\dfrac{2}{x}$
   (E) $\dfrac{-2}{x^3}$

Questions 54-58 are based on the assumption that two functions F(x) and G(x) are differentiable to any order over the set of all real numbers.

54. $\displaystyle\int_1^2 F(x)dx + \int_1^2 G(x)dx =$

   (A) $2\displaystyle\int_1^2 [F(x) + G(x)]dx$
   (B) $3\displaystyle\int_1^2 [F(x) + G(x)]dx$
   (C) $\displaystyle\int_1^2 [F(x) + G(x)]dx$
   (D) $\displaystyle\int_1^2 F(x)G(x)dx$
   (E) None of the above

55. $\displaystyle\int_{S=0}^{S=F(x)-b} G'(s)ds =$

(A) $G[F(x)] - G(b) - G(0)$  (D) $G[F(x) - b]$

(B) $G[F(x)] - G(0) - b$  (E) $\frac{1}{b}[G[F(x)] - G(0)]$

(C) $G[F(x) - b] - G(0)$

56. $\int_0^1 F(x)dx + \int_1^2 F(x)dx - \int_0^3 F(x)dx =$

(A) $\int_1^3 F(x)dx$  (D) $\int_2^3 F(x)dx$

(B) $-\int_1^3 F(x)dx$  (E) $-\int_0^3 F(x)dx$

(C) $-\int_2^3 F(x)dx$

57. $\frac{d}{dx}[G(x^2)F(x^3)] =$

(A) $2xG'(x^2)F(x^3) + F'(x^3)G(x^2)$

(B) $2xG'(x)F(x^3) + x^2F'(x^3)G(x^2)$

(C) $2xG'(x^2)F(x^3) + 3x^2F'(x^3)G(x^2)$

(D) $3x^2F'(x^3)G'(x^2)$

(E) $2xF'(x^3)G'(x^2)$

58. If $F(x)$ is decreasing at the points $x_1, x_2, x_3, \ldots, x_n$, which are not inflection points, then the slope of the function, (i.e. $F'(x)$) will be

(A) constant at these points

(B) negative at these points

(C) positive at these points

(D) increasing at these points

(E) undetermined at these points

59. Given $y = \ln \sinh x$, $D_x y =$

   (A) csch x
   (B) $\dfrac{\tanh x}{\sinh x}$
   (C) coth x
   (D) $\dfrac{1}{x \cosh x}$
   (E) ln cosh x

60. $\displaystyle\int_0^{\pi/2} e^{\ln(\cos^2 x + \sin^2 x)}\, dx =$

   (A) 0
   (B) $\dfrac{1}{2}$
   (C) $\dfrac{-1}{2}$
   (D) $\dfrac{\pi}{2}$
   (E) $\dfrac{-\pi}{2}$

61. If $i = \sqrt{-1}$ then

$$1 + (ix) + \frac{(ix)^2}{2!} + \frac{(ix)^3}{3!} + \ldots =$$

   (A) $\displaystyle\sum_{n=1}^{\infty} \frac{(ix)^n}{n!}$
   (B) $\dfrac{(ix)^{n-1}}{(n-1)!}$
   (C) $\cos x + i \sin x$
   (D) $\tan(ix)$
   (E) $\sin x + i \cos x$

62. Find: $\displaystyle\lim_{x \to 0} \frac{x}{1 - e^x}$

   (A) $\infty$
   (B) 1
   (C) 0
   (D) $-1$
   (E) $\dfrac{1}{2}$

63. If $A = \begin{bmatrix} 1 & 2 & 0 \\ 3 & -1 & 4 \end{bmatrix}$, then $AA^t =$

(A) $\begin{bmatrix} 1 & 0 & 1 \\ -1 & 2 & 3 \end{bmatrix}$  (D) $\begin{bmatrix} 5 & 1 \\ 1 & 26 \end{bmatrix}$

(B) $\begin{bmatrix} 1 & 3 \\ -1 & 4 \end{bmatrix}$  (E) an undefined product

(C) 46

64. For $|x| < 1$, $\dfrac{1}{1-x} =$

(A) $\sum_{n=0}^{\infty} x^n$  (D) $e^{\infty}$

(B) $1 - x + x^2 - x^3 + \ldots$  (E) $x!$

(C) $(1 + x)$

Directions: Choose the best answer for each question and mark the letter of your selection on the corresponding answer sheet. Assume every curve in this section has derivatives of all orders at each point of its domain unless otherwise indicated.

This is the graph of the function, f, to be used in questions 65-70.

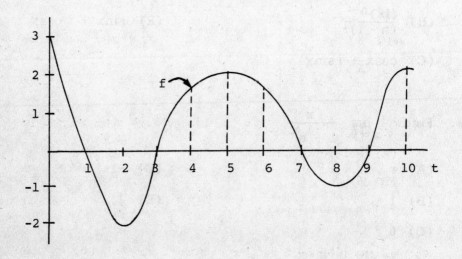

65. If the function g(t) is defined as $g(t) = \int_0^t f(a)\,da$, then g(3) is

(A) greater than 6

(B) between 0 and -2

(C) between 1 and 4

(D) 0

(E) less than -4

66. The function g(t) is defined as in the previous problem. The value of g(f(7)) is

(A) 7

(B) between 4 and 7

(C) 0

(D) g(7)

(E) between -4 and -7

67. For the previously defined function, g(t), a local maximum occurs at the value t =

(A) 0

(B) 3

(C) 5

(D) 7

(E) g(t) has no maxima

68. The function g(t) has a local minimum at the value t =

(A) 0

(B) 2

(C) 5

(D) 9

(E) g(t) has no minima

69. On the interval $0 \leq t \leq 10$, the average value of f(t) is

(A) between 0 and 1

(B) 0

(C) between 0 and -1

(D) g(-1)

(E) between -1 and -2

70. In the interval $5 \leq t \leq 8$, which of the following is true?

I. $\frac{df}{dt} < 0$ for all t

II. $\frac{dg}{dt} < 0$ for all t

III. $f(t) \cong -t^2$

(A) I only  (D) II and III only
(B) I and II only  (E) I, II and III
(C) II only

Two particles, $P_1$ and $P_2$, are released at a minimum distance from each other and travel in circles with equal radii at equal speeds. Both particles travel clockwise.

Questions 71-80 are based on this information and on the following graph, which plots the distance between the particles as a function of time.

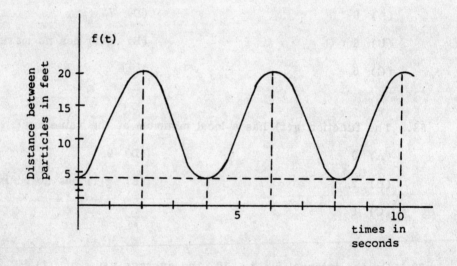

71. On the time interval from $0 \leq t \leq 10$, how many times do the particles simultaneously intersect the line joining the centers of their circular paths?

(A) zero  (D) four
(B) two  (E) six
(C) three

72. The area bounded by the path of $P_1$ is

    (A) $4\pi$

    (B) $8\pi$

    (C) $16\pi$

    (D) $20\pi$

    (E) $32\pi$

73. Let $\theta$ be the angle between the line joining $P_1$ and $P_2$ and the horizontal line joining the two centers of the circular paths of the particles. What is $\tan\theta$ at time $t = 5$?

    (A) $\frac{2}{3}$

    (B) $\frac{3}{4}$

    (C) $\frac{8}{7}$

    (D) $\frac{15}{17}$

    (E) $\frac{2}{5}$

74. The angular velocity of particle $P_2$ is

    (A) 25 rpm

    (B) 15 rpm

    (C) 20 rpm

    (D) 36 rpm

    (E) 40 rpm

75. On the interval from $0 \leq t \leq 10$, at how many points does the tangent to the curve cross the curve?

    (A) zero

    (B) four

    (C) six

    (D) two

    (E) three

76. For which of the following values of t is $f''(t)$ greater than zero?

    (A) $t = 2$

    (B) $t = 3$

    (C) $t = 8$

    (D) $t = 5$

    (E) $t = 10$

If a function g(x) is defined as follows:

$$g(x) = \int_0^x f(t)\,dt$$

where f(t) is the function whose graph was given previously.

77. g(0) is equal to

    (A) 4
    (B) 0
    (C) $\frac{29}{2}$
    (D) 20
    (E) $\frac{14}{3}$

78. The previously defined function g is an increasing function

    (A) on some subintervals only
    (B) on no subintervals
    (C) when f is a maximum only
    (D) when f is a minimum only
    (E) on all subintervals

79. If the curve f is rotated once around the t-axis, then the volume enclosed over one period of the function is approximately

    (A) 830π
    (B) 944π
    (C) 426π
    (D) 360π
    (E) 1268π

80. If the function g is defined as previously, then $\frac{dg}{dt} = 0$ at

    (A) t = 0,2,4,6,8,10
    (B) t = 0
    (C) t = 1,3,5,7,9
    (D) all values of t
    (E) no values of t

For Questions 81-85.

Acceleration $\left(\dfrac{ft}{sec^2}\right)$

Particle A travels on a path with the above acceleration with respect to time.

Particle A is initially at rest and 12 feet above the origin of some coordinate system.

81. At time t = 3 seconds, the velocity of A is approximately

   (A) 30
   (B) 50
   (C) 60
   (D) 80
   (E) 100

82. At time t = 10 seconds the position of A with respect to the origin is

   (A) 342 ft
   (B) 450 ft
   (C) 762 ft
   (D) 886 ft
   (E) 1054 ft

83. At what time does A reach its maximum velocity?

   (A) 5
   (B) 10

(C) 15

(D) 20

(E) 25

84. Over the interval $0 \leq t \leq 30$ the displacement of A in feet is

(A) 4,050

(B) 3,600

(C) 5,350

(D) 4,880

(E) 6,450

85. What is the total distance in feet that A travels in the interval $0 \leq t \leq 30$?

(A) 8,100

(B) 3,600

(C) 5,350

(D) 6,450

(E) 4,050

Directions: Choose the best answer for each question and mark the letter of your selection on the corresponding answer sheet.

86. Two wires are attached to a corner fence post with the wires making an angle of 90° with each other. If each wire pulls on the post with a force of 5 pounds, what is the approximate resultant force acting on the post?

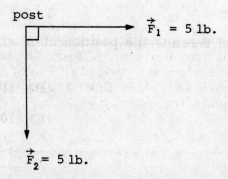

(A) 5 pounds

(B) 6 pounds

(C) 7 pounds

(D) 8 pounds

(E) 9 pounds

87. An engineer uses an ohmmeter to measure resistance of an electromagnet. A spark occurs when one of the leads of the ohmmeter is disconnected. This is due to

(A) Faraday's law

(B) Ohm's law

(C) Kirchoff's law

(D) the Thompson effect

(E) the Seebeck effect

88. One (1) gram of aspirin, empirical formula $C_9H_8O_4$, is dissolved in 200 mL of solution. How many molecules of aspirin are in 1.0L of solution? (Assume Avogadro's number, $N_A = 6 \times 10^{23}$, atomic weights: C = 12, H = 1, O = 16)

(A) $3.0 \times 10^{16}$ molecules/L

(B) $1.5 \times 10^{25}$ molecules/L

(C) $1.7 \times 10^{19}$ molecules/L

(D) $20 \times 10^{23}$ molecules/L

(E) $1.7 \times 10^{22}$ molecules/L

89. The core of a transformer is made of soft iron because soft iron is

(A) a good conductor of electricity

(B) malleable

(C) a natural magnet

(D) permeable to magnetic lines of force

(E) retentive to magnetic lines of force

90. A liquid flows at a velocity of 2 m/s in a constant-area pipe of cross-sectional area 0.8 m². A liquid jet enters the pipe through $A_i$ of 0.3 m² at a velocity, $V_i$, of 10 m/sec. Assuming uniform pressure and water flow at 1 and that the flow is frictionless, calculate the velocity at 2.

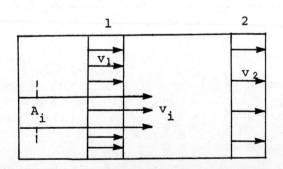

(A)  5 m/sec   (D)  12 m/sec
(B)  6 m/sec   (E)  5.75 m/sec
(C)  11 m/sec

91. The outlet of a tank of water is changed as shown below from a distance d to a distance 4d from the surface of the water. This has which of the following effects on the velocity of the water leaving the tank?

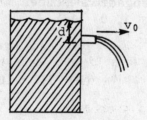
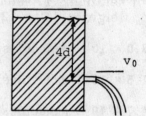

(A) increases the velocity by 2
(B) increases the velocity by 4
(C) increases the velocity only if the diameter of the outlet is increased
(D) decreases the velocity due to lower pressure
(E) has no effect on the velocity

92. What is the equivalent resistance, $R_{eq}$, of the circuit below?

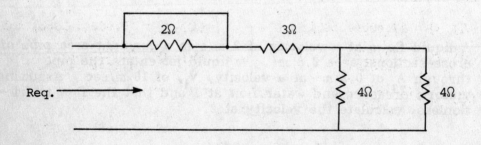

(A)  7Ω
(B)  5Ω
(C)  13Ω
(D)  11Ω
(E)  6/5 Ω

93. The speed of a DC motor, with other factors remaining constant, is correctly described by which of the following?

   (A) The speed varies with the frequency of voltage supply.

   (B) The speed is inversely proportional to the counter emf.

   (C) The speed depends on the number of poles only.

   (D) The speed is directly proportional to the magnetic flux per pole and the counter emf.

   (E) The speed is directly proportional to the counter emf and inversely proportional to the magnetic flux per pole.

94. The horn of a car seems to have a higher pitch when it approaches an observer. This occurs because

   (A) the density of the air has decreased

   (B) the elasticity of the air has decreased

   (C) the temperature of air has decreased

   (D) the horn echoes

   (E) the number of waves per second reaching the observer has increased

95. A car of mass 2000 Kg traveling at 7 m/s has its speed reduced to 2 m/s by a brake mechanism. Neglecting all other effects, such as changes in potential energy, how many calories of heat are transferred to the brake mechanism? (Note: 1 cal = 4.1858J)

   (A) $Q = 24,000(4.1858)$ cal
   (B) $Q = \frac{-45,000}{4.1858}$ cal
   (C) $Q = \frac{25,000}{4.1858}$ cal
   (D) $Q = 25,000(4.1858)$ cal
   (E) $Q = \frac{48,000}{4.1858}$ cal

96. Two similar cars, A and B, are connected rigidly together and have a combined mass of 4 kg. Car C has a mass of 1 kg. Initially, A and B have a speed of 5 m/sec and C is at rest as shown in the figure.

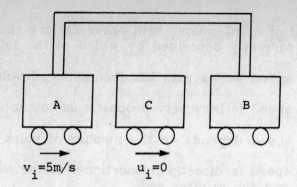

$v_i = 5\text{m/s}$   $u_i = 0$

Assuming a perfectly inelastic collision between A and C, the final speed of the system is

(A) 1 m/sec  (D) 3 m/sec
(B) 2 m/sec  (E) 4 m/sec
(C) 2.5 m/sec

97. The equilibrium-constant expression for the dissociation of water into hydrogen and oxygen is given by

$$K_{eq} = \frac{[H_2]^2[O_2]}{[H_2O]^2}$$

For the reversed reaction of hydrogen and oxygen into water, the equilibrium-constant is

(A) inverted, $1/K_{eq}$  (D) the square-root, $\sqrt{K_{eq}}$
(B) the same, $K_{eq}$  (E) doubled, $2K_{eq}$
(C) the negative, $-K_{eq}$

98. Your reaction time can be calculated by catching a ruler which is dropped between your thumb and first finger. If when you catch the ruler the marking shows that it has dropped 4 inches, then your reaction time is approximately

(A) $\frac{1}{6}$ sec  (D) $\frac{\sqrt{3}}{8}$ sec
(B) $\frac{1}{7}$ sec  (E) $\frac{\sqrt{2}}{8}$ sec
(C) $\frac{1}{8}$ sec

99. Which of the following represents the temperature variation of $c_v$ (constant volume heat capacity) for a copper solid?

(A)
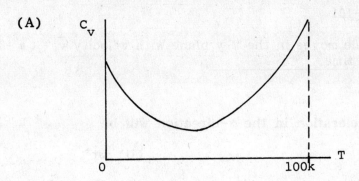

(B)
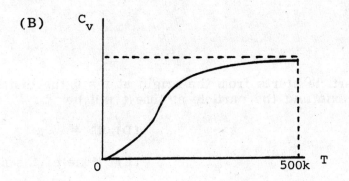

(C)
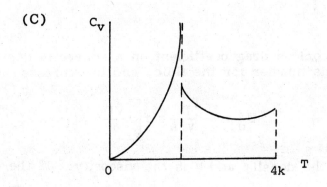

(D)
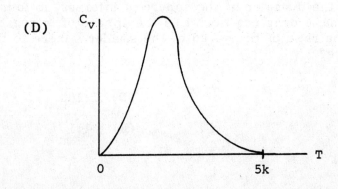

(E) None of the above

Questions 100-101

A particle moves in the x-y plane with velocity $\vec{v} = t^3 \mathbf{i} + 3t^2 \mathbf{j}$ where t is the time.

100. The acceleration in the x-direction will be

(A) $t^4$

(B) $t^4/4$

(C) $t^3$

(D) $3t^2$

(E) None of the above

101. If the particle starts from the origin at t = 0, the distance between x-axis and the particle at time t will be

(A) $3t^2$

(B) $t^3$

(C) 6t

(D) $t^4/4$

(E) None of the above

102. The dimensionless drag coefficient on a sphere is dependent on the Reynolds number for the fluid, and is expressed as follows:

$$C_d = \frac{F_D}{\frac{1}{2}\rho v^2 A} = f\left(\rho \frac{VD}{\mu}\right),$$

where $\rho$ is the density and $\mu$ is the viscosity. If the drag coefficient for a sphere, which has a diameter which is $\frac{1}{10}$ the length of the diameter of the sphere of interest, is found to have the same drag coefficient as the sphere of interest, then what is the ratio of the speed of the smaller sphere to the larger sphere?

(A) 10:1

(B) 1:10

(C) 100:1

(D) 1:100

(E) 1:1

103. A box starts sliding at a constant velocity down the ramp of a truck inclined at 60° while the truck is still moving. If an observer outside the truck sees the box moving down vertically, and the speed of the box with respect to the moving ramp is 4 m/sec, then what is the speed of the truck?

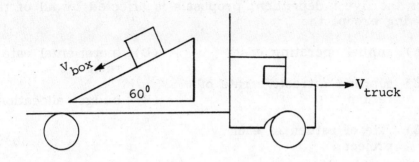

(A) 2 m/sec

(B) 4 m/sec

(C) $2\sqrt{3}$ m/sec

(D) $4/\sqrt{3}$ m/sec

(E) $2/\sqrt{3}$ m/sec

104. The melting point of solid argon is higher than the melting point of solid helium. Which of the following is responsible for this property?

(A) The bond length between the argon atoms is greater than the bond length between the helium atoms.

(B) Van der Waals Forces in molecular solids are stronger with increasing size of the atoms involved.

(C) The argon atoms are more electronegative than the helium atoms.

(D) Solid argon has an fcc lattice structure.

(E) The atomic radius for argon is much greater than the atomic radius for helium.

105. Which of the following is not a mathematical statement of the ideal gas law for two states, 1 and 2, of an ideal gas?

(A) $\dfrac{P_1 V_1}{T_1} = \dfrac{P_2 V_2}{T_2}$

(B) $\dfrac{P_1 T_2}{T_1} = \dfrac{P_2 V_2}{V_1}$

(C) $\dfrac{V_1}{P_2 T_1} = \dfrac{V_2}{P_1 T_2}$

(D) $P_1 V_1 T_2 = P_2 V_2 T_1$

(E) (A)-(D) are all correct

106. The ABC Milling Company is considering five proposals for new equipment and plant modifications. The project selection process for the five independent proposals is affected by all of the following except the

(A) annual operating costs

(B) minimum attractive rate of return

(C) rate of return of each project

(D) incremental rate of return

(E) budget allocation

107. The diagram below shows a gun and a target at the same elevation. The gun is fired at a departure angle $\alpha_0$ with muzzle velocity u. R is the position of the target and is less than $R_{max}$, the maximum range. In order to hit the target, the departure angle $\alpha_0$ should be

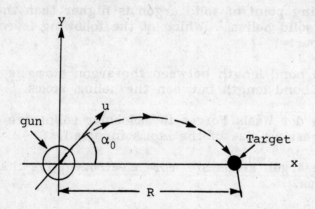

(A) $\sin^{-1}(RR_{max})$

(B) 45°

(C) $\frac{1}{2}\sin^{-1}(R/R_{max})$

(D) $2\tan^{-1}(R/R_{max})$

(E) $\sin^{-1}(R_{max}/R)$

108. The intrinsic carrier concentration for silicon is approximately $10^{10}$ carriers/cm³. If an impurity concentration of $10^{14}$ donor atoms/cm³ is added, the carrier concentration is very close to

(A) $n = 10^6$ atoms/cm$^3$, $P = 10^{14}$ atoms/cm$^3$

(B) $n = 2 \times 10^4$ atoms/cm$^3$, $P = \frac{1}{2} \times 10^6$ atoms/cm$^3$

(C) $n = 10^4$ atoms/cm$^3$, $P = 10^6$ atoms/cm$^3$

(D) $n = 2 \times 10^6$ atoms/cm$^3$, $P = 2 \times 10^{14}$ atoms/cm$^3$

(E) $n = \frac{1}{2} \times 10^{14}$ atoms/cm$^3$, $P = \frac{1}{2} \times 10^{14}$ atoms/cm$^3$

109. In steady state, which of the following would have the largest rate of heat transferred from a heat source of 100°C to an environment of air of 10°C?

(A) Decreasing the temperature of air by half

(B) Increasing the surface area by two

(C) Covering the heat source with cotton fiber

(D) Placing the heat source in a vacuum

(E) Decreasing the surface area by half

110. A 1 in. × 1 in. × 1 in. copper cube was heated to 90°C. It was thrown into a lake of 10°C. After steady state is reached, how much heat was released? (Note that $C_{copper}$ = 0.386 kJ/kg°K)

(A) 3.86 kJ/kg        (D) 30.9 kJ/kg

(B) 19.9 kJ/kg        (E) 34.74 kJ/kg

(C) 27.4 kJ/kg

111. The open-tube manometer shown in the figure is used to measure the gauge pressure of a tank (not shown). Which of the following is correct?

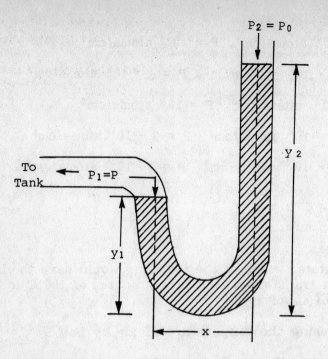

(A) The pressure, P, is found independent of atmospheric pressure, $P_0$.

(B) The gauge pressure, $(P - P_0)$, cannot be determined without knowing the value x.

(C) The product of the density of the liquid, $\rho$, and the acceleration constant, g, relate the distance, $(y_2 - y_1)$ to gauge pressure $(P - P_0)$.

(D) The gauge pressure $(P - P_0)$ is independent of the distance $(y_2 - y_1)$.

(E) The dimensions of the tank must be known.

112. The mass of a rectangular bar of iron is 200 grams. Its dimensions are 2cm × 2cm × 10cm. The specific gravity of this iron bar is: (assume that the density of water is 1 g/cm$^3$).

(A) 4           (D) 7

(B) 5           (E) 8

(C) 6

113. If the number of gas molecules in a constant volume vessel is increased, the effect is: (Note: assume ideal gas behavior.)

(A) decreased pressure

(B) increased pressure and increased temperature

(C) decreased temperature

(D) increased pressure and decreased temperature

(E) decreased pressure and increased temperature

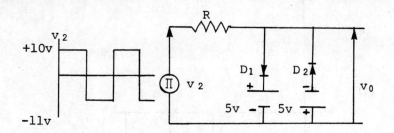

114. A square wave having a peak voltage of 10V is applied to the above circuit. If the diodes are ideal, which of the following figures represents the output?

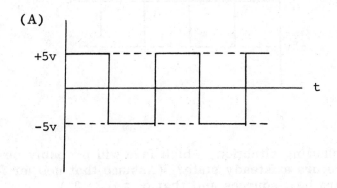

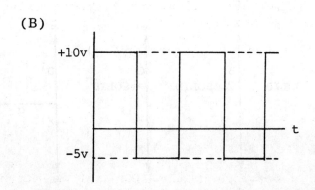

(C)

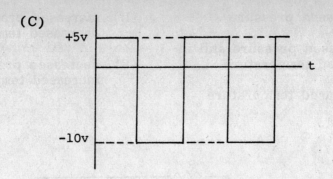

(D)

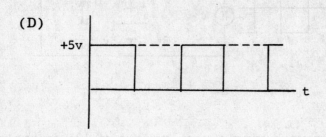

(E)

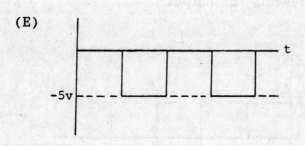

115. For the following situation, which face will probably be higher in temperature at steady state? (Assume that neither force B nor C are heat sources and that $q_1 = q_2 > 0$.)

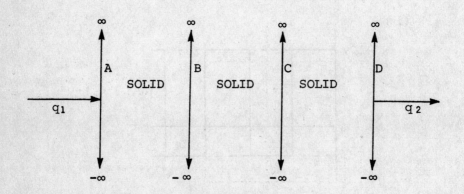

(A) A  (D) D
(B) B  (E) insufficient information
(C) C

116. Which of the following phenomena illustrates the particle property of light?

    (A) Electron diffraction
    (B) Interference
    (C) The photoelectric effect
    (D) Radiation
    (E) Polarization

117. The addition of certain alloying elements such as manganese or nickel to steel increases its hardenability when quenched because

    (A) it decreases the rate of transformation of austenite to pearlite
    (B) it decreases the rate of transformation of austenite to martensite
    (C) it increases the rate of transformation of austenite to pearlite
    (D) it increases the rate of transformation of pearlite to bainite
    (E) it decreases the rate of transformation of martensite to bainite

Questions 118-119

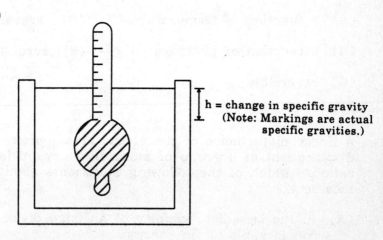

h = change in specific gravity
(Note: Markings are actual specific gravities.)

A hydrometer is placed in a tank of oil to measure its specific gravity. The hydrometer sinks to a reading of h. The volume of the hydrometer that is submerged under the oil is V. (Atmospheric pressure is negligible; specific weight of water is $\gamma$.)

118. Which of the following quantities are equal for the hydrometer shown?

    (A) The weight of the hydrometer and the weight of the fluid displaced

    (B) The pressure force on the submerged portion of hydrometer and the pressure force on the portion open to the atmosphere

    (C) The weight of the hydrometer and the weight of the fluid surrounding the hydrometer

    (D) The pressure force on the portion of the hydrometer open to the atmosphere and the weight of the fluid displaced

    (E) None of the above

119. The total force acting on the fluid surrounding the hydrometer is

    (A) $h \dfrac{\gamma}{V}$

    (B) $\dfrac{h}{\gamma V}$

    (C) $\dfrac{h}{\gamma} V$

    (D) $h \gamma V$

    (E) $\dfrac{\gamma}{h} V$

120. The heat transfer during an adiabatic process is

    (A) a function of temperature
    (B) a function of pressure
    (C) reversible
    (D) irreversible
    (E) zero

121. A linear disturbance of the atomic arrangement caused by the displacement of a group of atoms in a crystal is called a dislocation. Which of the following statements about dislocations is incorrect?

    (A) In the immediate vicinity of an edge dislocation there is considerable strain energy.

(B) The Burgers vector describes both the magnitude and direction of the relative displacement.

(C) Both edge and screw dislocations may occur together in a crystal.

(D) Dislocations may be caused by thermal stresses, growth accidents or collapse of vacancies.

(E) Dislocations do not interact with vacancies or other defects in a crystal.

122. In thermal equilibrium, which of the following is always true?

(A) The emissivity is small and the absorbance is large.

(B) The rate of emission of radiant energy is equal to the rate of absorption.

(C) The emissivity and absorbance are zero.

(D) The rate of emission and the rate of absorption of radiant energy are zero.

(E) The ratio of emissivity to absorbance is one.

123. How much heat is given off when 4 kg of water cools from 80°C to 10°C?

(A) 70 kcal

(B) 35 kcal

(C) 7000 kcal

(D) 280 kcal

(E) 240 kcal

124. An engineer wishes to build a trapezoidal open channel for water. The Chezy constant for the conduit was found experimentally to be 100. It is required that the discharge for the channel is 10 cfs. The channel must be built on an incline of slope .001. If water is in contact with 800 ft. of the channel surface, then its cross-sectional area must be

(A) 40 ft²

(B) 20 ft²

(C) 30 ft²

(D) 25 ft²

(E) 50 ft²

125. The first of two alternatives for the local water supply requires an initial investment of $8,000,000 and an annual operating expense of $25,000 for a dam that will last infinitely long. The second alternative requires 10 wells at an initial cost of $50,000 with an average life expectancy of 13 years and an annual operating cost of $5,000 per well. If money is valued at 5% and the uniform annual cost factor for thirteen years is 0.1, then the city would

   (A) save $25,000 annually if it built the dam

   (B) save $7,500,000 on the initial investment if it built the wells

   (C) save $75,000 annually if it built the dam

   (D) save $3,000,000 on the capital cost if it built the wells

   (E) save $6,500,000 on the capital cost if it built the wells

126. If a polymer such as rubber is to remain rigid at high temperatures, which of the following must be true?

   I. The possibility for branching must exist between the polymer molecules.

   II. Some of the bonds in the molecule must be ionic.

   III. Cross-linking between linear polymers must be facilitated with the presence of an intermediate atom such as sulfur.

   (A) II only         (D) II and III only

   (B) I and III only  (E) I, II, and III

   (C) I and II only

127. For a mass on the end of a spring displaced from equilibrium (and ignoring any air resistance), the period of the resultant harmonic motion is related to which of the following?

   (A) mass and distance displaced

   (B) mass and force-constant of the spring

   (C) mass and distance displaced and force-constant of spring

(D) distance displaced and force-constant of spring

(E) mass only

128. Which of the phase diagrams indicates a eutectic system with partial solubility in the solid state?

(A)
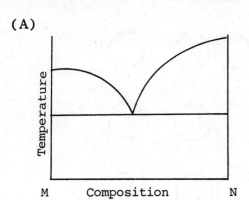

(B)
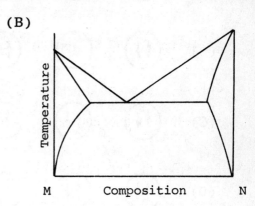

(C)
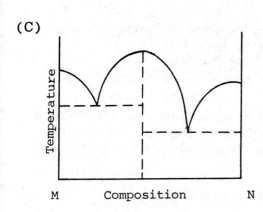

(D)
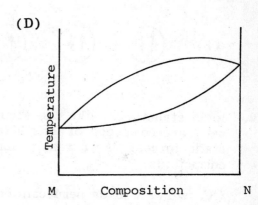

(E)
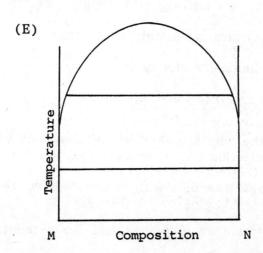

129. Which of the following orbital configurations cannot exist because of violation of the Pauli Exclusion Principle?

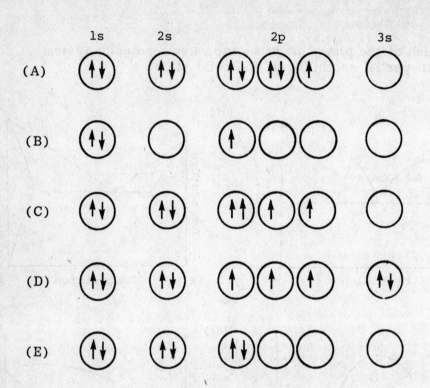

130. Ionic structures, like the structure of sodium chloride or rock-salt, are composed of ions strongly held together by electrostatic forces. As a result, ionic compounds compared to other compounds,

   (A) serve well as semi-conductor material

   (B) have fairly low melting and boiling points

   (C) don't dissociate completely in solution

   (D) yield mechanically strong crystals

   (E) contain large interionic distances

Directions: Choose the best answer for each question and mark the letter of your selection on the corresponding answer sheet.

Note: The natural logarithm of x will be denoted by $\ln x$.

131. If

$$u \cos v - x = 0 \text{ and}$$

$$u \sin v - y = 0$$

where u and v are defined as implicit functions of x and y, then $\frac{\partial u}{\partial x}$ is given by:

(A) $\frac{\sin u}{v}$

(B) $\cos v$

(C) $\frac{\cos u}{v}$

(D) $\frac{-\sin v}{u}$

(E) $\frac{-\sin u}{v}$

132. Let the graphs $g_1$, $g_2$, $g_3$ correspond to the equations:
$g_1$: $y - x^2 = 9$
$g_2$: $x^2 + y^2 = 81$
$g_3$: $y = x$

The graph(s) which intersect(s) only the y-axis is(are)

(A) $g_1$

(B) $g_2$

(C) $g_3$

(D) $g_2$ and $g_1$

(E) $g_1$ and $g_3$

133. Given an n × n matrix A, the characteristic polynomial of A is given by

(A) $P(\lambda) = \det |A - \lambda A|$

(B) $P(\lambda) = \det |\lambda - I_n|$

(C) $P(\lambda) = \det |A - \lambda I_n|$

(D) $P(\lambda) = \det |A^T - \lambda A|$

(E) $P(\lambda) = \det |A - \lambda|$

134. $\int_0^{\pi/2} (\sin^5 x) \cos^3 x \, dx =$

(A) $\frac{1}{24}$

(B) $\frac{-1}{24}$

(C) $\frac{11}{24}$

(D) $\frac{10}{24}$

(E) $\frac{5}{24}$

135. Green's Theorem in the plane states:

(A) $\oint_C Pdx + Qdy = \iint_R \left(\frac{\partial Q}{\partial y} - \frac{\partial P}{\partial x}\right) dA$

(B) $\oint Pdy + Qdx = \iint_R \left(\frac{\partial Q}{\partial x} + \frac{\partial P}{\partial y}\right) dA$

(C) $\int Pdy + Qdx = \iint \left(-\frac{\partial Q}{\partial x} - \frac{\partial P}{\partial y}\right) dA$

(D) $\oint_C Pdx + Qdy = \iint_R \left(\frac{\partial Q}{\partial x} - \frac{\partial P}{\partial y}\right) dA$

(E) $\oint_C Pdy + Qdx = \iint_R \left(\frac{\partial Q}{\partial y} + \frac{\partial P}{\partial x}\right) dA$

136. A conic section is an ellipse if

(A) $e \to \infty$

(B) $0 < e < 1$

(C) $e > 1$

(D) $e = 0$

(E) $e < 0$

137. The family of curves given by $f(x) = ax^2$ are

(A) ellipses

(B) lines through $(0, a)$

(C) parabolas with vertices $(0, 0)$

(D) hyperbolas

(E) quadratic curves with exactly two zeros

138. The polynomial $\dfrac{x^2 - 1}{(x + 1)(x^2 + 4x - 5)}$ has non-removable singularities at $x =$

(A) $-1, -3$

(B) $-1$

(C) $-5, 0$

(D) $4$

(E) $-5$

139. $\sum_{n=0}^{\infty} \frac{x^n}{n!} =$

(A) $e^x$

(B) $\infty$

(C) 0

(D) $\ln x$

(E) $(1 + x)^n$

140. The sequence

$$x_{n+1} = x_n - \frac{x_n^3 - 30}{3x_n^2}$$

converges to:

(A) 0

(B) $\infty$

(C) $30^{\frac{1}{3}}$

(D) $\sqrt{30}$

(E) 2

141. For the boundary value problem

$$y'' - 4\lambda y' + 4\lambda^2 y = 0 \qquad (1)$$

with the boundary conditions

$$y(0) + y'(0) = 0, \quad y'(1) = 0 \qquad (1')$$

the only real solution that can satisfy the differential equation and the boundary conditions is a trivial solution, $y \equiv 0$. This problem has

(A) two real eigenvalues

(B) one real and one complex eigenvalue

(C) no eigenvalues

(D) two complex eigenvalues

(E) more than two eigenvalues

142. If the equation $e^x \cos 3y + \ln y (\sin x^2) = 3y$ defines $y$ implicitly as a function of $x$ then $\frac{dy}{dx} =$

(A) $\dfrac{e^x \cos 3y + 2x \ln y \cos x^2}{3(1 + e^x \sin 3y) - \dfrac{\sin x^2}{y}}$

(B) $\dfrac{e^x \cos 3y - 2x \ln y \cos x^2}{\dfrac{\sin x^2}{y} + 3(1 - e^x \sin 3y)}$

(C) $\dfrac{3e^x \cos 3y - \ln y \cos x^3}{\sin x^2 - \dfrac{\sin x^2}{y}}$

(D) $\dfrac{3y \cos 3y + 2x \cos x^2}{\sin 3y + 3}$

(E) $\dfrac{\dfrac{\sin x^2}{y} - 3e^x \sin 3y}{e^x \cos 3y - \ln y \cos x^2}$

143. A curve passing through the origin has a slope of 2x at any point on the curve. The equation of the curve is

(A) $y = x^2$

(B) $y = 2x + C$, where C is a constant

(C) $y^2 = 2x$

(D) $x^2 + y^2 = 2$

(E) $2x + y = 2$

144. Consider a triangle with vertices (1,0), (1,1), (0,0). If $C_1$ is the line connecting vertices (0,0) and (1,0) then

$$\int_{C_1} y^2 \, dx + x^2 \, dy =$$

(A) 0

(B) 1

(C) -1

(D) $\pi$

(E) C

# THE GRADUATE RECORD EXAMINATION ENGINEERING TEST

## MODEL TEST I

# ANSWERS

| | | | | | |
|---|---|---|---|---|---|
| 1. | C | 21. | D | 41. | C |
| 2. | A | 22. | C | 42. | C |
| 3. | E | 23. | A | 43. | E |
| 4. | C | 24. | C | 44. | E |
| 5. | B | 25. | B | 45. | A |
| 6. | D | 26. | C | 46. | C |
| 7. | D | 27. | C | 47. | C |
| 8. | E | 28. | A | 48. | B |
| 9. | A | 29. | B | 49. | B |
| 10. | C | 30. | D | 50. | E |
| 11. | D | 31. | D | 51. | C |
| 12. | D | 32. | C | 52. | C |
| 13. | E | 33. | A | 53. | C |
| 14. | E | 34. | B | 54. | C |
| 15. | C | 35. | D | 55. | C |
| 16. | C | 36. | A | 56. | C |
| 17. | C | 37. | A | 57. | C |
| 18. | B | 38. | C | 58. | B |
| 19. | A | 39. | D | 59. | C |
| 20. | A | 40. | B | 60. | D |

| | | | | | |
|---|---|---|---|---|---|
| 61. | C | 89. | D | 117. | A |
| 62. | D | 90. | A | 118. | A |
| 63. | D | 91. | A | 119. | D |
| 64. | A | 92. | B | 120. | E |
| 65. | B | 93. | E | 121. | E |
| 66. | C | 94. | E | 122. | B |
| 67. | D | 95. | B | 123. | D |
| 68. | D | 96. | E | 124. | B |
| 69. | A | 97. | A | 125. | E |
| 70. | A | 98. | B | 126. | B |
| 71. | E | 99. | B | 127. | B |
| 72. | C | 100. | D | 128. | B |
| 73. | A | 101. | B | 129. | C |
| 74. | B | 102. | A | 130. | D |
| 75. | B | 103. | A | 131. | B |
| 76. | C | 104. | B | 132. | A |
| 77. | B | 105. | E | 133. | C |
| 78. | E | 106. | D | 134. | A |
| 79. | A | 107. | C | 135. | D |
| 80. | E | 108. | C | 136. | B |
| 81. | B | 109. | B | 137. | C |
| 82. | C | 110. | D | 138. | E |
| 83. | D | 111. | C | 139. | A |
| 84. | A | 112. | B | 140. | C |
| 85. | E | 113. | B | 141. | D |
| 86. | C | 114. | A | 142. | A |
| 87. | A | 115. | A | 143. | A |
| 88. | E | 116. | C | 144. | A |

# THE GRADUATE RECORD EXAMINATION ENGINEERING TEST

## MODEL TEST I

## DETAILED EXPLANATIONS OF ANSWERS

**1. (C)**
In an isothermal process for an ideal gas, the internal energy does not change since it depends only on the temperature. Only by a slow change in pressure and volume can the work done by the system be equal to the heat entering the system. The simultaneous decrease in pressure and increase in volume means the performance of positive work by the system. Since we have $\Delta W = \Delta Q$ for isothermal processes, the heat entering the system is also positive. Therefore, the system must have received some heat.

**2. (A)**
The v-i characteristic of a typical non-linear resistance is shown below.

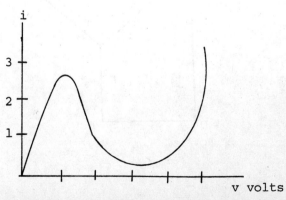

A current-controlled non-linear resistor is one whose terminal voltage is a single-valued function of its terminal current, i.e. v = f(i), where v and i are the terminal voltage and terminal current, respectively.

A voltage-controlled non-linear resistor is one whose terminal current is a single-valued function of its terminal voltage, i.e. i = g(v).

As seen from the v-i characteristics of the non-linear resistance, a change in coil voltage produces a change in the resistance since R = V/I, and for distinct points on the curve this is different.

## 3. (E)
According to Poiseville's law the total volume of flow per unit time is given by

$$\frac{dV}{dt} = \frac{\pi}{8} \frac{R^4}{\eta} \frac{p_1 - p_2}{L}$$

Therefore, the volume flow rate is inversely proportional to the viscosity coefficient and length of the tube, while it is directly proportional to the radius of the tube. The viscosity of a liquid will decrease with an increasing temperature. Therefore, decreasing the length of the tube and increasing the temperature will increase the flow rate the most.

## 4. (C)
When section BEFCB is removed, we are left with the following:

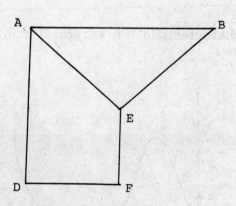

To find the centroid ($\bar{X}, \bar{Y}$) of this composite area we use:

$$\bar{X} = \frac{\Sigma \bar{x}_i A_i}{\Sigma A_i} \qquad \bar{Y} = \frac{\Sigma \bar{y}_i A_i}{\Sigma A_i}$$

We break the figure into three areas. Note that the center of mass of a triangle is the point of intersection of its medians. This, in addition to the fact that the medians intersect at one-third distance between the vertex and the side (closer to the side), can be used to evaluate the $\bar{X}$ and $\bar{Y}$ coordinates of the centroid of the triangle.

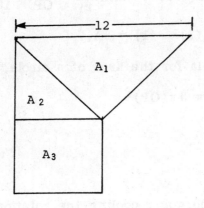

$$\bar{X} = \frac{\bar{x}_1 A_1 + \bar{x}_2 A_2 + \bar{x}_3 A_3}{A_1 + A_2 + A_3}$$

$$= \frac{6(36) + 2(18) + 3(36)}{36 + 18 + 36} = \frac{10(36)}{5(18)}$$

$$\bar{X} = 4$$

$$\bar{Y} = \frac{\bar{y}_1 A_1 + \bar{y}_2 A_2 + \bar{y}_3 A_3}{A_1 + A_2 + A_3}$$

$$= \frac{10(36) + 8(18) + 3(36)}{36 + 18 + 36} = \frac{(17)(36)}{5(18)} = \frac{34}{5} = 6.8$$

Thus the centroid is at (4, 6.8).


5. (B)
The area of a circle is $A = \pi r^2$. For two circles of areas $A_1$ and $A_2$, the ratio of their areas is:

$$\frac{A_2}{A_1} = \frac{\pi r_2^2}{\pi r_1^2} = \left(\frac{r_2}{r_1}\right)^2$$

Therefore we have

$$A_2 = A_1 \left(\frac{r_2}{r_1}\right)^2.$$

The area of Track 2 is then:

$$\text{Area} = A_2 - A_1 = A_1\left(\frac{r_2}{r_1}\right)^2 - A_1 = A_1\left[\left(\frac{r_2}{r_1}\right)^2 - 1\right]$$

Now if a man runs at a constant speed v, the distance he runs in time t will be d = vt. Using this, we conclude that:

$$OQ = 20v, \quad OP = 10v, \quad \text{and} \quad \frac{r_2}{r_1} = \frac{OQ}{OP} = \frac{20v}{10v} = 2$$

Then Area = $A_1[(2)^2 - 1] = 3A_1$.

Using the formula for the area of a circle, we see that:

$$\text{Area} = 3\pi(OP)^2$$

6. (D)
For a glass undergoing cooling, the rotation and translation of the atoms and ions or molecules are slowed down until they effectively cease once they reach the glass transition temperature. Liquids such as glass having complex molecular structure and high viscosities will not crystallize and instead become frozen in the glassy state. Once they are frozen, the glass will be stiff enough to support its own weight without deforming and becomes rigid.

7. (D)
A balanced equation means that the number of atoms of each element is the same on both sides of the chemical equation. For this problem there should be 3 carbon atoms (C), 8 hydrogen atoms (H) and 10 oxygen atoms (O). A, B and C have incorrect products and have unequal numbers of oxygen, hydrogen and carbon, respectively. E has unequal numbers of hydrogen (6) and oxygen (9) in the products.

8. (E)
This is a simple problem in which the value of I or V is substituted in the given equation to calculate the value of

the other variable.

Choice E is the right solution. This is verified as follows:

Substitute V = 10 into $I = \left(\dfrac{V}{5}\right)^2$. We get

$$I = \left(\dfrac{10}{5}\right)^2 = 4 \text{ mA}.$$

Substitute V = 20. We get

$$I = \left(\dfrac{20}{5}\right)^2 = 16 \text{ mA}$$

The above two values of current can be checked against their respective voltages on the given curve.

## 9. (A)

In this problem, the train, the boy, and the ball move with the same horizontal velocity. In the reference frame of the moving train, the boy perceives the ball as if he were in a stationary train. From his point of view the ball does not have a horizontal velocity, so it will not have a horizontal movement with respect to him (as it does, from the point of view of another boy standing on the ground). Therefore, a ball thrown vertically upward will come vertically downward and land in the boy's hands.

## 10. (C)

To find the value of the current i, we have to apply Kirchoff's current law.

Kirchoff's current law: For each junction in a network, the algebraic sum of all currents entering the junction is equal to the algebraic sum of all currents leaving the junction.

The word "algebraic" implies that the signs of the current are to be considered while adding. Applying Kirchoff's law to the given figure we get:

$$2i + 2t = 6 - 3 \sin \pi t$$

i.e.

$$i = \dfrac{6 - 3 \sin \pi t - 2t}{2}$$

Substituting t = 0.5 we get:

$$i = \frac{6 - 3\sin 90 - 1}{2}$$

$$= \frac{6 - 3 - 1}{2}$$

$$= 1$$

## 11. (D)

The given circuit is a balanced Wheatstone bridge circuit. A circuit such as the one given is said to be balanced if

$$\frac{R_{AB}}{R_{AD}} = \frac{R_{BC}}{R_{DC}}$$

Now $R_{AB} = 10\Omega$, $R_{AD} = 30\Omega$, $R_{BC} = 5\Omega$ and $R_{DC} = 15\Omega$.

So $\frac{10}{30} = \frac{5}{15}$ and hence the bridge is balanced.

If the bridge is balanced then I = 0A and V = 0 volts. This is proved as follows:

If I = 0, $I_1 = I_3$ and $I_2 = I_4$ and if V = 0, $V_1 = V_2$ and $V_3 = V_4$.

So, $\qquad I_1 R_{AB} = I_2 R_{AD}$ and $I_3 R_{BC} = I_4 R_{DC}$ \hfill (1)

Substituting $I_3 = I_1$ and $I_4 = I_2$ into (1) we get

$$I_1 R_{BC} = I_2 R_{DC}$$

or

$$I_1 = \frac{I_2 R_{DC}}{R_{BC}} \qquad (2)$$

Solving for $I_2$ in (1) we get

$$I_2 = \frac{I_1 R_{AB}}{R_{AD}}$$

Substituting this $I_2$ in (2) we get

$$I_1 = \left(\frac{I_1 R_{AB}}{R_{AD}}\right) \frac{R_{DC}}{R_{BC}}$$

i.e.
$$\frac{R_{AB}}{R_{AD}} = \frac{I_1}{I_1} \times \frac{R_{BC}}{R_{DC}}$$

$$\frac{R_{AB}}{R_{AD}} = \frac{R_{BC}}{R_{DC}} \qquad \text{(bridge is balanced).}$$

## 12. (D)

The impulsive torque acting on both disks is the same, since they are internal to the system and cancel according to Newton's third law. The angular momentum of each disk is changed under the action of the same torque

$$J_0 = I\omega_1 - I\omega_0$$

$$J_0' = I'\omega_1 - I'\omega_0'$$

But since the sum of both torques is equal to zero, the change in the angular momenta of both disks is equal to zero:

$$T' + T' = 0 \Rightarrow \int T dt + \int T' d't = 0 \Rightarrow$$

$$J_0 + J_0' = 0$$

$$I\omega_1 - I\omega_0 + I'\omega_1 - I'\omega_0 = 0$$

therefore angular momentum is conserved

$$I\omega_1 + I'\omega_1 = I\omega_0 + I'\omega_0$$

## 13. (E)

The smallest variation in readings among different gas thermometers is obtained at low pressures. By reducing the gas in the $O_2$ thermometer, the temperature reading will be reduced since the pressure is proportional to the temperature. The variation can also be reduced if the pressure of both gases is reduced to a common pressure. Therefore both B and D are correct.

## 14. (E)

The kinetic theory of gases states that all matter is composed of molecules which are in constant random motion. These molecules collide with each other and with the container wall. When molecules collide with the wall, the effect of numerous collisions is producing a constant force against the wall. This force is the result of a change in momentum for the molecules.

## 15. (C)

Convection is the transfer of heat by means of displacement of liquid or gas which is in turn due to the difference of densities. When the water is heated in the furnace it expands and therefore has smaller density. The greater pressure of the cold water from the radiator forces the hot water from the furnace to the radiator while the cold water moves to the furnace and is heated. Hence, the transfer of heat is through free convection. Forced convection would require a pump or other mechanical device to move the water.

## 16. (C)

Water is one of the few materials that contracts upon melting. For materials that expand upon melting, any temperature increase must be accompanied by an increase in pressure in order to maintain solid-liquid equilibrium. For water, however, any decrease in temperature must be accompanied by an increase in pressure. For any real substance, the phase diagram is a projection of a PVT surface onto a constant volume plane.

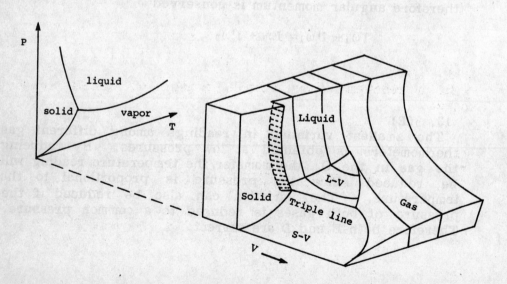

In order to have a complete understanding about the situation, we have to consider the three-dimensional PVT surface where water and ice can be in equilibrium. As shown in the figure, if we want to stay on the solid-liquid surface, an increase of pressure must be accompanied by a decrease in temperature and vice-versa. On the other hand, we can move upward on the surface (which means increasing the pressure and decreasing the temperature) while tending to both right and left, which causes an increase or decrease in volume. Therefore a change in volume is feasible, and we are not restricted to a constant volume. Consequently, both A and C are correct.

17. (C)
The speed of sound in any media is proportional to the compressibility of the media. The governing equation is:

$$C^2 \alpha \left(\frac{\partial P}{\partial \rho}\right)_S$$

where   C = velocity of sound

P = pressure

$\rho$ = density

$\frac{\partial P}{\partial \rho}$ for liquid is definitely higher than that of gas. Therefore, the speed of sound in liquid is greater than the sonic velocity; i.e. the Mach Number is greater than 1. (Mach number is defined to be $\frac{\text{velocity}}{\text{speed of sound in air}}$)

18. (B)
In a reaction, mass must be conserved. Therefore we have:

$$NO + \tfrac{1}{2}O_2 \rightarrow NO_2$$

$$NO: \frac{\text{Weight}}{\text{Molecular weight}} = \frac{15g}{16 + 14} = \frac{1}{2}$$

$$NO_2: \frac{\text{Weight}}{\text{Molecular weight}} = \frac{X}{16 + 16 + 14} = \frac{X}{46}$$

$$\frac{1}{2} = \frac{X}{46} \; ; \; X = 23g$$

19. (A)
The mass of an atom is approximately a whole number,

since each proton and neutron contributes approximately one unit of atomic mass to an atom. The fact that many atomic weights are far from being whole numbers is due to the fact that most elements exist as mixtures of two or more kinds of atoms of different atomic masses but of similar chemical properties. The difference between the atomic masses is due to the different number of neutrons. Chlorine is such an element; it has an atomic weight of 35.453, existing as two kinds of atoms having masses very close to whole numbers 35 and 37. Both of the atoms have an atomic number of 17.

Such atoms having the same atomic number, same number of protons and different masses are called isotopes.

20. (A)
In an isochoric process all energy changes in the gas take place at constant volume. Since $\Delta V = 0$, the thermal equivalent of the work, W, done by the system on its surroundings is also zero. Also, the net flow of heat into the system, Q, equals the change in the internal energy of the system, $\Delta U$.

Hence, in an isochoric process, the thermal energy transferred to the gas is completely converted to internal energy, resulting in an increase in the gas temperature.

21. (D)
Water is usually measured in units of volume which are $L^3$. Thus, the dimensional formula is $L^3 T^{-1}$.

22. (C)
$\left(P + \frac{a}{V^2}\right)(V - b) = nRT$ is the correct choice. This expression is known as Van der Waals' equation of state. The constants a and b are determined experimentally for each gas. The term $a/V^2$ is used to express the reduced pressure due to the mutual attraction of the gas molecules when the density is very high and molecules are close together. The term b is a volume correction due to the fact that the molecules have finite size, thus reducing the volume available to them.

23. (A)

For motion to start, the moment couple on the drum must be equal to the friction force of the rod against the drum, multiplied by the drum radius. ($\Sigma M_o = 0$).

Since the rod is in equilibrium, we can take torques about the fulcrum to find the normal force of the drum on the rod.

$$\Sigma \tau_{fulcrum} = 0 = -100 \text{ lb}(4 \text{ ft}) + N(2 \text{ ft}) \qquad (1)$$

$$N = 200 \text{ lb}$$

From N, one can find the frictional force f by writing:

$$f = \mu N = 0.1(200) \qquad (2)$$

$$f = 20 \text{ lb}$$

Therefore, the torque produced by the couple is

$$T = 20 \text{ lb} \times 1 \text{ ft} = 20 \text{ ft-lb}$$

24. (C)

The cooling coil of the refrigerator accepts heat from the air in the room, but the same amount of heat is emitted by the condenser. There is no net heat exchange in this process. However, the work done by the compressor appears as heat, and therefore the temperature of the room goes up. The following is the energy diagram of a refrigerator. As might be concluded, the heat $Q_2$ is extracted from the heat reservoir at lower temperature (inside the refrigerator) and is transferred to the higher temperature reservoir (room). In addition, some work supplied from a source of energy (electricity in this case) is again transferred to the high temperature reservoir as heat. Although we have a heat transfer from the inside of the refrigerator to the outside, it doesn't affect the room temperature since the doors of the refrigerator are open. However, the additional heat generated by the consumption of electrical energy will raise the room temperature.

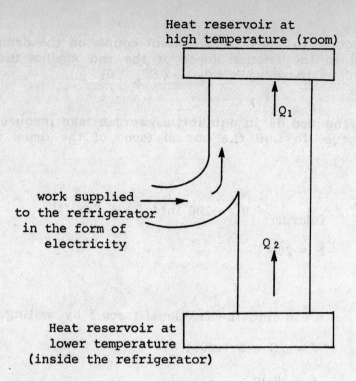

## 25. (B)

The resonant circuit is a combination of R, L, C elements having a normal frequency characteristic as shown in the figure.

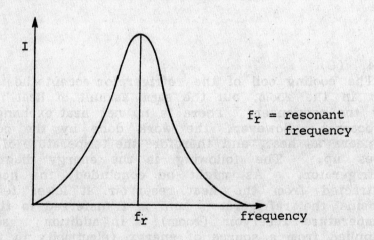

$f_r$ = resonant frequency

At resonant frequency $f_r$, for a given voltage, current through the circuit is at a maximum and hence the power absorbed by the circuit is at a maximum (note that the phase angle of this case is zero) and we have:

$$P = RI_e^2,$$

since $I_e$ is maximum, P is maximum too.

In an ac series circuit, the current is given by

$$I = \frac{V}{\sqrt{R^2 + (X_L - X_C)^2}} = \frac{I}{Z} \qquad (1)$$

where V is the voltage, R the resistance, $X_L$ is the inductive reactance and $X_C$ is the capacitive reactance. $\sqrt{R^2+(X_L-X_C)^2}$ is Z, the impedance of the circuit.

Meanwhile $X_L = 2\pi fL$ where L is the inductance and $X_C = 1/(2\pi fC)$ where C is the capacitance. $X_L$ and $X_C$ vary with the frequency of the driving voltage. So the current I in the circuit is a function of the oscillator frequency.

When $X_L = X_C$, the denominator (1) is minimum and so the current becomes maximum. This occurs when $|X_L|=|X_C|$, thus $X_L = X_C$ only when the magnitudes are equal. The frequency when $|X_L|=|X_C|$ is called the resonance frequency and is given by

$$f_r = \frac{1}{2\pi\sqrt{LC}}$$

## 26. (C)
For any reversible process the change in entropy is given as

$$\Delta S = \int_1^2 \frac{dQ}{T}.$$

Since in any adiabatic process no heat is allowed to enter or leave the system, Q = 0. Therefore the change in entropy

$$\int_1^2 \frac{dQ}{T}$$

is zero (dQ = 0).

## 27. (C)
Solar energy, transmitted by radiation, warms the ground during the day. At night, the ground radiates energy back

into space. Clouds slow down the process, so cloudy nights are warmer.

28. (A)

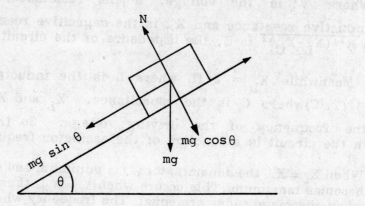

The figure above shows the distribution of forces when an object is placed on an inclined plane. As seen from the diagram, the force needed to push the object is $mg \sin\theta$. This force differs by a factor of $\sin\theta$ from the force that would be necessary otherwise. Since the maximum of $\sin\theta$ is one, the force required to push an object up the incline is usually reduced. This method is only useful if the frictional force retarding the object's movement up the incline is relatively small.

29. (B)
Any ball that is thrown straight up in the air with an initial velocity $v_o$ will have its speed reduced to 0 linearly by the gravitational force. It will then fall back to the ground and have its speed linearly increased back to the speed with which it was thrown.

30. (D)
Energy density for a plane wave:

$$u = \varepsilon_0 E^2$$

The Poynting vector (energy flow rate) is

$$S = u\frac{C}{n}$$

$$S = uC = C\varepsilon_0 E^2,$$

since n, the coefficient of refraction, is one for outer space.

$$E^2 = \frac{S}{C\varepsilon_0} = \frac{26.55 \times 10^2 \text{ W/m}^2}{(3 \times 10^8 \text{m/sf})(8.85 \times 10^{-8} C^2 N^{-1} m^{-2})}$$

$$E = \frac{S}{C\varepsilon_0} = 10 NC^{-1}$$

$$E = 10 \frac{V}{m}$$

### 31. (D)

To evaluate the angular acceleration of the disk which is expected to be negative, we first have to evaluate the total frictional torque acting on it. Since the disk has a purely rotational movement, the frictional forces acting on it at any point of contact with the surface will be tangent to the circle the point is rotating along. To evaluate the total torque, we have to integrate over the infinitesimal contributions provided by the points of contact. Therefore, we can write:

$$\Gamma = \int_0^R r\, dF_f$$

while:

$$dF_f = \mu dF_n$$

$$dF_n = \frac{(Mg)}{(\pi R^2)} ds$$

$$ds = 2\pi r\, dr \quad \Rightarrow$$

$$\Rightarrow \Gamma = \int_0^R r\mu \frac{(Mg)}{(\pi R^2)} 2\pi r\, dr$$

$$= \frac{2\pi \mu Mg}{\pi R^2} \int_0^R r^2 dr$$

$$= \frac{2\mu Mg}{R^2} \frac{R^3}{3}$$

$$\Gamma = 2\mu Mg \frac{R}{3}$$

$$\Gamma = I\alpha = \tfrac{1}{2}MR^2\alpha = 2\mu Mg \frac{R}{3}$$

$$\alpha = \frac{4\mu g}{3R}$$

## 32. (C)

The differential equation for the modeling of the system is as follows:

$$\underset{\text{restoring force}}{-kx} - \underset{\substack{\text{damping}\\\text{term}}}{b\frac{dx}{dt}} + \underset{\substack{\text{driving}\\\text{force}}}{F_m \cos \omega''t} = m\frac{d^2x}{dt^2}$$

The solution is

$$X = \frac{F_m \sin(\omega''t - \phi)}{\sqrt{m^2(\omega''^2 - \omega^2)^2 + b^2\omega''^2}}$$

in which $\omega$ = natural undamped frequency, $F_m$ is the amplitude of the driving force, $\omega''$ is the frequency of the driving force and b is the damping constant.

As one can see from the above, when $\omega = \omega''$, the oscillation will take place with a finite constant amplitude equal to $F_m/b\omega$ (and therefore undamped). $\omega$ is not the resonant frequency of the system since the value of $\omega''$ for which the amplitude would be maximum is less than $\omega$ (although close to it when b is very small). Therefore resonance will not occur.

## 33. (A)

Write a KCL equation using the lower node as a reference node.

$$\underbrace{(1 - 2 + 4) \times 10^{-3}}_{\text{current sources}} = \underbrace{v(60\mu\Omega + 40\mu\Omega)}_{\text{branch currents}}$$

Solving for v,

$$v = \frac{3 \times 10^{-3}}{100 \times 10^{-6}} = 30\text{V}.$$

In order to find the energy absorbed, find the power

and integrate with respect to time, remembering that when energy is absorbed, the current must flow out of the negative side of the source.

For the 1-mA source, note that the current flows out of the positive side. This means the source absorbs negative power and energy.

$$E = \int_{t_0}^{t} p\, dt = -\int_{0}^{3600} (1\text{ mA})(30\text{v})\, dt = -108 \text{ J}.$$

## 34. (B)
The lithium isotope is unstable because of the unequal pairing of the neutrons and electrons. This causes the nuclear force to be less than the electrostatic repulsive force. The magnitude of the nuclear force is dependent on the number of nucleon pairs. Therefore the lithium isotope with a greater number of protons than nucleon pairs will result in a greater repulsive interaction.

## 35. (D)
The tension in the cord could be found by determining the net force due to gravity and buoyancy.

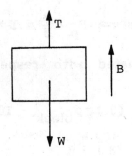

$$\vec{W} = \rho_{block} gV = 0.1g\, \rho_{block}$$

$$\vec{B} = \rho_{liq} gV$$

$$= 100g$$

$$\vec{T} = \vec{W} - \vec{B}$$

$$\vec{T} = 0.1g\rho_{block} - 100g$$

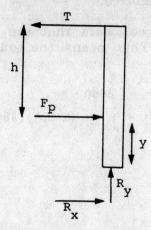

The total force, $F_p$, exerted by water pressure on the beam is equal to the force exerted if the pressure at any point of the beam is equal to the pressure at its middle point. Thus we can write:

$$F_p = \rho g\left(\frac{h}{2}\right) \text{ Area} = 1\times 10^3 \times g \times \frac{8}{2} \times 0.2$$

$$= 800 \text{ N}$$

The torque exerted on the beam by the liquid is equal to the torque exerted by $F_p$ if it is acting at 2/3 of the distance from the surface of liquid. (It can be proved by integration.) Therefore we can write:

$$\Sigma\Gamma = 0 \Rightarrow F_p \times \frac{8}{3} = T \times 10$$

(Note that $\Gamma$ is evaluated with respect to point R.) So we have:

$$800g \times \frac{8}{3} = (0.1g\rho_{block} - 100g) \times 10$$

$$\Rightarrow \rho_{block} = 3133.\bar{3} \text{ } \frac{kg}{m^3}$$

36. (A)

The elements of the column next to the inert gases, the halogens, lack one electron to form a closed shell, $np^6$. Since this is the most stable configuration, these elements will form a compound with other atoms such as H which can provide the extra electron to close the shell.

37. (A)

Suppose that instead of liquid, we had a cube of solid. When this solid is resting, two equal and opposite forces are acting upon it. Because these two external forces are balanced, the solid remains stationary. But nonetheless, the solid will experience stress and strain. (The unit of this stress is lb/in$^2$.) Now suppose this solid was falling freely without air friction. In the latter case, the solid experiences only one force and consequently accelerates. Because there are no opposing forces, stress is not induced; therefore, the solid is stress-free. Similarly a liquid is stressed when it's stationary. This stress is called the fluid pressure. (Note that the unit of stress is also lb/in$^2$.) Furthermore, when this liquid has no opposing force acting upon it, which is the effect of viscous-free free fall, the liquid is stress-free or pressure-free.

38. (C)

The forces acting on the block are its weight, mg, and the tension, T, in the string, pulling upwards. The block is at rest relative to the elevator, so they both have the same acceleration. For an elevator at rest, taking upward acceleration as being positive, we can write the sum of the forces on the block as:

$$\Sigma F = ma_{elev} = T - mg$$

$$0 = T - mg$$

$$T = mg = 9.8m \text{ Newtons}$$

As the elevator accelerates downwards:

$$\Sigma F = -ma_{elev} = T - mg$$

$$T = m(g - a_{elev})$$

and the tension in the string can only decrease from 9.8m Newtons. Therefore the string can break only for upward accelerations of the elevator. In this case:

$$\Sigma F = ma_{elev} = T - mg$$

$$a_{elev} = \frac{T - mg}{m}$$

When the tension in the string reaches 50m Newtons, the string will break. The acceleration of the elevator will then be:

$$a_{elev} = \frac{50m - gm}{m}$$

$$= 50 - g$$

$$= 50 - 9.8$$

$$= 40.2 \text{ m/s}^2$$

### 39. (D)

Because the temperature of a human hand is higher than room temperature, heat is transferred from the hand to the two metals when the man touches them. The rate of heat transfer depends upon the thermal conductivities of the two metals. Heat is transferred faster when the thermal conductivity is higher. Therefore, the faster the transfer of heat from the hand to the metal, the colder the metal feels.

### 40. (B)

Choice B shows the external characteristics of a separately excited d.c. generator. The decrease in terminal voltage is due to:

1. Armature reaction: This reduces the amount of effective field flux which results in a decrease in generated emf.

2. Armature circuit resistance: This causes a voltage drop.

Choice A shows external characteristics of a self-excited d.c. generator, choice C that of a series d.c. generator, choice D, that of an over-compounded d.c. generator and choice E shows external characteristics of a flat compounded d.c. generator.

### 41. (C)

The intensity distribution for double slit interference where the slit width is negligible would be

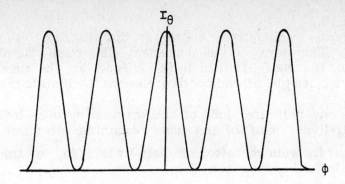

The intensity distribution for a single slit system where the slit width is not negligible would be

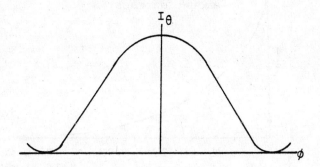

For the double slit system in question, the intensity distribution would be obtained by superimposing the diffraction pattern on the interference pattern. If one of the slits is covered, there will be no interference effects and only the single slit diffraction would remain.

42. (C)
An electron inside a metal is firmly bound to the atom. In order to pluck an electron, energy has to be used. The minimum amount of energy needed to remove an electron from a particular material is said to be the work function (W) of that material.

If we increase the frequency of radiation, the kinetic energy of the photoelectron increases. If the photon emitted by the light has a greater energy than the work function, the electron is ejected from the material with a kinetic energy equal to the difference of the photon energy (hν) and the work function of the material,(i.e. $E_k = hf - W$.)

The plot of the kinetic energy of the electron against the frequency of radiation is a straight line as shown by curve (C). The frequency at which the curve starts on the x- axis is called the threshold frequency, which is the frequency at which the photon has enough energy to overcome the work function of the material.

## 43. (E)

Different radioactive elements disintegrate at different rates. The unit of measurement for such disintegration is called the half-life, which is defined as the time for half of a given sample of radioactive element to disintegrate.

After one half-life, half of the original mass is left. After two half-lives, half of the mass remaining after one half-life is left. In general after T half-lives, $(\frac{1}{2})^T$ of the original mass is left.

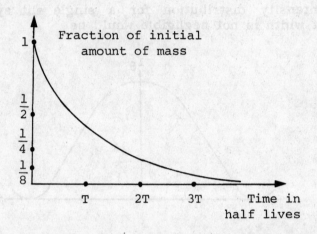

If we plot the curves of fraction of initial amount of mass to the time in half-lives, we get a curve as shown above which is similar to curve (E). This is a negative exponential curve, since the $(\frac{1}{2})^T$ can be written as $(e^{-T\ln 2})$.

## 44. (E)

The capacitive reactance $(X_c)$ is the resistance to the flow of an alternating current as a result of capacitance and is measured in ohms.

It is mathematically expressed as

$$X_c = \frac{1}{2\pi fc}$$

where f is the frequency in cycles/sec and c is the capacitance in farads. It is clear from the above relation that with an increase in frequency, the capacitive reactance decreases. This is represented by curve (E).

**45. (A)**
For the first year, with the 10% interest rate on an $800 deposit, $80 will be accumulated; these $80 will then be withdrawn by the depositor, leaving the original $800 in the bank.

The next year, a new $80 will again be accumulated and, as before, withdrawn by the depositor. This will repeat for the remainder of the 5 year period, at the end of which only the principal of $800 will be left.

**46. (C)**

$$\int_{-1}^{+1} (x^2 - 4)\,dx$$

$$= \int_{-1}^{+1} x^2\,dx - 4\int_{-1}^{+1} dx$$

$$= \left.\frac{x^3}{3}\right]_{-1}^{+1} - 4x\Big]_{-1}^{+1}$$

$$= \tfrac{1}{3}[(1)^3 - (-1)^3] - 4[(1) - (-1)]$$

$$= \tfrac{2}{3} - 8$$

$$= -\tfrac{22}{3}$$

**47. (C)**
We are dealing with separate rolls of a balanced die. The 2 rolls are independent, therefore we invoke the following multiplication rule: The probability of getting any particular combination in two or more independent trials will be the product of their individual probabilities. The probability of getting a 5 on any single toss is $\tfrac{1}{6}$ and by the multiplication rule

$$P(5 \text{ and } 5) = \tfrac{1}{6} \cdot \tfrac{1}{6} = \tfrac{1}{36}.$$

Note also that the problem could have been stated as follows: What is the probability of rolling 2 balanced dice simultaneously and getting a 5 on each?

**48. (B)**
Let X = the number of heads observed in 4 tosses of a fair coin. X is binomially distributed if we assume that each toss is independent. If the coin is fair, $p = \Pr$ (a head is observed on a single toss) $= \frac{1}{2}$.

Thus Pr (at least two heads in 4 tosses)

$$= \Pr(X \geq 2) = \sum_{x=2}^{4} \binom{4}{x} \left(\frac{1}{2}\right)^x \left(\frac{1}{2}\right)^{4-x}$$

$$= \binom{4}{2}\left(\frac{1}{2}\right)^2\left(\frac{1}{2}\right)^2 + \binom{4}{3}\left(\frac{1}{2}\right)^3\left(\frac{1}{2}\right) + \binom{4}{4}\left(\frac{1}{2}\right)^4\left(\frac{1}{2}\right)^0$$

$$= \frac{6}{16} + \frac{4}{16} + \frac{1}{16} = \frac{11}{16}.$$

**49. (B)**

$$\int_0^\pi (1 + \sin x)\,dx = \int_0^\pi dx + \int_0^\pi \sin x\, dx = x - \cos x \Big|_0^\pi$$

$$= \pi - \cos \pi - (0 - \cos 0).$$

$$= \pi + 1 - (0 - 1)$$

$$= \pi + 2.$$

**50. (E)**
If the determinant of a matrix [A] is zero, $[A]^{-1}$ does not exist and the matrix [A] is said to be singular. In our case, each of the matrices (A) through (D) has a determinant of zero; e.g.,

$$\begin{vmatrix} 3 & 1 \\ 6 & 2 \end{vmatrix} = 3(2) - 1(6) = 6 - 6 = 0$$

Therefore, they are singular and none of them have inverses, except for the matrix (E) whose determinant is non-zero; i.e.

$$\begin{vmatrix} 5 & 2 \\ 2 & 1 \end{vmatrix} = 5(1) - 2(2) = 5 - 4 = 1$$

51. (C)
We have the formula

$$\frac{e^{i\theta} + e^{-i\theta}}{2} = \cos\theta$$

Therefore, for $\theta = \cos^{-1}x$

$$\frac{e^{i(\cos^{-1}x)} + e^{-i(\cos^{-1}x)}}{2} = \cos(\cos^{-1}x)$$

$$= x$$

52. (C)
The mode is the observation that appears most often in the sample. Both 5 and 7 appear twice in the data while the other observations appear only once. Both observations, 5 and 7, are modes. This sample is called "bimodal."

53. (C)
Substitute $x^2 = u$, then $\frac{du}{dx} = 2x$.

Now $\frac{d^2}{dx^2}(\ln x^2) = \frac{d}{dx}\left(\frac{d}{dx}\ln x^2\right)$

$$= \frac{d}{dx}\left(\frac{d}{dx}(\ln u)\right)$$

$$= \frac{d}{dx}\left(\frac{1}{u} \cdot \frac{du}{dx}\right)$$

$$= \frac{d}{dx}\left(\frac{1}{x^2} \cdot 2x\right)$$

$$= \frac{d}{dx}\left(\frac{2}{x}\right) = -\frac{2}{x^2}$$

**54. (C)**

If $F(x)$ and $G(x)$ are integrable on $[a,b]$, then,

$$\int_a^b [F(x) + G(x)]dx = \int_a^b F(x)dx + \int_a^b G(x)dx$$

In our case, $a = 1$, $b = 2$, thus the answer is (C).

**55. (C)**

$$\int_{s=0}^{s=F(x)-b} G'(s)ds = G(s) \Big]_0^{F(x)-b}$$

$$= G[F(x) - b] - G(0)$$

**56. (C)**

From the figure,

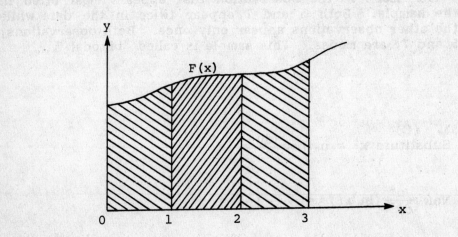

$$\int_0^1 F(x)dx + \int_1^2 F(x)dx = \int_0^2 F(x)dx$$

$$\Rightarrow \int_0^1 F(x)dx + \int_1^2 F(x)dx - \int_0^3 F(x)dx$$

$$= \int_0^2 F(x)dx - \int_0^3 F(x)dx$$

$$= -\left[\int_0^3 F(x)dx - \int_0^2 F(x)dx\right]$$

$$= -\int_2^3 F(x)dx$$

**57. (C)**
Let $x^2 = t$, $x^3 = s$, then $G(x^2) = G(t) = u$ and $F(x^3) = F(s) = v$. Now

$$\frac{d}{dx}(uv) = v\frac{du}{dx} + u\frac{dv}{dx}$$

$$\therefore \frac{d}{dx}[G(t)F(s)] = F(s)G'(t)\frac{dt}{dx} + G(t)F'(s)\frac{ds}{dx}$$

Now $x^2 = t$, therefore

$$2x\,dx = dt \quad \text{or} \quad \frac{dt}{dx} = 2x$$

and $x^3 = s$, therefore

$$3x^2\,dx = ds \quad \text{or} \quad \frac{ds}{dx} = 3x^2$$

Therefore on substitution,

$$\frac{d}{dx}[G(x^2)F(x^3)] = F(x^3)G'(x^2) \cdot 2x + G(x^2)F'(x^3) \cdot 3x^2$$

$$= 2xG'(x^2)F(x^3) + 3x^2F'(x^3)G(x^2)$$

**58. (B)**
Consider a portion AB of the curve.

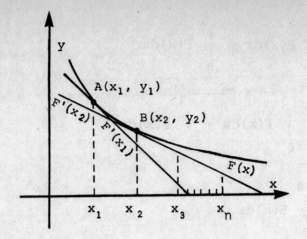

$$\text{Slope of AB} = \frac{y_2 - y_1}{x_2 - x_1}$$

since $y_1 > y_2$, we have $y_2 - y_1 < 0$

Therefore the slope of AB is negative. Hence,

$F'(x_1)$, $F'(x_2)$,...,$F'(x_n)$ are negative.

## 59. (C)

To find

$$D_x y = \frac{dy}{dx},$$

we use the differentiation formula:

$$\frac{d}{dx} \ln u = \frac{1}{u} \frac{du}{dx},$$

with $u = \sinh x$. Applying this formula, we have:

$$D_x y = \frac{1}{\sinh x} \cdot \cosh x = \frac{\cosh x}{\sinh x}$$

$$= \coth x.$$

## 60. (D)

$$\int_0^{\pi/2} e^{\ln(\cos^2 x + \sin^2 x)} dx$$

$$= \int_0^{\pi/2} e^{\ln 1}\, dx$$

$$= \int_0^{\pi/2} e^0\, dx$$

$$= \int_0^{\pi/2} 1\, dx$$

$$= x \Big]_0^{\pi/2} = \frac{\pi}{2} - 0 = \frac{\pi}{2}$$

## 61. (C)

$$1 + ix + \frac{(ix)^2}{2!} + \frac{(ix)^3}{3!} + \cdots$$

$$= e^{ix}$$

$$= \cos x + i \sin x$$

## 62. (D)

Because, for the numerator

$$\lim_{x \to 0} x = 0, \text{ and } \lim_{x \to 0} (1 - e^x) = 0,$$

for the denominator, the function takes the form 0/0, to which L'Hospital's rule may be applied. Therefore,

$$\lim_{x \to 0} \frac{x}{1 - e^x} = \lim_{x \to 0} \frac{1}{-e^x} = \frac{1}{-1} = -1$$

Alternate Method:

Since $e^x = 1 + x + \frac{x^2}{2!} + \frac{x^3}{3!} + \cdots$

we have $\lim_{x \to 0} \frac{x}{1 - e^x}$

$$= \lim_{x \to 0} \frac{x}{1 - \left[1 + x + \frac{x^2}{2!} + \frac{x^3}{3!} + \ldots\right]}$$

$$= \lim_{x \to 0} \frac{x}{1 - 1 - \left[x + \frac{x^2}{2!} + \frac{x^3}{3!} + \ldots\right]}$$

$$= \lim_{x \to 0} \frac{x}{-x\left[1 + \frac{x}{2!} + \frac{x^2}{3!} + \ldots\right]}$$

$$= -\frac{1}{1 + \left[\frac{0}{2!} + \frac{(0)^2}{3!} + \ldots\right]} = -\frac{1}{1+0} = -1$$

### 63. (D)

Observe that if A is an m × n matrix, $A^t$ is an n × m matrix. Hence, the products $AA^t$ and $A^tA$ are always defined.

$$A = \begin{bmatrix} 1 & 2 & 0 \\ 3 & -1 & 4 \end{bmatrix}$$

then,

$$A^t = \begin{bmatrix} 1 & 3 \\ 2 & -1 \\ 0 & 4 \end{bmatrix}$$

$$AA^t = \begin{bmatrix} 1 & 2 & 0 \\ 3 & -1 & 4 \end{bmatrix} \begin{bmatrix} 1 & 3 \\ 2 & -1 \\ 0 & 4 \end{bmatrix}$$

$$= \begin{bmatrix} 1 \cdot 1 + 2 \cdot 2 + 0 \cdot 0 & 1 \cdot 3 + 2 \cdot (-1) + 0 \cdot 4 \\ 3 \cdot 1 + (-1) \cdot 2 + (4) \cdot 0 & 3 \cdot 3 + (-1) \cdot (-1) + 4 \cdot 4 \end{bmatrix}$$

$$= \begin{bmatrix} 1+4+0 & 3-2+0 \\ 3-2+0 & 9+1+16 \end{bmatrix} = \begin{bmatrix} 5 & 1 \\ 1 & 26 \end{bmatrix}$$

### 64. (A)

$$\frac{1}{(1-x)} = (1-x)^{-1}$$

Expanding by binomial theorem

$$(1-x)^{-1} = 1 - \frac{(-1)x}{1!} + \frac{(-1)(-1-1)x^2}{2!} + \ldots$$

$$= 1 + x + \frac{(-1)(-2)}{2}x^2 + \ldots$$

$$= 1 + x + x^2 + \ldots$$

$$= \sum_{n=0}^{\infty} x^n$$

**65. (B)**
The function $g(t)$, which is not shown explicitly, is defined as the integral on the interval from zero to t. That is:

$$g(t) = \int_0^t f(a)\,da$$

Then:

$$g(3) = \int_0^3 f(a)\,da.$$

$f(a)$ is not stated directly, but is shown.

$g(t)$ is thus the area bounded by the curve f. We can use numerical methods to approximate the value of $g(3)$.

By using Simpson's rule:

$$\int_a^b f(x)\,dx \approx \frac{b-a}{3n}[f(x_0) + 4f(x_1) + 2f(x_2) + 4f(x_3) + \ldots + 4f(x_{n-1}) + f(x_n)]$$

Applying this to the figure:

$$b - a = 3,$$

$$n = 3,$$

$$f(x_0) = 3, \quad f(x_1) = 0, \quad f(x_2) = -2, \quad f(x_3) = 0.$$

Hence:

$$\int_0^3 f(a)\,da \approx \frac{3}{3(3)}[3 + 4(0) + 4(-2) + 0]$$

$$= \frac{1}{3}[3 - 8]$$

$$= -\frac{5}{3}$$

This is an approximate value and it lies between 0 and -2.

66. (C)
The function g is defined as previously:

$$g(t) = \int_0^t f(a)\,da.$$

From the figure, it is seen that $f(7) = 0$. Thus $g(f(7)) = g(0)$.

$$g(0) = \int_0^0 f(a)\,da.$$

The length of the interval is equal to 0.

The integral of a function over an interval of length zero is zero.

Thus $g(0) = 0$.

67. (D)
The function $g(t) = \int_0^t f(a)\,da$ is interpreted as the area under or above the function f.

On an interval where f is positive for all t, the area under f is defined as positive.

On an interval where f is negative for all t, the area bounded by f is the "negative" area.

The critical points of the area as a function of t occur when f crosses the t-axis ($f(t) = 0$). On the interval specified above, the area bounded by f is increasing when f is positive and decreasing when f is negative.

A critical point P is a maximum when f(x) is increasing for x < P and decreasing for x > P. This translates into the given function f going from positive to negative.

The only point on the graph satisfying this condition is where t = 7.

**68. (D)**
As stated previously, when f is positive on a given interval, the area bounded by f increases as t increases. As soon as f takes on negative values, the area bounded by f begins to decrease.

A value, x, that is a local minimum has the property that in a certain interval containing x, the value of the function at x is less than the value of the function at all other points in the interval. At a local minimum, the concavity of the function changes from concave downward to concave upward. That is, the function changes from decreasing to increasing.

In the given function g, this happens when f crosses the t-axis from negative to positive. The only choice given that satisfies this criterion is t = 9.

**69. (A)**
The average value of a function defined on an interval [a,b] is:

$$\frac{1}{b-a} \int_a^b f(x)\,dx,$$

where a and b are the limits of the interval.

The interval given is $0 \le t \le 10$. To find the average value of f we must evaluate

$$\frac{1}{10} \int_0^{10} f(a)\,da,$$

which is the same as $\frac{1}{10} g(10)$.

Since the relation for f is not directly stated, we must find the average using numerical methods such as Simpson's rule.

With 10 equal intervals between 0 and 10 we have:

$$\int_0^{10} f(a)\,da \approx \frac{10}{3(10-0)}\left[f(x_0) + 4f(x_1) + 2f(x_2) + 4f(x_3) + 2f(x_4)\right.$$
$$+ 4f(x_5) + 2f(x_6) + 4f(x_7) + 2f(x_8) + 4f(x_9)$$
$$\left.+ f(x_{10})\right]$$

$$= \frac{10}{30}\left[f(0) + 4f(1) + 2f(2) + 4f(3) + 2f(4) + 4f(5) + 2f(6) + 4f(7) + 2f(8) + 4f(9) + f(10)\right]$$

Substituting values from the figure:

$$\int_0^{10} f(a)\,da \approx \frac{1}{3}\left[3 + 0 - 4 + 0 + 2 + 8 + 2 + 0 - 2 + 0 + 2\right]$$

$$= \frac{1}{3}[11].$$

Multiplying by $\frac{1}{10}$ we have:

$$\frac{1}{10}\left(\frac{11}{3}\right) = \frac{11}{30}$$

This is the approximate average value of f, which is between 0 and 1.

70. **(A)**

This problem is best solved by testing each statement. The first statement says that $\frac{df}{dt}$ is negative on the interval. This means that f is a decreasing function on the interval. From the figure, this is verified to be true.

The second statement says that $\frac{dg}{dt}$ is negative. But g is decreasing only when f takes on negative values; on the interval $5 \leq t \leq 8$, f takes on both positive and negative values. Therefore, g is both increasing and decreasing in the interval. From this we know that $\frac{dg}{dt}$ takes on both positive and negative values. Statement II is thereby false.

The third statement gives a general relation for the

function in the interval $5 \leq t \leq 8$. From basic algebra we know that $-t^2$ must be a negative number for all t. Statement III is false because f takes on both positive and negative values in the interval.

Thus, Statement I is the only true statement.

71. (E)
The particles and their paths are illustrated in the following figure:

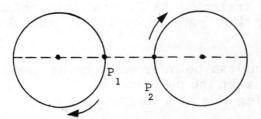

This is the path that the two particles must traverse as shown by the constant fluctuation in the graph of the distance between them.

Initially, at $t = 0$, the particles are at a minimum distance from each other. At the same time the particles are simultaneously intersecting the line joining the centers of the two circles. Thus, when the distance between the two particles is a minimum the line between them coincides with the line between the centers of their respective circles.

The other time when both particles simultaneously intersect the line between the two centers is when the distance between them is a maximum. This occurs in the following diagram.

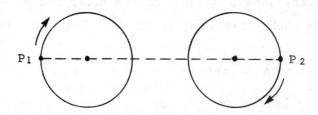

It is easily verified that the lines between the particles and the centers are coincident.

Thus, to find the total number of times over the given time interval that both particles intersect the line joining the centers, we must count the number of maxima and minima in this interval.

Looking at the graph, in the interval $0 \le t \le 10$ there are three minima and three maxima.

Therefore, both particles are coincident on the line six times over the interval.

72. (C)

The path traversed by $P_1$ is a circle. Therefore the area bounded by this path is $\pi r^2$. We must find the radius of the circle.

This radius can be inferred from the graph. The minimum distance between the particles is 4. The maximum distance is 20. Sketching this in a diagram, we have:

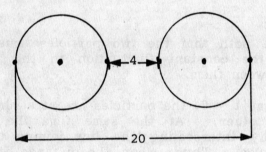

Since the two paths are equal, the diameters of their circles are equal.

We subtract the distance of 4 from 20 and get 16, which is the length of both diameters combined.

Thus, the length of one of the diameters is $\frac{16}{2} = 8$. The radius is half this length, or 4. Now we can determine the area:

$$A = \pi r^2$$
$$= \pi 4^2$$
$$= 16\pi.$$

73. (A)

To answer this question, we must know the position of each of the two particles on their respective paths. This can be inferred from the graph. Each particle moves a quarter-cycle each second. At t = 4, the particles are at a minimum distance from each other. One second later, t = 5, the particles move a quarter of their circular paths as shown in the following diagram.

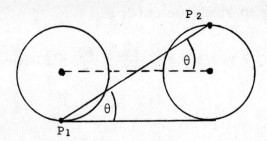

The angle θ is labelled as prescribed. The angle between the line connecting $P_1$ and $P_2$ and the line parallel to the line conecting the centers of the circles is also θ.

Thus we may construct a right triangle:

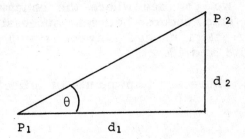

From this diagram, $\tan \theta = \dfrac{d_2}{d_1}$.

We see that $d_2$ is the diameter of the circle, which is eight.

$d_1$ is equal to the minimum distance, 4, plus the two radii of the circles: $d_1 = 4 + 4 + 4 = 12$.

Thus, $\tan \theta = \dfrac{d_2}{d_1} = \dfrac{8}{12} = \dfrac{2}{3}$.

74. (B)
Looking at the graph, one revolution can correspond to both particles being released from a minimum distance of

4 and returning to this position.

From the graph, this action takes 4 seconds. Thus, the particle moves with an angular velocity of

$$\frac{1 \text{ rev}}{4 \text{ sec}} \quad \text{or} \quad \frac{1}{4} \text{ rev/sec.}$$

Converting to rpm's we obtain:

$$\frac{1}{4} \text{ rev/sec} \times \frac{60 \text{ sec}}{1 \text{ min}} = \frac{60}{4} = 15 \text{ rpm.}$$

75. (B)
The point at which the tangent line crosses the curve to which it is tangent is called a point of inflection.

At a point of inflection, the concavity of the curve changes from concave up to concave down or vice-versa. From the given graph, the exact inflection points cannot be determined.

However, we can see where the concavity of the graph changes. This occurs between successive local extrema. We see there is a change between t = 0 and 2, 2 and 4, 6 and 8 and 8 and 10.

Therefore, there are four points of inflection.

76. (C)
f"(t) represents the rate of change of the derivative f'(t) of f. A function whose graph is concave up in some interval has a derivation which is initially negative and increases to a positive value. f' is increasing, hence f" is positive. When the graph is concave down the derivative is initially positive and decreases to a negative value. f' is decreasing, hence f" is negative. At a point where f' is changing from an increasing function to a decreasing one or vice-versa, f" is zero. This is called an inflection point. We are asked for a point where f" is greater than zero. This occurs where the graph is concave up; thus t = 8 is the appropriate point.

77. (B)
When the function g is defined as

$$g(x) = \int_0^x f(t)\,dt$$

then $g(0)$ is

$$\int_0^0 f(t)\,dt.$$

When the limits of integration are the same, that is, when the difference between the upper and lower limit is zero, the value of the integral is zero.

$$\int_x^x f(t)\,dt = 0, \quad x \text{ being any real number.}$$

Thus

$$\int_0^0 f(t)\,dt \quad \text{equals zero.}$$

78. **(E)**
Since $g(x)$ represents the area bounded by $f$, the area is positive only when $f$ is positive and negative only when $f$ is negative.

When a function has positive and negative values, the area that it bounds increases and decreases. It increases when $f$ is positive and decreases when $f$ is negative.

Looking at the given function, we observe that for all $t$ in the interval, $f$ is positive. This means that over all subintervals, $g$ is increasing (the area under $f$ is increasing). Thus $g$ is an increasing function on all subintervals.

79. **(A)**
When a function $f$, which is continuous on an interval $a \leq x \leq b$, is rotated around the x-axis, the volume that is generated between $x = a$ and $x = b$ is given by the relation:

$$V = \int_a^b \pi(f(x))^2 dx.$$

f(t) is not given explicitly. However, the volume of revolution can be obtained approximately by applying Simpson's rule. Noting that the period of the function is 4, the equation we apply is:

$$V = \pi \int_0^4 [f(t)]^2 dt = \frac{\pi}{3}[f^2(x_0) + 4f^2(x_1) + 2f^2(x_2) + 4f^2(x_3) + f^2(x_4)].$$

Substituting values from the curve,

$$V \approx \frac{\pi}{3}[4^2 + 4(14.4)^2 + 2(20)^2 + 4(14.4)^2 + 4^2]$$

$$\approx \frac{\pi}{3}[16 + 829.44 + 800 + 829.44 + 16]$$

$$\approx \frac{\pi}{3}[2488.88]$$

$$\approx 830\pi.$$

80. (E)
The function g is defined as

$$g(x) = \int_0^x f(t) dt$$

By the Fundamental Theorem of Calculus, we can differentiate this function with respect to x and obtain the function f(t). That is,

$$\frac{dg}{dx} = \frac{d}{dx} \int_0^x f(t) dt = f(t).$$

Thus we may look at the graph of f to locate where it crosses the t-axis (f(t) = 0). We find that this occurs at no points in the interval.

81. (B)

From the graph we have enough data to find the relation for the acceleration of particle A.

Particle A is accelerating uniformly. The relation for the acceleration as a function of time is the equation of a straight line.

From the graph, the slope is determined from the points (0,18) and (20,0):

$$\frac{0-18}{20-0} = \frac{-18}{20} = \frac{-9}{10}$$

Now use slope-intercept form with the point (0,18) giving the intercept. The acceleration is:

$$a(t) = 18 - \frac{9}{10}t.$$

We are asked for the velocity at time $t = 3$. The velocity is the integral of the acceleration.

In order to find the velocity at a particular time, it is necessary to find the velocity as a function of time. Thus the following equation is used:

$$v = \int a\,dt$$

With $a = 18 - \frac{9}{10}t$ we have

$$v = \int \left(18 - \frac{9}{10}t\right) dt$$

$$= 18t - \frac{9}{20}t^2 + C_1$$

It was given that at time $t = 0$, $v = 0$. Solving for $C_1$:

$$0 = 18(0) - \frac{9}{20}(0)^2 + C_1 \quad \text{or} \quad 0 = C_1.$$

Therefore, $v(t) = 18t - \frac{9}{20}t^2$. Substituting $t = 3$:

$$v(3) = 18(3) - \frac{9}{20}(3)^2$$

$$= 54 - \frac{81}{20}$$

which is approximately $54 - 4$ or $50$.

**82. (C)**
To find the position of A at a particular time we must first find the position of A as a function of time.

From the graph we know:

$$a(t) = 18 - \frac{9}{10}t$$

Integrating, we obtain velocity:

$$v(t) = \int a \, dt = \int \left(18 - \frac{9}{10}t\right) dt$$

$$v(t) = 18t - \frac{9}{20}t^2 + C_1$$

Integrating again we obtain position:

$$s(t) = \int v \, dt = \int \left(18t - \frac{9}{20}t^2 + C_1\right) dt$$

$$s(t) = 9t^2 - \frac{3}{20}t^3 + C_1 t + C_2$$

We are given $v(0) = 0$, so $C_1 = 0$.

Also, we are given $s(0) = 12$, so $C_2 = 12$. Therefore the position function is

$$s(t) = 9t^2 - \frac{3}{20}t^3 + 12$$

and

$$s(10) = 9(10)^2 - \frac{3}{20}(10)^3 + 12$$

$$= 900 - 150 + 12$$

$$= 762.$$

**83. (D)**
From the previous problem we have the velocity as a function of time:

$$v(t) = 18t - \frac{9}{20}t^2.$$

We wish to find when the maximum velocity occurs. We differentiate v and obtain

$$v'(t) = a = 18 - \frac{9}{10}t.$$

Setting this equal to zero and solving for t:

$$18 - \frac{9}{10}t = 0,$$

$$t = 20.$$

To check that this is indeed a maximum we find v":

$$v''(t) = \frac{-9}{10}.$$

Since v" < 0, a maximum does occur at t = 20.

84. (A)
The total displacement of particle A over the interval $\leq t \leq 30$ is simply the difference between the final position and the initial position. That is, the total displacement is equal to s(30) - s(0).

The relation for position as a function of time was found earlier to be:

$$s(t) = 9t^2 - \frac{3}{20}(t^3)^3 + 12$$

$$s(30) = 9(30)^2 - \frac{3}{20}(30)^3 + 12$$

$$= 8100 - 4050 + 12 = 4062,$$

$$s(0) = 12$$

Thus, s(30) - s(0) = 4062 - 12 = 4050.

85. (E)
The displacement of a particle depends only on the initial and final positions of the particle. The distance traveled, however, depends on the path of the particle.

The procedure for determining distance traveled is to find the displacement of the particle in each direction that it travels and sum up these displacements. In order to determine a change in direction, we look for the times when the velocity is equal to zero.

The equation for v(t) was found earlier to be:

$$v(t) = 18t - \frac{9}{20}t^2$$

This equation is set to zero:

$$18t - \frac{9}{20}t^2 = 0$$

The equation is readily simplified to:

$$t^2 - 40t = 0$$

or $\quad t(t - 40) = 0$

The velocity is zero at $t = 0$ and $t = 40$. $t = 0$ corresponds to the starting motion of A. $t = 40$ is outside the interval specified in the question. Thus there are no reversals in direction during the interval and the distance traveled is equivalent to the displacement. The displacement was found previously to be

$$s(30) - s(0) = 4062 - 12 = 4050.$$

86. (C)

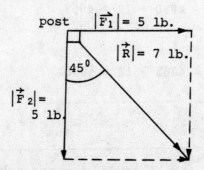

If we use the fact that the component vectors are at right angles to each other, we can write

$$R^2 = 5^2 + 5^2$$

whence R = 7 pounds approximately at 45° to each wire.

87. (A)
Whenever the number of flux lines of force threading

through a coil is changed, an emf is induced in that coil. The amount of emf induced depends on the rate at which the number of flux lines through the coil is changing. This is Faraday's law of electromagnetic induction.

A coil of wire can induce an emf in itself. This follows directly from the fact that when a current flows in a coil, the current causes a flux through the same coil. If there is a change in the current, the flux also changes and an emf is induced in the coil.

In the case of the electromagnet, when the ohmmeter is connected, current in the electromagnet is accompanied by a strong magnetic field. Each loop in the coil interacts with the magnetic field produced by other loops in the same coil due to self induction. When the ohmmeter lead is disconnected, the current rapidly falls to zero and the magnetic field in the coil undergoes a sudden decrease. By Faraday's law, due to a changing magnetic field, a voltage is induced in the electromagnet. The rapidly collapsing magnetic field with its store of energy induces an emf large enough to develop an arc across the leads of the ohmmeter. For this reason, electromagnets are connected to a circuit that provides a bypass for the induced voltage.

88. (E)
First, find the

$$\text{molar mass of aspirin} = ((12 \times 9) + (1 \times 8) + (16 \times 4)) \text{g/mol}$$

$$= (108 + 8 + 64) \text{g/mol}$$

$$= 180 \text{ g/mol}$$

To convert 1g of aspirin to moles of aspirin:

$$1\text{g } C_9H_8O_4 \times \frac{1 \text{ mol } C_9H_8O_4}{180\text{g } C_9H_8O_4} = \frac{1}{180} \text{ mol } C_9H_8O_4$$

Next, find the molarity of the solution:

$$M = \frac{\text{moles of solute}}{\text{solution volume in liters}} = \frac{1/180 \text{ moles solute}}{0.2 \text{ liters of solution}}$$

$$M = \frac{1}{180 \times 0.2} = \frac{1}{36} = 0.028$$

Thus there are 0.028 moles of $C_9H_8O_4$ in 1L of solution.

$$\therefore \quad 0.028 \text{ moles of } C_9H_8O_4 \times \frac{6 \times 10^{23} \text{ molecules } C_9H_8O_4}{1 \text{ mole } C_9H_8O_4}$$

$$= 0.168 \times 10^{23} \text{ molecules } C_9H_8O_4/L$$

Answer: $1.7 \times 10^{22}$ molecules $C_9H_8O_4/L$.

### 89. (D)
Soft iron has a high permeability. The permeability of a material is the ratio of the number of flux lines passing through it to the number of flux lines that would be present in the same space. The space is a vacuum and the material is kept in a magnetic field.

### 90. (A)
By continuity considerations, the flow rate at 1 must equal the flow rate at 2. Thus

$$v_1 A_1 + v_i A_i = v_2 A_2$$

$$v_2 = \frac{v_1 A_1 + v_i A_i}{A_2}$$

where $A_1$ is the difference between the area of the pipe ($A_2$) and the jet ($A_i$)

$A_2 = .8 m^2 \quad v_1 = 2 \text{ m/s}$
$A_i = 0.3 m^2 \quad v_i = 10 \text{ m/s}$
$A_1 = .5 m^2$

$$v_2 = \frac{(2)(.5) + (10)(.3)}{(.8)}$$

$$v_2 = (1 + 3)/.8 = 5$$

$$v_2 = 5.0 \text{ m/s}$$

### 91. (A)
The speed of efflux, v, can be derived from Bernoulli's equation as follows:

$$p + \rho g h + \tfrac{1}{2} \rho v^2 = \text{constant}$$

for unchanged pressure,

$$\rho gh = \tfrac{1}{2}\rho v^2 \quad \text{and} \quad 2gh = v^2$$

or $v = \sqrt{2gh}$ which is Torricelli's law, where g is the gravitational constant and h = d which is the distance from the surface of the fluid to the outlet.

From Torricelli's law, the speed of efflux and the square root of the distance are proportionally related, $v \propto k\sqrt{d}$. So the correct answer is A.

## 92. (B)

Note that the 2Ω resistor has no effect in the circuit, because there is a short circuit above it. Hence, the true circuit looks like the following.

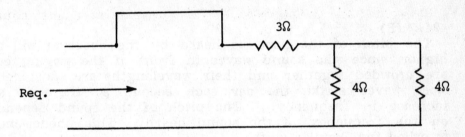

Two 4Ω resistors are in parallel. This resistance is in series with the 3Ω resistor.

$$\therefore \quad R_{eq} = \frac{4 \times 4}{4 + 4} + 3 = 5 \,\Omega.$$

## 93. (E)

The speed of a DC motor is given by the following equation:

$$N = \frac{E_a}{k\phi}$$

where
  $N$ = speed of the motor
  $E_a$ = counter emf
  $\phi$ = resultant magnetic flux per pole
  $k$ = a constant involving number of poles, parallel paths through armature, choice of units, etc.

Hence, the speed is directly proportional to the counter emf and inversely proportional to the resultant magnetic flux.

It is useful to express the speed with respect to the terminal voltage as

$$N = \frac{E_t - I_a R_a}{k\phi}$$

where
- $E_t$ = terminal voltage
- $I_a$ = armature current
- $R_a$ = armature resistance
- $\phi$ = flux per pole

and usually $I_a R_a$ is very small compared with $E_t$.

## 94. (E)

The pitch of the horn as heard by the observer will be higher since the sound waves in front of the moving car are crowded together and their wavelengths are shortened. More waves strike the ear each second so there is an increase in frequency. The pitch of the sound depends on the frequency of the sound heard. This phenomenon is called the Doppler shift.

## 95. (B)

This problem is solved using the law of conservation of energy. The energy lost by the car appears in the form of heat in the brake mechanism.

The change in kinetic energy is equal to the heat energy given off.

$$\tfrac{1}{2}mv_f^2 - \tfrac{1}{2}mv_i^2 = Q$$

$$Q = \tfrac{1}{2}(2000)\left[(2)^2 - (7)^2\right] = -45,000 \text{ Joules}$$

$$Q_{(cal)} = -45,000 J \left(\frac{1 \text{ cal}}{4.1858 J}\right)$$

## 96. (E)

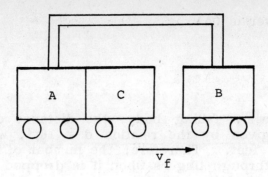

A perfectly inelastic collision means that the two colliding bodies stick together and move with the same velocity after the collision, as shown in the figure. From the Principle of Conservation of Linear Momentum, we may write

Total Momentum Before Collision = Total Momentum After Collision.

Thus,

$$4(v_i) + 0 = (4 + 1)(v_f). \qquad (1)$$

Solving for the final velocity, $v_f$:

$$v_f = \frac{4}{5} v_i = \frac{4}{5} (5 \text{ m/sec}) = 4 \text{ m/sec}.$$

**97. (A)**
The balanced chemical reaction for the dissociation of water into hydrogen and oxygen is

$$2H_2O \rightleftarrows 2H_2 + O_2$$

so the equilibrium-constant expression is

$$K_{eq} = \frac{[H_2]^2[O_2]}{[H_2O]^2}$$

where the concentration of the products appear in the numerator and the concentration of the reactants appears in the denominator. For the reverse reaction, the reactants are now products and vice-versa, so that the expression is

$$K'_{eq} = \frac{[H_2O]^2}{[H_2]^2[O_2]} = \frac{1}{K_{eq}}$$

which is the inverse of $K_{eq}$.

So the answer is (A).

**98. (B)**
For a freely falling body, the distance d, travelled in t sec. is given by the relation d = 16t², where d is the distance in feet. Therefore, the number of inches of ruler that pass through fingers when it is dropped is proportional to your reaction time.

From $\quad d = 16t^2,\quad$ we get

$$4 \text{ inches} = \frac{1}{3}\text{ ft.} = 16t^2$$

or $\quad t = \sqrt{\frac{1}{3 \times 16}} \simeq \frac{1}{7} \text{ sec.}$

**99. (B)**
According to the law of Dulong and Petit, at constant volume the average molar heat capacities for all metals except the lightest are approximately the same and equal to 3R. However, at low temperatures, the specific heat begins to deviate considerably from 3R and tends to 0 as the temperature goes to 0.0°K. The only curve that satisfies these criteria is (B).

**100. (D)**
Since $\vec{a} = \frac{d\vec{v}}{dt}$

$$a = \frac{d}{dt}(t^3\vec{i} + 3t^2\vec{i}) = 3t^2\vec{i} + 6t\vec{i}$$

The x - component of $\vec{a}$ is $3t^2$.

**101. (B)**
The velocity,

$$\vec{v} = t^3\vec{i} + 3t^2\vec{j} = \frac{d\vec{r}}{dt}$$

Therefore the displacement,

or
$$\vec{v} = \int \vec{V}\, dt$$
$$\vec{v} = \int (t^3 \vec{i} + 3t^2 \vec{j})\, dt$$
$$\vec{v} = \frac{t^4}{4}\vec{i} + \frac{3t^3}{3}\vec{j} + c$$

Therefore at $t = 0$, $r = 0$.

$$\therefore\ \vec{v} = \frac{t^4}{4}\vec{i} + t^3 \vec{j} = \vec{v}_x + \vec{v}_y \tag{A}$$

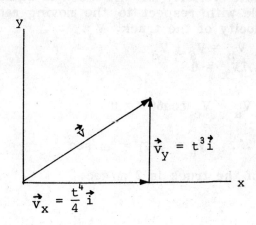

Now from the figure and from equation (A), the distance of the particle from the y-axis is $= t^3$.

## 102. (A)

Since the drag coefficient for both spheres is the same, the Reynolds number must be the same:

$$Re_1 = Re_2$$

$$\frac{\rho V_1 \left(\frac{1}{10} D\right)}{\mu} = \frac{\rho V_2 (D)}{\mu}$$

$$\frac{V_1}{V_2} = 10$$

## 103. (A)

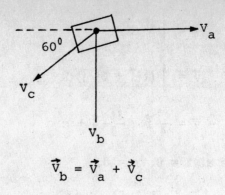

$$\vec{V}_b = \vec{V}_a + \vec{V}_c$$

The absolute velocity of the block is equal to the velocity of the block with respect to the moving ramp, $V_c$, summed with the velocity of the truck, $V_a$.

$$V_b = V_a + V_c$$
$$\Rightarrow \Sigma V_x = 0$$

$$V_a - V_c \cos 60° = 0$$

$$\therefore \quad V_a = 4(\tfrac{1}{2}) = 2 \text{ m/sec}$$

The speed of the truck is 2 m/sec.

## 104. (B)

The Van der Waals bonds in molecular solids become stronger with the increasing size of the atoms involved. Van der Waals attraction between atoms such as helium arises from an instantaneous electric dipole moment in one atom which induces a dipole moment in the other atom. With more electrons in the argon atom, larger instantaneous dipoles and induced dipoles are possible. Therefore, the attraction will be stronger. The larger mass and surface area of the argon atom also contribute to the strength of the bond.

## 105. (E)

For an ideal gas,

$$PV = nRT$$

so,

$$\frac{PV}{T} = nR = \text{constant}$$

So for two states,

$$\frac{P_1 V_1}{T_1} = \frac{P_2 V_2}{T_2}$$

which can be manipulated to arrive at the mathematical expressions (A) through (D). Therefore (E) is the correct answer.

## 106. (D)

For an independent project, its acceptance or rejection is independent of that of any other project. The rate of return needs to be computed for every project. This rate is dependent on the annual operating costs. Those projects whose rates of return are less than the minimum annual rate of return are rejected. The incremental rate of return is determined only in a situation where the projects are mutually exclusive, and at most one of them may be accepted.

## 107. (C)

The x-component of the force on the projectile = 0, thus the acceleration in the x-direction, $a_x = 0$.

For the y-direction

$$a_y = \frac{F_y}{m} = \frac{-mg}{m} = -g$$

so that

$$v_x = u\cos\alpha_0$$

and $\quad v_y = u\sin\alpha_0 - gt$

We now use the equations of motion with constant velocity and constant acceleration to find the position of the projectile at any time t. Thus

$$x = u_x t = u\cos\alpha_0 t \tag{1}$$

$$y = u^y t - \tfrac{1}{2}gt^2 = (u\sin\alpha_0)t - \tfrac{1}{2}gt^2 \tag{2}$$

When the projectile hits the target, $y = 0$.

Therefore, from equation (2)

$$(u\sin\alpha_0)t_2 - \tfrac{1}{2}gt_2^2 = 0$$

so that, $t_2 = \dfrac{2u}{g}\sin\alpha_0$ = time taken for projectile to hit the target.

The range,

$$R = v_x t_2 = (u\cos\alpha_0)\left[\dfrac{2u}{g}\sin\alpha_0\right] = \dfrac{u^2}{g}\sin 2\alpha_0 \qquad (3)$$

when $\dfrac{u^2}{g}\sin 2\alpha_0$ is maximum, $R = R_{max}$. i.e. .This occurs when $2\alpha_0 = 90°$. Therefore,

$$R = R_{max} = \dfrac{u^2}{g}$$

and equation (3) becomes

$$R = R_{max}\sin 2\alpha_0$$

from which $\alpha_0 = \tfrac{1}{2}\sin^{-1}(R/R_{max})$.

108. (C)
$n_i$ = Intrinsic carrier concentration = $10^{10}$ atoms/cm$^3$

$N_d$ = Impurity concentration of donors = $10^{14}$ atoms/cm$^3$.

Since $N_d \gg n_i$

$$n = N_d = 10^{14} \text{ atoms/cm}^3$$

$$P = \dfrac{n_i^2}{N_d} = \dfrac{(10^{10})^2}{N_d} = 10^6 \text{ atoms/cm}^3.$$

109. (B)
Because the heat is being transferred to the environment through convection, Newton's Law of Cooling is applicable:

$$q = hA\,\Delta T\;\left[\dfrac{\text{Joule}}{\text{sec}}\right]$$

where        h = Heat transfer coefficient
$\Delta T$ = Temperature difference
A = Surface area

Observe from the above equation that q would increase if h, A or $\Delta T$ increased; by doubling the area, q would indeed double.

### 110. (D)

$q = C \Delta T$

$= 0.386 \frac{kJ}{kg °K} (80°K)$

$q = 30.9 \frac{kJ}{kg}$

### 111. (C)

The open-tube manometer measures gauge pressure. The density of liquid, $\rho$, is considered constant because liquids are nearly incompressible. Taking p and g as constants, and with the knowledge that pressure is the same at all points of the same depth, we can write: p2 - p1 = p g(y2 - y1). This leads us to conclude (C) is correct and (D) and (E) are incorrect. Since only the height matters, (B) is wrong. Since we need to know P2 in order to know P1, (A) is also wrong.

### 112. (B)

Specific gravity for solids is the ratio of the density of the solid to the density of water (approximately 1 gram per cubic centimeter). The specific gravity of iron is then

$$S = \frac{P_i}{P_\omega}$$

But

$$P_i = \frac{200 \text{ gm}}{2 \times 2 \times 10 \text{ cm}^3} = 5 \text{ gm/cm}^3$$

Since $P_\omega = 1 \text{ gm/cm}^3$

$$S = \frac{5 \text{ gm/cm}^3}{1 \text{ gm/cm}^3} = 5$$

113. (B)

Since the pressure is equal to the change in momentum multiplied by the number of collisions per unit time, increasing the number of molecules will increase the number of collisions per unit time; therefore, the pressure will increase. According to Charles' Law: $\frac{P_1}{T_1} = \frac{P_2}{T_2}$ = constant.

For constant volume, when the pressure goes up so does the temperature.

114. (A)

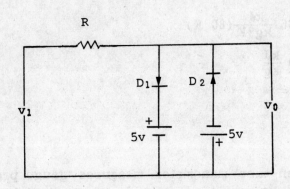

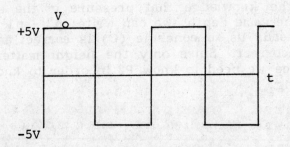

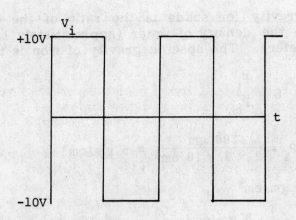

In the figure shown, when $v_i$ is less than 5 V, both diodes are reverse biased, and $v_0 = v_i$. As soon as $v_i$ is slightly more positive than 5 V, diode $D_1$ is forward biased (diode $D_2$ remains reverse biased). The output is clamped (maintained) at +5 V and remains clamped at that value until $v_i$ becomes less than 5 V.

When $v_i$ is between +5 V and -5 V, both diodes are reverse biased again, and $v_0 = v_i$. As soon as $v_i$ becomes less than -5 V, $D_2$ conducts ($D_1$ is reverse biased), and the output is now clamped to -5 V. It remains clamped at this value until $v_i$ is greater than -5 V.

### 115. (A)
Since neither face B nor C are heat sources, and since positive heat always goes from higher temperature to lower temperature, face A must have the highest temperature.

### 116. (C)
The photoelectric effect illustrates the particle property of light by showing that light acts as if it were "quantized"; a quantum of light is called a photon.

The photoelectric effect was explained by Einstein. He stated that incident light exists in concentrated bundles called photons, the energy of a photon being given by $E = h\nu$, where $\nu$ is the frequency and $h$ is Planck's constant

### 117. (A)
Hardenability is the ease with which martensite can be produced in the steel. In any critical cooling rate, austenite will be transformed into pearlite before it is transformed into martensite. Alloying elements form solid solutions with austenite and thereby stabilize austenite. By decreasing the rate of transformation of austenite to pearlite, more martensite can be formed at slower cooling rates.

## 118. (A)

A hydrometer works by displacing the fluid in which it is submerged. The instrument sinks in the fluid until the weight of the fluid it displaces is equal to its own weight.

## 119. (D)

According to Newton's Law of Action and Reaction, the total force acting on the fluid surrounding the hydrometer is equal to the total force acting on the hydrometer by the fluid.

Furthermore, equilibrium dictates that this force is equal and opposite to the total weight of the hydrometer. Since the hydrometer's weight is equal to the weight of the displaced fluid, the total force acting on the fluid surrounding the hydrometer is equal to the weight of the displaced fluid. Note that hydrometers are carefully calibrated so that the marking which the hydrometer sinks to denotes the specific gravity of the fluid. In this problem, we were given the specific gravity as h.

Since specific gravity is defined as

$$\frac{\text{specific weight of fluid}}{\text{specific weight of water}}$$

we can thus find the specific weight of the fluid as

Specific weight of the fluid = (specific weight of water) × (specific gravity)

or

Specific weight of the fluid = (specific weight of water) × (h)

Since the specific weight of water was given as $\gamma$:

Specific weight of the fluid = $\gamma h$

Since weight = $mg = \rho V g = \rho g V$ = (specific weight) × (Volume)

we can find the weight of the fluid as:

W = (specific weight) × (displaced volume)

W = $h \gamma V$

## 120. (E)
An adiabatic process is defined as a process in which no heat transfer takes place between the system and its surroundings.

## 121. (E)
Vacancies are primary causes of dislocations.

## 122. (B)
The definition of thermal equilibrium is that state of a system wherein the rate of emission is equal to the rate of absorption of radiant energy. This definition makes no claims about the system's particular ability to emit or absorb radiant energy (i.e. its emissivity or absorptivity). Though the rate of emission and rate of absorption may be zero (hence equal) this does not have to be the case. Thus (B) is the correct response.

## 123. (D)
The ratio of the heat Q supplied to a body, to its corresponding temperature rise $\Delta T$ is called the heat capacity C of the body, $\frac{Q}{\Delta T}$.

The heat capacity per unit mass of a body, called specific heat, is $c = \frac{Q}{m \Delta T} \quad \frac{cal}{g \, ^\circ C}$

Therefore, $\quad Q = cm \Delta T$

$$= (4000)(1.00)(70^\circ C)$$

$$= 280,000 \text{ cal}$$

$$= 280 \text{ Kcal.}$$

## 124. (B)
Given the Chezy constant, the Chezy relation for open-channel flow is used:

$$Q = C\sqrt{RS} \; A$$

115

where $R = \dfrac{A}{P} = \dfrac{\text{cross-sectional area}}{\text{wetted perimeter}}$, $S = $ slope

This formula is rearranged:

$$Q^2 = C^2 RSA^2$$

$$Q^2 = C^2 \dfrac{AS}{P}(A^2)$$

$$\dfrac{PQ^2}{C^2 S} = A^3$$

Substituting knowns:

$$\dfrac{(800)(10)^2}{(100)^2(.001)} = A^3 = 8000$$

Taking the cube root: $A = 20$ ft$^2$.


## 125. (E)

The capitalized cost refers to the present-worth of the project that will last forever:

Capitalized cost = Equivalent uniform annual cost (EUAC)/interest rate

Thus:

$$PW_{dam} = 8{,}000{,}000 + \dfrac{25{,}000}{0.05} = 8{,}500{,}000$$

$$EUAC_{wells} = \text{EUAC of investment} + \text{annual operating costs}$$

$$= 50{,}000(10)(.1) + 5{,}000(10)$$

$$= 100{,}000$$

$$PW_{wells} = \dfrac{100{,}000}{0.05} = \$2{,}000{,}000$$

Therefore, \$6,500,000 will be saved if this alternative is chosen.


## 126. (B)

In order for a polymer to become more rigid, it must become more and more viscous through entanglements.

These entanglements interfere with plastic deformation. When a polymer chain branches, it is able to resist plastic deformation. An atom such as sulfur can serve as an anchor point between adjacent molecules of rubber. Cross-linking will prevent adjacent molecules from sliding by their neighbors and thus maintain rigidity.

127. (B)
The correct answer is (B) because for simple harmonic motion, frequency (and period) are related only to the mass and the spring constant, not displacement.

128. (B)
An eutectic system is made up of two components and must satisfy the condition that at one point, the eutectic point, the liquid state must be completely insoluble in the solid state. In order to accomodate partial solubility in the solid state, a solid solution should exist above the eutectic point along with the liquid solution for each component. This solid solubility should increase with decreasing temperature above the eutectic point and decrease with decreasing temperature below the eutectic point.

129. (C)
In studying the electronic structure of a many-electron atom, Wolfgang Pauli came up with an exclusion principle which can be stated as follows: No two electrons in the same atom can have the same quantum state. So, with the notation used, one orbital can hold an electron of positive spin, ⊕

and two electrons occupying one orbital, one with positive and one with negative spin, (↑↓)

This is violated in (C).

130. (D)
In solid form, ionic compounds will not conduct electricity. The reason for this is that ionic forces are sufficiently strong to hold the compound in a crystal lattice. This strong force also makes the crystal mechanically strong.

## 131. (B)
Differentiating we have,

$$\cos v \frac{\partial u}{\partial x} - u \sin v \frac{\partial v}{\partial x} - 1 = 0, \text{ and}$$

$$\sin v \frac{\partial u}{\partial x} + u \cos v \frac{\partial v}{\partial x} - 0 = 0.$$

These two linear equations in $\frac{\partial u}{\partial x}$ and $\frac{\partial v}{\partial x}$ may be solved by Cramer's rule,

$$\frac{\begin{vmatrix} 1 & -u \sin v \\ 0 & u \cos v \end{vmatrix}}{\begin{vmatrix} \cos v & -u \sin v \\ \sin v & u \cos v \end{vmatrix}} = \frac{u \cos v}{u} = \cos v = \frac{\partial u}{\partial x}$$

as long as u is not zero.

## 132. (A)
The sketch is

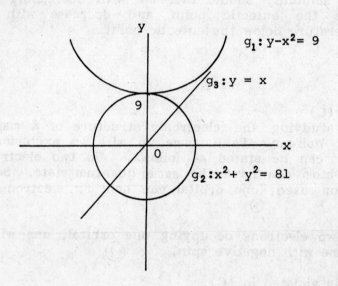

$g_2$ and $g_3$ intersect both y and x axes.

## 133. (C)
The characteristic polynomial of any matrix $A_{n \times n}$ is the polynomial in $\lambda$ whose roots are the eigenvalues. This polynomial is obtained by evaluating $\det |A - \lambda I_n|$.

## 134. (A)

When there are powers of trigonometric functions, we have to split them so

$$\sin^5 x(\cos^2 x) = (\sin x)\cos^3 x(\sin^4 x)$$
$$= (\sin x)\cos^3 x(1 - \cos^2 x)^2$$
$$= \sin x[\cos^3 x - 2\cos^5 x + \cos^7 x]$$

With the substitution $u = \cos x$, we get $du = -\sin x\, dx$.

$$\int_0^{\frac{\pi}{2}} \sin^5 x \cos^3 x\, dx = \int_0^{\frac{\pi}{2}} (-u^7 + 2u^5 - u^3)\, du$$

$$= \left. \frac{-u^8}{8} + \frac{u^6}{3} - \frac{u^4}{4} \right|_0^{\frac{\pi}{2}}$$

Then replace $u$ with $\cos x$ and evaluate to get $\frac{1}{24}$

Alternate Method:

Let $\sin x = u$, $\quad x = 0, \quad u = 0$

$(\cos x)dx = du \quad x = \frac{\pi}{2}, \quad u = 1$

$$\therefore \int_0^{\frac{\pi}{2}} \sin^5 x(\cos^3 x)\, dx = \int_0^{\frac{\pi}{2}} (\sin^5 x)\cos^2 x(\cos x)\, dx$$

$$= \int_0^{\frac{\pi}{2}} \sin^5 x(1 - \sin^2 x)\cos x\, dx$$

$$= \int_0^1 u^5(1 - u^2)\, du$$

$$= \int_0^1 (u^5 - u^7)\, du = \left. \frac{u^6}{6} - \frac{u^8}{8} \right|_0^1$$

$$= \frac{1}{6} - \frac{1}{8} = \frac{1}{24}$$

**135. (D)**
Recall Green's theorem:

For $F:R^2 \to R^2$ (where $R^2$ denotes n-dimensional cartesian space) such that $F(x,y) = P(x,y) + Q(x,y)$, with P and Q having continuous first partial derivatives in some simple region R with boundary C, then

$$\oint_C Pdx + Qdy = \iint_R \left( \frac{\partial Q}{\partial x} - \frac{\partial P}{\partial y} \right) dA$$

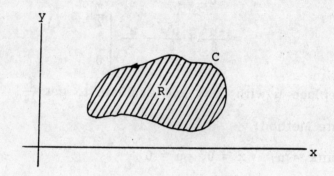

C is traversed in the counterclockwise direction.

**136. (B)**
Given an equation which determines a conic section, if it can be written as:

$$\frac{(x - h)^2}{a^2} + \frac{(y - k)^2}{b^2} = r^2$$

then it may be an ellipse, or a circle if $a = b = 1$.

The eccentricity e is $e = \frac{\sqrt{a^2 - b^2}}{a}$. When $0 < e < 1$, then $a = b$, and we have an ellipse.

**137. (C)**
Below is a sketch of a family of curves $f(x) = ax^2$

for values of $-2 \leq a \leq 2$; they are all parabolas with vertex $(0,0)$.

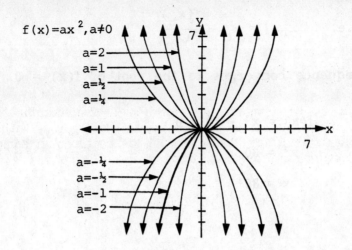

**138. (E)**
The polynomial

$$\frac{x^2 - 1}{(x + 1)(x^2 + 4x - 5)}$$

can be written as

$$\frac{(x + 1)(x - 1)}{(x + 1)(x + 5)(x - 1)} = \frac{1}{(x + 5)}$$

which has only one singularity at $x = -5$.

The singularities at $x = -1$ and $x = 1$ are removable singularities.

**139. (A)**
The series $\sum_{n=0}^{\infty} \frac{x^n}{n!}$ if expanded is

$$1 + x + \frac{x^2}{2!} + \frac{x^3}{3!} + \frac{x^4}{4!} + \dots$$

which we immediately recognize as the series expansion of $e^x$.

## 140. (C)

We recognize the sequence as that of the Newton-Raphson procedure

i.e.  $x_{n+1} = x_n - \dfrac{f(x_n)}{f'(x_n)}$

This sequence converges to the root of $f(x) = 0$. In our case,

$$f(x) = x^3 - 30 = 0$$

$\therefore \quad x^3 = 30$

$$x = 30^{\frac{1}{3}}$$

## 141. (D)

We first solve the differential equation (1). Then we apply the boundary conditions to the solution $y = f(x, \lambda)$ and proceed to find the real values of $\lambda$.

Assume the solution of (1) is in the form $y = e^{mx}$. Substituting this and its derivatives into (1) we find

$$(m^2 - 4\lambda m + 4\lambda^2)e^{mx} = 0$$

or, cancelling $e^{mx}$, since it is never zero,

$$m^2 - 4\lambda m + 4\lambda^2 = 0.$$

This yields $m = 2\lambda$. Since the general solution of a second order differential equation must contain two linearly independent solutions, the general solution of (1) is

$$y = c_1 e^{2\lambda x} + c_2 x e^{2\lambda x}. \tag{2}$$

Applying the boundary conditions, (1') to (2) we find

$$(1 + 2\lambda)c_1 + c_2 = 0 \tag{3}$$

$$(1 - 2\lambda)c_1 + (-2\lambda)c_2 = 0. \tag{3'}$$

Equations (3), (3') have a non-trivial solution, i.e., $c_1 \neq 0$, $c_2 \neq 0$ only if,

$$\begin{vmatrix} 1 + 2\lambda & 1 \\ 1 - 2\lambda & -2\lambda \end{vmatrix}$$

$$= 4\lambda^2 + 1 = 0. \tag{4}$$

From (4), $4\lambda^2 = -1$, $\lambda^2 = -\frac{1}{4}$, $\lambda = \pm i(\frac{1}{2})$.

These eigenvalues are complex. This problem has no real eigenvalues. The only real solution is $y \equiv 0$.

**142. (A)**
Recall that for $F(x,y) = 0$ an implicit function, $\frac{dy}{dx} = \frac{-Fx}{Fy}$

so with $F(x,y) = e^x \cos 3y + \ln y (\sin x^2) - 3y = 0$,

$$\frac{dy}{dx} = -\frac{F_x}{F_y}$$

$$= \frac{-\{e^x \cos 3y + (2x)\ln y (\cos x^2)\}}{\{-3e^x \sin 3y + \frac{\sin x^2}{y} - 3\}}$$

$$= \frac{e^x \cos 3y + 2x(\ln y)(\cos x^2)}{3(1 + e^x \sin 3y) - \frac{\sin x^2}{y}}$$

**143. (A)**

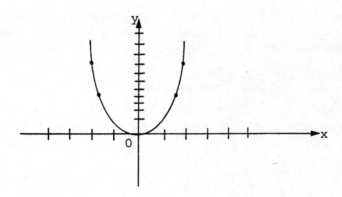

We know that the slope m is equal to $y' = \frac{dy}{dx}$. Thus $y' = 2x$. Integrating, we get

$$y = x^2 + C.$$

The curve passes through the origin ($x = 0, y = 0$). Therefore $C = 0$. Hence $y = x^2$.

144. (A)

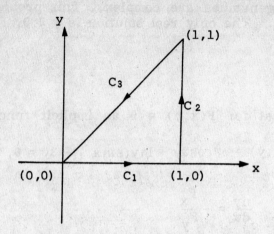

To compute $\int_{C_1} y^2\,dx + x^2\,dy$, compute y and dy.

Along $C_1$, $y = 0$ and $dy = 0$; therefore

$$\int_{C_1} y^2\,dx + x^2\,dy = \int_{C_1} 0 \cdot dx + x^2 \cdot 0 = 0.$$

# The Graduate Record Examination in
# ENGINEERING

Test 2

# GRE ENGINEERING TEST 2
## ANSWER SHEET

61. Ⓐ Ⓑ Ⓒ Ⓓ Ⓔ
62. Ⓐ Ⓑ Ⓒ Ⓓ Ⓔ
63. Ⓐ Ⓑ Ⓒ Ⓓ Ⓔ
64. Ⓐ Ⓑ Ⓒ Ⓓ Ⓔ
65. Ⓐ Ⓑ Ⓒ Ⓓ Ⓔ
66. Ⓐ Ⓑ Ⓒ Ⓓ Ⓔ
67. Ⓐ Ⓑ Ⓒ Ⓓ Ⓔ
68. Ⓐ Ⓑ Ⓒ Ⓓ Ⓔ
69. Ⓐ Ⓑ Ⓒ Ⓓ Ⓔ
70. Ⓐ Ⓑ Ⓒ Ⓓ Ⓔ
71. Ⓐ Ⓑ Ⓒ Ⓓ Ⓔ

72. Ⓐ Ⓑ Ⓒ Ⓓ Ⓔ
73. Ⓐ Ⓑ Ⓒ Ⓓ Ⓔ
74. Ⓐ Ⓑ Ⓒ Ⓓ Ⓔ
75. Ⓐ Ⓑ Ⓒ Ⓓ Ⓔ
76. Ⓐ Ⓑ Ⓒ Ⓓ Ⓔ
77. Ⓐ Ⓑ Ⓒ Ⓓ Ⓔ
78. Ⓐ Ⓑ Ⓒ Ⓓ Ⓔ
79. Ⓐ Ⓑ Ⓒ Ⓓ Ⓔ
80. Ⓐ Ⓑ Ⓒ Ⓓ Ⓔ
81. Ⓐ Ⓑ Ⓒ Ⓓ Ⓔ
82. Ⓐ Ⓑ Ⓒ Ⓓ Ⓔ

83. Ⓐ Ⓑ Ⓒ Ⓓ Ⓔ
84. Ⓐ Ⓑ Ⓒ Ⓓ Ⓔ
85. Ⓐ Ⓑ Ⓒ Ⓓ Ⓔ
86. Ⓐ Ⓑ Ⓒ Ⓓ Ⓔ
87. Ⓐ Ⓑ Ⓒ Ⓓ Ⓔ
88. Ⓐ Ⓑ Ⓒ Ⓓ Ⓔ
89. Ⓐ Ⓑ Ⓒ Ⓓ Ⓔ
90. Ⓐ Ⓑ Ⓒ Ⓓ Ⓔ
91. Ⓐ Ⓑ Ⓒ Ⓓ Ⓔ
92. Ⓐ Ⓑ Ⓒ Ⓓ Ⓔ
93. Ⓐ Ⓑ Ⓒ Ⓓ Ⓔ

94. Ⓐ Ⓑ Ⓒ Ⓓ Ⓔ
95. Ⓐ Ⓑ Ⓒ Ⓓ Ⓔ
96. Ⓐ Ⓑ Ⓒ Ⓓ Ⓔ
97. Ⓐ Ⓑ Ⓒ Ⓓ Ⓔ
98. Ⓐ Ⓑ Ⓒ Ⓓ Ⓔ
99. Ⓐ Ⓑ Ⓒ Ⓓ Ⓔ
100. Ⓐ Ⓑ Ⓒ Ⓓ Ⓔ
101. Ⓐ Ⓑ Ⓒ Ⓓ Ⓔ
102. Ⓐ Ⓑ Ⓒ Ⓓ Ⓔ
103. Ⓐ Ⓑ Ⓒ Ⓓ Ⓔ
104. Ⓐ Ⓑ Ⓒ Ⓓ Ⓔ

105. Ⓐ Ⓑ Ⓒ Ⓓ Ⓔ
106. Ⓐ Ⓑ Ⓒ Ⓓ Ⓔ
107. Ⓐ Ⓑ Ⓒ Ⓓ Ⓔ
108. Ⓐ Ⓑ Ⓒ Ⓓ Ⓔ
109. Ⓐ Ⓑ Ⓒ Ⓓ Ⓔ
110. Ⓐ Ⓑ Ⓒ Ⓓ Ⓔ
111. Ⓐ Ⓑ Ⓒ Ⓓ Ⓔ
112. Ⓐ Ⓑ Ⓒ Ⓓ Ⓔ
113. Ⓐ Ⓑ Ⓒ Ⓓ Ⓔ
114. Ⓐ Ⓑ Ⓒ Ⓓ Ⓔ

115. Ⓐ Ⓑ Ⓒ Ⓓ Ⓔ
116. Ⓐ Ⓑ Ⓒ Ⓓ Ⓔ
117. Ⓐ Ⓑ Ⓒ Ⓓ Ⓔ
118. Ⓐ Ⓑ Ⓒ Ⓓ Ⓔ
119. Ⓐ Ⓑ Ⓒ Ⓓ Ⓔ
120. Ⓐ Ⓑ Ⓒ Ⓓ Ⓔ
121. Ⓐ Ⓑ Ⓒ Ⓓ Ⓔ
122. Ⓐ Ⓑ Ⓒ Ⓓ Ⓔ
123. Ⓐ Ⓑ Ⓒ Ⓓ Ⓔ
124. Ⓐ Ⓑ Ⓒ Ⓓ Ⓔ

125. Ⓐ Ⓑ Ⓒ Ⓓ Ⓔ
126. Ⓐ Ⓑ Ⓒ Ⓓ Ⓔ
127. Ⓐ Ⓑ Ⓒ Ⓓ Ⓔ
128. Ⓐ Ⓑ Ⓒ Ⓓ Ⓔ
129. Ⓐ Ⓑ Ⓒ Ⓓ Ⓔ
130. Ⓐ Ⓑ Ⓒ Ⓓ Ⓔ
131. Ⓐ Ⓑ Ⓒ Ⓓ Ⓔ
132. Ⓐ Ⓑ Ⓒ Ⓓ Ⓔ
133. Ⓐ Ⓑ Ⓒ Ⓓ Ⓔ
134. Ⓐ Ⓑ Ⓒ Ⓓ Ⓔ

135. Ⓐ Ⓑ Ⓒ Ⓓ Ⓔ
136. Ⓐ Ⓑ Ⓒ Ⓓ Ⓔ
137. Ⓐ Ⓑ Ⓒ Ⓓ Ⓔ
138. Ⓐ Ⓑ Ⓒ Ⓓ Ⓔ
139. Ⓐ Ⓑ Ⓒ Ⓓ Ⓔ
140. Ⓐ Ⓑ Ⓒ Ⓓ Ⓔ
141. Ⓐ Ⓑ Ⓒ Ⓓ Ⓔ
142. Ⓐ Ⓑ Ⓒ Ⓓ Ⓔ
143. Ⓐ Ⓑ Ⓒ Ⓓ Ⓔ
144. Ⓐ Ⓑ Ⓒ Ⓓ Ⓔ

# THE GRADUATE RECORD EXAMINATION ENGINEERING TEST

## MODEL TEST II

Time: 170 Minutes
      144 Questions

Directions: Choose the best answer for each question and mark the letter of your selection on the corresponding answer sheet.

1. The two isotherms represent possible states of an ideal gas at temperatures T and T + dT. For which of the given processes is the change in internal energy the same?

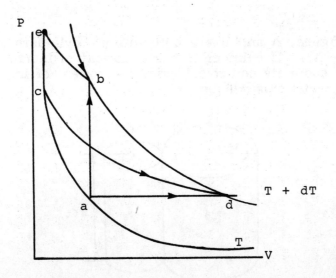

(A) ab, ad and cd     (D) ab and eb

(B) eb and cd         (E) All of the above

(C) ad, cd and eb

2. The resistance of a metallic conductor can be reduced by

   (A) increasing the length of the conductor.

   (B) increasing the temperature.

   (C) cooling the conductor.

   (D) coating the conductor with an insulator.

   (E) connecting a resistor in series with the conductor.

3. Two spheres of equal weight are dropped simultaneously from the same height but in separate baths. The sphere dropped in the bath of water reaches the bottom before the sphere dropped in the bath of oil because

   (A) the terminal velocity of the sphere in the bath of oil is less than that of the sphere in the bath of water.

   (B) the specific gravity of oil is greater than 1.

   (C) water has a greater specific heat.

   (D) water has a greater viscosity.

   (E) both (A) and (B)

4. In the figure, A and B are both uniform cylinders with density 100 slug/ft$^3$. If cylinder B has a moment of inertia of 800$\pi$ slug-ft$^2$ about its centroidal axis, its moment of inertia about A's centroidal axis will be

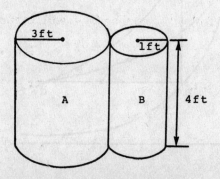

(A)  800 π slug-ft².   (D)  3200 π slug-ft².

(B)  1600 π slug-ft².  (E)  7200 π slug-ft².

(C)  2400 π slug-ft².

5. A man walks towards the edge of a cliff until he can just see the top of a pole past the edge of the cliff. The pole is 10 feet high and the man's eyes are 6 feet above the top of the cliff. The man is standing 3 feet from the edge of the cliff, while the pole is 15 ft. from the edge of the cliff. If the geometry is as shown above, what is the height of the cliff above the ground?

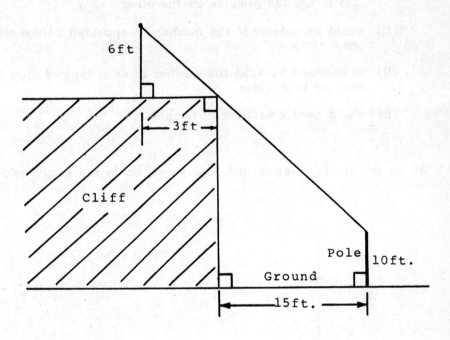

(A)  10 ft.    (D)  18 ft.

(B)  25 ft.    (E)  30 ft.

(C)  40 ft.

6. Which of the following processes will strengthen glass?

(A)  annealing    (D)  sintering

(B)  prestressing  (E)  tempering

(C)  firing

7. The equation describing the reaction of calcium carbonate (limestone) and hydrochloric acid is

$$CaCO_3 + 2HCl \rightarrow CaCl_2 + CO_2 + H_2O.$$

This equation

(A) is not balanced because the number of reactants is less than the number of products.

(B) is incorrect because the amount of chlorine, Cl, on one side is not the same as on the other.

(C) would be correct if the number '2' appeared before all the components.

(D) is balanced because the number of each type of atom is the same on both sides.

(E) would need a catalyst to happen.


8. A current $i_{AB}$ varies with time as shown in the graph below.

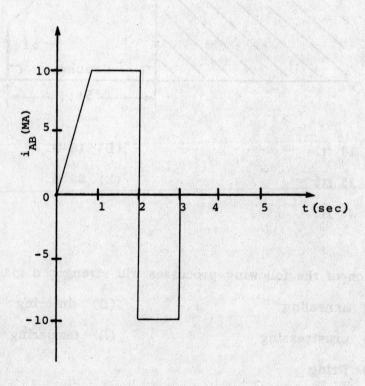

How much charge in coulombs passes through the wire in the interval t = 0 to 4 sec?

(A) 0.050 C

(D) 0.005 C

(B) 0.025 C

(E) −0.005 C

(C) 0.015 C

9. Two trains having average speeds of 25 km/hr and 15 km/hr respectively, are moving toward each other on straight parallel tracks. A bird that can fly with an average speed of 80 km/hr, flies off the first train when the trains are separated by a distance of 40 km. Upon touching the second train, the bird immediately flies back and continues this process until trains pass each other. The total distance covered by the bird is

(A) 25 km.

(D) 40 km.

(B) 15 km.

(E) none of the above.

(C) 80 km.

10. The destruction (by open circuit) of which resistor in the circuit given below would cause the greatest increase in the power supplied by the current source?

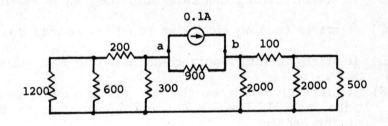

(A) 500 Ω

(D) 200 Ω

(B) 100 Ω

(E) 1200 Ω

(C) 900 Ω

11. How many independent loops are there in the network shown?

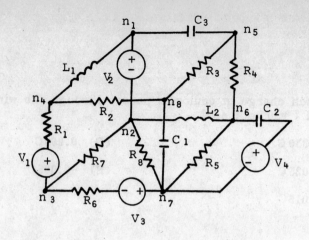

(A) 7        (D) 10

(B) 14       (E) 9

(C) 8

12. A man is initially standing still on the rim of a small circular platform that is at rest. He starts moving counterclockwise along the rim of the platform and then steps radially toward the center along a spiral path around the center of the platform with a constant linear velocity. Neglecting the time it takes the man to step closer to the center and any dissipative frictional effects, what happens to the platform?

(A) It starts turning counterclockwise and then speeds up.

(B) It starts turning clockwise and then slows down.

(C) It starts turning clockwise and then speeds up.

(D) It starts turning counterclockwise and then slows down.

(E) It stays at rest when the man starts moving counterclockwise and only turns counterclockwise after he steps closer to the center.

13. A liquid column mercury thermometer and a constant volume gas thermometer both record temperatures for a tank of water. The temperature recorded by the constant volume gas thermometer is greater than that recorded by the liquid column mercury thermometer. If both thermometers are then brought in contact

(A) both thermometers will show the same temperature readings.

- (B) the temperature readings of both thermometers will not change.

- (C) the temperature recorded by the constant volume gas thermometer will decrease but still remains greater than the liquid column thermometer.

- (D) the temperature recorded by the liquid column thermometer will increase to a reading greater than that of the constant volume gas thermometer.

- (E) none of the above.

14. Brownian movement refers to

    - (A) the slightly upward curving of the edges of the surface of water in glass.

    - (B) the slightly downward curving of the edges of the surface of mercury in water.

    - (C) the irregular motion of small particles suspended in a fluid.

    - (D) the occurrence of convection currents in a gas or liquid.

    - (E) the stretching of a body beyond its elastic limit.

15. In order to maintain the constancy of temperature inside a double-walled vessel the amount of heat transfer between the container and its surroundings must be minimized. The best design specifications for a double-walled vessel is to

    - (A) create a vacuum between two walls of mirrored glass.

    - (B) fill the region between two walls of mirrored glass with water.

    - (C) create a vacuum between two walls covered with lampblack.

    - (D) fill the region between two lampblack covered walls with water.

    - (E) create a vacuum between two walls of cast iron steel.

16. Above the critical temperature

    - (A) no clear distinction is evident between a gas and a liquid.

(B) a solid, liquid and vapor are in equilibrium.

(C) a solid sublimes into a gas.

(D) a liquid vaporizes into a gas.

(E) a liquid and vapor are in equilibrium.

17. Light is incident at 90° on a cubic container of water filled with undissolved sugar. The proportion of sugar in the water gradually increases from top to bottom of the container. The variation in the density of the sugar-water solution will produce a linear increase in the index of refraction from top to bottom. Which of the following represents the path of light in the sugar-water solution?

(A)

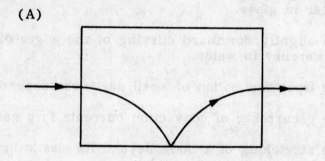

(B)

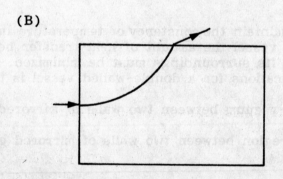

(C)

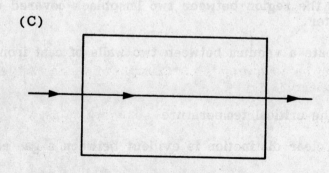

(D)

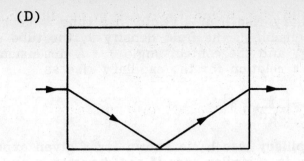

(E)

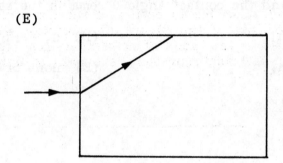

18. When 30 g of sodium hydroxide is neutralized by 36 g of hydrochloric acid, how many moles of water will it yield? (Molecular weights: HCl = 18, NaOH = 20)

   (A) 3.0 mol         (D) 4.0 mol

   (B) 1.5 mol         (E) .5 mol

   (C) 2.0 mol

19. Which of the following is most directly related to the chemical behavior of an atom?

   (A) Atomic weight           (D) Number of electrons in outermost shell
   (B) Natural atomic weight
                               (E) Mass number
   (C) Molecular weight

20. Which of the following is an example of a thermodynamic intensive property?

   (A) Temperature     (D) Energy

   (B) Mass            (E) Entropy

   (C) Volume

21. The capillary rise h of a liquid in a tube is dependent on the surface tension γ, the fluid density ρ, the tube diameter d, gravity g, and the contact angle θ. A dimensionless analysis gives us a relation for the capillary rise as

$$\frac{h}{d} = F\left(\frac{\gamma}{\rho g d^2}, \theta\right)$$

If the capillary rise $h_1$ is known for a given experiment, what will $h_2$ be in a similar case if the diameter d and surface tension γ are reduced by half of the original value, the density ρ is doubled and the contact angle θ remains the same?

(A) $h_2 = \tfrac{1}{2}h_1$

(B) $h_2 = 2h_1$

(C) $h_2 = h_1$

(D) $h_2 = \tfrac{1}{4}h_1$

(E) none of the above

22. An isolated container is filled with gas. It has two chambers of equal volume which are separated by a small door. Initially, the door was open, but then a small "demon" was placed in the container. The demon is smart and nimble to close and open the door very rapidly so that only faster molecules are allowed to enter one chamber and only slow molecules the other. Which statement is true about such a setup?

(A) It is in violation of the second law of thermodynamics because the volume remains the same.

(B) It is in violation of the second law of thermodynamics because the system performs work on the container.

(C) The second law of thermodynamics is satisfied.

(D) It is in violation of the second law of thermodynamics because it results in a temperature difference between the chambers.

(E) none of the above

23. A car turns on a road banked at 45° of radius ρ = 200 ft. Knowing that the coefficient of static friction between the tires and pavement μ = .6 and that the car weights 64.4 lbs., for what range of speed will a lateral friction force be exerted up the incline on its wheels? Note: g = 32 ft/s².

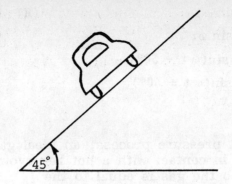

(A)  v > 80

(B)  v < 100

(C)  v < 80

(D)  100 > v > 80

(E)  v > 100

24. Water and vapor are in equilibrium in an insulated container. If a small quantity of vapor is released from the container

   (A) the temperature of the liquid goes up.

   (B) the temperature of the liquid goes down.

   (C) the vapor in the container gets superheated.

   (D) enthalpy of the system goes up.

   (E) internal energy of the system rises.

25. For the given pairs of voltages and currents, determine the pair which best describes an inductive circuit given below:

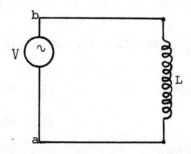

(A)  i = 7 sin($\omega$t - 10°)
     v = 14 sin($\omega$t - 100°)

(B)  i = 4 sin($\omega$t + 170°)
     v = 36 sin($\omega$t + 80°)

(C)  i = 15 sin ω t
     v = 10 sin ω t

(D)  i = 555 sin(ω t - 50°)
     v = 12 sin(ω t + 40°)

(E)  i = 40 sin(ω t + 50°)
     v = 200 sin(ω t - 40°)

26. In a constant pressure process, an ideal gas is enclosed in a cylinder and in contact with a hot reservoir. The net energy transferred to the gas is equal to the

   (A) work done on the gas.

   (B) change in enthalpy.

   (C) $nc_p \Delta t$.

   (D) heat transferred in a constant volume process that produces an equivalent temperature change.

   (E) change in entropy.

27. When alcohol is applied to bare skin, there is a cold sensation which lasts as long as alcohol is present on the skin. This is caused by

   (A) alcohol affecting the nerves under the skin.

   (B) the fact that alcohol is always cooler than the atmosphere.

   (C) the absence of hydrogen bonds in alcohol.

   (D) the transfer of heat from the skin to the alcohol.

   (E) none of the above.

28. All of the following properties are inherent to a given material except

   (A) flexural strength.          (D) Poisson's ratio.

   (B) density.                    (E) yield point.

   (C) Young's modulus.

29. The graph below could represent the variation in

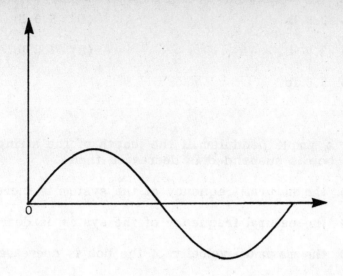

(A) the tangential component of acceleration with respect to the angle subtended by a body whirling in a vertical circle.

(B) the potential energy with respect to the displacement from a spring.

(C) the centripetal force acting on a vehicle with respect to the angle of banking for a circular track.

(D) the restoring force acting on a body suspended by a spring with respect to the displacement from the equilibrium position.

(E) both (A) and (C)

30. A hypothetical spaceship of the future sails through space by reflecting solar radiation from a mirrored surface of area $A_1$. If the intensity of the light is $I_1$ then the average force that is exerted by the incoming light is   (Note: c = Velocity of light)

(A) $\frac{I_1 A_1}{c}$

(B) $\frac{I_1 c}{A_1}$

(C) $\frac{2I_1 c}{A_1}$

(D) $I_1 A_1 c$

(E) $\frac{2I_1 A_1}{c}$

31. If the coefficient of sliding friction for steel on ice is 0.05, what force is required to keep a man weighing 150 pounds moving at constant speed along the ice?

(A) 2.5 lb.

(B) 3.0 lb.

(C) 5.0 lb.

(D) 6.0 lb.

(E) 7.5 lb.

32. For a simple pendulum if the length of the string from which the bob is suspended is decreased then

   (A) the natural frequency of the system is increased.

   (B) the natural frequency of the system is decreased.

   (C) the maximum velocity of the bob is decreased.

   (D) the maximum acceleration of the bob is increased.

   (E) both (A) and (D)

33. The value of the resistance R in the circuit given below should be

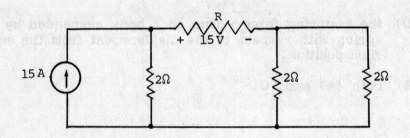

   (A) 15 Ω

   (B) 10 Ω

   (C) 6 Ω

   (D) 3 Ω

   (E) 1 Ω

34. In the beta decay of a neutron the kinetic energy of the electron emitted is significantly less than the maximum kinetic energy available from the decay process. This difference could be accounted for by

   (A) a greater recoil energy for the proton emitted.

   (B) the gamma ray emission of an excited nucleus.

(C) the emission of a positron.

(D) the emission of an antineutrino.

(E) both (A) and (B)

35. A 4 ft. wide weightless gate is hinged at point B and rests against a smooth wall at point A. If the center of pressure of the force acts 0.5 ft. below the middle of the gate, the reaction force at A is

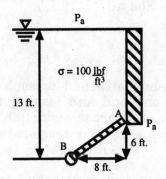

(A) $6 \times 10^4$ lbf.

(B) $2.25 \times 10^4$ lbf.

(C) $3 \times 10^4$ lbf.

(D) $5 \times 10^4$ lbf.

(E) $5 \times 10^3$ lbf.

36. Nitrogen's outermost subshell is a half-filled 2p, occupied by only electrons with spins in the same direction. This phenomenon is best described by

(A) Pauli Exclusion Principle.

(B) Hund's Rule.

(C) Heisenberg's Uncertainty Principle.

(D) The Periodic Law.

(E) None of the above

37. A flat automobile tire weighs 5 lb. How much would the tire weigh after it is inflated to 25 psia?

(A) 0 lb.

(B) 5 lb.

(C) 25 lb.

(D) 30 lb.

(E) insufficient information

38. Two iron blocks of different weights are placed on a frictionless inclined plane and are allowed to slide. They will

   (A) have different accelerations but the same forces acting on them.

   (B) have the same acceleration but different forces acting on them.

   (C) have equal potential energy at any point on the plane.

   (D) have equal kinetic energy at any point on the plane.

   (E) none of the above

39. Two rods, with ends attached to each other, are conducting heat entering at one end and leaving the other at a constant rate. If both have the same cross-sectional areas and lengths, which one will experience a greater temperature difference across its two ends?

   (A) The one whose end is attached to the source at higher temperature

   (B) The one which has higher thermal condutivity

   (C) The one which has lower thermal conductivity

   (D) The one which has lower heat capacity

   (E) The one which has higher heat capacity

40. A graph of a voltage source $V_{ab} = 10tu(t)$ is given below.

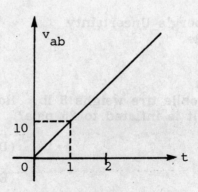

If a 2-Henry inductance with i(o) = 3 is connected to $V_{ab}$, then which of the following graphs represents the current in the inductance?

(A)
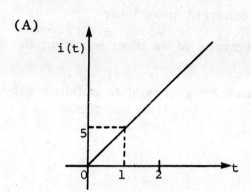

(B)
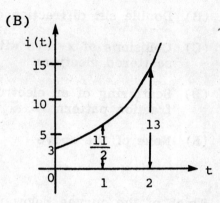

(C)
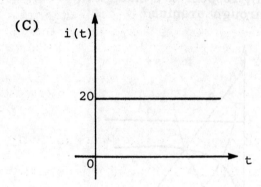

(D)
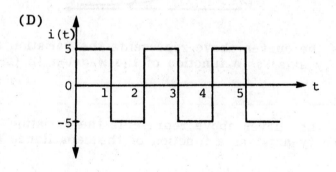

(E)
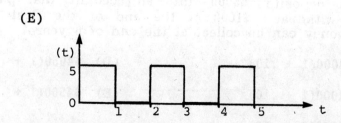

41. Which of the following experiments would provide additional confirmation of the quantum nature of light?

   (A) Diffraction of x-rays by a crystal of NaCl

   (B) Double slit diffraction of a coherent neon laser

   (C) Collisions of x-rays with electrons of an atom resulting in scattered electrons

   (D) Scattering of an electron beam by a crystal to obtain a diffraction pattern

   (E) None of the above

42. Which of the curves below represents the variation of nuclear mass (y-axis) with respect to a change in atomic number (x-axis) from hydrogen through uranium?

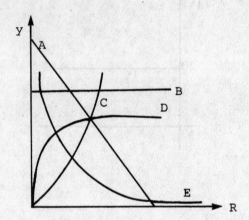

43. Which of the curves above represents the variation of photon momentum (y-axis) as a function of the wavelength (x-axis)?

44. Which of the curves above represents the variation of capacitive reactance (y-axis) as a function of the capacitance (x-axis)?

45. A man deposits $5000 into an account that pays 10% annually. If he withdraws $1000 at the end of the first year, how much money can he collect at the end of 5 years?

   (A) $4000(1 + .10)^4$

   (B) $4000(1 + .10)^5$

   (C) $4500(1 + .10)^4$

   (D) $4050(1 + .10)^4$

   (E) $4500(1 + .10)^5$

Directions: Choose the best answer for each question and mark the letter of your selection on the corresponding answer sheet.

Note: The natural logarithm of x will be denoted by ln x.

46. The integral

$$\int_1^2 x(x + 2x^2)\,dx$$

has the value:

(A) $\frac{65}{6}$

(B) $\frac{21}{2}$

(C) 8

(D) $\frac{22}{3}$

(E) $\frac{59}{6}$

47. We have three light bulbs, each of which has a probability of $e^{-kt}$ of having failed after t hours. The probability that none fails after $t_0$ hours is

(A) $1 - e^{-kt_0}$

(B) $e^{-kt_0}$

(C) $e^{-3kt_0}$

(D) $e^{-2kt_0}$

(E) $e^{3kt_0}$

48. A bag contains 4 white balls, 6 black balls, 3 red balls, and 8 green balls. If one ball is drawn from the bag, find the probability that it will be either white or green.

(A) $\frac{6}{19}$

(B) $\frac{3}{8}$

(C) $\frac{4}{7}$

(D) $\frac{1}{2}$

(E) $\frac{1}{7}$

49. $\displaystyle\int_0^{\pi/2} \tan^2 \frac{x}{2}\,dx =$

(A) $2 + \frac{\pi}{2}$

(B) $2 - \frac{\pi}{2}$

(C) $3 - \frac{\pi^2}{4}$

(D) $\frac{\pi}{6}$

(E) $3 + \frac{\pi}{4}$

50. Matrix D is an orthogonal matrix.

    $D = \begin{bmatrix} A & B \\ C & 0 \end{bmatrix}$

    The value of $|B|$ is

    (A) 0

    (B) $\sqrt{1/2}$

    (C) 1

    (D) -A

    (E) 1/2

51. For $0 < x < \frac{\pi}{2}$

    $$\frac{\cos \frac{x}{2} \cos \frac{x}{2} + \sin \frac{x}{2} \sin \frac{x}{2}}{\cos \frac{x}{2} \cos \frac{x}{2} - \sin \frac{x}{2} \sin \frac{x}{2}} =$$

    (A) cosx

    (B) sinx

    (C) secx

    (D) cscx

    (E) cotx

52. A survey asking for the number of times toast is burned during one week was distributed to eight randomly selected households. The survey yielded the following results:

    2, 3, 0, 3, 4, 1, 3, 0.

    The range, variance and standard deviation for this data set in their respective order is,

    (A) 2, 4 and -2

    (B) 3, 2 and -1

    (C) 4, 1 and 2

    (D) 3, 2 and $\sqrt{3}$

    (E) 4, 2 and $\sqrt{2}$

53. If $y = \ln(\tan x)$, $\frac{dy}{dx} =$

(A) $\csc x \sec x$

(B) $2\sin x \cos x$

(C) $\csc^2 x$

(D) $\csc x \cos x$

(E) $\frac{\tan x}{\sec^2 x}$


Questions 54-58 are based on the following assumption:

Let $F(x)$ and $G(x)$ be functions differentiable to any order over the set of all real numbers.

54. $\int_2^3 F(x)dx - \int_3^2 G(x)dx =$

(A) $\int_2^3 [F(x) + G(x)]dx$

(B) $\int_2^3 [F(x) - G(x)]dx$

(C) $2\int_3^2 [F(x) + G(x)]dx$

(D) $-\int_2^3 [F(x) + G(x)]dx$

(E) $2\int_1^3 [F(x) - G(x)]dx$

55. $\int_a^{2a} F'(x - a)dx =$

(A) F(a) - F(o)  
(B) F(2a) - F(a)  
(C) 2[F(a) - F(o)]  
(D) F'(a)F'(o)  
(E) None of the above  

56. $\int_1^2 F(x)dx - \int_1^3 F(x)dx =$

(A) $\int_2^3 F(x)dx$

(B) $\int_1^3 F(x)dx$

(C) $-\int_1^3 F(x)dx$

(D) $-\int_1^2 F(x)dx$

(E) $-\int_2^3 F(x)dx$

57. $\dfrac{d}{dx}\left[\dfrac{F(x)}{G(x^2)}\right] =$

(A) $\dfrac{2xF'(x)}{G'(x^2)}$

(B) $G(x)F'(x) - 2xF(x)G'(x^2)$

(C) $\dfrac{G(x^2)F'(x) + 2xF(x)G'(x^2)}{G(x^2)}$

(D) $\dfrac{G(x)F'(x) - F(x)G'(x^2)(2x)}{[G'(x^2)]^2}$

(E) $\dfrac{G(x^2)F'(x) - F(x)G'(x^2)(2x)}{[G(x^2)]^2}$

58. If $F(x) = 3x^3 + 4x + 7$ the relative maxima and minima of $F(x)$ are

(A) $\dfrac{+2}{3}$ and $\dfrac{-2}{3}$ respectively

(B) $\frac{+7}{4}$ and $\frac{-4}{3}$ respectively

(C) $\frac{4}{3}$ and $\frac{-4}{3}$ respectively

(D) 0

(E) Do not exist

59. If $y = \frac{1}{\ln x}$, what is $\frac{dy}{dx}$?

(A) $\frac{1}{x^2 \ln x}$

(B) $-\frac{1}{x(\ln x)^2}$

(C) $\frac{1}{x(\ln x)}$

(D) $x \ln x - x$

(E) None of the above

60. $\int_0^\pi \frac{(\csc^2 x)e^{ax}}{1 + \cot^2 x} \, dx =$

(A) $e^{-\pi a} \sin ax$

(B) $(e^{\pi a} - 1)\cot x$

(C) $a(e^\pi - 1)$

(D) $\frac{1}{a}(1 - e^{-\pi a})$

(E) $\frac{1}{a}(e^{\pi a} - 1)$

61. If n is any positive integer, then

$(n - 1)(n - 2)(n - 3)\ldots 3 \cdot 2 \cdot 1 =$

(A) $n!$

(B) $(n - 1)!$

(C) $(n - 1)^n$

(D) 6

(E) $e^{(n - 1)}$.

62. Find $\lim_{x \to 0} \frac{x \cos x - \sin x}{x}$.

(A) -1

(B) 1

(C) 0

(D) $\pi$

(E) $\infty$

63. If $A = \begin{pmatrix} 1 & 1 & 0 \\ 1 & 0 & 1 \end{pmatrix}$ and $B = \begin{pmatrix} 1 \\ 0 \\ 1 \end{pmatrix}$, the product of A and B is,

(A) $\begin{pmatrix} 1 & 0 \\ 0 & 1 \end{pmatrix}$

(B) $\begin{pmatrix} 1 \\ 0 \end{pmatrix}$

(C) $\begin{pmatrix} 1 \\ 2 \end{pmatrix}$

(D) $\begin{pmatrix} 0 \\ 0 \end{pmatrix}$

(E) $\begin{pmatrix} 1 \\ 1 \\ 1 \end{pmatrix}$

64. $\sum_{n=0}^{\infty} \frac{2^n}{n!} =$

(A) 2

(B) $\frac{1}{2}$

(C) 1

(D) $\infty$

(E) $e^2$

Directions: Choose the best answer for each question and mark the letter of your selection on the corresponding answer sheet. Assume that every function in this section has derivatives of all order at each point of its domain unless otherwise indicated.

Curve g(x) to be used in 65-70:

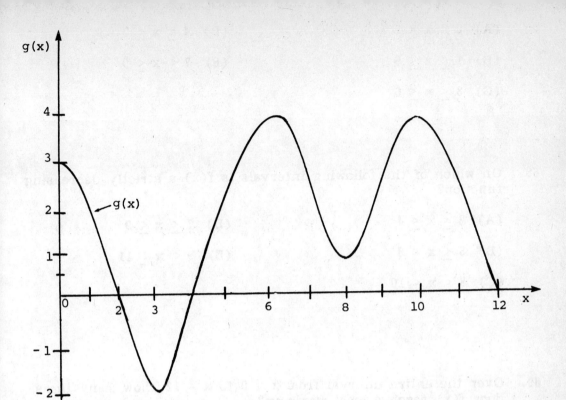

$$g(x) = \int_{-\infty}^{x} f(a)\,da$$

65. Given the function g(x), the function f(x) crosses the x-axis at x =

   (A) 1
   (B) 2
   (C) 4
   (D) 6
   (E) 9

66. f(x) ≥ 0 for all x in which of the following intervals?

   (A) 0 ≤ x ≤ 2
   (B) 1 ≤ x ≤ 5
   (C) 3 ≤ x ≤ 6
   (D) 4 ≤ x ≤ 7
   (E) 7 ≤ x ≤ 9

67. On what interval is f(x) nonpositive?

(A) $0 \leq x \leq 2$
(B) $1 \leq x \leq 5$
(C) $3 \leq x \leq 6$
(D) $4 \leq x \leq 7$
(E) $7 \leq x \leq 9$

68. On which of the following intervals is f(x) a strictly decreasing function?

(A) $2 \leq x \leq 4$
(B) $3 \leq x \leq 6$
(C) $8 \leq x \leq 10$
(D) $7 \leq x \leq 9$
(E) $9 \leq x \leq 11$

69. Over the entire interval from x = 0 to x = 12, how many times does f(x) reach a local maximum?

(A) 1
(B) 2
(C) 3
(D) 4
(E) 5

70. The area bounded by g(x) over $0 \leq x \leq 12$ is a minimum at x =

(A) 1
(B) 2
(C) 4
(D) 8
(E) 12

Figure for questions 71-80.

Two hands move in a clockface. Particle A is at the end of the smaller hand and B is at the end of the larger. The curve below is the distance between A and B as a function of time on the interval $0 \leq t \leq 14$.

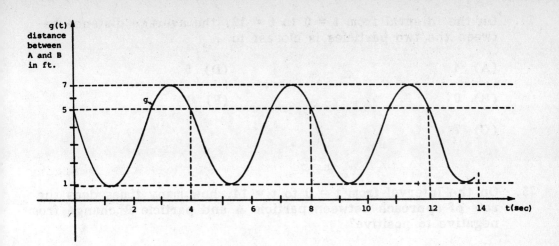

Particle A, on the smaller hand, has angular velocity $\frac{1}{8}$ rpm.

71. On the interval from t = 0 to t = 14, the number of times the two clock hands overlap is

    (A) one  
    (B) two  
    (C) three  
    (D) four  
    (E) five

72. How long is the larger clock hand?

    (A) 3  
    (B) 4  
    (C) 7  
    (D) 8  
    (E) 15

73. If the small hand of a clock is moving at 1/8 RPM, how fast must the large hand move to have the maxima and minima conditions satisfied?

    (A) 14.75 RPM  
    (B) 14.875 RPM  
    (C) 15.00 RPM  
    (D) 15.125 RPM  
    (E) 15.25 RPM

74. On the interval from t = 0 to t = 12, the average distance between the two particles is closest to

   (A) 4

   (B) 0

   (C) 7

   (D) 5

   (E) 1

75. On the interval from t = 0 to t = 14, how many times does the rate of approach between particle A and particle B change from negative to positive?

   (A) one

   (B) two

   (C) three

   (D) four

   (E) seven

76. If we define a function f as:

   $$f = \frac{dg}{dt}$$

   then $|f|$ is a maximum at how many points from t = 0 to t = 14?

   (A) one

   (B) two

   (C) three

   (D) four

   (E) seven

77. With the function f defined as before, on which of the following intervals is f strictly increasing?

   (A) $3 \leq t \leq 4$

   (B) $1 \leq t \leq 3$

   (C) $5 \leq t \leq 6$

   (D) $6 \leq t \leq 8$

   (E) $12 \leq t \leq 13$

78. A function h is defined as:

   $$h(x) = \int_0^x g(t)dt,$$

where g is the distance function depicted in the figure and $x \geq 0$. At which of the following points is h(x) an increasing function?

(A) t = 5

(B) t = 7

(C) t = 0

(D) all of the above

(E) none of the above

79. With h(x) defined as in the last problem, $D_x h(x)$ on $0 \leq x \leq 14$ is

(A) periodic

(B) strictly increasing

(C) strictly decreasing

(D) not monotonic on any subinterval

(E) none of the above

80. If h(x) is defined as before, h(4) is closest to

(A) 26

(B) 17

(C) 52

(D) 8

(E) 34

Figure for Questions 81-85.

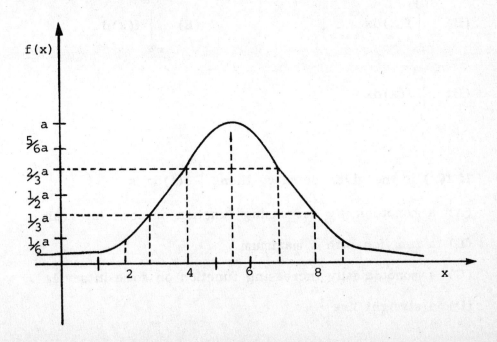

X is a gaussian random variable with density function shown above.

81. The standard deviation of X is closest to

    (A) 1
    (B) 4
    (C) 3
    (D) 6
    (E) 2

82. With f(x) so defined, $\int_{-\infty}^{\infty} f(x)\, dx$ is equal to

    (A) $2a_x$
    (B) $(a_x - \sigma_x)/\sigma_x$
    (C) $\sigma_x^2/(a_{x^2} - a_x^2)$
    (D) 2
    (E) infinity

83. What is the probability that random variable X is not greater than 2?

    (A) $\int_{2}^{\infty} f(x)dx$
    (B) $\int_{-\infty}^{0} f(x)dx$
    (C) $\int_{-2}^{\infty} f(x)dx$
    (D) $\int_{-\infty}^{2} f(x)dx$
    (E) $\int_{-\infty}^{\infty} f(x)dx$

84. If f(x) is the given density, then $\int^{x} f(t)dt$ is

    (A) a monotonically decreasing function on some intervals
    (B) a function with 1 maximum
    (C) a monotonically increasing function on some intervals
    (D) a straight line

(E) equal to 1 everywhere

85. The skew of $f_x(x)$ is equal to

(A) 0

(B) 3

(C) 9

(D) 7

(E) 12

Directions: Choose the best answer for each question and mark the letter of your selection on the corresponding answer sheet.

86. The resultant of a 13-newton force and a 8-newton force acting on the same object is 5 newtons. The angle between the forces is

(A) 0°

(B) 30°

(C) 45°

(D) 90°

(E) 180°

87. A student, while trying to measure the resistance of a lamp on a 120V line by the voltmeter-ammeter method of measuring resistance, connects the ammeter in parallel with the lamp and then the voltmeter in series with the lamp-ammeter circuit. The following result could occur:

(A) the ammeter is damaged

(B) the voltmeter is damaged

(C) the lamp is damaged

(D) the lamp may not light

(E) the student is successful in measuring the resistance

88. In an automobile battery, the chemical reaction below generates electricity.

$$Pb(c) + PbO_2(c) + 2H_2SO_4(aq) \rightarrow 2PbSO_4(c) + 2H_2O$$

What mass of Pb is consumed when 6.06g $PbSO_4$ is formed? (Assume the following atomic weights: Pb = 207, O = 16, S = 32.)

(A) 1.1g  (D) 20g
(B) 1.7g  (E) 25.5g
(C) 2.07g

89. For an induction motor, an increase in rotor resistance leads to

    (A) an increase in starting torque
    (B) an increase in motor efficiency
    (C) an increase in the starting current
    (D) a decrease in the power factor
    (E) an increase in speed

90. For the nozzle shown below with the given values, which of the following equations will be helpful in solving for the flow rate? (Neglect viscous effects.)

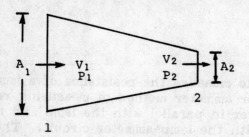

$P_1 = P \quad A_1 = 2A \quad v_1 = v$
$P_2 = 0 \quad A_2 = A$

I. Continuity equation $A_1 V_1 = A_2 V_2$

II. Bernoulli's equation $\frac{P_1}{\rho g} + \frac{V_1^2}{2g} = \frac{P_2}{\rho g} + \frac{V_2^2}{2g}$

III. Momentum equation $p_1 A_1 - p_2 A_2 = mV_1 + mV_2$

(A) I only
(B) I or II
(C) I and II and $\rho$ must be close to 1
(D) III only
(E) I, II and III

91. A liquid passes through a venturi tube as represented in the

160

figure below. The pressure difference between the two ends is a function of which of the following?

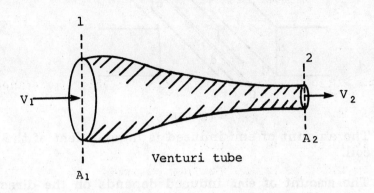

Venturi tube

(A) flow rates, $v_1$ and $v_2$

(B) cross-sectional $A_1$ and $A_2$

(C) density of the passing liquid

(D) correction coefficient if the viscosity is large

(E) all of the above

92. Calculate the value of R, for the circuit below,

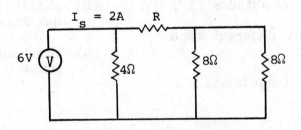

(A) 16Ω

(B) 8Ω

(C) 32Ω

(D) 4Ω

(E) 6Ω

93. The figure below represents a simple two pole single-coil generator. The coil is rotated at a constant speed. Which of the following correctly describes a property about the emf produced in the coil?

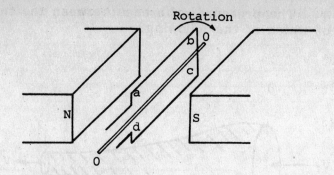

(A) The amount of emf induced is independent of the size of the coil.

(B) The amount of emf induced depends on the direction of rotation of the coil.

(C) The amount of emf induced has a maximum, when the coil is parallel to the magnetic poles.

(D) The amount of emf induced has a minimum, when the coil is parallel to the magnetic poles.

(E) The amount of emf induced has a minimum, when the coil is perpendicular to the magnetic poles.

94. Two sound waves produce beats when they have:

(A) equal amplitudes

(B) sightly different amplitudes

(C) equal frequencies

(D) slightly different frequencies

(E) equal number of overtones

95. What amount of heat is generated if a 1 kg metal block falls a distance of 10 m? (Assume all the energy of the block appears as heat)

(A) 9.8J

(B) 10J

(C) 98J

(D) $10^3$J

(E) $10^4$J

96. Two masses, $m_1$ = a kg and $m_2$ = b kg have velocities $\vec{u}_1$ = b m/sec in the +x direction and $\vec{u}_2$ = a m/sec in the +y direction.

They collide and stick together, moving at an angle to the horizontal equal to

(A) 0°

(B) 30°

(C) 45°

(D) 60°

(E) 90°

97. Heterogeneous reactions include reactants and products which are not all in the same phase. The equilibrium-constant expression does not for these reactions, include the concentrations of solids and liquids since the concentrations are constant in those phases. If (g) denotes a gas and (s) denotes a solid, then the equilibrum-constant expression for the decomposition of solid $NaHCO_3$ given in the reaction

$$2NaHCO_3(s) \rightleftarrows Na_2CO_3(s) + CO_2(g) + H_2O(g)$$

is

(A) $\dfrac{[CO_2(g)]}{[H_2O(g)]}$

(B) $[CO_2(g)][H_2O(g)]$

(C) $\dfrac{[CO_2(g)][Na]^2}{[H_2O(g)]}$

(D) $\dfrac{[NaHCO_3(s)]}{[Na_2CO_3(s)]}$

(E) $\dfrac{[CO_3]^2}{[H_2O(g)] + [CO_2(g)]}$

98. A man weighing 128 lb. is standing on the floor of an elevator that is going up at 4 ft/s². If a scale were placed on the floor how much would he weigh? Note: gravitational acceleration is 32.2 ft/s².

(A) 128 lb.

(B) 144 lb.

(C) 112 lb.

(D) 16 lb.

(E) None of the above

99. If two gases of masses $m_1$ and $m_2$ are at the same temperature $T_0$, what is the ratio of the root mean-square speeds of molecules of the two different gases?

$$\left(\dfrac{V_1 \text{ rms}}{V_2 \text{ rms}}\right)$$

Note: the root mean square speed is the average molecular speed ($\sqrt{\overline{v^2}}$)

(A) $\dfrac{m_2^2}{m_1^2}$

(B) $\dfrac{m_1}{m_2}$

(C) $\sqrt{\dfrac{m_2}{m_1}}$

(D) $\dfrac{m_2}{m_1}$

(E) $\sqrt{\dfrac{m_1}{m_2}}$

**Questions 100-101.**

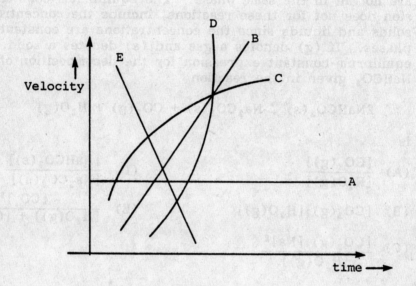

100. Which of the curves in the above figure describes the motion of an object with no acceleration?

   (A) C  
   (B) E  
   (C) B  
   (D) D  
   (E) A  

101. Which of the curves in the figure above describes the motion of a freely falling object? (Neglect drag).

   (A) B  
   (B) A  
   (C) D  
   (D) C  
   (E) E

102. A liquid impinges on the frictionless vane shown below at A.  A force must be exerted to hold the vane stationary.  The ratio of the magnitude of the horizontal to the magnitude of the vertical component of this force is

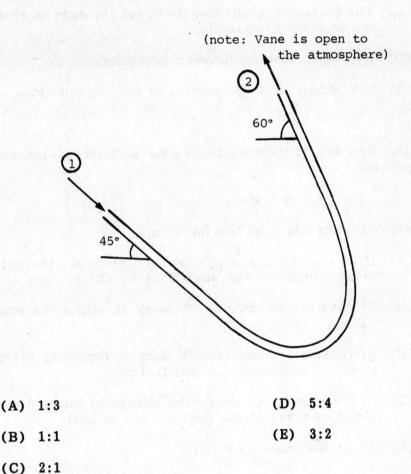

(note: Vane is open to the atmosphere)

(A)  1:3

(B)  1:1

(C)  2:1

(D)  5:4

(E)  3:2

103. A balloon will rise at a speed of 60 m/sec straight up in no wind. What would be the approximate magnitude of the balloon's velocity with a wind speed of 20 m/sec perpendicular to the upward velocity of the balloon?

(A)  40 m/sec

(B)  60 m/sec

(C)  63 m/sec

(D)  70 m/sec

(E)  80 m/sec

104. Ice is less dense than water at the melting temperature. Which of the following is usually cited as the reason for this phenomenon?

(A) The change in the length of the hydrogen-oxygen bond upon melting.

(B) The increased electron pair repulsion upon melting.

(C) The increased electronegativity of the oxygen atom with the increased temperature.

(D) Hydrogen bonding between molecules.

(E) the paramagnetic properties of the oxygen atom.

105. The first law of thermodynamics for a closed system can be expressed

$$\Delta E = q - w$$

which implies which of the following?

(A) If no change in energy during a reaction, the net change in heat balances the work done by the system.

(B) At constant volume, no PV work simplifies the equation to $\Delta E = q$.

(C) At constant volume, no PV work is done and the reaction is either exothermic or endothermic.

(D) In the absence of work, the change in internal energy is a direct measure of the gain or loss of heat.

(E) All of the above (A)-(D)

106. Two bridges are being considered for construction. One is expected to last indefinitely without any major overhaul, while the other bridge requires an overhaul every 25 years. In order to choose between these two alternatives, all of the following factors need to be determined except

(A) the capitalized cost of the annual operating expenses

(B) the future worth of the annual operating expenses

(C) the equivalent uniform annual cost for the overhaul process

(D) the capitalized costs of the overhaul process

(E) the present worth of the bridges

107. A military supply plane is supplying materials on a war front. When it is flying at a constant velocity it drops pack B two seconds later than pack A, both without parachutes. Assuming air resistance to be negligible, before pack A lands on the ground, it will be:

   (A) ahead of pack B

   (B) behind pack B

   (C) 9.8 meters in front of pack B

   (D) directly under pack B

   (E) either ahead of or behind pack B but not directly under it

108. The electrical conductivity of an n-type semiconductor crystal, with other factors remaining constant, depends on which of the following?

   (A) Size and the structure of the semiconductor crystal

   (B) The number of electrons that are lost to impurity acceptor atoms

   (C) Number of acceptor atoms and the electron mobility

   (D) Number of donor atoms and the hole mobility

   (E) Number of donor atoms and the electron mobility

109. Which of the following would not increase the rate of heat transferred from a heater pipe?

   (A) Insulating with materials whose thickness is below that of critical thickness for insulation

   (B) Blowing air over it

   (C) Providing fins

   (D) Submerging it in another media of higher density

   (E) Putting the heater pipe within another whose thermal conductivity is smaller in number and 2 inches thick.
   Note that Fourier Law of Conduction is: $q = -kA \frac{\partial T}{\partial X}$.

110. Find the efficiency of a Carnot Cycle if the temperature of the source and sink is 100°C and -273°C respectively.

(A) 0%  (D) 75%
(B) 10%  (E) 100%
(C) 50%

111. The cross-section of a tube of water is shown below. Which of the following statements is incorrect?

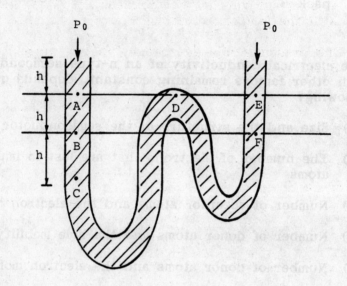

(A) The pressure at A, D, and E are equal and less than that of B.

(B) The pressure difference between A and B is linearly related to h.

(C) The pressure difference between A and C is dependent of the density of water.

(D) The pressure difference between A and B is less than the pressure difference between B and C.

(E) The pressure is not as great at D as it is at F.

112. What is the mass of a 10 ft³ material having a specific gravity of 13? Assume the density of water is 62.4 lbm/ft³.

(A) 130×62.4 lbm/ft³ lbm  (D) 130 - 62.4 lbm/ft³ lbm
(B) 130/62.4 lbm/ft³ lbm  (E) 130 lbm
(C) 130 + 62.4 lbm/ft³ lbm

113. Boyle's law states that

   (A) the volume of a sample of gas varies inversely with pressure under constant temperature.

   (B) at constant pressure, a volume of gas varies directly with temperature °K.

   (C) at constant volume, the pressure of a gas varies directly with the temperature.

   (D) at constant volume, the pressure of a gas varies indirectly with the temperature.

   (E) the pressure of a gas cannot change unless it begins at STP.

114. The transistor in the configuration below is

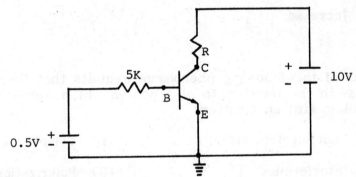

Transistor Characteristics:
$h_{FE} = 100$
$V_{BE} = 1\ V$
$I_{CBO} = 0\ A$

   (A) saturated                (D) open circuited

   (B) cutoff                   (E) forward biased

   (C) operating in the active region

115. For the situation below, what would happen to the average temperature at face C if the thermal conductivity of solid II was increased?

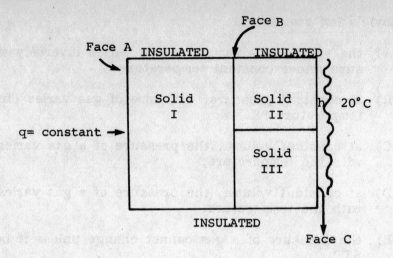

(A) No change

(B) Becomes 20°C

(C) Increase

(D) Decrease

(E) Insufficient information

116. Which of the following phenomena exhibits that there is an increase in the wavelength of a photon which was scattered by colliding with an electron?

   (A) Photoelectric effect
   (B) Interference
   (C) Electron diffraction
   (D) Compton effect
   (E) Polarization

117. Annealing wrought steel above its critical temperature range will primarily increase the toughness of the steel because

   (A) it will relieve the stresses brought about by rapid cooling
   (B) it will result in the formation of spheroids in the amtrix
   (C) it will bring about the refinement of a coarse grain size
   (D) it will increase the hardness of the steel through the formation of martensite
   (E) both (A) and (C)

Questions 118-119.

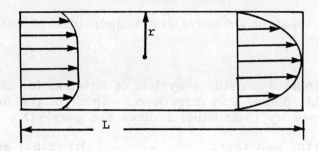

The diagram above represents the velocity distribution for viscous flow in a cylindrical pipe. The flow is laminar for a length L of the pipe for an incompressible fluid of viscosity coefficient $\eta$.

118. Which of the following is true for the fluid traversing the region L?

   (A) The inviscid region for the fluid increases from the center to the wall.

   (B) The region of viscous flow remains the same throughout the length L.

   (C) The total velocity over the cross-sectional area remains the same throughout the length L.

   (D) The flow rate increases while traversing the length L.

   (E) None of the above.

119. The total force acting on the fluid throughout the control volume of length L is

   (A) 0
   (B) $\eta(2\pi rL)\frac{dv}{dr}$
   (C) $\Delta p\, \pi r^2$
   (D) $\Delta p\, 2\pi rL$
   (E) $\eta \pi r^2 \frac{dv}{dr}$

120. An isentropic process is defined as

   (A) an adiabatic and reversible process

   (B) an adiabatic but not reversible process

(C) reversible but not adiabatic process

(D) a constant enthalpy process

(E) a constant pressure and temperature process

121. Miller indices provide a system of notation for describing crystallographic planes and directions. Which of the following planes, expressed by their Miller indices are parallel?

(A) (110) and (011)

(B) (101) and (111)

(C) (101) and (201)

(D) (103) and (310)

(E) (012) and (123)

122. The measure of specific heat of a material relates the quantity of heat required to change the temperature of a given mass to the mass of that material. At ordinary temperature ranges (or for a small temperature range) the specific heat, C, may be considered constant. If aluminum has a higher specific heat (.217 cal/gm°C) than copper (.093 cal/gm°C), then

(A) copper is less massive than aluminum

(B) it takes more energy in the form of heat to change the temperature of aluminum than copper of equal mass

(C) aluminum can be kept at a higher temperature than copper of equal mass

(D) $C_{Al}/C_{Cu}$ is higher at higher temperajtures than lower temperatures

(E) All of the above (A)-(D)

123. How many calories are needed to change 50g of water at 30°C to steam at 90°C? (Assume heat of vaporization: 540 $\frac{cal}{g}$ )

(A) 1,000 cal

(B) 30,000 cal

(C) 10,000 cal

(D) 5,000 cal

(E) 500 cal

124. A rectangular prism has dimensions 8 feet high, 15 feet long and 2 feet wide. Its gross weight is 6 tons and floats in a liquid of specific weight 100 lb/ft$^3$. The prism is then tilted so that it floats in the diagonal planes. The bouyant force exerted by the liquid in the tilted position is

   (A) 10,000 lb
   (B) 12,000 lb
   (C) 15,000 lb
   (D) 18,000 lb
   (E) 24,000 lb

125. The Acme Milling Company is evaluating three proposals to boost its production. Proposals A, B, and C require $55,000, $85,000, and $91,000 initial investments with the operating costs savings of $18,000/year, $24,000/year, and $27,000/year, respectively. All three proposals will be evaluated for a ten-year period. If the minimum attractive rate of return is 25% ((A/P, 25%,10) = .3) which of the three proposals is acceptable?

   (A) Proposal A, Proposal B, Proposal C
   (B) Proposal A only
   (C) Proposal B and Proposal C
   (D) Proposal B only
   (E) Proposal A and Proposal B

126. If a liquid polymer is to supercool and not crystallize during cooling below the melting temperature then which of the following must be true?

   I. Cross-linking must be facilitated with the presence of a bifunctional group.
   II. A single polymer molecule must fit in one cell.
   III. The polymer chains must be large or branched.

   (A) II only
   (B) I and III only
   (C) I only
   (D) III only
   (E) I and II only

127. For a mass suspended by a spring, at maximum displacement (k = spring constant, A is displacement amplitude)

(A) kinetic energy is zero, potential energy is $\frac{1}{2}kA^2$

(B) potential energy is zero, kinetic energy is $\frac{1}{2}kA^2$

(C) both kinetic and potential energy contribute equally to give a total energy of $\frac{1}{2}kA^2$

(D) potential energy is less than kinetic energy

(E) kinetic energy and potential energy are increasing

128. Curve A represents the continuous cooling transformation curve for carbon steel. Which of the following curves represents the continuous-cooling-transformation curve for a low alloy steel that stabilizes austenite?

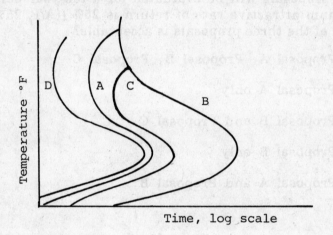

(A) Curve C only
(B) Curve D only
(C) Curve B only
(D) both Curve C and Curve B
(E) both Curve D and Curve C

129. Which of the following orbital configurations represent a possible excited state?

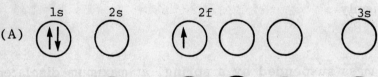

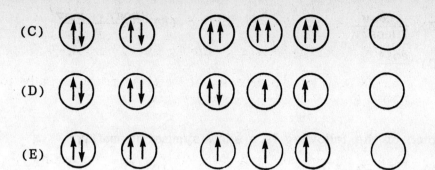

130. Ionic, metallic and most intermolecular bonds form structures that are different from those formed by covalent bonds in that

    (A) the former are directional and form loosely packed structures

    (B) the formers tend to form densely packed, highly coordinated structures and the latter is less densely packed

    (C) the bond length of the latter is always smaller than the intermolecular radii of the formers

    (D) the formers deal with microscopic forces while the latter deals with macroscopic forces

    (E) the formers are the only structures that are not artificially manufactured


Directions: Choose the best answer for each question and mark the letter of your selection on the corresponding answer sheet.

Note: The natural logarithm of x will be denoted by $\ln x$.


131. If

$$u^2 \cos v - x = 0$$

$$u^2 \sin v - y = 0$$

define u and v as implicit functions of x and y, then $\frac{\partial u}{\partial x}$ is given by

(A) $\frac{\cos v - \sin v}{2u}$      (B) $\frac{\cos v}{u}$

(C) $\dfrac{\cos v}{2u\cos 2v}$

(E) $\dfrac{\cos v - \sin v}{u}$

(D) $\dfrac{\cos v}{2u}$

132. Which of the following is a skew symmetric matrix?

(A) $\begin{bmatrix} 0 & -2 & 5 \\ 2 & 0 & 6 \\ -5 & -6 & 0 \end{bmatrix}$

(D) $\begin{bmatrix} 1 & 5 & 2 \\ 6 & 3 & 1 \\ 2 & 4 & 0 \end{bmatrix}$

(B) $\begin{bmatrix} 0 & 1 & 3 \\ 1 & 0 & 5 \\ 3 & 5 & 0 \end{bmatrix}$

(E) $\begin{bmatrix} 0 & 3 & 3 \\ 2 & 0 & 2 \\ 1 & 1 & 0 \end{bmatrix}$

(C) $\begin{bmatrix} 1 & 2 & 3 \\ 4 & 5 & 6 \\ 7 & 8 & 9 \end{bmatrix}$

133. Given the graphs $g_1: y - x^2 = 9$, $g_2: x^2 + y^2 = 81$, $g_3: y = x$ then $g_1$ intersects $g_2$ at

(A) (9,0)

(D) (81,9)

(B) (0,9)

(E) (0,-9)

(C) (9,81)

134. If $A = \{1,2,3,4,5\}$ $B = \{2,3,4,5,6\}$ then $A \cap B$ is

(A) $\{1,2,3,4,5\}$

(D) $\{2,3,4,5\}$

(B) $\{2,3,4,5,6\}$

(E) $\{2,4,6\}$

(C) $\{1,6\}$

135. $\displaystyle\int_C \dfrac{z^2}{z-3}\, dz \quad c: |z| = 1$, z complex, has value

(A) 0

(D) $6\pi i$

(B) $\pi i$

(E) $12\pi i$

(C) $3\pi i$

136. $\dfrac{d}{dx} \ln(5^{3x^2} \cos 5x) =$

  (A) $6x\ln 5 - 5\tan 5x$

  (B) $\dfrac{-\sin 5x + 6x\cos 5x}{5^{3x^2}}$

  (C) $\dfrac{5\cos 5x - 6\ln 5}{\cos 5x}$

  (D) $\dfrac{5\sin 5x + 6x\ln 5}{\cos 5x}$

  (E) $5\tan 5x - 6\sec 5x \ln 5$

137. The equation of a circle centered at $(-a, b)$ with diameter $r$ is

  (A) $(x + a)^2 + (y + b)^2 = r^2$

  (B) $(x - a)^2 + (y - b)^2 = r^2$

  (C) $[2(x + a)]^2 + [2(y - b)]^2 = r^2$

  (D) $(x - a)^2 + (y + b)^2 = r^2$

  (E) $(x - a)^2 + (y - b)^2 = r$

138. The general solution of the differential equation

  $$y'' - 4\lambda y' + 4\lambda^2 y = 0$$

  can be written as

  (A) $y = e^{2\lambda x}(c_1 + c_2 x)$

  (B) $y = ce^{2\lambda x}$

  (C) $y = e^{2\lambda x}(c_1 x + c_2 x^2)$

  (D) $y = cxe^{2\lambda x}$

  (E) $y = c_1 e^{2\lambda x} + c_2 xe^{\lambda x}$

139. $\sum\limits_{n=1}^{\infty} \dfrac{1}{n^k}$ diverges if

  (A) k is a constant

  (B) $k > 1$

  (C) $k < 1$

  (D) $k \leq 1$

  (E) $k \geq 1$

140. $\sum\limits_{k=1}^{\infty} (\sqrt{k+1} - \sqrt{k})$ is

  (A) convergent

  (B) divergent

(C) conditionally convergent  (E) uniformly convergent

(D) absolutely convergent

141. The expansion $1 + x + x^2 + \ldots$ approximates $(1-x)^{-1}$ if

(A) $x \neq 1$

(B) $-1 < x \leq 1$

(C) $|x| \neq 1$

(D) $|x| > 1$

(E) $|x| < 1$

142. $\lim\limits_{n \to \infty} \sum\limits_{k=0}^{n} \dfrac{1}{3^k} =$

(A) $\dfrac{1}{3}$

(B) $0$

(C) $1$

(D) $\dfrac{3}{2}$

(E) $\infty$

143. The graph of $4x + y + 2z = 8$ in three-dimensional Euclidian space intercepts the y-axis at a point how far from the origin?

(A) 4

(B) 2

(C) 8

(D) $\tfrac{1}{2}$

(E) 1

144. The area of the isosceles triangle formed by the equation $y = x$, the x-axis, and $x = a$, is

(A) $2a$

(B) $\dfrac{1}{2}$

(C) $\dfrac{3}{2}a^2$

(D) $\dfrac{a^2}{2}$

(E) $\dfrac{\sqrt{a}}{2}$

# THE GRADUATE RECORD EXAMINATION ENGINEERING TEST

## MODEL TEST II

## ANSWERS

| | | | | | |
|---|---|---|---|---|---|
| 1. | C | 21. | A | 41. | C |
| 2. | C | 22. | D | 42. | C |
| 3. | A | 23. | C | 43. | E |
| 4. | E | 24. | B | 44. | E |
| 5. | C | 25. | D | 45. | C |
| 6. | B | 26. | D | 46. | E |
| 7. | D | 27. | D | 47. | C |
| 8. | D | 28. | A | 48. | C |
| 9. | C | 29. | A | 49. | B |
| 10. | C | 30. | E | 50. | C |
| 11. | A | 31. | E | 51. | C |
| 12. | B | 32. | E | 52. | E |
| 13. | B | 33. | D | 53. | A |
| 14. | C | 34. | D | 54. | A |
| 15. | A | 35. | C | 55. | A |
| 16. | A | 36. | B | 56. | E |
| 17. | A | 37. | B | 57. | E |
| 18. | B | 38. | B | 58. | E |
| 19. | D | 39. | C | 59. | B |
| 20. | A | 40. | B | 60. | E |

| | | | | | |
|---|---|---|---|---|---|
| 61. | B | 89. | A | 117. | C |
| 62. | C | 90. | A | 118. | C |
| 63. | C | 91. | A | 119. | A |
| 64. | E | 92. | B | 120. | A |
| 65. | D | 93. | D | 121. | C |
| 66. | C | 94. | D | 122. | B |
| 67. | A | 95. | C | 123. | B |
| 68. | E | 96. | C | 124. | B |
| 69. | B | 97. | B | 125. | B |
| 70. | C | 98. | B | 126. | D |
| 71. | D | 99. | C | 127. | A |
| 72. | B | 100. | E | 128. | D |
| 73. | D | 101. | A | 129. | A |
| 74. | A | 102. | B | 130. | B |
| 75. | C | 103. | C | 131. | D |
| 76. | E | 104. | D | 132. | A |
| 77. | C | 105. | E | 133. | B |
| 78. | D | 106. | B | 134. | D |
| 79. | A | 107. | D | 135. | A |
| 80. | B | 108. | E | 136. | A |
| 81. | E | 109. | E | 137. | C |
| 82. | C | 110. | E | 138. | A |
| 83. | D | 111. | D | 139. | D |
| 84. | C | 112. | A | 140. | B |
| 85. | A | 113. | A | 141. | E |
| 86. | E | 114. | B | 142. | D |
| 87. | D | 115. | C | 143. | C |
| 88. | C | 116. | D | 144. | D |

# THE GRADUATE RECORD EXAMINATION ENGINEERING TEST

## MODEL TEST II

## DETAILED EXPLANATIONS OF ANSWERS

1. (C)
The internal energy for an ideal gas depends only on its temperature. It remains constant if the temperature is constant. Each isothermal is at a constant internal energy. Therefore, during the transfer from any point located on any of them to any point located on the other, the change in internal energy, dU, will be the same.

2. (C)
For most conductors, when the temperature is increased there is an increase in resistance due to the increased molecular movement within the conductor. This phenomenon impedes the flow of charge in the conductor. Hence, the resistance can be reduced by lowering the temperature since a decrease in temperature reduces the molecular movement within the conductor.

For most metallic conductors such as copper, the resistance increases linearly with an increase in temperature. In the case of semiconductors, the resistance decreases with increase in temperature because of increased availability of charge carriers.

The practical empirical relation expressing the temperature effect on resistance is given by $R_T = R_0 (1 + \alpha_0 T)$, where $R_T$ is the desired resistance at temperature T in centigrade degrees, $R_0$ is the resistance at 0°C, and $\alpha_0$ is the temperature coefficient based on 0°C.

3. (A)
A sphere falling in a viscous fluid experiences a viscous retarding force and a buoyant force. When the combination of these two forces equals the weight of the sphere the terminal velocity of the sphere is reached. The density of oil is somewhat less than the density of water. Therefore it has a lower specific gravity and exerts a smaller buoyant force on the sphere. Oil has a much greater viscosity than water and hence exerts a greater retarding force. The terminal velocity of the sphere in the bath of oil is attained sooner and therefore, less than that in the bath of water. The sphere with the greater terminal velocity will reach the bottom sooner.

4. (E)
To solve this problem we make use of the parallel axis theorem: $I_Z = I_G + md^2$ where $I_Z$ is the moment of inertia of a body of mass m with distance d from the z-axis, and $I_G$ is the moment of inertia of the body about an axis, passing through the center of mass and parallel to z-axis. We are told that $I_G$ of cylinder B is $800\pi$ slug-ft². Since its density is 100 slug/ft³, its mass is

$$m = pV = (100)(4\pi) = 400\pi$$

The distance d is easily shown to be 4 ft. Hence:

$$I_Z = I_G + md^2$$
$$= (800\pi) + (400\pi)(4)^2$$
$$= 7200\pi \text{ slug-ft}^2$$

5. (C)
If we draw a line from the top of the pole perpendicular to the side of the cliff, we will see that we have two similar

triangles. The right angle is common to both. The side of the cliff and the man are parallel, therefore they intersect the man's line of sight at equal angles. Since two angles of the triangles are equal, all three angles must be equal, and we indeed have similar triangles. They are shown below:

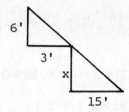

We can now write the following equation:

$$\frac{x}{6'} = \frac{15'}{3'}, \quad \text{or} \quad x = 30'$$

The height of the cliff is therefore 30 feet plus the height of the pole. Height = 40 feet.

6. (B)
Prestressing will increase the strength of the glass by introducing residual compressive stresses on the surface to counteract any internal or applied tensile stresses. One way to do this would be to cool the outside glass surfaces rapidly during heating so that the glass surface becomes rigid while the interior is still in the viscous state. When the interior cools and contracts it will cause a compressive stress on the outside surface. The compressive stress will counteract any tensile stress put on the glass, and thereby, increase the strength.

7. (D)
The balanced equation has equal number of atoms on both sides as shown. It doesn't necessarily need a catalyst. So only D is correct.

8. (D)
To find the charge in coulombs from t = 0 to 4 secs, we calculate and then add the charge in coulombs from t = 0

to 1 sec, t = 1 to 2 sec, t = 2 to 3 sec and t = 3 to 4 sec.

To find the charge for t = 0 to 1 sec, we first find equation for i(t). In this interval the graph is a straight line described by i(t) = 10t. The charge flow during this interval is given by

$$\Delta q = \int_0^1 10t \, dt = 10 \times \frac{t^2}{2} \Big|_0^1$$

$$= 5 \text{ mC} \quad \text{or} \quad 0.005 \text{ C}.$$

Similarly for the interval t = 1 to 2 we get i(t) = 10

$$\Delta q = \int_1^2 10 \, dt = 10t \Big|_1^2 = 10 \text{ mC or } 0.010 \text{ C}.$$

Similarly for the interval t = 2 to 3 we get i(t) = −10

$$\Delta q = \int_2^3 -10 \, dt = -10t \Big|_2^3 = -10 \text{ mC} \quad \text{or} \quad -0.010 \text{ C}.$$

Finally for the interval t = 3 to 4 we get i(t) = 0. Therefore Δq = 0.

Summing up the charge for the different intervals we get

$$\Sigma \Delta q = 0.005 + 0.010 - 0.010 + 0$$

$$= 0.005 \text{ coulombs}.$$

Notice that these integrals represent the areas under the respective curves. In cases where the curves are simple (straight lines), one need not perform integration to obtain the result. The areas can be found using simple geometric formulas, areas above x-axis being positive, and below the x-axis being negative.

9. (C)
In this case, the average speed of approach is the sum of the speeds of the two trains.

speed of approach = 25 + 15 = 40 km/hr

Therefore, it will take the two trains one hour to cover a distance of 40 km together. This means that the bird will spend a total of one hour in flight. Since its speed is 80 km/hr, it will cover 80 km by the time the two trains meet.

## 10. (C)

The power in a circuit is given by $P = VI$, so in order for the current source to supply increased power, the voltage across its terminals $V_{ba}$ must be at a maximum.

This will happen only if the equivalent resistance across its terminal is at a maximum since by Ohms Law we get $V = IR$.

The equivalent resistance for the complete circuit is found to be $360\,\Omega$. A larger equivalent resistance can be obtained by removing the $900\,\Omega$.

If we suppose that the equivalent resistance of the whole system excluding the $900\,\Omega$ resistor is $r$, then the resistance of the whole system will be: $\frac{900r}{900 + r}$. If we remove the 900 resistor from the network, the total resistance will be simply $r$, which is greater than $\frac{900r}{900 + r}$. Since the $900\,\Omega$ resistor is the nearest resistor to the terminals, its removal will have the greatest effect in increasing the equivalent resistance in the way shown above.

## 11. (A)

For the problem the following definitions will be helpful:

A branch is that part of a network which consists of a single component or a group of components connected in series. A point connecting two or more branches is said to be a node of a network. A loop is any closed path in a network. An independent loop is a loop whose KVL equation cannot be deduced from those of the other loops. In this way, the KVL equations related to these loops form a linear system of equations whose determinant is not zero and therefore has a unique answer. We can draw the planar equivalent of the given network (given below).

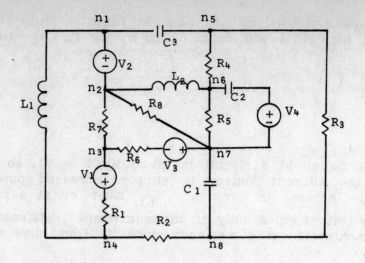

The following diagram shows the loops.

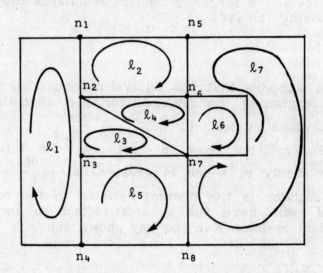

Since there are 8 nodes and 14 branches, there are 14 − (8 + 1) = 7 independent KVL equations. The equations are:

$\ell_1$: $v_1 - v_2 = v_{R_1} + v_{R_7} + v_{L_1}$

$\ell_2$: $\quad\quad v_2 = v_{C_3} + v_{R_4} + v_{L_2}$

$\ell_3$: $\quad\quad -v_3 = v_{R_6} + v_{R_8} - v_{R_7}$

$\ell_4$: $\quad\quad\quad 0 = v_{R_5} - v_{R_8} - v_{L_2}$

$\ell_5:\quad -v_1 + v_3 = v_{C_1} - v_{R_2} - v_{R_1} - v_{R_6}$

$\ell_6:\quad\quad -v_4 = v_{C_2} - v_{R_5}$

$\ell_7:\quad\quad v_4 = v_{R_3} - v_{C_1} - v_{C_2} = v_{R_4}$

## 12. (B)

The system of the man and the platform is initially at rest (therefore the angular momentum = 0). As soon as the man starts walking counterclockwise along the rim the platform must turn clockwise so as to conserve the zero angular momentum. When the man steps closer to the center he is decreasing his radial distance. Since angular momentum ($\vec{r}\;\;\vec{mv}$) is dependent on the length of the vector $\vec{r}$ (which is decreasing because the man is approaching the center), the decrease in the angular momentum component of the man must be compensated by a decrease in the angular momentum component of the platform ($I\omega$). In order to maintain a constant angular momentum of 0 for the system (note that the angular momentum of the man and that of the platform are opposite vectors whose sum is zero), the inertia of the platform remains constant and therefore its angular velocity must decrease.

## 13. (B)

When the temperature is recorded by a thermometer it means that the thermometer is in thermal equilibrium with the system with which it is in contact. If two thermometers are in contact with the same system but not with each other, they are in thermal equilibrium with that system. According to the zeroth law of thermodynamics two systems in thermal equilibrium with a third one are in thermal equilibrium with each other. Hence, there will be no transfer of heat when the two thermometers are brought in contact. Thus, the temperature readings will remain the same.

## 14. (C)

When solid particles are suspended in water, they are bombarded by the water molecules causing the suspended particles to move in a zigzag manner. At any given instant, larger numbers of water molecules may strike the particle from one side then the other and in this way the particle is given a push at different instances from different sides, resulting in an irregular motion of the particle, termed as Brownian movement.

## 15. (A)

The vessel must be designed so as to minimize heat transfer through conduction, convection and radiation. Glass is used since its thermal conductivity is lower than all metals. The surface of the glass is coated with silver to give a mirror finish. The mirror surface is a good reflector of heat and therefore a poor absorber. The region between the two walls should be evacuated so that no convection can take place through any medium.

## 16. (A)

As the liquid-vapor equilibrium curve approaches the critical point the liquid phase expands and the gas phase contracts. When the critical point is reached the molar volumes of liquid and gas become the same, so the distinction between the two separate phases vanishes.

## 17. (A)

According to the Huygen's principle every point of a wavefront may be considered the source of secondary wavelets which spread out in all directions with a speed equal to the speed of propagation of the waves. The new wave front is then found by constructing a surface tangent to the secondary wavelets. For an increasing index of refraction the speed of propagation of the waves will decrease. The secondary wavelets will be much smaller at slower wave speeds.

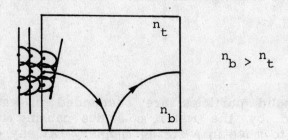

$n_b > n_t$

The Huygens construction for the path of light through

a sugar-water solution is as shown above. The majority of light will be reflected from the denser medium at the bottom of the container.

18. (B)
The balanced equation is:

$$NaOH + HCl \rightarrow NaCl + H_2O$$

The number of moles of NaOH at the start is

$$\frac{30 \text{ g}}{20 \text{ g/mol}} = 1.5 \text{ mol}$$

The balanced equation shows that the number of moles of HCl which is needed is 1.5.

$$1.5 \text{ mol} \times 18 \text{ g/mol} = 27 \text{ g HCl}$$

Therefore only 1.5 mol of water is created.

19. (D)
The chemical behavior of an atom depends on the outermost electrons surrounding the nucleus of the atom. The atomic number of an atom is equal to the number of protons in the nucleus of an atom. Protons are significant in the chemical behavior of an atom since they determine how many electrons surround the nucleus in a neutral atom. The number of electrons occupying the outermost electron orbital will determine the chemical stability of the atom. Hence, atomic number is most directly related to the chemical behavior of an atom.

20. (A)
Properties which are independent of the amount of matter are termed as intensive properties.

If matter existed in a uniform state, the temperature would be the same regardless of whether one were to describe all the matter or a part of it.

But other properties, such as energy or entropy, are defined for certain amounts of mass. For example if we divide a system into two parts, its energy will be divided as well, or in the case of entropy, the change of temperature due to the transfer of dQ depends on how much mass exists in the system.

## 21. (A)
For any similar case the original functional relationship still holds.

$$\frac{h_2}{d_2} = F\left(\frac{\gamma_2}{\rho_2 g d_2^2}, \theta_2\right)$$

But since

$$\frac{\gamma_2}{\rho_2 g (d_2)^2} = \frac{\frac{1}{2}\gamma_1}{2\rho_1 g (\frac{1}{2}d_1)^2} = \frac{\gamma_1}{\rho g d_1^2}$$

and $\theta_2 = \theta_1$

then

$$\frac{h_2}{d_2} = F\left(\frac{\gamma_1}{\rho_1 g d_1^2}, \theta_1\right) = \frac{h_1}{d_1}$$

Therefore since $h_1$ is known

$$\frac{h_2}{d_2} = \frac{h_1}{d_1}$$

$$h_2 = \frac{d_2}{d_1} h_1 = \frac{\frac{1}{2}d_1}{d_1} h_1 = \frac{1}{2}h_1$$

## 22. (D)
Such a process was envisioned by J.C. Maxwell and was debated over for many years. It is in clear violation of the Clausius statement of the second law of thermodynamics which is the following:"It is impossible to construct a device which operates in a cycle and whose sole effect is to transfer heat from a cooler body to a hotter body."

When the faster molecules (with high kinetic energy) enter the first chamber and slow molecules (with less kinetic energy) enter the second, gradually a temperature difference will be experienced due to the fact that molecules in the first chamber move faster and have higher energy.

## 23. (C)
Since the car is travelling in a horizontal circular path it has a centripetal acceleration directed toward the center

of $v^2/\rho$. If no lateral friction is to be exerted the reaction from the road is directed normal to the surface.

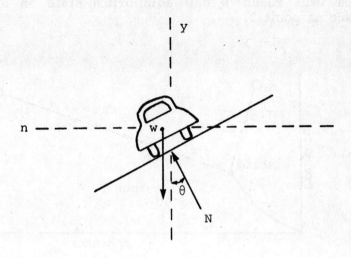

$$\Sigma F_y = 0 \;:\; N\cos\theta - W = 0$$

$$N = \frac{W}{\cos\theta}$$

$$\Sigma F_n = m\frac{v^2}{\rho}$$

$$N\sin\theta = \frac{W}{g}\frac{v^2}{\rho}$$

$$W\frac{\sin\theta}{\cos\theta} = \frac{W}{g}\frac{v^2}{\rho}$$

for $\theta = 45°$

$$\sin\theta = \cos\theta$$

$$1 = \frac{v^2}{g\rho}$$

$$v^2 = g\rho = 32(200)$$

$$v = 80$$

Therefore for speeds less than 80 ft/s a friction force will be exerted up the incline to keep the car from sliding down before the maximum friction force is reached.

## 24. (B)

When some of the vapor is removed, the temperature will decrease and some of the water will turn into vapor. The total internal energy as well as pressure will decrease. The system will reach a new equilibrium state on liquid-vapor surface at point 2.

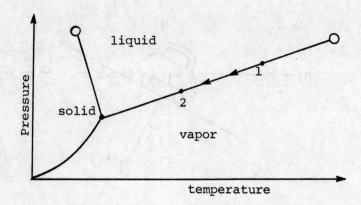

Another way to look at it is this: the vapor molecules have more kinetic energy than the liquid molecule. Therefore the effect of removing some of the vapor molecules, is to reduce the average kinetic energy of the system; the temperature is decreased.

## 25. (D)

To solve the problem the knowledge of the following characteristics of an a.c. circuit is useful.

1. For a purely resistive a.c. circuit, the voltage across and the current through the resistive element is always in phase as shown below.

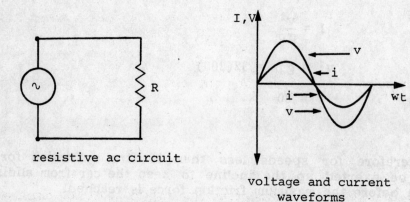

resistive ac circuit

voltage and current waveforms

This pair of voltage and current is represented by choice C.

2. For an a.c. circuit consisting of a capacitor, the current leads the voltage by 90° as shown below.

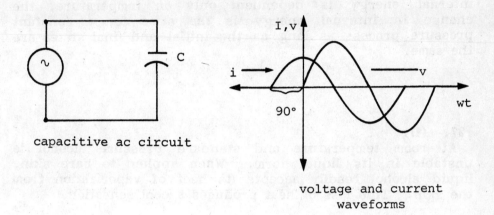

capacitive ac circuit

voltage and current waveforms

This pair of voltage and current is represented by choice A, B and E.

3. Finally, in an a.c. circuit consisting of an inductor, the voltage leads the current by 90° as shown below.

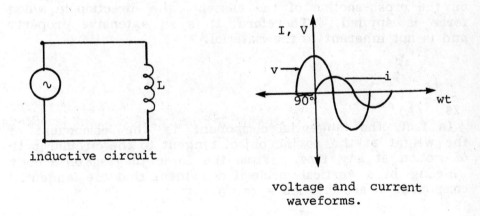

inductive circuit

voltage and current waveforms.

This pair of voltage and current is represented by choice D.

26. (D)
The net energy transferred in any process is equal to

the change in internal energy. Since internal energy is a state function and hence, path independent, the change in internal energy is the same for any number of processes as long as the initial and final states are the same. For a constant volume process the change in internal energy is equal to the heat added to the system because no work has been done on or by the system ($\Delta w = 0$). Since internal energy is dependent only on temperature, the change in internal energy is the same for a constant pressure process as long as the initial and final states are the same.

27. (D)

At room temperature and standard pressure alcohol is unstable in its liquid form. When applied to bare skin, liquid alcohol readily accepts its heat of vaporization from the skin. This loss of heat produces a cool sensation.

28. (A)

Any material's properties can be described as either extensive or intensive. An extensive property is one which depends on the amount of material present. An intensive property is one which depends on each specific material. In this case, the flexural strength of the material depends on the cross-section of the element, the direction in which force is applied. Therefore, it is an extensive property and is not inherent to the material.

29. (A)

In fact, the tangential component is the component of the weight of the moving object tangent to the circular path of motion at any time. From the force diagram on a body whirling in a vertical circle it is evident that the tangential component of acceleration = $g \sin \theta$.

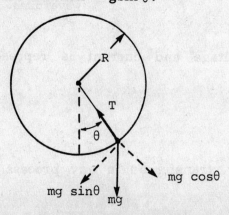

## 30. (E)

Average force is defined as follows:

$$F = \frac{\Delta p}{\Delta t}$$

The momentum delivered by the light when it is totally reflected is

$$p = \frac{2U}{c} \qquad \begin{array}{l} U \Rightarrow \text{energy} \\ c \Rightarrow \text{speed of light} \end{array}$$

The average energy U is:

$$U = I_1 A_1 \Delta t$$

Therefore the average force is:

$$F = \frac{2 I_1 A_1}{c}$$

## 31. (E)

To keep the man moving at constant velocity, we must oppose the force of friction tending to retard his motion with an equal but opposite force (see diagram).

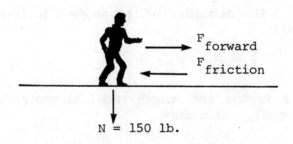

The force of friction is given by:

$$F = \mu_{kinetic} N$$

By Newton's Third Law

$$F_{forward} = F_{friction}$$

Therefore

$$F_{forward} = \mu_{kinetic} N$$

$$F_{forward} = (.05)(150 \text{ lb}) = 7.5 \text{ lb}.$$

**32. (E)**
The restoring force acting on the bob is $F = -mg \sin\theta$. For small oscillations we can assume:

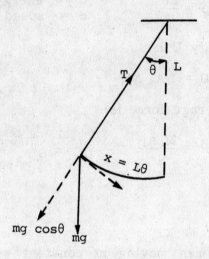

$$\sin\theta \approx \theta$$

According to the definition of the angle $\theta$ in terms of radians we can write:

$$\theta = \frac{x}{L},$$

in which x is the arc length and L is the radius (here the length of rope). Therefore:

$$F = -\left(\frac{mg}{L}\right) x$$

$$K = \frac{mg}{L}$$

$$T = 2\pi\sqrt{\frac{m}{k}} = 2\pi\sqrt{\frac{L}{g}}$$

$$\omega = \frac{1}{T} = \frac{1}{2\pi}\sqrt{\frac{g}{L}}$$

Therefore $\omega$ increases with increasing L.

$$x = A \cos(\omega t + \phi)$$

$$\frac{d^2x}{dt^2} = \omega^2 Ax$$

Therefore maximum acceleration = $\omega^2 A$

Therefore it increases with increasing fequency.

33. (D)
Combining the two $2\Omega$ resistance at the right of the circuit we get the following circuit.

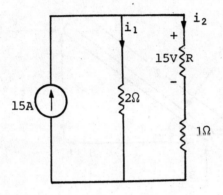

The current $i_2 = 15/R$.

Applying concept of current division we get

$$i_2 = \frac{15(2)}{2 + R + 1}$$

Substituting $i_2 = 15/R$ we get,

$$\frac{15}{R} = \frac{30}{R + 3}$$

i.e.   $30R = 15R + 45$

$15R = 45$

$$R = \frac{45}{15} = 3\,\Omega$$

34. (D)
In any decay process spin angular momentum must be conserved. The intrinsic spins of the proton, neutron and electron are all $\frac{1}{2}$. The decay of a neutron into an electron and proton would conserve charge but not spin. Any other

particle emitted would therefore have to be neutral and of spin ½. A photon-like particle of zero rest mass such as an antineutrino will conserve spin.

35. (C).
The length of the gate is 10 ft.

Therefore, $A = 40 \text{ ft}^2$

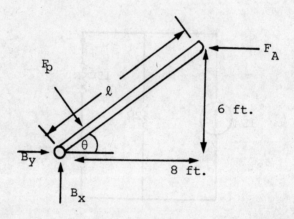

The hydrostatic force on the gate is

$$F_p = p_{CG}A = \rho g h_{CG} A$$

$$= 100 \frac{\text{lbf}}{\text{ft}^3} (10 \text{ ft.})(40 \text{ ft}^2)$$

This fact can be proved in the following way: if x is the distance of a point on the gate from the top of it, we may write:

$$F_p = \int_0^\ell \rho g[x\sin\theta + m](L dx) = \rho g L \int_0^\ell (x\sin\theta + m) dx$$

$$= \rho g L \left[ \sin \frac{\ell^2}{2} + m\ell \right]$$

$$= \rho g L \ell \left[ \frac{\ell}{2} \sin\theta + m \right]$$

$$= \rho g A\, h_{CG}$$

in which L is the width of the gate, $h_{CG}$ is vertical distance b between the middle of the gate and water surface, and m is the vertical distance from the top of the gate to the liquid surface.

$$F_p = 4 \times 10^4 \text{ lbf}$$

$$\Sigma M_B = 0$$

$$F_A(\ell \sin \theta) - F_p(h_{CG} - .5) = 0$$

$$F_A(6) = 4 \times 10^4 (4.5)$$

$$F_A = 4 \times 10^4 (.75)$$

$$= 3 \times 10^4$$

**36. (B)**
Hund's Rule summarizes what has been found experimentally: electrons that enter a subshell that contains more than one orbital (such as the p subshell) will spread out over the available orbitals with their spins in the same direction. Therefore the configuration of a nitrogen atom is

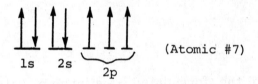

(Atomic #7)

**37. (B)**
Although the tire was inflated to 25 psia, all of that fluid pressure is distributed around the tire. In other words, all of the forces are balanced. Therefore, the weight of the tire has not increased. For further illustration, consider the common experience of blowing up a balloon. Anyone who has experienced it would know that a lot of pressure must be exerted to blow up a balloon. However, once the balloon is inflated, it can be bounced around with the greatest of ease; if the air pressure had affected the weight of the balloon, the inflated balloon could not have been bounced around so effortlessly.

**38. (B)**

If F is the net force acting on a body sliding down the inclined plane, then according to Newton's second law of motion

$$F = ma$$

where m is the mass and a is the acceleration of the body. Therefore,

$$m_1 = \frac{W_1}{g}$$

where $W_1$ is the weight of the first body and g is the gravitational acceleration

$$m_2 = \frac{W_2}{g}$$

where $W_2$ is the weight of the second body.

For the inclined plane $a = g \sin\theta$ where $\theta$ is the angle of inclination. Therefore,

$$F_1 = m_1 a = W_1 \sin\theta$$

$$F_2 = m_2 a = W_2 \sin\theta$$

It is obvious that the forces acting on the two iron blocks are different while acceleration is the same.


### 39. (C)

According to the formulated relationship between the rate of heat transfer in a solid body, $H = -kA \frac{dT}{dx}$, in which: k is thermal conductivity, x is the distance from one end, T is the temperature at that distance and H is the time rate of heat transfer, we can say:

Since all heat which enters one end leaves the end of the other, H is the same for both rods. In addition, since the cross-sectional areas are the same, we can assume a uniform increase (or decrease) of temperature in both rods.

So we can write:

$$K_1 \frac{\Delta T_1}{\Delta x} = K_2 \frac{\Delta T_2}{\Delta x}$$

lengths are equal, so we have:

$$K_1 \Delta T_1 = K_2 \Delta T_2$$

so if $\Delta T_2 > \Delta T_1$, then $K_1 > K_2$. So C is correct.

## 40. (B)

The current through an inductor is given as

$$i_L = i_L(0) + \frac{1}{L} \int_0^t v_L \, dT$$

where $i_L(0) = 3$, $v_L = 10t$, and $L = 2$ H. Hence,

$$i_L = 3 + \frac{1}{2} \int_0^t 10 \, T \, dT$$

$$i_L = 3 + \frac{1}{2} \left[ \frac{10 \, T^2}{2} \right]_0^t$$

$$i_L = \left( 3 + \frac{5}{2} t^2 \right) u(t) \text{ A}$$

This is represented by graph B.

This question can also be solved using the process of elimination. In the statement of the problem, it is given that the current at t = 0 is 3 amperes. Looking at the choices, there is only one graph indicating that the initial current in the inductance is 3 amperes. Logically, this must be the correct choice.

## 41. (C)

The collision of x-rays with electrons produces a loss in the energy of the radiation since electrons are being emitted. Since the energy of a photon is proportional to the frequency, a photon of light that collides with an electron will have a decreased frequency. Therefore, the experiment provides further confirmation of the quantum nature of light.

## 42. (C)

Since an atom is usually electrically neutral, the number of positively charged protons in its nucleus must equal the number of negatively charged electrons moving around the nucleus. The Zth element in the periodic table has Z electrons and therefore has Z protons in its nucleus. Z is called the atomic number. If there are also N neutrons in the nucleus, the total number of protons and neutrons is A = Z + N, where A is called the mass number. It is an exact integer. The nuclear mass always differs only very slightly from an integer, and the mass number, A, is this nearest integer.

From the above relation it is clear that an increase in the atomic number will increase the mass number and hence the nuclear mass. This is best represented by curve C.

### 43. (E)

To evaluate the momentum the photon, we start from the formula for the energy of electromagnetic waves: $p = \frac{E}{C}$. Since the photon has dual characteristics, we can use this formula in the following way:

$$E = h\nu, \quad E = pc \Rightarrow h\nu = pc$$

also we have:

$$\lambda\nu = c \quad \text{therefore:} \quad h\nu = p\lambda\nu \Rightarrow p = \frac{h}{\lambda}$$

where p and $\lambda$ are momentum and wavelength of the photon. This relation is expressed by curve E.

### 44. (E)

The capacitive reactance $(X_C)$ is the resistance to the flow of an alternating current as a result of capacitance and is measured in ohms.

It is mathematically expressed as

$$X_C = \frac{1}{2\pi fC}$$

where f is the frequency in cycles/sec and C is the capacitance in farads. It is clear from the above relation that as the capacitance is increased, the capacitive reactance decreases. This is represented by curve E.

### 45. (C)

The future worth factor equation which calculates the accumulated sum, when the principle, number of years and interest rate are given, is as follows:

$$F = p(1 + i)^n$$

where:  $p$ = principle
 $n$ = number of years
 $i$ = interest rate

Thus, the sum which will be accumulated in the account at the end of the first year is $500(1 + .10) = \$5500$. The withdrawal of $1000 will leave $4500 in the bank. There will be no further withdrawals for the next 4 years. Therefore, the amount that will remain in the account at the end of 4 years is $\$4500 \ (1 + .10)^4$.

46. (E)
$$\int_1^2 x(x + 2x^2)dx = \int_1^2 (x^2 + 2x^3)dx$$

$$= \left.\frac{x^3}{3}\right]_1^2 + \left.\frac{2}{4}x^4\right]_1^2$$

$$= \frac{1}{3}(8 - 1) + \frac{1}{2}(16 - 1)$$

$$= \frac{7}{3} + \frac{15}{2}$$

$$= \frac{59}{6}$$

47. (C)
For none to have failed at $T = t_0$, the three must have lifetimes greater than $t_0$. Since the lifetimes are independent, the probability of this is the product

$$Pr(T_1 \geq t_0) \times Pr(T_2 \geq t_0) \times Pr(T_3 \geq t_0)$$

$$= [Pr(T \geq t_0)]^3$$

$$= (e^{-kt_0})^3 = e^{-3kt_0}.$$

48. (C)
The probability that it will be either white or green is: P(a white ball or a green ball) = P(a white ball) + P(a green ball). This is true because if we are given two mutually exclusive events A or B, then P(A or B) = P(A) + P(B). Note that two events, A and B, are mutually exclusive events if their intersection is the null or empty set. In this case the intersection of choosing a white ball and of choosing a green ball is the empty set. There are no elements in common.

$$P \text{ (a white ball)} = \frac{\text{number of ways to choose a white ball}}{\text{number of ways to select a ball}}$$

$$= \frac{4}{21}$$

$$P \text{ (a green ball)} = \frac{\text{number of ways to choose a green ball}}{\text{number of ways to select a ball}}$$

$$= \frac{8}{21}$$

Thus,

$$P \text{ (a white ball or a green ball)} = \frac{4}{21} + \frac{8}{21} = \frac{12}{21} = \frac{4}{7}.$$


49. (B)
Using the identity: $\tan^2 x = \sec^2 x - 1$, we have:

$$\int_0^{\pi/2} \tan^2 \frac{x}{2} \, dx = \int_0^{\pi/2} \left( \sec^2 \frac{x}{2} - 1 \right) dx$$

$$= \int_0^{\pi/2} \sec^2 \frac{x}{2} \, dx - \int_0^{\pi/2} dx$$

$$= 2 \int_0^{\pi/2} \sec^2 \frac{x}{2} \cdot \frac{1}{2} \, dx$$

$$+ \int_0^{\pi/2} dx.$$

Applying the formula: $\int \sec^2 u \, du = \tan u + C$, and evaluating between $\frac{\pi}{2}$ and 0, we obtain:

$$\left[2\tan\frac{x}{2} - x\right]_0^{\pi/2} = 2\tan\frac{\pi}{4} - \frac{\pi}{2} = 2 - \frac{\pi}{2}$$

50. **(C)**
Orthogonal matrices have the following properties:

$$\det M = 1$$
$$M^{-1} = M^T$$

Hence:

$$D^{-1} = D^T$$

The inverse of D is

$$D^{-1} = \frac{1}{-BC}\begin{bmatrix} 0 & -B \\ -C & A \end{bmatrix}$$

The transpose of D is

$$D^T = \begin{bmatrix} A & C \\ B & 0 \end{bmatrix}$$

It is easy to see that

$$B = \frac{-C}{-BC} \Rightarrow B = \frac{1}{B} \quad B = \pm 1$$

Hence $|B| = 1$

51. **(C)**

$$\frac{\left(\cos\frac{x}{2}\right)\left(\cos\frac{x}{2}\right) + \left(\sin\frac{x}{2}\right)\left(\sin\frac{x}{2}\right)}{\left(\cos\frac{x}{2}\right)\left(\cos\frac{x}{2}\right) - \left(\sin\frac{x}{2}\right)\left(\sin\frac{x}{2}\right)}$$

$$= \frac{\cos^2\frac{x}{2} + \sin^2\frac{x}{2}}{\cos^2\frac{x}{2} - \sin^2\frac{x}{2}}$$

$$= \frac{1}{\cos x} = \sec x$$

**52. (E)**

The range is the difference between the largest and smallest observation is $4 - 0 = 4$.

The variance is the mean or average squared deviation from X. To compute the variance of this sample we use the formula

$$s^2 = \frac{\Sigma X^2 - n\overline{X}^2}{n}$$

To facilitate the computation we use the following table,

| X | $X^2$ |
|---|---|
| 2 | 4 |
| 3 | 9 |
| 0 | 0 |
| 3 | 9 |
| 4 | 16 |
| 1 | 1 |
| 3 | 9 |
| 0 | 0 |
| $\Sigma X = 16$ | $X^2 = 48$ |

Thus

$$\overline{X} = \frac{\Sigma X}{n} = \frac{16}{8} = 2 \quad \text{and}$$

$$s^2 = \frac{48 - 8(2)^2}{8} = \frac{48 - 8(4)}{8} = \frac{16}{8} = 2.$$

The standard deviation is

$$s = \sqrt{s^2} = \sqrt{2} = 1.414.$$

**53. (A)**

To find $\frac{dy}{dx}$, we use the differentiation formula:

$$\frac{d}{dx} \ln u = \frac{1}{u} \frac{du}{dx},$$

letting $u = \tan x$, $\frac{du}{dx} = \sec^2 u$.

Applying the formula, we obtain:

$$\frac{dy}{dx} = \frac{1}{\tan x} \sec^2 x = \frac{\cos x}{\sin x} \frac{1}{\cos^2 x}$$

$$= \frac{1}{\sin x} \frac{1}{\cos x} = \csc x \sec x.$$

54. (A)

$$\int_2^3 F(x)dx - \int_3^2 G(x)dx$$

$$= \int_2^3 F(x)dx + \int_2^3 G(x)dx \quad \left[\text{Since } \int_2^3 G(x)dx = -\int_3^2 G(x)dx\right]$$

$$= \int_2^3 [F(x) + G(x)]dx \quad \left[\text{Since } \int_a^b [F(x) + G(x)]dx = \int_a^b F(x)dx + \int_a^b G(x)dx\right]$$

55. (A)
Let

$$I = \int_a^{2a} F'(x - a)dx$$

Let $x - a = u$, therefore $dx = du$

at $x = a$, $u = a - a = 0$
at $x = 2a$, $u = 2a - a = a$

Therefore

$$I = \int_0^a F'(u)\,du = F(u)\Big]_0^a$$

$$= F(a) - F(0)$$

56. (E)

$$\int_1^2 F(x)\,dx - \int_1^3 F(x)\,dx$$

$$= -\int_2^1 F(x)\,dx - \int_1^3 F(x)\,dx$$

$$= -\left[\int_2^1 F(x)\,dx + \int_1^3 F(x)\,dx\right]$$

$$= -\int_2^3 F(x)\,dx \quad \left[\text{Since } \int_a^b F(x)\,dx + \int_b^c F(x)\,dx = \int_a^c F(x)\,dx\right]$$

57. (E)
Let $u = F(x)$ then $\frac{du}{dx} = F'(x)$ and $v = G(x^2)$ then $\frac{dv}{dx} = G'(x^2) \cdot \frac{d}{dx}(x^2)$ or $\frac{dv}{dx} = G'(x^2)(2x)$.

The formula is,

$$\frac{d}{dx}\frac{u}{v} = \frac{v\frac{du}{dx} - u\frac{dv}{dx}}{v^2}$$

$$\frac{d}{dx}\left[\frac{F(x)}{G(x^2)}\right] = \frac{G(x^2)F'(x) - F(x)G'(x^2)(2x)}{(G(x^2))^2}$$

58. (E)
Let $F(x) = y$

$$y = 3x^3 + 4x + 7, \text{ then } \frac{dy}{dx} = 9x^2 + 4 = 0, \; x^2 = -\frac{4}{9},$$

$$x = \pm\sqrt{-\frac{4}{9}} = \pm\frac{2i}{3}.$$

This is an imaginary quantity. Since these are not real roots, in this example y has neither relative maxima or minima.

59. (B)
To find $\frac{dy}{dx}$, we use the quotient rule, obtaining:

$$\frac{dy}{dx} = \frac{\ln x \, (0) - 1) \left(\frac{1}{x}\right)}{(\ln x)^2}$$

$$= \frac{-\frac{1}{x}}{(\ln x)^2}$$

$$= -\frac{1}{x(\ln x)^2}$$

60. (E)

$$\int_0^\pi \frac{(\csc^2 x)e^{ax}}{1 + \cot^2 x} \, dx$$

$$= \int_0^\pi \frac{e^{ax}}{\sin^2 x \left(1 + \frac{\cos^2 x}{\sin^2 x}\right)} \, dx$$

$$= \int_0^\pi \frac{e^{ax}}{(\sin^2 x + \cos^2 x)} \, dx$$

$$= \int_0^\pi e^{ax} \, dx = \frac{1}{a} e^{ax} \Big|_0^\pi$$

$$= \frac{1}{a}(e^{a\pi} - e^{a(o)})$$

$$= \frac{1}{a}(e^{\pi a} - 1)$$

**61. (B)**
Since $n(n - 1)(n - 2)\ldots 3\cdot 2\cdot 1 = n!$

Therefore replacing n by n - 1

$$(n - 1)(n - 2)(n - 3)\ldots 3\cdot 2\cdot 1 = (n = 1)!$$

**62. (C)**
This function is in the form 0/0. Consequently, applying L'Hospital's rule,

$$\lim_{x \to 0} \frac{x \cos x - \sin x}{x} = \lim_{x \to 0} \frac{-x \sin x}{1}$$

$$= \frac{0}{1} = 0.$$

Another approach is to recognize that, for small angles, $\sin x \cong x$, $\cos x \cong 1$. Thus the given expression becomes:

$$\lim_{x \to 0} \frac{x(1) - x}{x} = \frac{1 - 1}{1} = 0.$$

**63. (C)**

$$\begin{pmatrix} 1 & 1 & 0 \\ 1 & 0 & 1 \end{pmatrix} \begin{pmatrix} 1 \\ 0 \\ 1 \end{pmatrix}$$

$$= \begin{pmatrix} 1 \times 1 + 1 \times 0 + 0 \times 1 \\ 1 \times 1 + 0 \times 0 + 1 \times 1 \end{pmatrix}$$

$$= \begin{pmatrix} 1 + 0 + 0 \\ 1 + 0 + 1 \end{pmatrix}$$

$$= \begin{pmatrix} 1 \\ 2 \end{pmatrix}$$

64. (E)
The power series expansion for $e^x$, where x is any real number is

$$e^x = 1 + x + \frac{x^2}{2!} + \frac{x^3}{3!} + \ldots = \lim_{n \to \infty} \sum_{k=0}^{n} \frac{x^k}{k!}$$

Therefore

$$\sum_{n=0}^{\infty} \frac{2^n}{n!}$$

$$= \frac{2^0}{0!} + \frac{2^1}{1!} + \frac{2^2}{2!} + \frac{2^3}{3!} + \ldots$$

$$= 1 + 2 + \frac{2^2}{2!} + \frac{2^3}{3!} + \ldots$$

$$= e^2$$

65. (D)
The graph of the function g(x) is given; we are not given the graph of f(x).

However, we are given that $g(x) = \int_{-\infty}^{x} f(t)dt$. The function f(x) is obtained by applying the Fundamental Theorem of Calculus.

Differentiating both sides we obtain:

$$\frac{dg(x)}{dx} = f(x).$$

At the points where f(x) crosses the x-axis, f(x) = 0. Since f(x) = g'(x), we explore the local extrema of the graphed function g(x); that is, we look for the points where g'(x) = 0.

Looking at g(x) and the choices offered we see that at points x = 1,2,4 and 9 the slope of the tangent to g(x) is non-zero.

It is zero at x = 6; therefore f(x) crosses the x-axis at x = 6.

66. (C)
In this problem the Fundamental Theorem of Calculus is invoked once again:

$$\frac{d}{dx} \int_{-\infty}^{x} f(t)\,dt = \frac{dg(x)}{dx} = f(x).$$

We are asked for an interval where, for all values of x, $f(x) \geq 0$.

This is the same as finding an interval where the slope of the tangent of g(x) is non-negative for all values of x. That is, we look for an interval where g(x) is non-decreasing.

From the graph, we see that the only interval given in which g(x) is non-decreasing at any value of x is the interval $3 \leq x \leq 6$.

67. (A)
Since $f(x) = \frac{d}{dx} \int_{-\infty}^{x} f(t)\,dt = \frac{dg}{dx}$, we examine the graph of g(x) to see where the slope of its tangent is negative. This is the same as looking for points where g(x) is decreasing.

Examining the itervals provided we see that the only interval where g(x) is decreasing for all x is the interval of x between 0 and 2.

68. (E)
Wherever f(x) is decreasing the derivative of f(x) must be negative. The first derivative of f(x) is the same as the second derivative of g(x) since the first derivative of g(x) is the slope of the tangent to the curve, the second derivative is the rate of change of the slope of the tangent.

Thus, if the slope is increasing on an interval then $\frac{d^2 g(x)}{dx^2}$ will be positive and if the slope is decreasing then $\frac{d^2 g(x)}{dx^2}$ will be negative.

When the second derivative of a function is positive, the graph of the function is concave up; when the second derivative is negative, the graph is concave down. We are asked for an interval where g"(x) is negative so the graph should be concave down on that interval. Of the choices given, the interval $9 \leq x \leq 11$ satisfies this condition.

## 69. (B)

If the function f(x) is at a maximum then f'(x) must be zero. If $f(x) = g'(x)$, then at $f_{max}$ g"(x) = 0. A function's second derivative is equal to zero at its points of inflection; that is, at the points where its concavity changes. When a function's second derivative changes from incrasing to decreasing then the concavity of the function changes from concave down to concave up. This corresponds to a local maximum for $f(x)(= g'(x))$.

Looking at the graph of g(x), we see that the concavity of g changes from concave up to down twice. This corresponds to two local maxima for f(x).

## 70. (C)

The area bounded by g(x) is increasing whenever g(x) is positive and decreasing when g(x) is negative. g(x) is negative over the interval $2 \leq x \leq 4$. Thus g(x) decreases over $2 \leq x \leq 4$.

When x = 4, the area under g(x) is at a minimum because over the rest of the interval, g(x) is positive and the area it bounds is increasing.

## 71. (D)

When the two hands overlap, the distance between the two particles is at a minimum. Therefore, by looking at the curve and noting the number of minima in the interval, we can tell how many times the hands of the clock overlap.

Since there are four minima on the graph, the hands overlap four times in the interval.

## 72. (B)

By looking at the distance between particle A and particle

B at different times we can determine the lengths. We know the distance between the two particles when they are diametrically opposite to each other, i.e. the maximum distance between the particles.

We also know the minimum distance between the particles, which occurs when the two hands overlap.

Letting  L = length of the longer hand,

and  $\ell$ = length of the shorter hand,

we have:

$$L + \ell = 7$$

and  $$L - \ell = 1.$$

Adding these two equations we obtain

$$2L = 8,$$

$$L = 4.$$

Thus, the length of the longer hand is 4 ft.

### 73. (D)

If the small hand moves at 1/8 RPM, then in 4 seconds it travels the equivalent of 0.5 seconds on the face.

$$\frac{1}{8} \ RPM \times 4 \ sec. =$$

$$.125 \ RPM \times .066\overline{6} \ min. = .0083375 \ rotations$$

$$.0083375 \times 60 = .5025 \ sec.$$

The large hand would have to travel 60.5 seconds on the face in 4 seconds to achieve the minima and 30.25 seconds on the face in 2 seconds to achieve the maxima. Thus it would have to travel at 15.125 RPM.

$$\frac{60.5}{60} \ rev. = 1.083\overline{3}$$

$$\frac{1.0833 \ rev.}{4 \ sec.} \times \frac{60 \ sec}{min.} = 15.125 \ RPM$$

## 74. (A)

Note that the given curve is a periodic function. The average value of a periodic function is equivalent to the area bounded by the function over one period divided by the period. That is:

$$f_{AV} = \frac{1}{T} \int_0^t f(t)dt, \text{ where } T = \text{period.}$$

Often, a periodic function is symmetrical at a certain ordinate level; that is, it has equal portions above that level and equal portions below it. When this is the case, the level to which the function is symmetrical is the average value of the function.

Looking at the given distance curve, we see that the function ranges from $d = 7$ to $d = 1$ and is symmetric about $d = 4$. Thus, the average distance between the particles is 4 ft.

## 75. (C)

When we define two particles as approaching each other, we mean that the distance between the particles is negatively changing or decreasing.

Using this definition, an increase in approach corresponds to a decrese in distance between A and B. Thus, an "approach function" with respect to time would be opposite to the distance function in that everywhere the distance function is increasing the "approach function" would be decreasing.

The rate of approach is the rate at which the particles distance decreases. A positive rate of approach corresponds to a negative rate of change of the distance function. A negative rate of approach is the same as a positive rate of change of distance between particles. Thus, in reality, we are looking for the points where the rate of change of distance between particles changes from positive to negative.

Looking at the curve this occurs three times on the interval (corresponding to the three maxima).

## 76. (E)

Points of inflection are points where the rate of change of the tangent to a curve is zero. That is, the concavity of the curve changes from concave up to concave down or vice versa. Points of inflection are where the function f reaches a maximum or minimum value. Since the function is symmetrical the minimum is equal to the negative of the maximum of f.

Therefore, at every point of inflection, $|f|$ is at its maximum value.

From the curve we see that the function g changes concavity seven times.

### 77. (C)
The function f has been defined as the first derivative of the distance function g. To see whether a function is increasing or decreasing in a certain interval, we look at its derivative in that interval. If the derivative is positive the function is increasing; if it's negative it is decreasing.

Thus, to see if f is decreasing or increasing we must examine its derivative, which is the second derivative of the function g.

Examining the second derivative of a function is analagous to examining the concavity of the function. When a function is concave up the second derivative is positive. When a function is concave down its second derivative is negative.

From the graph of g, we see that only on the interval from $t = 5$ to $t = 6$ the function g is strictly concave upward.

### 78. (D)
The function h(x) is equivalent to the area bounded by the distance function g(t) on the interval from $t = 0$ to $t = x$ where $x \geq 0$.

We look at the curve on the given interval and observe that it is always positive and thus, bounds only positive area. As x increases, more positive area is bounded by g. Thus, on the given interval h(x) is always increasing; so it is increasing at all points from 0 to 14.

### 79. (A)
The operator $D_x$ takes the derivative of h(x). $D_x h(x)$

is equivalent to $\frac{d}{dx} h(x)$ or $\frac{d}{dx} \int_0^x g(t)dt$.

From the Fundamental Theorem of Calculus, this is equal to $g(x)$. This is simply $g(t)$ with the variable changed to $x$. Thus, looking at the graph of $g(t)$ on the interval $0 \leq t \leq 14$ is the same as $g(x)$ on the interval $0 \leq x \leq 14$.

We observe that $g(x)$ is not strictly increasing or decreasing and it <u>is</u> monotonic on subintervals. However, it is periodic (with period = 4). Thus, the answer is (A).

80. (B)
In this problem, we are asked to find the approximate area bounded by g from 0 to 4. One method for doing this is Simpson's rule.

Simpson's rule is stated as:

$$\int_a^b f(x)dx = \frac{b-a}{3n} [f(x_0) + 4f(x_1) + 2f(x_2) + 4f(x_3) + \ldots + 2f(x_{n-2}) + 4f(x_{n-1}) + f(x_n)]$$

In this problem, we have $b = 4$, $a = 0$, $n$ = number of intervals = 4. Thus, we have:

$$\int_0^4 f(x)dx = \frac{4}{12} [f(x_0) + \ldots + f(x_n)]$$

From the graph we obtain:

$f(x_0) = f(0) = 5$, $f(x_1) = f(1) \approx 3$, $f(x_2) = f(2) \approx 3$,

$f(x_3) = f(3) \approx 6$, $f(x_4) = f(4) = 5$.

Hence

$h(4) \approx \frac{1}{4(3)} [5 + 12 + 6 + 24 + 5]$

$= \frac{52}{12} 4 = \frac{52}{3} \approx 17$

81. (E)

We are told that the curve is a gaussian density function. When the function decays to approximately 0.6 times its maximum value, then the function is being evaluated at either $a_x - \sigma_x$ or $a_x + \sigma_x$ where $a_x$ is the mean of X and $\sigma_x$ is the standard deviation.

Since the function is symmetrical about a point between $x = 5$ and $x = 6$, this is where the mean lies.

Next, we see where the function is approximately 0.6 of its maximum. The maximum is $a$. $0.6a$ lies between $2/3a$ and $1/2a$.

Following a vertical path from the function between $\frac{2}{3}a$ and $\frac{1}{2}a$ down to the x-axis, we see that the corresponding f interval is between $x = 3$ and $x = 4$ or between $x = 7$ and $x = 8$. These represent deviations from the mean.

Estimating from $|3 < x < 4| - |5 < x < 6|$ or $|7 < x < 8| - |5 < x < 6|$ we see that the standard deviation of the curve is approximately 2.

82. (C)
f(x) is defined to be a probability density function, therefore, the area that it bounds from negative to positive infinity, that is, $\int_{-\infty}^{\infty} f(x)dx$ is equal to 1.

One is not given in the choices. Choices (D) and (E) can be eliminated because they do not equal 1. We shall prove that (C) is the correct choice.

$\sigma_x^2$, or variance, is defined as $E[(X - a_x)^2]$, where $E[N]$ is defined as the "expected" value or average value of random variable N.

$E[(X - a_x)^2]$ may be expanded:

$E[X^2 - 2Xa_x + a_x^2]$.

The expected value of a sum is equal to the sum of expected values. Thus, we have:

$\sigma_x^2 = E[X^2] - E[2Xa_x] + E[a_x^2]$.

The average of a constant is the constant, so:

$$\sigma_x^2 = E[X^2] - 2a_x^2 + a_x^2$$
$$= a_{x^2} - a_x^2.$$

Dividing both sides by $a_{x^2} - a_x^2$ gives us

$$\frac{\sigma_x^2}{a_{x^2} - a_x^2} = 1 \text{ which equals } \int_{-\infty}^{\infty} f(x)dx.$$

**83. (D)**
One property of a probability density function is that the area it bounds over a certain interval is equal to the probability that its random variable lies in that interval. In other words:

$$P\{x_1 < X \le x_2\} = \int_{x_1}^{x_2} f_x(x)dx.$$

In this question, $x_2$ is equal to 2. $x_1$ equals $-\infty$ because it was specified that x not be greater than 2. Therefore the interval is $-\infty < X \le 2$ and the probability is

$$P\{-\infty < X \le 2\} = \int_{-\infty}^{2} f(x)dx.$$

**84. (C)**

The integral $\int_{-\infty}^{x} f(t)dt$ is defined as the probability distribution fucntion of Random variable X. The distribution function of a random variable is defined as:

$$F_x(x) = P\{X \le x\}$$

where X is the random variable. We see that as x increases the probability that X is less than x will increase. Therefore the function $F_x(x)$ is a monotonically increasing function on certain intervals. It is not necessarily a straight line, and it is never a decreasing function. It is also not a function with just 1 maximum, such as f(x).

### 85. (A)

The skew of a density function is defined as the third central moment of the density function. It is expressed as: $E[(X - \overline{X})^3]$, where E stands for the expected value of $(X - \overline{X})^3$. When the density function $f_x(x)$ is symmetric about its mean, then the density function has zero skew, the skew being a measure of the density function's asymmetry.

### 86. (E)

When two forces act in opposite directions on an object, the resultant force produced is equal to the difference between the two forces.

Since in the given problem, the resultant force is 5-newtons, which is equal to the difference of the two given forces, the angle between the two forces is 180°. Also the direction of the resultant is the direction of the larger of the two forces, in our case the 13-newton force.

### 87. (D)

First of all, the student has made a mistake in circuit connection for measuring the resistance by the voltmeter ammeter method. The right method is that the ammeter should be in series and the voltmeter across the lamp.

For this student's set-up, very little current will flow through the lamp. The reason for this, is two fold: One, ammeters have very small resistances. When connected in parallel with the lamp, it acts almost as a short circuit. Two, voltmeters have very large resistances ($\sim 10^6 \Omega$). Therefore, when the voltmeter is connected in series with the lamp, it acts almost as an open circuit.

### 88. (C)

From the equation of the chemical reaction we see that

1 mol Pb   produces   2 mol $PbSo_4$

or 207 g Pb produces $2(207 + 32 + (4 \times 16))$ g PbSo$_4$

i.e. 207 g Pb produces 606 g PbSo$_4$

Thus

$$6.06 \text{ g PbSo}_4 \times \frac{207 \text{ g Pb}}{606 \text{ g PbSo}_4} = 2.07 \text{ g Pb}$$

Answer: 2.07 g Pb


### 89. (A)

The curves below show the speed-torque characteristics for a wound rotor induction motor for different values of rotor resistance.

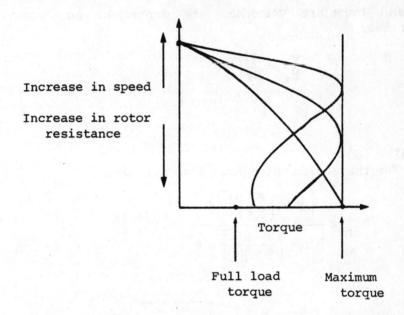

As the rotor resistance is increased, the maximum torque occurs at lower speeds. It gives a high starting torque, a high slip at normal load and hence a high rotor loss with a corresponding decreased efficiency. It decreases the starting current and increases the power factor, the power factor increasing faster than the starting current decreasing, thus leading to a better starting torque.


### 90. (A)

In (I) there is only one unknown variable, $v_2$, so it can be used. Equation (III) has left us with another variable, m, which is no help. Likewise, equation (II) needs the value of the density.

## 91. (A)
Bernoulli's equation

$$\frac{P_1}{\gamma} + \frac{V_1^2}{2g} + Z_1 = \frac{P_2}{\gamma} + \frac{V_2^2}{2g} + Z_2 + h_{1-2}$$

shows that all the pressure differences depends on all the listed parameters. Hence:

$$\frac{P_1 - P_2}{\gamma} = \frac{1}{2g}(V_2^2 - V_1^2) + Z_2 - Z_1 + h_{1-2}$$

$$P_1 - P_2 = \frac{\gamma}{2g}(V_2^2 - V_1^2) + \gamma(Z_2 - Z_1) + h_{1-2}\gamma$$

From continuity we have:

$$\rho A_1 V_1 = \rho A_2 V_2$$

and therefore velocities are dependent on cross-sectional areas.

$$\frac{V_1}{V_2} = \frac{A_2}{A_1}$$

## 92. (B)
For the equivalent circuit shown below,

$$R_{eq} = \frac{\left[R + \left(\frac{8 \times 8}{8 + 8}\right)\right]4}{\left[R + \left(\frac{8 \times 8}{8 + 8}\right)\right] + 4} = \frac{4R + 16}{R + 8}$$

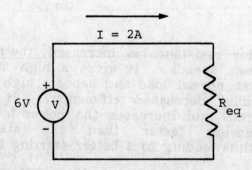

Now use Ohm's Law $V = IR_{eq}$

$$\therefore R_{eq} = \frac{4R + 16}{R + 8} = \frac{V}{I} = \frac{6}{2} = 3$$

Solving the equation we obtain,

$$4R + 16 = 3R + 24$$

$$\therefore \quad R = 8\,\Omega.$$

**93. (D)**
For the figure shown, the amount of emf induced depends on the rate at which the flux is changing through the coil when the coil is turned at a constant speed, the coil edges intersects the line at forces, generating certain amount of emf inside the coil. This generated emf continuously changes with the position of the coil about the magnetic poles. When the coil is parallel to the magnetic poles, the amount of field lines intersected by the coil is a minimum, thereby producing a minimum emf. Similarly when the coil is perpendicular to the magnetic poles, during its constant rotation, it cuts the maximum number of field lines and the generated emf has a maximum at that instant of time.

**94. (D)**
Beats are defined as a series of alternate reinforcements and cancellations produced by the interference of two sets of superimposed waves of slightly different frequencies.

The number of beats is equal to the difference in the frequencies of the two sound waves. If the difference in frequency is greater than 10 Hz, it is not readily detected by the ear. If the two sound waves have the same frequency, there would be no beats.

**95. (C)**
The principle of conservation of energy will be used to solve this problem. Initially, the block possesses potential energy equal to mgh, where m is the mass of the block, g is the gravitational acceleration, and h is the distance from the ground level. When the block is dropped, the potential energy is decreased while the kinetic energy is increased until the instant before impact when the potential energy is at its minimum, zero, and the kinetic energy is at its maximum. Since there were no losses, the potential energy is transferred completely to kinetic energy. Thus, maximum kinetic energy is mgh. At the time of impact, all of the kinetic energy is changed into heat. Therefore,

there are mgh joules of heat generated.

$$\text{Energy} = mgh = (1 \text{ kg})(9.8 \text{m/sec}^2)(10\text{m}) = 98\text{J}$$

## 96. (C)

The total x and y components of linear momentum must be conserved after the collision. The mass of the body resulting after the collision is

$$m = m_1 + m_2$$

and the velocity $\vec{v}$ is inclined at angle to the x-axis. We know that the total momentum vector is unchanged, and we can write down the x and y components of momentum.

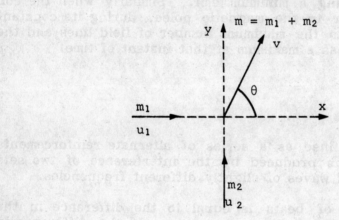

|  | INITIAL MOMENTUM | FINAL MOMENTUM |
|---|---|---|
| x component | $m_1 u_1$ | $(m_1 + m_2)v \cos\theta$ |
| y component | $m_2 u_2$ | $(m_1 + m_2)v \sin\theta$ |

Thus $m_1 u_1 = (m_1 + m_2)v \cos\theta$

$$m_2 u_2 = (m_1 + m_2)v \sin\theta$$

$$\therefore \tan\theta = \frac{m_2 u_2}{m_1 u_1}$$

$$= \frac{b \times a}{a \times b} = 1$$

Hence $\theta = 45°$.

## 97. (B)

Since the equilibrium-constant expression for the reaction is the ratio of concentrations of products to reactants,

$$K'_{eq} = \frac{[Na_2CO_3(s)][CO_2(g)][H_2O(g)]}{[NaHCO_3(s)]^2}$$

With the information that the solids have constant concentrations, their ratio can be incorporated into the equilibrium constant, so

$$K_{eq} = [CO_2(g)][H_2O(g)]$$

which is (B).

## 98. (B)

By Newton's third law the force exerted on one man by the floor of the elevator is counteracted by an equal and opposite force R exerted by the man on the floor. Therefore, the scale will measure the weight W of the man plus the force, R, exerted by him on the floor.

$$R = m a_{elev}$$

$$R = \frac{W}{g} a_{elev} = \frac{128 \text{ lbs}}{32.0 \frac{ft}{s^2}} (4 \frac{ft}{s^2})$$

$$R = 16 \text{ lbs.}$$

Therefore the reading on the scale is

$$W + R = 128 + 16 = 144 \text{ lbs.}$$

## 99. (C)

The kinetic energy per mole of the molecules of an ideal gas is proportional to the temperature.

$$\frac{1}{2} m \bar{v}^2 = \frac{3}{2} RT$$

Therefore at the same temperature to the translational kinetic energy of both gases are equal,

$$\frac{3}{2} RT_0 = \frac{1}{2} m_1 \bar{v}_1^2 = \frac{1}{2} m_2 \bar{v}_2^2$$

$$\frac{\overline{v_1}^2}{\overline{v_2}^2} = \frac{m_2}{m_1}$$

$$\frac{\sqrt{\overline{v_1}^2}}{\sqrt{\overline{v_2}^2}} = \frac{v_{1\,rms}}{v_{2\,rms}} = \sqrt{\frac{m_2}{m_1}}$$

**100. (E)**

Acceleration $= \frac{dv}{dt}$ is the slope of the velocity vs. time graph. When acceleration is zero, $\frac{dv}{dt} = 0$.

So that the velocity stays constant at all times, curve A gives this information.

**101. (A)**

When the object is not far away from the earth's center, the acceleration due to gravity g is constant or nearly so. Thus

$$\frac{dv}{dt} = constant$$

i.e. the velocity varies linearly with time.

**102. (B)**

This problem is an application of the principle of conservation of linear momentum. Actually, the equation that is used is the simplification of the momentum equation:

$$\Sigma F = \rho Q (V_2 - V_1)$$

where  p = density

Q = discharge

$V_2$ = out velocity of liquid

$V_1$ = in velocity of liquid

The component equation for the horizontal (x) direction is:

$$\Sigma F_x = \rho A (V_{2x}^2 - V_{1x}^2)$$

The velocity magnitude at the inlet is V and is also V at the outlet.

Thus $V_{1x} = V\cos 45° = \frac{\sqrt{2}}{2} V$.

Similarly,

$$V_{2x} = \cos 60° = -V(\tfrac{1}{2})$$

Hence: $\Sigma F_x = \rho A(V_{2x}^2 - V_{1x}^2)$

$$= \rho A \left[\tfrac{1}{4}V^2 - \tfrac{2}{4}V^2\right]$$

$$= -\rho AV^2(\tfrac{1}{4})$$

In the y direction:

$$V_{1y} = -V\sin 45° = -V\frac{\sqrt{2}}{2}$$

$$V_{2y} = \sin 60° = V\sqrt{\tfrac{3}{2}}$$

Thus $\Sigma F_y = \rho A(V_{2y}^2 - V_{1y}^2)$

$$= \rho A \left[\tfrac{3}{4}V^2 - \tfrac{2}{4}V^2\right]$$

$$= \rho A(\tfrac{1}{4})V^2$$

Taking the ratio of their magnitudes:

$$\frac{|\Sigma F_x|}{|\Sigma F_y|} = \frac{\rho AV^2(\tfrac{1}{4})}{\rho AV^2(\tfrac{1}{4})} = 1:1$$

103. (C)
 The correct answer, adding vectors as shown below, is (C) − 63 m/sec.

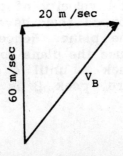

$$v_B = \sqrt{60^2 + 20^2} = \sqrt{4000} \approx 63 \text{ m/sec}.$$

### 104. (D)
Hydrogen bonding causes an open network structure in ice which collapses when the ice melts. Only part of the hydrogen bonds of ice are broken when it melts. The clusters of hydrogen bonded molecules that remain break up as the temperature increases and the volume continues to shrink.

### 105. (E)
The expression

$$\Delta E = q - w$$

for a closed system equates the change in internal energy for a system is made up of the net change in heat added to or subtracted from (endothermic or exothermic) the system and the net work done by the system during a reaction. All of the statements are correct and applicable.

### 106. (B)
When comparing the two alternatives the present worth must be determined for each of them. The capitalized cost represents the present total cost of financing and maintaining a given alternative for an indefinite period of time. The capitalized costs need to be determined for the annual operating expenses and the overhaul process. However, the future worth of the annual operating expenses cannot be obtained for an indefinite period of time.

### 107. (D)
Suppose the horizontal velocity of the plane is v. When pack A and B are dropped they have the same horizontal velocity as that of the plane. Hence pack A is directly beneath the plane because the plane covers exactly the same distance as that of pack A until A hits the ground two seconds later. Therefore, pack B is directly beneath the plane as well.

Hence, both the packs are directly beneath the plane at all times.

The following figure illustrates the movement of the two packs at different instants of time.

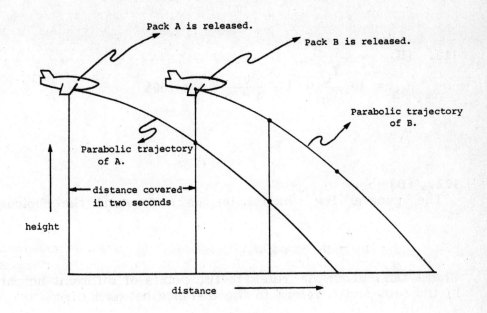

**108. (E)**
In an n-type semiconductor crystal, the density of electrons in the conduction band is essentially the density of donor atoms in the crystal, and hence the electrical conductivity is given by the relation

$$\sigma_n \sigma_n = Ndq\,\mu_n$$

where
- $\mu_n$ = Electrical conductivity of an n-type semiconductor
- Nd = Number of donor atoms per cubic meter
- $\mu_n$ = electron mobility
- q = charge of an electron, which is a constant.

Hence the conductivity closely depends on the number of donor atoms and the electron mobility.

**109. (E)**
Because the outer pipe is smaller in thermal conductivity, it serves effectively as an insulator. Furthermore, since

its thickness is 2 inches, it can be assumed that it's beyond the critical thickness; the critical thickness refers to the thickness of the insulator below which the rate of heat transfer would actually increase. Typical values for this thickness are 1 or 2 mm.

## 110. (E)

$$\eta_{th} = 1 - \frac{T_L}{T_H} = 1 - \frac{0}{3+3} = 1 = 100\%$$

## 111. (D)

The general law which helps us understand the choices is

$$p_2 - p_1 = \rho g h$$

where the subscripts refer to two points of different height in the tube and h refers to the distance between them.

The product of the density, $\rho$, with the acceleration constant, g, is called the weight density and is constant. So the pressure difference, $p_2 - p_1$ is linearly related to height difference, h. For equal values of h, $(p_2 - p_1)$ values will be equal.

## 112. (A)

The mass of the material is given by

$$\text{mass} = (\text{mass density})(\text{volume}) \quad (1)$$

We need to find the mass density first.

mass density of material = specific gravity × mass density of water

$$= 13 \times 62.4 \text{ lb/ft}^3.$$

Substituting this in (1) we get

mass = $13 \times 62.4 \text{ lb/ft}^3 \times 10$

= $130 \times 62.4 \text{ lb/ft}^3$ lbm

## 113. (A)

Boyle's Law states:

The volume of a given mass of gas held at constant temperature is inversely proportional to the pressure under which it is measured.

$$V \propto \frac{1}{P}$$

or $PV = k$

The second equation means that the product of the pressure of a given mass of gas times its volume is always constant if the temperature does not change.

## 114. (B)

$$\text{Base current } I_B = \frac{V_{BB} - V_{BE}}{R_B} = \frac{.5 - 1}{5k} = -.1 \text{ mA}$$

Since the base current that flows through the base-emitter junction is negative and given that the reverse saturation current $I_{CBO} = 0$, then the transistor is cut off.

## 115. (C)

Solid I, II and III can be thought of as three resistors, $R_1$, $R_2$ and $R_3$:

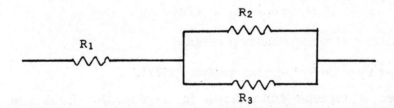

where the analogy of electric current is q and the analogy of electric potential is temperature. When the thermal conductivity of solid II is increased, the effect is to reduce $R_2$ since thermal conductivity and resistance are inversely proportional. $R_2$ and $R_3$ are in parallel so the equivalent parallel resistance decreases as $R_2$ decreases, and the total resistance from face A to face C is reduced. Since the heat flow is constant and obeys the equation $q = \Delta T/R_{TOT}$ — if $R_{TOT}$ decreases, so must $\Delta T$. Face A − face C = $\Delta T$ and face C must increase in temperature.

## 116. (D)

A.H. Compton explained that when a beam of x-rays of sharply defined wavelength fell on a graphite block, the photons striking the free electrons transferred some of its energy to the electron with which it collided. The photon must then have a lower energy and so a lower frequency which implies a larger wavelength and hence the shift in wavelength which is known as the Compton shift.

## 117. (C)

Full annealing consists of heating steel to a temperature above its critical range, holding it there for a sufficient period of time and slowly cooling it. Heat treatment of the steel to a temperature below the critical temperature will relieve internal stresses produced by rapid cooling or produce spheroids in a territe matrix. Annealing above the critical temperature will refine the coarse granular condition of the steel.

## 118. (C)

The continuity equation tells us that:

$$\rho_1 V_1 A_1 = \rho_2 V_2 A_2.$$

In this problem, $\rho_1 = \rho_2$ and $A_1 = A_2$ throughout the length, L, of the tube. Therefore V is constant from one cross-sectional area to another over the length, L.

## 119. (A)

When a viscous fluid flows in a pipe the fluid near the walls of the pipe become retarded while the fluid in the center line increases in velocity. This occurs if the flow rate is to remain constant. Integrating the velocity distribution over the cross-sectional area the net force acting on the fluid is $F = (p_1 - p_2) \pi r^2$. Since the flow rate does not change the force due to the pressure must just balance the viscous retarding force

$$F = \eta A \frac{dv}{dr} = \eta 2\pi rL \frac{dv}{dr}$$

Therefore, the sum of the forces is equal to 0.

**120. (A)**
The equation for the change in entropy for a process is given as

$$ds = \left(\frac{\delta q}{T}\right)_{rev}$$

where q is the heat supplied to the system and T, the temperature.

If the process is isentropic it follows that the process is adiabatic and reversible.

**121. (C)**
Miller indices specify a plane by the integral common denominator of the values of the three coordinates of its location.

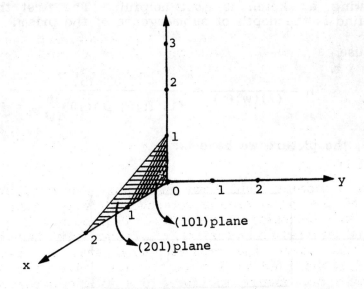

In fact these planes overlap.

122. (B)
For a solid:
$$q = C \Delta T$$

Rearranging,
$$\Delta T = \frac{q}{C}$$

For the same $\Delta T$, as C decreases, q must increase. Likewise, as C increases, q must decrease.

123. (B)
To change 50g of water from 30°C to water at 90°C = $m \Delta T$ = 50g × 60°C = 3000 calories. To change 50g of water at 90°C to steam at 90°C =

$$m(\text{heat of vaporization} = 540 \frac{cal}{g}) = 50g \times 540 \frac{cal}{g}$$

$$= 27,000 \text{ cal.}$$

Total = 27,000 + 3000 = 30,000 cal.

124. (B)
Drawing a sketch is most helpful: The first thing we must find is the depth of submergence of the prism.

We use

$$h = \frac{W}{(\ell)(w)(\gamma)} = \frac{12,000 \text{ lb.}}{(15 \text{ ft})(2 \text{ ft})(100\frac{lb}{ft^3})} = 4 \text{ ft.}$$

Hence, the picture we have is:

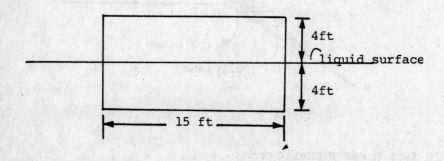

When the prism is rotated until it floats in the diagonal plane

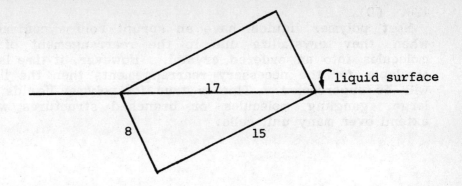

The buoyant force is:

$$F_B = \gamma V,$$

where V is the volume submerged (displaced).

$$V = (\text{Area})(w)$$

$$A = \tfrac{1}{2}(bh) = \tfrac{1}{2}(15)(8) = 60 \text{ ft}^2$$

$$V = Aw = (60)(2) = 120 \text{ ft}^3$$

and $\qquad F_B = (120)(100) = 12{,}000$ lb.

**125. (B)**
In order to determine the viability of each proposal, the net present value of each of them needs to be determined at the minimum attractive rate of return. If the net present value is greater than or equal to zero in any given proposal, then the proposal is acceptable.

$$NPV_A = -55{,}000 + 18{,}000(P/A, \ 25\%, \ 10)$$

$$= -55{,}000 + 18{,}000 \ \tfrac{1}{.3}$$

$$= \$+5{,}000$$

$$NPV_B = -85{,}000 + 24{,}000 \ \tfrac{1}{.3}$$

$$= \$-5{,}000$$

$$NPV_C = -91{,}000 + 27{,}000 \ \tfrac{1}{.3}$$

$$= \$-1{,}000$$

Therefore only proposal A is an acceptable one ($NPV_A > 0$).

126. (D)
Most polymer liquids have an abrupt volume contraction when they crystallize due to the rearrangement of the molecules into an ordered crystal. However, if time is not available for the necessary rearrangements then the liquid will be supercooled. This is true for polymer liquids with large, gangling molecules or branched structures which extend over many unit cells.

127. (A)
A mass suspended from a spring with spring constant, k, will oscillate in simple harmonic motion when losses from air friction and the like are neglected. The potential energy, the energy due to the mass' location, is

$$PE = \tfrac{1}{2}kx^2 \qquad PE = \tfrac{1}{2}kA^2$$

where x is the distance from the mass' starting point at highest amplitude.

The kinetic energy, the energy due to the mass' motion, is, $$KE = \tfrac{1}{2}mv^2$$

At full amplitude, A, the spring is stretched furthest and the mass stops moving before changing direction. At this point,

$$KE = \tfrac{1}{2}m(\phi)^2 = \phi$$

So all the energy is potential and the kinetic energy is zero.

128. (D)
The effect of alloying elements such as nickel or chromium is to delay the transformation of austenite to pearlite and thus shift the s curve to the right. The effect of nickel is to lower the critical transformation temperature to produce longer transformation times. Chromium will raise the critical transformation temperature but produce transformation times longer than nickel.

129. (A)
The orbital configurations in (B), (C) and (E) are impossible because they violate the Pauli exclusion principle by having two electrons in an orbital with the same spin

( ↑↑ ). The configuration represented in (D) is in the ground state. Only (A) represents a possible excited state.

130. (B)
The ionic bonds form very densely packed and highly coordinated structures. Covalent bonds are weaker attractions, so form less densely packed structures.

131. (D)
Differentiate to yield,

$$2u \cos v \frac{\partial u}{\partial x} - u^2 \sin v \frac{\partial v}{\partial x} - 1 = 0,$$

$$2u \sin v \frac{\partial u}{\partial x} + u^2 \cos v \frac{\partial v}{\partial x} - 0 = 0.$$

These two linear equations in $\frac{\partial u}{\partial x}$ and $\frac{\partial v}{\partial x}$ may be solved by Cramer's rule,

$$\frac{\partial u}{\partial x} = \frac{\begin{vmatrix} 1 & -u^2 \sin v \\ 0 & u^2 \cos v \end{vmatrix}}{\begin{vmatrix} 2u \cos v & -u^2 \sin v \\ 2u \sin v & u^2 \cos v \end{vmatrix}} = \frac{u^2 \cos v}{2u^3(\cos^2 v + \sin^2 v)} = \frac{\cos v}{2u}$$

$$(u \neq 0)$$

132. (A)
A skew symmetric matrix $A_{n \times n}$ is a matrix with $A^T = -A$, where $A^T$ is the transpose of matrix A.

The matrix

$$\begin{bmatrix} 0 & -2 & 5 \\ 2 & 0 & 6 \\ -5 & -6 & 0 \end{bmatrix}^T = \begin{bmatrix} 0 & 2 & -5 \\ -2 & 0 & -6 \\ 5 & 6 & 0 \end{bmatrix}$$

is skew symmetric.

133. (B)
The sketch is

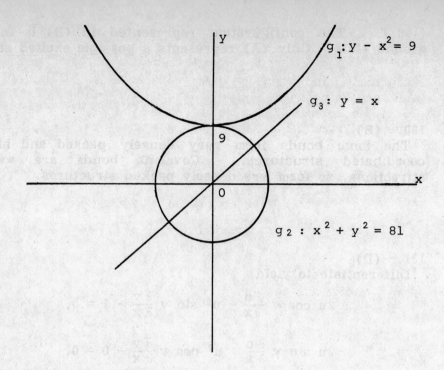

We can see that $g_1$ intersects $g_2$ at $(0,9)$.

Analytical Method:

The point of intersection of $g_1$ and $g_2$ can be found mathematically.

$$y - x^2 = 9$$

$$x^2 + y^2 = 81$$

Adding these two equations,

$$y^2 + y - 90 = 0$$

Solving, $\quad y = 9, -10.$

134. (D)

The intersection of two sets a and b is the set of all elements that belong to both a and b; that is, all elements common to a and b. In this problem, if

$$a = \{1, 2, 3, 4, 5\} \quad \text{and} \quad b = \{2, 3, 4, 5, 6\},$$

then

$$a \cap b = \{2, 3, 4, 5\}.$$

135. (A)

The function $f(z) = \dfrac{z^2}{z-3}$ is analytic everywhere in the complex plane except at the point $z = 3$. However, since this point is neither on nor contained in the circle $|z| = 1$, $f$ is analytic both on and within C so by the Cauchy-Goursat Theorem

$$\int_C \frac{z^2}{z-3}\,dz = 0.$$

136. (A)
If $f$ and $g$ are two functions of $x$, and $h = fg$

$$\frac{d}{dx}\ln(h) = \frac{1}{h}\frac{dh}{dx} = \frac{1}{fg}\frac{d}{dx}(fg) = \frac{1}{fg}\left(f\frac{dg}{dx} + g\frac{df}{dx}\right)$$

The rule can be applied with $f = 5^{3x^2}$, $g = \cos 5x$ to give

$$\frac{1}{5^{3x^2}\cos 5x}[(-5 \times 5)^{3x^2}\sin 5x + \cos 5x \cdot 5^{3x^2} \cdot \ln 5 \cdot 6x]$$

$$= 6x\ln 5 - 5\tan 5x$$

137. (C)
The general equation of a circle is

$$(x - a)^2 + (y - b)^2 = c^2$$

whose center is at $(a,b)$ and radius $c$. The given circle is centered at $(-a,b)$ and has diameter $r$, so the radius is $\dfrac{r}{2}$ the equation is

$$[x - (-a)]^2 + (y - b)^2 = \left(\frac{r}{2}\right)^2$$

$$= (x + a)^2 + (y - b)^2 = \frac{r^2}{4}$$

$$= [2(x + a)]^2 + [2(y - b)]^2 = r^2$$

138. (A)
Assume the solution is of the form $y = e^{mx}$ and substitute

for y. Thus, we obtain the characteristic equation

$$m^2 - 4\lambda m + 4\lambda^2 = 0.$$

Solving for m by the quadratic formula we obtain $m = 2\lambda$, a double root. The general solution is therefore

$$y = c_1 e^{2\lambda x} + c_2 x e^{2\lambda x}$$

### 139. (D)
With the integral test, consider

$$\int_1^t \frac{dx}{x^k} = \int_1^t x^{-k} dx = \frac{x^{1-k}}{1-k} \bigg|_1^t = \frac{t^{1-k} - 1}{1 - k}$$

at $k = 1$, $\quad \int_1^t \frac{dx}{x^k} = \int_1^t \frac{dx}{x} = \ln x \bigg|_1^t = \ln t$

and $\quad \lim_{t \to \infty} \ln t = \infty$

if $k < 1$

$$\lim_{t \to \infty} \frac{t^{1-k} - 1}{1 - k} = \infty$$

if $k > 1$

$$\lim_{t \to \infty} \frac{t^{1-k} - 1}{1 - k} = \frac{1}{k - 1}$$

So the series diverges for $k \leq 1$.

### 140. (B)
Applying the limit test $\lim_{k \to \infty} (\sqrt{k+1} - \sqrt{k})$

$$= \lim_{k \to \infty} (\sqrt{k+1} - \sqrt{k}) \left[ \frac{\sqrt{k+1} + \sqrt{k}}{\sqrt{k+1} + \sqrt{k}} \right] = \lim_{k \to \infty} \frac{1}{\sqrt{k+1} + \sqrt{k}}$$

$$= 0,$$

so it may or may not converge.

The kth partial sum

$$S_k = u_1 + \ldots + u_k = (\sqrt{2} - \sqrt{1}) + (\sqrt{3} - \sqrt{2}) + \ldots$$
$$\ldots + (\sqrt{k+1} - \sqrt{k}) = \sqrt{k+1} - \sqrt{1}$$

$\lim_{k \to \infty} S_k = \infty$, the series diverges.

## 141. (E)
Recall

$$\frac{1}{1-x} = 1 + x + x^2 + \ldots$$

if $|x| < 1$, or $-1 < x < 1$.

A result of the theorem: If $c \neq 0$, $|x| < 1$ and the geometric series

$$\sum_{n=1}^{\infty} cx^{n-1} = c + cx + cx^2 + \ldots + cx^{n-1} + \ldots$$

converges to the sum $S = \frac{c}{1-x}$. Setting $c = 1$ gives the result.

## 142. (D)
The given series is a geometric series. We have

$$\lim_{n \to \infty} \sum_{k=0}^{n} \frac{1}{3^k} = \lim_{n \to \infty} \sum_{k=0}^{n} \left(\frac{1}{3}\right)^k$$

Now to proceed further, we consider the geometric series

$$\sum_{n=0}^{n} x^n,$$

which converges if $|x| < 1$. That is $R = 1$ and

$$\sum_{n=0}^{\infty} x^n = \frac{1}{(1-x)}$$

where $|x| < 1$.

$$\therefore \lim_{n \to \infty} \sum_{k=0}^{n} \left(\frac{1}{3}\right)^k = \frac{1}{1 - \frac{1}{3}}$$

$$= \frac{1}{\frac{2}{3}}$$

$$= \frac{3}{2}$$

**143. (C)**

The given equation represents a plane in a 3-dimensional rectangular coordinate system. The intercepts of the plane are given by the points $(x',0,0)$, $(0,y',0)$, and $(0,0,z')$, where the non-zero coordinates signifies the axis with which the plane intersects.

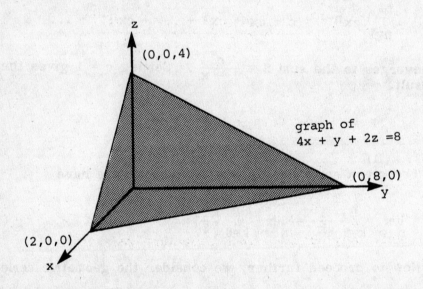

Therefore at the y-axis,

$$x = z = 0 \quad \text{and} \quad y = 8.$$

Hence the point of interception on the y-axis is 8 units from the origin.

**144. (D)**
Looking at the diagram, it is evident that the required area is to be evaluated between x = 0 and x = a. These values are the limits of the integral which give us the required area. The area is equal to the integral of the upper function minus the lower function. In this problem, the area is between y = x as the upper function and y = 0 (the x-axis) as the lower function. Therefore,

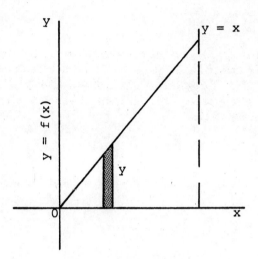

$$A = \int_0^a (x) - (0)\,dx = \int_0^a x\,dx$$

$$A = \frac{x^2}{2}\Big]_0^a = \frac{a^2}{2},$$

a result easily checked from elementary geometry.

# The Graduate Record Examination in
# ENGINEERING

## Test 3

# GRE ENGINEERING TEST 3
## ANSWER SHEET

1. Ⓐ Ⓑ Ⓒ Ⓓ Ⓔ
2. Ⓐ Ⓑ Ⓒ Ⓓ Ⓔ
3. Ⓐ Ⓑ Ⓒ Ⓓ Ⓔ
4. Ⓐ Ⓑ Ⓒ Ⓓ Ⓔ
5. Ⓐ Ⓑ Ⓒ Ⓓ Ⓔ
6. Ⓐ Ⓑ Ⓒ Ⓓ Ⓔ
7. Ⓐ Ⓑ Ⓒ Ⓓ Ⓔ
8. Ⓐ Ⓑ Ⓒ Ⓓ Ⓔ
9. Ⓐ Ⓑ Ⓒ Ⓓ Ⓔ
10. Ⓐ Ⓑ Ⓒ Ⓓ Ⓔ
11. Ⓐ Ⓑ Ⓒ Ⓓ Ⓔ
12. Ⓐ Ⓑ Ⓒ Ⓓ Ⓔ
13. Ⓐ Ⓑ Ⓒ Ⓓ Ⓔ
14. Ⓐ Ⓑ Ⓒ Ⓓ Ⓔ
15. Ⓐ Ⓑ Ⓒ Ⓓ Ⓔ
16. Ⓐ Ⓑ Ⓒ Ⓓ Ⓔ
17. Ⓐ Ⓑ Ⓒ Ⓓ Ⓔ
18. Ⓐ Ⓑ Ⓒ Ⓓ Ⓔ
19. Ⓐ Ⓑ Ⓒ Ⓓ Ⓔ
20. Ⓐ Ⓑ Ⓒ Ⓓ Ⓔ

21. Ⓐ Ⓑ Ⓒ Ⓓ Ⓔ
22. Ⓐ Ⓑ Ⓒ Ⓓ Ⓔ
23. Ⓐ Ⓑ Ⓒ Ⓓ Ⓔ
24. Ⓐ Ⓑ Ⓒ Ⓓ Ⓔ
25. Ⓐ Ⓑ Ⓒ Ⓓ Ⓔ
26. Ⓐ Ⓑ Ⓒ Ⓓ Ⓔ
27. Ⓐ Ⓑ Ⓒ Ⓓ Ⓔ
28. Ⓐ Ⓑ Ⓒ Ⓓ Ⓔ
29. Ⓐ Ⓑ Ⓒ Ⓓ Ⓔ
30. Ⓐ Ⓑ Ⓒ Ⓓ Ⓔ
31. Ⓐ Ⓑ Ⓒ Ⓓ Ⓔ
32. Ⓐ Ⓑ Ⓒ Ⓓ Ⓔ
33. Ⓐ Ⓑ Ⓒ Ⓓ Ⓔ
34. Ⓐ Ⓑ Ⓒ Ⓓ Ⓔ
35. Ⓐ Ⓑ Ⓒ Ⓓ Ⓔ
36. Ⓐ Ⓑ Ⓒ Ⓓ Ⓔ
37. Ⓐ Ⓑ Ⓒ Ⓓ Ⓔ
38. Ⓐ Ⓑ Ⓒ Ⓓ Ⓔ
39. Ⓐ Ⓑ Ⓒ Ⓓ Ⓔ
40. Ⓐ Ⓑ Ⓒ Ⓓ Ⓔ

41. Ⓐ Ⓑ Ⓒ Ⓓ Ⓔ
42. Ⓐ Ⓑ Ⓒ Ⓓ Ⓔ
43. Ⓐ Ⓑ Ⓒ Ⓓ Ⓔ
44. Ⓐ Ⓑ Ⓒ Ⓓ Ⓔ
45. Ⓐ Ⓑ Ⓒ Ⓓ Ⓔ
46. Ⓐ Ⓑ Ⓒ Ⓓ Ⓔ
47. Ⓐ Ⓑ Ⓒ Ⓓ Ⓔ
48. Ⓐ Ⓑ Ⓒ Ⓓ Ⓔ
49. Ⓐ Ⓑ Ⓒ Ⓓ Ⓔ
50. Ⓐ Ⓑ Ⓒ Ⓓ Ⓔ
51. Ⓐ Ⓑ Ⓒ Ⓓ Ⓔ
52. Ⓐ Ⓑ Ⓒ Ⓓ Ⓔ
53. Ⓐ Ⓑ Ⓒ Ⓓ Ⓔ
54. Ⓐ Ⓑ Ⓒ Ⓓ Ⓔ
55. Ⓐ Ⓑ Ⓒ Ⓓ Ⓔ
56. Ⓐ Ⓑ Ⓒ Ⓓ Ⓔ
57. Ⓐ Ⓑ Ⓒ Ⓓ Ⓔ
58. Ⓐ Ⓑ Ⓒ Ⓓ Ⓔ
59. Ⓐ Ⓑ Ⓒ Ⓓ Ⓔ
60. Ⓐ Ⓑ Ⓒ Ⓓ Ⓔ

61. Ⓐ Ⓑ Ⓒ Ⓓ Ⓔ
62. Ⓐ Ⓑ Ⓒ Ⓓ Ⓔ
63. Ⓐ Ⓑ Ⓒ Ⓓ Ⓔ
64. Ⓐ Ⓑ Ⓒ Ⓓ Ⓔ
65. Ⓐ Ⓑ Ⓒ Ⓓ Ⓔ
66. Ⓐ Ⓑ Ⓒ Ⓓ Ⓔ
67. Ⓐ Ⓑ Ⓒ Ⓓ Ⓔ
68. Ⓐ Ⓑ Ⓒ Ⓓ Ⓔ
69. Ⓐ Ⓑ Ⓒ Ⓓ Ⓔ
70. Ⓐ Ⓑ Ⓒ Ⓓ Ⓔ
71. Ⓐ Ⓑ Ⓒ Ⓓ Ⓔ
72. Ⓐ Ⓑ Ⓒ Ⓓ Ⓔ
73. Ⓐ Ⓑ Ⓒ Ⓓ Ⓔ
74. Ⓐ Ⓑ Ⓒ Ⓓ Ⓔ
75. Ⓐ Ⓑ Ⓒ Ⓓ Ⓔ
76. Ⓐ Ⓑ Ⓒ Ⓓ Ⓔ
77. Ⓐ Ⓑ Ⓒ Ⓓ Ⓔ
78. Ⓐ Ⓑ Ⓒ Ⓓ Ⓔ
79. Ⓐ Ⓑ Ⓒ Ⓓ Ⓔ
80. Ⓐ Ⓑ Ⓒ Ⓓ Ⓔ
81. Ⓐ Ⓑ Ⓒ Ⓓ Ⓔ
82. Ⓐ Ⓑ Ⓒ Ⓓ Ⓔ

83. Ⓐ Ⓑ Ⓒ Ⓓ Ⓔ
84. Ⓐ Ⓑ Ⓒ Ⓓ Ⓔ
85. Ⓐ Ⓑ Ⓒ Ⓓ Ⓔ
86. Ⓐ Ⓑ Ⓒ Ⓓ Ⓔ
87. Ⓐ Ⓑ Ⓒ Ⓓ Ⓔ
88. Ⓐ Ⓑ Ⓒ Ⓓ Ⓔ
89. Ⓐ Ⓑ Ⓒ Ⓓ Ⓔ
90. Ⓐ Ⓑ Ⓒ Ⓓ Ⓔ
91. Ⓐ Ⓑ Ⓒ Ⓓ Ⓔ
92. Ⓐ Ⓑ Ⓒ Ⓓ Ⓔ
93. Ⓐ Ⓑ Ⓒ Ⓓ Ⓔ
94. Ⓐ Ⓑ Ⓒ Ⓓ Ⓔ
95. Ⓐ Ⓑ Ⓒ Ⓓ Ⓔ
96. Ⓐ Ⓑ Ⓒ Ⓓ Ⓔ
97. Ⓐ Ⓑ Ⓒ Ⓓ Ⓔ
98. Ⓐ Ⓑ Ⓒ Ⓓ Ⓔ
99. Ⓐ Ⓑ Ⓒ Ⓓ Ⓔ
100. Ⓐ Ⓑ Ⓒ Ⓓ Ⓔ
101. Ⓐ Ⓑ Ⓒ Ⓓ Ⓔ
102. Ⓐ Ⓑ Ⓒ Ⓓ Ⓔ
103. Ⓐ Ⓑ Ⓒ Ⓓ Ⓔ
104. Ⓐ Ⓑ Ⓒ Ⓓ Ⓔ

105. Ⓐ Ⓑ Ⓒ Ⓓ Ⓔ
106. Ⓐ Ⓑ Ⓒ Ⓓ Ⓔ
107. Ⓐ Ⓑ Ⓒ Ⓓ Ⓔ
108. Ⓐ Ⓑ Ⓒ Ⓓ Ⓔ
109. Ⓐ Ⓑ Ⓒ Ⓓ Ⓔ
110. Ⓐ Ⓑ Ⓒ Ⓓ Ⓔ
111. Ⓐ Ⓑ Ⓒ Ⓓ Ⓔ
112. Ⓐ Ⓑ Ⓒ Ⓓ Ⓔ
113. Ⓐ Ⓑ Ⓒ Ⓓ Ⓔ
114. Ⓐ Ⓑ Ⓒ Ⓓ Ⓔ
115. Ⓐ Ⓑ Ⓒ Ⓓ Ⓔ
116. Ⓐ Ⓑ Ⓒ Ⓓ Ⓔ
117. Ⓐ Ⓑ Ⓒ Ⓓ Ⓔ
118. Ⓐ Ⓑ Ⓒ Ⓓ Ⓔ
119. Ⓐ Ⓑ Ⓒ Ⓓ Ⓔ
120. Ⓐ Ⓑ Ⓒ Ⓓ Ⓔ
121. Ⓐ Ⓑ Ⓒ Ⓓ Ⓔ
122. Ⓐ Ⓑ Ⓒ Ⓓ Ⓔ
123. Ⓐ Ⓑ Ⓒ Ⓓ Ⓔ
124. Ⓐ Ⓑ Ⓒ Ⓓ Ⓔ

125. Ⓐ Ⓑ Ⓒ Ⓓ Ⓔ
126. Ⓐ Ⓑ Ⓒ Ⓓ Ⓔ
127. Ⓐ Ⓑ Ⓒ Ⓓ Ⓔ
128. Ⓐ Ⓑ Ⓒ Ⓓ Ⓔ
129. Ⓐ Ⓑ Ⓒ Ⓓ Ⓔ
130. Ⓐ Ⓑ Ⓒ Ⓓ Ⓔ
131. Ⓐ Ⓑ Ⓒ Ⓓ Ⓔ
132. Ⓐ Ⓑ Ⓒ Ⓓ Ⓔ
133. Ⓐ Ⓑ Ⓒ Ⓓ Ⓔ
134. Ⓐ Ⓑ Ⓒ Ⓓ Ⓔ
135. Ⓐ Ⓑ Ⓒ Ⓓ Ⓔ
136. Ⓐ Ⓑ Ⓒ Ⓓ Ⓔ
137. Ⓐ Ⓑ Ⓒ Ⓓ Ⓔ
138. Ⓐ Ⓑ Ⓒ Ⓓ Ⓔ
139. Ⓐ Ⓑ Ⓒ Ⓓ Ⓔ
140. Ⓐ Ⓑ Ⓒ Ⓓ Ⓔ
141. Ⓐ Ⓑ Ⓒ Ⓓ Ⓔ
142. Ⓐ Ⓑ Ⓒ Ⓓ Ⓔ
143. Ⓐ Ⓑ Ⓒ Ⓓ Ⓔ
144. Ⓐ Ⓑ Ⓒ Ⓓ Ⓔ

# THE GRADUATE RECORD EXAMINATION ENGINEERING TEST

## MODEL TEST III

Time: 170 Minutes
      144 Questions

Directions: Choose the best answer for each question and mark the letter of your selection on the corresponding answer sheet.

1. For an internal combustion engine (Otto Cycle), the cycle begins with the adiabatic compression of air in the cylinder, followed by the heating of air at constant volume through spontaneous ignition, the adiabatic expansion of air back to the starting volume of the cylinder and ends with

   (A) the cooling of air at constant pressure

   (B) the cooling of air at constant volume

   (C) the isothermal expansion of air

   (D) both (A) and (C)

   (E) both (B) and (C)

2. The electric circuit shown below consists of a resistance and a source. If the potential difference of the source were decreased, the total heat developed in the circuit would

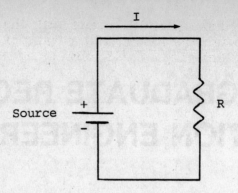

(A) decrease initially but then increase

(B) increase initially but then decrease

(C) increase

(D) decrease

(E) remain the same

3. A non-viscous liquid flows through a pipe that has a Venturi tube as shown below. If the Venturi tube acts as a meter to measure the mass flow rate and velocity of the fluid then one could determine that

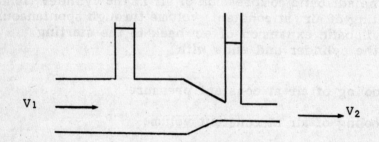

(A) the pressure has decreased and the velocity decreased

(B) the velocity has increased and the mass flow rate has increased

(C) the pressure has decreased and the velocity increased

(D) the velocity has decreased and the mass flow rate has remained constant

(E) the pressure has increased and the velocity decreased

4. In the figure, disks A and B rotate with the same angular velocity and weigh the same. If the ratio of the radii of A and B is 2:1, then the ratio of their kinetic energies is

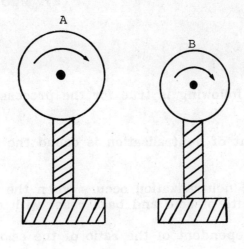

(A) 1:2

(B) 2:1

(C) 1:1

(D) 4:1

(E) 1:4

5. The geometric construction below can be used to prove that

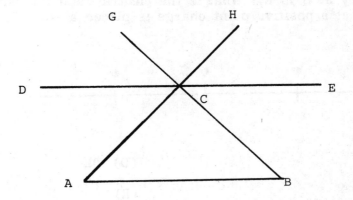

(A) the sum of two complementary angles is 90°

(B) the sum of two supplementary angles is 180°

(C) the sum of all the angles of a triangle is 180°

(D) the triangle ABC is a right triangle

(E) both (C) and (D)

6. All of the following ceramic materials are isotropic except

   (A) $SiO_4$
   (B) $SiO_2$
   (C) $MgSiO_3$
   (D) $Al_2O_3$
   (E) $MgO$

7. Which of the following is true for the process of acid-base titration?

   I. The point of neutralization is called the equivalence point.

   II. Complete neutralization occurs when the same number of equivalents of acid and base react with one another.

   III. It is independent of the ratio of the amounts of acid to base.

   (A) I only
   (B) III only
   (C) I and II only
   (D) II and III only
   (E) I, II and III

8. Based on the given diagram, assume that the electric field intensity at B is E. What is the electric field intensity of D? Note that a positive point charge is placed at A.

   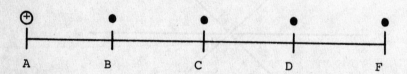

   (A) 4E
   (B) 3E
   (C) $\frac{1}{3}E$
   (D) 9E
   (E) $\frac{1}{9}E$

9. A pendulum is hung from the roof of a railroad car which is accelerating with constant acceleration a as shown in the figure below.

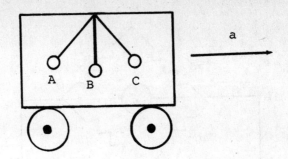

While the car is accelerating, the pendulum will

(A) stay in position A

(B) stay in position B

(C) stay in position C

(D) oscillate between A and C

(E) none of the above

10. In the circuit given below, what should be the value of x such that a charge of 600 coulombs enters the positive terminal of the 100V source in 1 min?

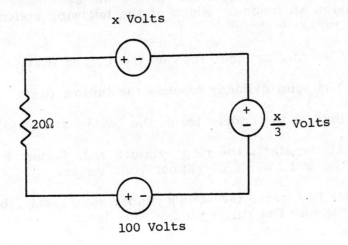

(A) 200V

(B) 100V

(C) 110V

(D) 240V

(E) 225V

11. The bridge in the circuit given below is balanced when the value of the unknown resistance is

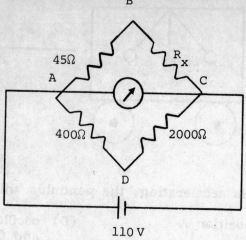

(A) 200Ω

(B) 225Ω

(C) 250Ω

(D) 275Ω

(E) 300Ω

12. A solid and a ring cylinder having the same mass and radius roll down an incline. Which of the following statements about their motion is true?

   (A) The ring cylinder reaches the bottom first.

   (B) The solid cylinder reaches the bottom first.

   (C) Both the cylinders reach the bottom at the same time.

   (D) At the start, the ring cylinder rolls faster, but towards the end the solid cylinder rolls faster.

   (E) At the start, the solid cylinder rolls faster, but towards the end the ring cylinder rolls faster.

13. Which of the following thermometers would be unreliable for measuring temperatures between 0°C to 10°C?

   (A) liquid column of mercury

   (B) liquid column of water

   (C) constant volume of He gas

   (D) thermocouple whose test junction is between two copper wires

(E) Both (B) and (D)

14. When water is carefully poured into a container having a very small hole in its base, the water does not leak out. This is possible due to the property of

    (A) adhesiveness
    (B) cohesiveness
    (C) capillarity
    (D) Brownian movement
    (E) surface tension

15. A spherical body at a temperature $T_1$ is surrounded by walls at a temperature $T_2$. At what rate must energy be supplied in order to keep the temperature of the body constant if $T_1 > T_2$?

    (A) $4\pi r^2 e \sigma T_1^4$
    (B) $k4\pi r(T_2)$
    (C) $4\pi r^2 e \sigma (T_1^4 - T_2^4)$
    (D) $4\pi r^2 e \sigma (T_2^4 - T_1^4)$
    (E) $k4\pi r(T_2 - T_1)$

16. Which of the following isothermal paths designated by temperatures would result in a phase change from vapor to liquid without the appearance of a meniscus?

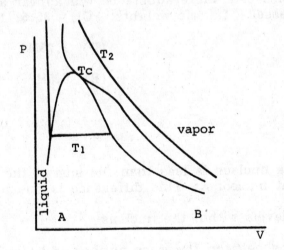

    (A) $T_2$ only
    (B) $T_1$ and $T_C$
    (C) $T_1$ only
    (D) $T_C$ and $T_2$
    (E) $T_C$ only

17. If the output of the circuit below is 8V then R must be adjusted to:
    (Note: A and B are ideal op-amps)

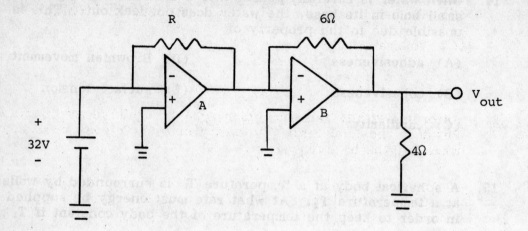

(A) 6Ω
(B) 10Ω
(C) 12Ω
(D) 16Ω
(E) 24Ω

18. 0.25L of 0.2M solution of silver nitrate has reacted with a solution containing 29.25 grams of salt (NaCl). Determine how many moles of a white substance will appear at the bottom of the vessel. (atomic weights: Cl = 35.5, Na = 23)

(A) 0.05M
(B) 1.5M
(C) 0.8M
(D) 1.25M
(E) 0.5M

19. The mass of a nucleus is less than the sum of the masses of its constituent nucleons. This difference is due to

(A) energy levels within the nucleus

(B) difference between the mass number and atomic number

(C) ionization potential

(D) binding energy

(E) inaccuracy of nucleon measuring device

20. Which of the following is an example of a thermodynamic extensive property?

   (A) Pressure
   (B) Temperature
   (C) Volume
   (D) Molar heat capacity
   (E) Specific volume

21. According to Stefan's law which must be modified by Quantum Physics, the total radiant intensity of a black body over all wave lengths is given as

$$\int_0^\infty R\,d\lambda = \sigma T^4.$$

If the radiant intensity is measured as $R(\lambda) = \frac{8\pi}{\lambda^4} kT \frac{c}{4}$, what are the units of the Stefan-Boltzman constant $\sigma$ in MKS notation?
Note: Boltzman's constant k is measured as $\frac{J}{K}$.

   (A) $\frac{J}{K^4}$
   (B) $\frac{W}{m^4 K^4}$
   (C) $\frac{W}{m^2}$
   (D) $\frac{W}{m^2 K^4}$
   (E) $\frac{W K^4}{m^2}$

22. The diagram below represents which of the following cycles?

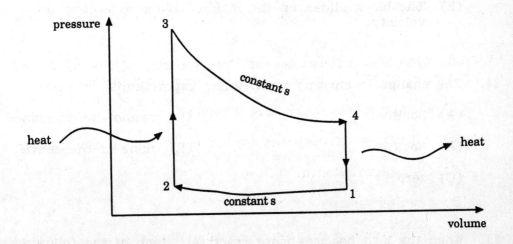

(A) Otto cycle
(B) Diesel cycle
(C) Gas-Turbine cycle
(D) Vapor Refrigeration cycle
(E) Creb's cycle

23. A 200 lb. force acts on the 500 lb. block along the inclined plane. The coefficients of friction between the block and plane are $\mu_s = 0.30$ and $\mu_k = 0.25$. Which of the following best describes the status of the block with the application of the 200 lb. force?

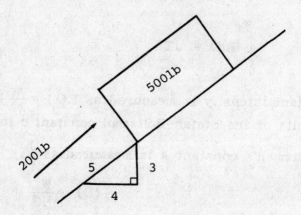

(A) The block remains at rest on the inclined plane.

(B) The block slides down the inclined plane at a constant velocity.

(C) The block accelerates up the inclined plane.

(D) The block accelerates down the inclined plane.

(E) The block slides up the inclined plane at a constant velocity.

24. The change in entropy when water vaporizes is

(A) positive
(B) negative
(C) zero
(D) cannot be determined
(E) none of the above

25. When the load becomes more reactive, which of the following is true?

(A) The power factor decreases

(B) The power factor increases

(C) The average power delivered increases

(D) No change in the power factor

(E) No change in the average power delivered

26. The efficiency for a Carnot engine that is operating between a cold reservoir at a temperature $T_C$ and a hot reservoir at a temperature $T_H$ is dependent upon

   (A) the heat capacity of the working substance

   (B) only the temperatures of the two reservoirs

   (C) the reservoir temperatures and the heat capacity of the working substance

   (D) the reservoir temperatures and the volume change during heat absorption

   (E) the volume change during heat absorption and the volume change during heat rejection

27. A hollow ball is carefully lowered into the middle of a pool of water. When released, the ball remained suspended in the water. Which of the following would best explain this behavior?

   (A) The surface turbulence that had caused the ball to sink doesn't exist in the middle of the water.

   (B) The turbulence at the bottom of the water acts to counter the weight of the ball.

   (C) The weight of the water on top of the ball negates the pressure of the water acting on the bottom of the ball.

   (D) The above phenomena could not occur; the ball should remain on the surface of the water.

   (E) None of the above

28. A wooden sphere of mass 1 kg is suspended on a string which is 1 meter long. A bullet of mass 50 grams is shot at the

sphere with a velocity $v_0$ and becomes embedded in it. Because of the impact, the sphere is raised a distance of 0.2m above the horizontal. What is $v_0$ of the bullet?

Note: assume $g = 10 \frac{m}{sec^2}$

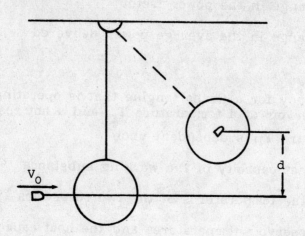

(A) $26 \frac{m}{sec}$

(B) $35 \frac{m}{sec}$

(C) $42 \frac{m}{sec}$

(D) $45 \frac{m}{sec}$

(E) $51 \frac{m}{sec}$

29. The inverse proportionality in the graph below could not represent the variation

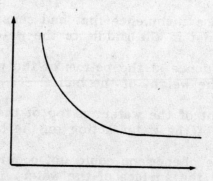

(A) of pressure with respect to volume for an ideal gas undergoing an adiabatic process

(B) of the electrostatic potential of a point charge with respect to distance from the charge

(C) of voltage for a discharging capacitor with respect to time

(D) of pressure with respect to volume for an ideal gas undergoing an isothermal process

(E) both (A) and (C)

30. Two parallel planes, 4 meters apart, cut a conic region which has a light bulb at its vertex. If the vertex is 2 meters above the upper plane at which the average light intensity is $1.1 \times 10^{-2}$ J/s·m$^2$, what is the average light intensity at the lower plane?

(A) $1.2 \times 10^{-5}$ J/m$^2$sec
(B) $3.3 \times 10^{-2}$ J/m$^2$sec
(C) $9.9 \times 10^{-2}$ J/m$^2$sec
(D) $3.6 \times 10^{-3}$ J/m$^2$sec
(E) $1.2 \times 10^{-3}$ J/m$^2$sec

31. A force of 1000 Newtons is needed to pull a sled weighing 1200 lb. along a horizontal surface at uniform speed by means of a rope which makes an angle of 30° with the horizon. What is the coefficient of friction?

(A) $\frac{\sqrt{3}}{4}$
(B) $\frac{1}{\sqrt{3}}$
(C) $\frac{5\sqrt{3}}{7}$
(D) $\frac{\sqrt{3}}{2}$
(E) $\frac{\sqrt{3}}{10}$

32. An object hangs on a vertical spring and displaces the spring an amount $y_0$ from its unstretched position. If the object is set into oscillation by stretching it an additional amount $y_1$, when it returns to its stretched equilibrium position the velocity is

(A) a maximum

(B) zero

(C) a minimum not equal to zero

(D) increasing

(E) decreasing

33. In the circuit diagram given below, the value of $i_1$ is

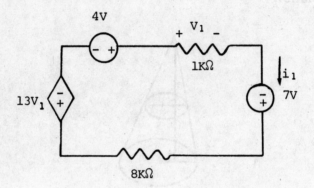

(A) 0.9 mA

(B) 0.6 mA

(C) 0.5 mA

(D) 0.2 mA

(E) 0.1 mA

34. The number of particles emitted from a reaction with energy $E_{Y_0}$ decreases with an increase in the energy $E_{Y_0}$ of the particle. This could be accounted for by

(A) the lower reaction probability of leaving the nucleus Y in a lower excited state

(B) the greater reaction probability of leaving the nucleus Y in a lower excited state

(C) the greater reaction probability of leaving the nucleus Y in a greater excited state

(D) the greater reaction probability of leaving the nucleus in a lower excited state

(E) none of the above

35. In the given situation, what is the weight of the gate? $\rho$ is the mass density of the water.

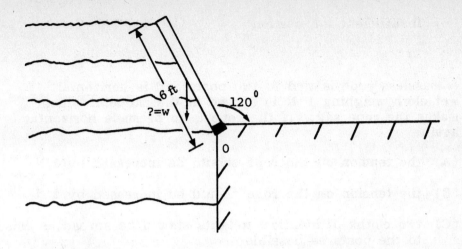

(A) 24ρg

(B) 12ρg

(C) 24√3ρg

(D) 12√3ρg

(E) √3ρg

36. The periodic table is not useful in predicting which of the following properties of an unknown element?

   (A) Number of chemical bonds usually formed

   (B) Formula of its chloride

   (C) Density

   (D) Melting point

   (E) All of the above

37. An elevator passenger was drinking juice from a cup by pouring it into his mouth, when the elevator started to descend. Assuming that the elevator descends downward at the same acceleration rate as gravity, what will happen to the passenger who tries to continue to drink while the elevator is still descending but has not started to slow down? (Neglect air friction)

   (A) drinks at the same rate

   (B) can't drink, because the juice stopped flowing

   (C) drinks slower, because the juice flows slower

   (D) drinks faster, because the juice flows faster

(E) insufficient information

38. A massless rope is tied to two posts and is horizontal. A wet cloth weighing 1 N is spread at the midpoint which makes the rope sag. If the rope is to be made horizontal again,

   (A) the tension on the rope should be increased by 1 N

   (B) the tension on the rope should be increased by 2 N

   (C) two cloths of identical weights should be spread as close to the posts as possible

   (D) supports on either side of the cloth should be provided

   (E) it is impossible to make the rope completely horizontal again

39. Your finger sticks to an ice tray just taken from the refrigerator. Which factor has more effect on this phenomenon?

   (A) The inside temperature of the freezer (Suppose it's working properly)

   (B) The humidity of the air

   (C) The heat capacity of both your finger and the tray

   (D) The thermal conductivity of the tray

   (E) The surface area of the tray

40. Which of the following graphs expresses the relationship of the distance d between two negative charges and the force between them?

(A)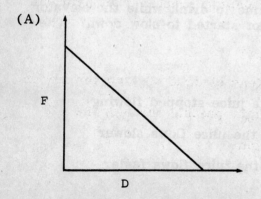

(B)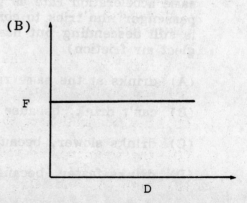

(C)

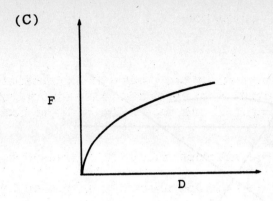

(D)

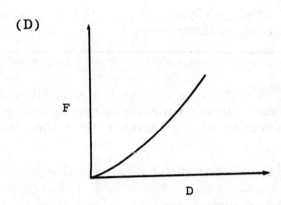

(E)

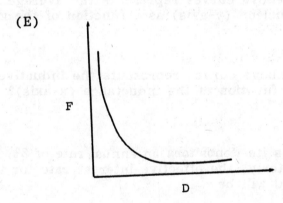

41. If the width of a slit producing a diffraction pattern for light is decreased, which of the following properties of light will become less certain?

   (A) the momentum of the light

   (B) the position of the photons of light striking the screen

   (C) the localization of the light passing through the slit

   (D) the frequency of the light

   (E) both (A) and (B)

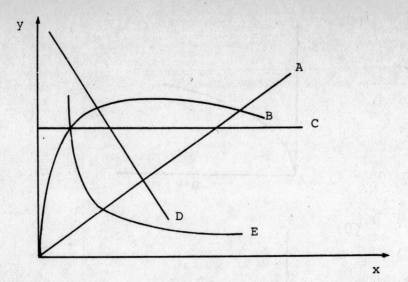

42. Which of the above curves represents the intensity of light (y-axis) as a function of the distance from the source (x-axis)?

43. Which of the above curves represents the average binding energy per nucleon (y-axis) as a function of the mass number (x-axis)?

44. Which of the above curves represents the inductive reactance (y-axis) as a function of the inductance (x-axis)?

45. If a bank pays its depositors an annual rate of 8%, compounded quarterly, the effective interest rate for the corresponding period will be

   (A) 2%
   (B) 8%
   (C) $2(1 + 0.02)^4$%
   (D) $8(1 + 0.02)^4$%
   (E) 2.25%

Directions: Choose the best answer for each question and mark the letter of your selection on the corresponding answer sheet.

Note: The natural logarithm of x will be denoted by $\ln x$.

46. $\int_{-1}^{1} (x^{4/3} + 4x^{1/3})dx =$

(A) 10           (D) $\frac{5}{3}$

(B) $\frac{6}{7}$   (E) $\frac{8}{3}$

(C) $\frac{1}{7}$

47. The number of permutations of the letters in the word BANANA is,

   (A) 36      (D) 60

   (B) 42      (E) 52

   (C) 76

48. Suppose that the probability of parents having a child with blond hair is $\frac{1}{4}$. If there are four children in the family, what is the probability that exactly half of them have blond hair?

   (A) $\frac{27}{128}$   (D) $\frac{44}{126}$

   (B) $\frac{37}{135}$   (E) $\frac{10}{135}$

   (C) $\frac{1}{126}$

49. $\int_0^{\pi/4} \sin^2 x \, dx =$

   (A) $\frac{\pi + 2}{8}$   (D) $\frac{\pi - 2}{8}$

   (B) $\frac{\pi - 4}{8}$   (E) $\frac{\pi + 1}{8}$

   (C) $\frac{2 - \pi}{8}$

50. If $A = \begin{bmatrix} 2 & -1 \\ 1 & 0 \\ -3 & 4 \end{bmatrix}$ and $B = \begin{bmatrix} 1 & -2 & -5 \\ 3 & 4 & 0 \end{bmatrix}$, then AB =

(A) $\begin{bmatrix} -1 & -8 & -10 \\ -1 & -2 & 5 \\ 9 & 22 & 15 \end{bmatrix}$ 
(D) $\begin{bmatrix} 0 & 0 & -10 \\ 1 & -2 & -5 \\ 0 & 21 & -15 \end{bmatrix}$

(B) $\begin{bmatrix} -1 & -8 & -10 \\ 1 & -2 & -5 \\ 9 & 22 & 15 \end{bmatrix}$ 
(E) $\begin{bmatrix} 0 & -8 & -10 \\ 1 & -2 & -5 \\ 9 & 21 & 15 \end{bmatrix}$

(C) $\begin{bmatrix} 1 & -8 & -10 \\ 0 & -2 & -5 \\ 9 & 21 & 15 \end{bmatrix}$

51. For $0 < x < 2\pi$,

$$\frac{\tan x \cot x + \cos x \sec x}{\cos x + \sin x \tan x} =$$

(A) sec x

(D) tan x

(B) cos x

(E) 2cos x

(C) 2sin x

52. A manufacturer of outboard motors receives a shipment of shearpins to be used in the assembly of its motors. A random sample of ten pins is selected and tested to determine the amount of pressure required to cause the pin to break. When tested, the required pressures to the nearest pound are 19, 23, 27, 19, 23, 28, 27, 28, 29, 27. The measures of central tendency are

|  | Mean | Median | Mode |
|---|---|---|---|
| (A) | 25, | 27, | 27 |
| (B) | 28, | 27, | 28 |
| (C) | 19, | 27, | 27 |

(D) 27, 19, 19

(E) 24, 28, 28

53. If $y = e^{x^2/4}$, what is $\frac{dy}{dx}$?

(A) $2xe^{\frac{x^2}{4}}$

(D) $\frac{2x}{e^{\frac{x^2}{4}}}$

(B) $x^2 e^{\frac{x^2}{4}}$

(E) $\frac{x}{2} e^{\frac{x^2}{4}}$

(C) $\frac{e^{\frac{x^2}{4}}}{2x}$

Questions 54 to 58 are based on the following assumption:

Let F(x) and G(x) be functions differentiable to any order over the set of all real numbers.

54. $\int_1^2 F(x)dx + \int_2^3 G(x)dx =$

   (A) $2\int_1^3 [F(x) + G(x)]dx$

   (B) $\int_1^3 [F(x) + G(x)]dx$

   (C) $3\int_1^2 [F(x) + G(x)]dx$

   (D) $\int_2^3 [F(x) + G(x)]dx$

   (E) None of the above

55. If F(x) is continuous over the range (a ≤ x ≤ b) and G(y) is continuous over the range c ≤ y ≤ d such that H(x,y) = F(x)G(y) is also continuous, then

$$\int_a^b \int_c^d F'(x)G'(y)dxdy =$$

   (A) F(a - b)G(c - d)

   (B) F'(a - b)G'(c - d)

   (C) [F(b) - F(a)][G(c) + G(d)]

   (D) [F(b) - F(a)][G(d) - G(c)]

   (E) [F(b)G(d) - F(a)G(c)]

56. The expression

$$\int_0^1 F(x)dx + \int_2^0 F(x)dx + \int_1^3 F(x)dx$$

269

can be simplified to

(A) $\int_{1}^{3} F(x)dx$

(D) $\int_{2}^{3} F(x)dx$

(B) $-\int_{1}^{2} F(x)dx$

(E) None of the above

(C) $-\int_{1}^{3} F(x)dx$

57. The derivative of $F(x) = \sqrt{x^2 + 1}$ is

(A) $2x\sqrt{x^2 + 1}$.

(D) $\dfrac{x}{\sqrt{x^2 + 1}}$

(B) $x(x^2 + 1)^{\frac{3}{2}}$

(E) None of the above

(C) $\dfrac{x^2}{(x^2 + 1)^{-\frac{1}{2}}}$

58. If $F(x) = 2x^2 - 8x + 6$ then $F(x)$ has a relative minimum for $x =$

(A) 2

(D) 3

(B) -8

(E) not necessarily any of the values from (A) to (D)

(C) 6

59. If $y = \left[e^{-\frac{1}{x}}\right]^2$, $dy/dx =$

(A) $-\dfrac{2e^{\frac{2}{x}}}{x^2}$

(D) $-2x\, e^{\frac{2}{x}}$

(B) $\dfrac{e^{\frac{2}{x}}}{2x^2}$

(E) $2x^2 e^{\frac{2}{x}}$

(C) $-\dfrac{[e^{\frac{1}{x}}]^2}{2x}$

60. If $F(s) = \int_0^\infty f(t)e^{-st}dt$, then $F(s)$ is called,

   (A) the Fourier transform of $f(t)$

   (B) the Laplace transform of $f(t)$ represented by $L\{f(t)\}$

   (C) the exponential function of $f(t)$

   (D) Bessel's function

   (E) Legendre's polynomial

61. If $x^2 + y^2 = 2$, $xy = -1$, then $x^3 + y^3 =$

   (A) 2

   (B) 3

   (C) 4

   (D) 0

   (E) -1

62. $\lim\limits_{x \to \infty} \dfrac{x^2 + 10}{6x^2 + 2} =$

   (A) 10

   (B) $\dfrac{1}{2}$

   (C) 5

   (D) $\dfrac{1}{6}$

   (E) undefined

63. If $A = \begin{pmatrix} 1 & 2 \\ 2 & 1 \\ 1 & 1 \end{pmatrix}$ then $A^{-1}$ is

   (A) $\begin{pmatrix} 1 & -2 \\ -2 & 1 \\ 1 & 2 \end{pmatrix}$

   (B) $\begin{pmatrix} 1 & 1 \\ 2 & -2 \\ 1 & 2 \end{pmatrix}$

   (C) $\begin{pmatrix} 2 & 3 \\ 3 & 1 \\ 2 & 4 \end{pmatrix}$

   (D) $\begin{pmatrix} 1 & 4 \\ 3 & 2 \\ 2 & 5 \end{pmatrix}$

   (E) undefined

64. If $S_n = 1 - t + t^2 - t^3 + \ldots + (-1)^n t^n$ where $|t| < 1$ then
$\lim_{n \to \infty} S_n =$

(A) 1

(B) $\dfrac{1}{1+t}$

(C) $\infty$

(D) $1 - t$

(E) $\frac{1}{2}$

Directions: Choose the best answer for each question and mark the letter of your selection on the corresponding answer sheet. Assume that every function in this section has derivatives of all order at each point of its domain unless otherwise indicated.

Function f(t) for problems 65-70.

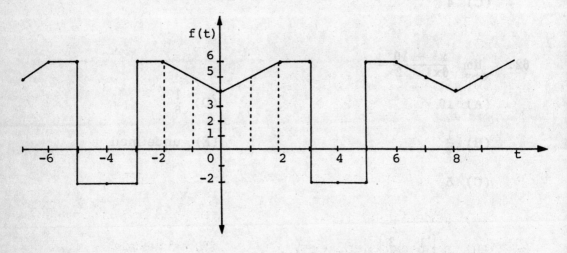

65. If f(t) is to be expanded into a convergent trigonometric series, which of the following is a sufficient condition that ensures this?

(A) f(t) is both positive and negative on certain intervals.

(B) f'(t) = 0 for at least one value of t on the periodic interval.

(C) f(t) bounds a finite area on the periodic interval.

(D) f(t) is a smooth curve.

(E) none of the above

66. When f(t) is expanded into a trigonometric series, its constant term coefficient is equal to

(A) 1

(B) $\frac{7}{2}$

(C) 5

(D) $\frac{14}{3}$

(E) $\frac{16}{3}$

67. When f(t) is expanded into a trigonometric series, its cosine coefficients are equal to

(A) $\frac{8}{\pi n} \left[ \left( \cos \frac{n\pi}{2} - 1 \right) /n\pi + 2\sin \frac{3}{4}\pi n \right]$  n = 2, 4, 6, ....

(B) $\frac{2}{\pi} \left[ \cos \frac{n\pi}{8} + \frac{\pi n}{8} \sin \frac{n\pi}{8} \right]$  n = 1, 3, 5, ....

(C) $\frac{8}{\pi n} \left[ \sin n\pi - \cos \frac{3}{4}\pi n \right]$  n = 2, 4, 6, ....

(D) $(1 - (-1)^n)/n\pi + \cos \frac{n\pi}{8}$  n = 1, 2, 3, ....

(E) 0

68. With f(t) defined as in the graph, then $\frac{df(t)}{dt}$ must be

(A) periodic and even

(B) nonperiodic and even

(C) odd and periodic

(D) odd and nonperiodic

(E) sinusoidal

69. The function f(t) is multiplied by sin t giving a new function g(t). Which of the following is true?

(A) The cosine coefficients of the Fourier series of g are all non-zero.

(B) g(t) will be nonperiodic.

(C) g(t) is an even function.

(D) The cosine coefficients of the Fourier series of g are all zero.

(E) g(t) is a smooth function.

70. The rms value of f(t) is equal to

(A) $\dfrac{36\sqrt{2}}{5}$

(B) $72\sqrt{3}$

(C) $\dfrac{\sqrt{3}}{12}$

(D) 36

(E) $\sqrt{\dfrac{68}{3}}$

The following information will be used to answer questions 71-75.

Particle K travels with constant tangential acceleration along a circular path of radius R = 10 m. Its initial and final velocities are 5 m/s and 40 m/s, respectively. If the distance covered by the particle during the above interval is 120 m, then:

71. The number of seconds that have elapsed is

(A) $\dfrac{16}{3}$ s

(B) $\dfrac{5}{4}$ s

(C) 4 s

(D) 10 s

(E) $3\dfrac{1}{3}$ s

72. The magnitude of the tangential acceleration of particle K is

(A) 8 m/s$^2$

(B) 9 m/s$^2$

(C) $\dfrac{3}{16}$ m/s$^2$

(D) $7\dfrac{8}{17}$ m/s$^2$

(E) $6\dfrac{9}{16}$ m/s$^2$

73. At the end of the time interval when K travels 120 m, the magnitude of its normal acceleration component is

(A) 100 m/s²  (D) 180 m/s²

(B) 130 m/s²  (E) 185 m/s²

(C) 160 m/s²

74. After particle K has traveled for 2 seconds the total angle turned through by line OK is: (O is the center of the circle)

(A) 3 rad  (D) $2\frac{5}{16}$ rad

(B) $5\frac{1}{3}$ rad  (E) $3\frac{3}{16}$ rad

(C) $1\frac{2}{3}$ rad

75. Over the time interval, the equation of displacement as a function of time is

(A) $s(t) = \frac{7}{2}t^2 + 18t + 3$  (D) $s(t) = 3t^2 + 5t + 4$

(B) $s(t) = \frac{105}{32}t^2 + 5t$  (E) $s(t) = \frac{49}{8}t^2 + 13t$

(C) $s(t) = \frac{38}{3}t^3 + \frac{14}{5}t$

Particle P starts at vertex A of a non-regular pentagon and moves along the perimeter stopping at each vertex B, C, D and E for 1 second. The entire traversal of the perimeter takes 12 seconds. The following graphs are the distances of P from A and C during the 12-second trip.

The above explanation and the following curves are related to questions 76-80.

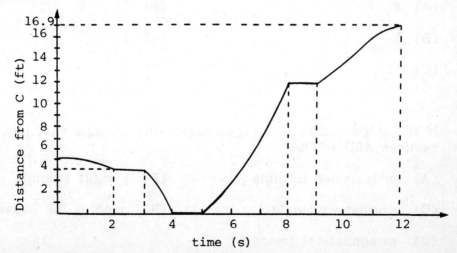

275

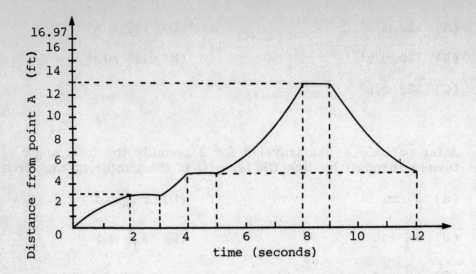

76. What is the distance between points A and D?

    (A) 5                    (D) 13
    (B) 4                    (E) 16
    (C) 12

77. What is the distance between points C and D?

    (A) 4                    (D) 13
    (B) 5                    (E) 16
    (C) 12

78. What is the distance between points A and C?

    (A) 4                    (D) 13
    (B) 5                    (E) 16
    (C) 12

79. If the angle ∠CDE is a right angle, the triangle that has vertices AED will be

    (A) an isosceles triangle      (D) a right triangle
    (B) a scalene triangle         (E) none of the above
    (C) an equilateral triangle

80. What is the area of △ABC?

   (A) 3 (D) 14
   (B) 6 (E) 16
   (C) 10

Questions 81-85.

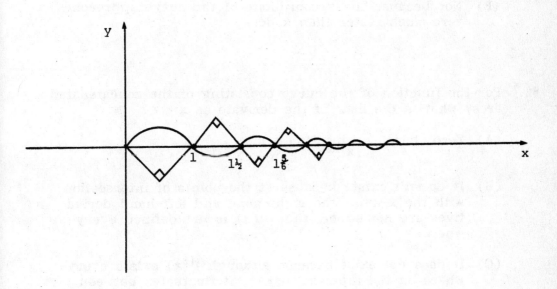

The above curve has been drawn by the following method:

The distance between the nth point and (n + 1)th point on the x-axis is equal to 1/(n + 1). Therefore the first point on the x-axis has the abscissa 1, the second has the abscissa 1 + ½, and the nth point has the abscissa $\sum_{m=1}^{n} \frac{1}{m}$.

The curve consists of arcs which are quarters of circles. Therefore the central angle is 90°. (as shown in the figure)

Now answer the following questions.

81. If we continue the construction of the curve for an unlimited number of points, will the x-coordinate of the points on the curve approach infinity? I.e., is the domain of the function specified by the curve equal to $R^+$?

   (A) No, because this curve does not specify a function.

(B) Yes, because the number of times the curve has cut the x-axis is infinity.

(C) No, because the sum of the lengths of the segments cut by the curve on the x-axis does not approach infinity as the number of points does.

(D) Yes, because we have: $\sum_{n=1}^{\infty} \frac{1}{n} = \infty$ and this series is equal to the abscissa of the nth point on the x-axis as n approaches infinity.

(E) No, because the y-coordinate of the curve approaches zero much faster than x does.

82. For the function of the curve consisting of the concatenated arcs, what is the limit of the derivate as $x \to \infty$?

(A) Zero, because $\lim_{x \to \infty} f(x) = 0$.

(B) It doesn't exist, because at the points of intersection with the x-axis, the right-hand and left-hand derivatives are not equal, thus f'(x) is not defined everywhere.

(C) It does not exist because although f'(x) exists everywhere on the interval $[0, +\infty]$, it fluctuates between $[-1,1]$ in the intervals bounded by any two points of intersection with x-axis, no matter how close are the two points to each other (especially as $x \to \infty$).

(D) Zero, because the x-axis is the asymptote of the curve as $x \to \infty$, and we have: $\lim_{x \to \infty} \frac{f(x)}{x} = 0$.

(E) It does exist and has a value between 1 and -1.

83. If f is the function represented by the curve, will the integral $\int_{0}^{\infty} f(x)dx$ be convergent?

(A) Yes, because the convergence of the integral is guaranteed by the convergence of the series $\sum_{0}^{\infty} \frac{(-1)^n}{n}$, and the series is convergent because the absolute value of its terms decrease and each term has an opposite sign from the previous one.

(B) Yes, because the series $\sum_{0}^{\infty} \frac{(-1)^n}{n^2}$ is convergent due to having terms of opposite sign and decreasing absolute value. The convergence of the series is sufficient for the convergence of the integral.

(C) No, because the series $\sum_{0}^{\infty} \frac{(-1)^n}{n^2}$ is not convergent and so neither is the integral.

(D) Yes, because the integral is the sum of positive and negative decreasing terms.

(E) No, because the sum of the (2n - 1)th and 2nth terms (n = 1,2,3,...) is always positive. Therefore the sum of the infinite number of these positive terms will diverge.

84. What is the minimum value of the function f?
Let the arc height be $\ell$.

(A) zero

(B) doesn't exist

(C) 1

(D) $-1 + \frac{\sqrt{2}}{2} \ell$

(E) $\frac{(-\sqrt{2} + 1)}{2} \ell$

85. Which of the following curves is the approximate representation of the function: $g(x) = \int_{0}^{x} f(t)\,dt$ while $f(t)$ is the function whose curve was given before?

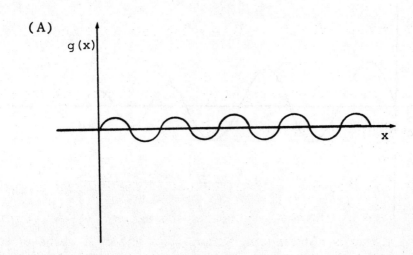

(B)

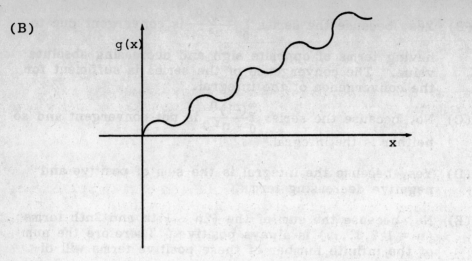

(C)

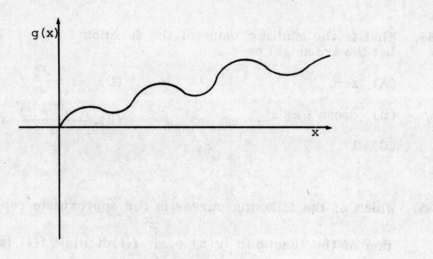

(D)

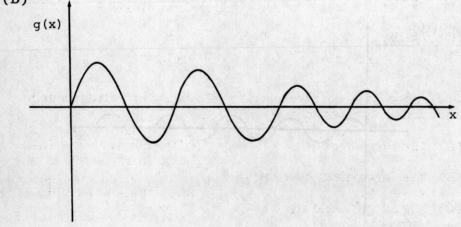

(E) None of the above

Directions: Choose the best answer for each question and mark the letter of your selection on the corresponding answer sheet.

86. A uniform wooden beam of weight 100 lb. is lying on a horizontal floor. A carpenter raises one end of it until the beam is inclined at 60° to the horizontal. He maintains it in this position by exerting a force at right angles to the beam while he waits for his mate to arrive to lift the other end. What is the magnitude of the force he exerts?

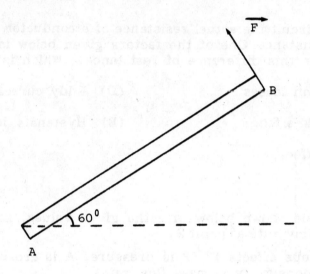

(A) 15 lb.

(B) 25 lb.

(C) 35 lb.

(D) 40 lb.

(E) 50 lb.

87. The speed of a dc series motor increases to a dangerously high value at no load due to

(A) the large torque

(B) the large field current

(C) the large armature current

(D) the very low field flux

(E) the low IR drop

88. When ethylene is burned in air, ethylene oxide is formed by the reaction below.

$$C_2H_4 + \tfrac{1}{2}O_2 \rightarrow C_2H_4O$$

(Atomic weights: C = 12.0, O = 16.0, H = 1.0)

If 66 g of $C_2H_4O$ is produced from 56 g of $C_2H_4$, the percent yield is

(A) 133%  
(B) 30%  
(C) 18%  
(D) 75%  
(E) 92%

89. In an a.c. circuit the actual resistance of a conductor differs from the d.c. resistance. One of the factors given below is not responsible for this difference of resistance. Which is it?

(A) Radiation losses  
(B) Seebeck effect  
(C) Skin effect  
(D) Eddy currents  
(E) Hystensis losses

90. For the nozzle shown below with the given values, solve for the volume flow rate at point 2.

(Ignore viscous effects.) (P is pressure, A is cross-section area, $\rho$ is density, v is mass flow rate)

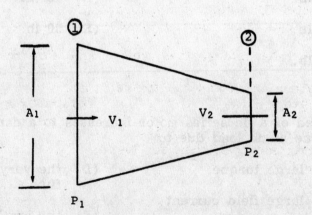

$P_1 = P$  
$P_2 = 0$  
$v_1 = v$  

$A_1 = 2A$  
$A_2 = A$

(A) $vA$ m³/s by the conservation equation

(B) $2vA$ m³/s by the continuity equation

(C) $\rho\left[1 - \left(\frac{A_2}{A_1}\right)^2\right]$ m³/s by the continuity equation

(D) $\left(\sqrt{\frac{2P}{\rho}}\right)$ m³/s by Bernoulli's equation

(E) cannot be calculated without more information

91. $V_1, V_2$    flow rates
     $A_1, A_2$    surface areas
     l          length

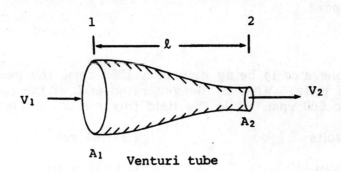

Venturi tube

A liquid passes through a venturi tube, represented in the above figure. The pressure difference between the two ends is a function of which of the following?
(Assume frictionless flow, i.e. tube wall is frictionless)

(A) flow rates and surface areas

(B) surface areas and the tube's length.

(C) the tube's length and the correction coefficient for low velocity

(D) flow rates and the tube's length

(E) surface area only

92. The DC current I, for the circuit is equal to

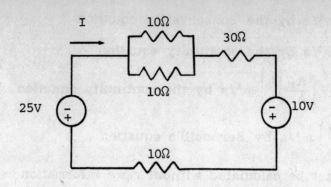

(A) $\frac{1}{3}$ ampere     (D) $-\frac{1}{3}$ ampere

(B) $\frac{2}{3}$ ampere     (E) $-\frac{2}{3}$ ampere

(C) 1 ampere

93. When a generator is being driven at 1200 rpm, the generated emf is 130 volts. What is the generated emf, if the speed is reduced to 600 rpm, while the field flux remains unchanged?

(A) 260 volts     (D) 60 volts

(B) 65 volts      (E) 13 volts

(C) 5538 volts

94. To prevent a bridge from being destroyed, soldiers marching on a bridge are asked to break their steps. Which of the following is responsible for such a danger?

(A) Beats         (D) Shock wave

(B) Resonance     (E) Sonic boom

(C) Reverberation

95. Two volts are applied across a copper wire of 5Ω that is inside an adiabatic container which contains 0.5 kg of water. What is the rise in the temperature of the water in 1 minute? Note: Specific heat of water (1 cal · $g^{-1}(C°)^{-1}$) 1 cal = 4.186J

(A) $\Delta T = \dfrac{2400}{4.186}\,°C$  (D) $\Delta T = \dfrac{24}{4.186}\,°C$

(B) $\Delta T = \dfrac{0.0960}{4.186}\,°C$  (E) $\Delta T = \dfrac{96}{4.186}\,°C$

(C) $\Delta T = (96)(4.186)\,°C$

96. A man weighing 150 lb. is at rest on roller skates. He throws a 7.5 lb. stone in a positive horizontal direction with a velocity of 20 ft/sec. What is the velocity with which the man moves?

(A) 1 ft/sec  (D) 2 ft/sec

(B) -1 ft/sec  (E) -2 ft/sec

(C) -1.5 ft/sec

97. The equilibrium constant, $K_c$, for the nitrogen fixation reaction for the production of ammonia

$$3H_2(g) + N_2(g) \rightleftarrows 2NH_2(g)$$

is

(A) $\dfrac{[NH_2]^2}{[H_2]^3[N_2]}$  (D) $\dfrac{[H_2]^3[N_2]}{[NH_2]^2}$

(B) $\dfrac{3[H_2] + [N_2]}{2[NH_2]}$  (E) $[H_2]^3[N_2]$

(C) $\dfrac{2[NH_2(g)]}{3[H_2(g)][N_2(g)]}$

98. Two blocks are connected to opposite ends of a light spring that exerts a force of 18N when the blocks are pulled apart. The mass ratio of the first block to the second is 3:1. When the two blocks are released, and if the acceleration of the first block is 6 m/s², what is the acceleration of the second block?

(A) $\dfrac{18}{4}\,m/s^2$  (D) $18\,m/s^2$

(B) $2\,m/s^2$  (E) $6\,m/s^2$

(C) $9\,m/s^2$

99. The speed distribution for a sample of gas containing N molecules is given by

$$N(v)dv = 4\pi N \left(\frac{m}{2}\pi kT\right)^{\frac{3}{2}} v^2 e^{-\frac{mv^2}{2kt}} dv$$

where N(v)dv represents the number of molecules in the gas sample having speeds between v and v + dv at temperature T. If the temperature of the gas sample is raised, which of the following represents the distribution of speeds of the same gas sample at two different temperatures?
Note: $\bar{v}$ is the average speed.

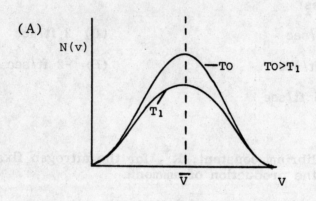

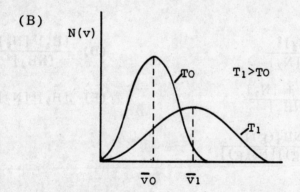

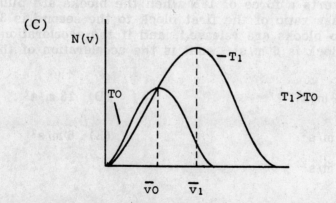

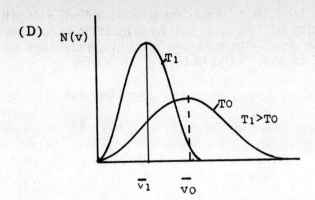

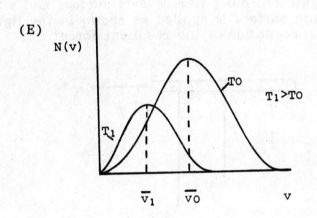

Questions 100-101.

A point moves in the x-y plane in such a way that its position vector at time t is given by the equation $\vec{s} = 3\sin t\,\vec{i} + 3t^2\,\vec{j}$.

100. The x component of the velocity of the point at any time t is

(A) $-3\cos t$

(B) $3\sin t$

(C) $6t$

(D) $t^3$

(E) $3\cos t$

101. The y-component of the acceleration of the point at any time t is

(A) $-3\cos t$

(B) $3\cos t$

(C) $t^3$

(D) $6t$

(E) $6$

102. Water flows in a very wide, open channel at a depth of 3m and a velocity of 100 m/s. A hydraulic jump occurs and its magnitude is 3m. Approximately what percentage of the initial energy is lost? (Assume $g = 10$ m/s$^2$)

(A) 74%

(B) 80%

(C) 15%

(D) 63%

(E) 54%

103. Three weights are hung from a level surface and a force parallel with the surface is applied as shown in the figure below. What is the magnitude of the resultant force?

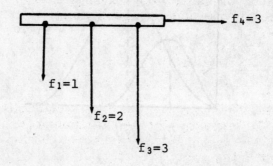

(A) 3

(B) $\sqrt{18}$

(C) 6

(D) $\sqrt{45}$

(E) 9

104. NaCl can dissolve readily in $H_2O$ but not in $C_6H_6$ (benzene). Which of the following is usually cited as the reason for this occurrence?

(A) $H_2O$ is a polar solvent while $C_6H_6$ is not.

(B) The number of hydrogen bonds in $H_2O$ is less than the number in $C_6H_6$.

(C) $C_6H_6$ cannot maintain a crystalline structure upon cooling.

(D) $H_2O$ has a greater probability for the existence of isotopes of both hydrogen and oxygen.

(E) Both (A) and (D)

105. The first law of thermodynamics expresses the conservation of energy in a closed system by

$$\Delta E = q - w.$$

Which of the following is a correct result of this equation?

(A) At constant volume, the change in internal energy is zero.

(B) In the absence of work done by the system, the change in internal energy is a measure of the gain or loss of heat.

(C) If there is no change in energy during a reaction, the net change in work done by the system is always zero.

(D) At constant pressure, the internal energy change is equal to only the net change in heat.

(E) Both (A) and (D)

106. A sale of a piece of equipment is being considered. If this asset is sold for a price which is equivalent to the updated book value of the asset, then all of the following accounts will be affected with the exception of

(A) accumulated depreciation

(B) depreciation expense

(C) gain on the disposal of the fixed asset

(D) book value

(E) both (C) and (D)

107. A motorbike starts from rest and accelerates at a rate of 4 m/s² for 10 seconds and then decelerates at 8 m/s² until it stops. The total distance covered is

(A) 100m

(B) 200m

(C) 300m

(D) 500m

(E) 600m

108. A constant-thickness integrated resistor is fabricated from a semiconductor 1 millimeter in length and width and having a resistance of R ohms. Then a similar resistor having a width of 3 millimeters and a length of 1 millimeter has a resistance of

(A) R  (D) 6R
(B) 3R  (E) $\frac{1}{3}$R
(C) $\frac{1}{4}$R

109. A cannon ball is heated to 100°C. Through a lumped parameter transient analysis, which of the following would allow the cannon ball to maintain its temperature the longest time?

    (A) Increasing the surface area

    (B) Decreasing the volume

    (C) Choose a cannon ball with low density

    (D) Choose a cannon ball with high density

    (E) Choose a cannon ball with low specific heat (Cp)

110. Water at 40°C enters a 2-inch diameter tube at an inlet velocity of 3 in/sec. If the tube that exits is 4 inches in diameter, find the exit velocity.

    (A) $\frac{1}{4}$ in/sec  (D) 1 in/sec
    (B) $\frac{1}{2}$ in/sec  (E) 2 in/sec
    (C) $\frac{3}{4}$ in/sec

111. A stepped pool of water is shown below. Which is true?

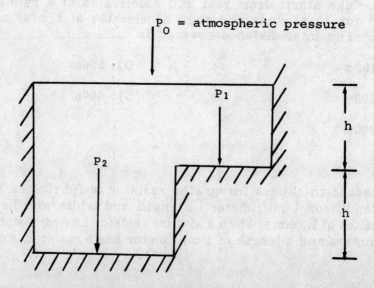

(A) The pressures $P_2$ and $P_1$ cannot be calculated without knowing the surface area of the pool.

(B) The pressure $P_2$ is twice that of $P_1$.

(C) The difference $(P_2 - P_1)$ is linearly related to h.

(D) The pressure $P_2$ is a simple product $2P_0h$.

(E) The pressures $P_2$ and $P_1$ are equal if $P_0 = 0$.

112. A block of ice, density = 0.92 g/cm$^3$, floats in fresh water. The percentage of ice that lies above the water surface is

(A) 100%

(B) 92%

(C) 50%

(D) 46%

(E) 8%

113. When the absolute pressure of a gas is halved at constant temperature, the volume of the gas is

(A) doubled

(B) halved

(C) quadrupled

(D) quartered

(E) unchanged

114. Which of the following circuits is configured by a common emitter transistor?

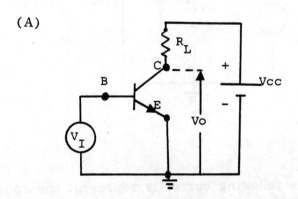

(B)

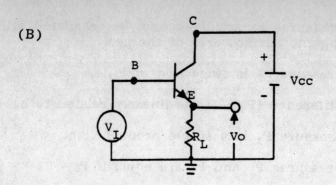

(C)

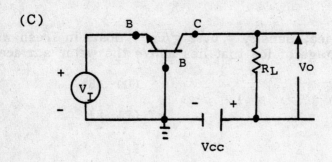

(D)

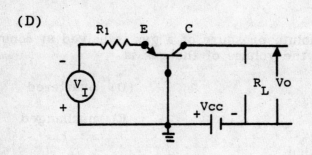

(E)

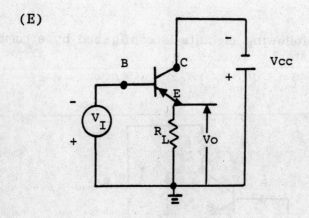

115. Which of the following correctly represent the Fourier's Law of Conduction?

(A) $q = h\Delta T$

(B) $\dfrac{q}{L} = h\Delta T$

(C) $\dfrac{q}{A} = h\Delta T$

(D) $\dfrac{q}{L} = k\dfrac{\partial T}{\partial x}$

(E) $\dfrac{q}{A} = -k\dfrac{\partial T}{\partial x}$

116. Which of the following phenomena illustrates the quantum nature of electron energy states of an atom?

    (A) The scattering of alpha particles by a thin foil of gold

    (B) The scattering of an electron beam by a crystal to obtain a diffraction pattern

    (C) The buildup of an intense beam in a laser through the interaction of an emitted photon with an excited atom

    (D) The periodic drop in current moving through a mercury vapor at certain voltage levels of a steadily increasing outside voltage

    (E) Both (D) and (C)

117. Precipitation hardening of a $CuAl_2$ alloy increases the yield strength of the alloy because

    (A) dislocation movements are more easily facilitated

    (B) the strain fields in the matrix are reduced by the presence of precipitates

    (C) contraction of the alloy is prevented

    (D) the precipitates are able to coalesce into larger particles

    (E) the precipitates exert large elastic strain fields on the solution matrix

A simple open manometer is used to measure the pressure of a fluid of density $\rho_1$. The U-tube contains a homogeneous liquid of density $\rho_2$ that is open to the atmsphere.

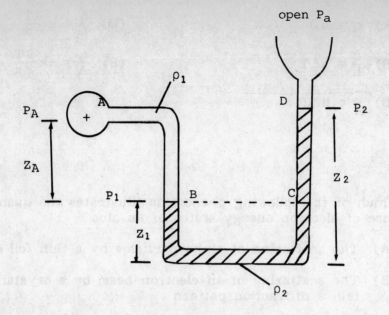

118. Which of the following quantities are equal for the U-tube manometer?

    (A) The gauge pressure from point A to point B and the gauge pressure from point D to point C.

    (B) The absolute pressure at point B and the gauge pressure from point C to point D.

    (C) The gauge pressure from point B to point C and the absolute pressure at point D.

    (D) The absolute pressure at point B and the absolute pressure at point C.

    (E) Both (C) and (D)

119. The net force acting on the fluid element between point C and and D with a cross-sectional area A is

    (A) 0

    (B) $(p_A + \rho_1 g(z_A) - p_a)A$

    (C) $(\rho_2 g(z_2 - z_1) - p_a)A$

    (D) $(\rho_2 g(z_1 - z_2) - p_a)A$

    (E) $(\rho_1 g(z_A) - p_a)A$

120. For an open system, entropy change during a process is

(A) less than zero or zero

(B) zero

(C) zero or greater than zero

(D) indeterminate

(E) dependent on the pressure

121. For molybdenum, which has a body-centered cubic lattice structure, the number of atoms per unit cell is

(A) 1

(B) 2

(C) 4

(D) 6

(E) 10

122. The spark-ignition internal combustion engine is often approximated by the Otto cycle by way of the following assumptions:

1. There is a fixed mass of air and air is an ideal gas.

2. All processes are internally reversible.

3. Air has a constant specific heat.

From this approximation we learn that, even in the actual engine, thermal efficiency is increased by increasing the compression ratio. But an important way in which the actual engine deviates from the ideal cycle is:

(A) The specific heats of the actual gases increase with an increase in temperature.

(B) There will be considerable heat transfer between the gases in the cylinder and the cylinder walls.

(C) There will be irreversibilities associated with pressure and temperature gradients.

(D) Work is required to charge the cylinder with air and exhaust the products of combustion.

(E) All of the above

123. How much heat is required to raise the temperature of 1 kg of water from 10°C to 100°C (specific heat of water = 1 cal/g°C)?

(A) 46 cal

(B) 90 cal

(C) 46 kcal

(D) 90 kcal

(E) 100 kcal

124. Water flows into a tank horizontally at 1 m³/s. There is a hole in the bottom of the tank which is 40 cm in diameter. If the flow is steady, by approximately how much is the apparent weight of the tank changed?
(Note: $\rho_{water}$ = 1000 kg/m³)

(A) 7 kN

(B) -2 kN

(C) -8.3 kN

(D) 1.5 kN

(E) 900 N


125. The APB Corp. is considering a proposal to replace its current equipment with an improved version. The annual cost of maintaining the present equipment is $400 and its present worth is $800. The new equipment will cost $10,000 with an annual operating cost of $400 and a lifetime of 22 years. The current machine is expected to last for 11 more years, before there will arise a need to purchase an improved version. Assuming a zero salvage value and a minimum attractive rate of return of 15% as well as an increase of 15%/year for the new equipment, the APB Corp. would

(A) save $9,200 by remaining with the present equipment

(B) save $800 by purchasing the new equipment

(C) save $\dfrac{\$10,000}{(1+.15)}$ by purchasing the new equipment

(D) save $\$9,200 - \dfrac{\$10,000}{(1+.15)^{11}}$ by remaining with the present equipment

(E) save $\$10,000(1-(1+.15)^{11})$ by purchasing new equipment


126. The elastic modulus for an elastomer such as rubber will increase with increasing temperature above the glass transition temperature due to which of the following reasons?

   I. The molecules are in a highly kinked conformation at higher temperatures.

   II. The molecules will align themselves in the direction of any applied stress.

   III. The molecules will distort the matrix at higher temperatures.

   (A) II only

   (B) I and III only

(C) III only  (E) I, II, and III

(D) I only

127. For a mass connected to a spring and oscillating in simple harmonic motion, to double the period of oscillation, which of the following must be done?

(A) Increase mass by a factor of 2

(B) Increase mass by a factor of 4

(C) Increase displacement by a factor of $2\pi$

(D) Increase spring constant by a factor of 2

(E) Decrease mass by a factor of 2

128. Which of the following diagrams below indicates the variation of hardness with aging time for three different aging temperatures? Note: $T_3 > T_2 > T_1$

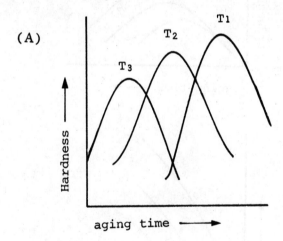

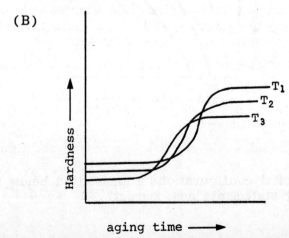

(C)
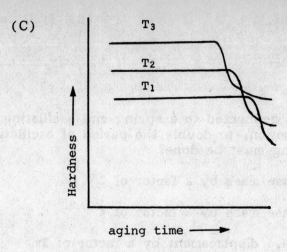

(D)
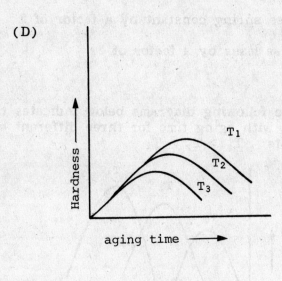

(E)
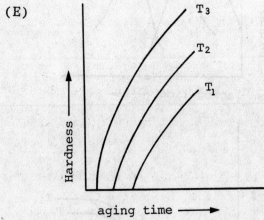

129. From the orbital configurations represented below which of the following statements are correct?

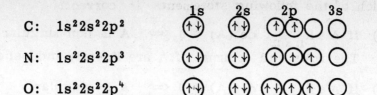

C: $1s^2 2s^2 2p^2$

N: $1s^2 2s^2 2p^3$

O: $1s^2 2s^2 2p^4$

I. As Hund's role predicts, the three p electrons in Nitrogen take up each of the 2p orbitals.

II. The two p electrons in carbon, as shown, violate the Pauli principle.

III. The carbon electrons, as shown, are in an excited state.

IV. All three atoms are shown in the ground state.

(A) I and IV

(B) I and II

(C) II and III

(D) I, II, and III

(E) I, II, III and IV

130. Aluminum has a higher atomic packing factor than molybdenum by about 6%. This means that

(A) aluminum is heavier than molybdenum

(B) the ratio of the volume of the unit cell to the volume of the atoms per unit cell is lower in aluminum

(C) the ratio of the number of atoms per unit cell to size of the unit cell is roughly equal in both

(D) the distances between atoms are equal for both

(E) molybdenum has more atoms per unit cell than aluminum

Directions: Choose the best answer for each question and mark the letter of your selection on the corresponding answer sheet.

Note: The natural logarithm of x will be denoted by $\ln x$.

131. If $x^2 + y^2 = 16$, find $\frac{dy}{dx}$ as an implicit function of x and y.

(A) $\frac{dy}{dx} = -y$

(B) $\frac{dy}{dx} = -\frac{y}{x}$

(C) $\frac{dy}{dx} = -\frac{x}{y}$

(D) $\frac{dy}{dx} = (x + y)$

(E) $\frac{dy}{dx} = \frac{1}{2(x+y)}$

132. Which of the following statements is correct?

   (A) If $A \in R_{n \times n}$, $\det(A) \neq 0 \iff$ A is non-singular $\iff$ The rows and columns of A are linearly independent.

   (B) If $A \in R_{n \times n}$, $\det(A) \neq 0 \iff$ A is singular.

   (C) If $A \in R_{n \times n}$, $\det(A) \neq 0 \iff$ A is non-singular $\iff$ The rows of A are linearly dependent.

   (D) If $A \in R_{n \times n}$, $\det(A) \neq 0 \iff$ A is non-singular $\iff$ A has one zero row.

   (E) If $A \in R_{n \times n}$, $\det(A) \neq 0 \iff$ A is non-singular $\iff$ A is tridiagonal.

133. If $g_1: x = y^2$
   $g_2: 5x + 3y = 8$
   $g_3: y = -1$

   $g_1$ and $g_2$ intersect at

   (A) $\left(\frac{64}{25}, -\frac{8}{5}\right)$ and $(1,1)$
   (B) $(1,-1)$ and $\left(\frac{64}{25}, -\frac{16}{10}\right)$
   (C) $(0,-1)$ and $(1,1)$
   (D) $\left(\frac{64}{25}, -\frac{8}{5}\right)$, and $(0,1)$
   (E) none of the above

134. The solid generated by revolving a circle about a line in its plane which lies outside the circle is the

   (A) Torus
   (B) Cylinder
   (C) Cone
   (D) Sphere
   (E) Ellipsoid

135. $\int_{-1}^{1} |3 - x^3| \, dx =$

   (A) 0
   (B) 6
   (C) $-\frac{1}{2}$
   (D) $\frac{1}{4}$
   (E) $\frac{3}{2}$

136. Given $f(x,y) = x^2 + y^3$, the directional derivative of f at (-1,3) in the direction A of maximal increase is

   (A) $3\sqrt{95}$
   (B) $2\sqrt{100}$
   (C) $\sqrt{733}$
   (D) 0
   (E) 50

137. $\lim_{n\to\infty} S_n$, where $S_n$ is the nth partial sum of the geometric series with first term a, common ratio r where $|r| < 1$ =

   (A) $\dfrac{a - ar^n}{1 - r^n}$
   (B) $\dfrac{a - ar}{1 - r^n}$
   (C) $\dfrac{a}{1 - r}$
   (D) $\dfrac{a - r^n}{1 - r}$
   (E) 0

138. $\displaystyle\int_{\pi}^{e} \pi e\, dx =$

   (A) 1
   (B) $\dfrac{1}{\sqrt{2}}$
   (C) $\pi e[e - \pi]$
   (D) $e\pi[\pi - e]$
   (E) 0

139. $e^{i\pi} =$

   (A) 1
   (B) -1
   (C) 0
   (D) $\sqrt{2}$
   (E) $\dfrac{1}{\sqrt{2}}$

140. The condition that $\lim_{k\to\infty} u_k = 0$ for the convergence of a series $\Sigma\, u_k$ is a condition that is

(A) sufficient

(B) necessary

(C) necessary and sufficient

(D) sufficient but not necessary

(E) necessary but not sufficient

141. The differential equation $y' + \left(\frac{4}{x}\right)y = x^4$ has as an integrating factor

(A) $\frac{x}{y}$

(B) $y^2$

(C) $\frac{y'}{y}$

(D) $x$

(E) $x^4$

142. $\lim\limits_{k\to\infty} \sum\limits_{n=1}^{k} \frac{1}{(2n-1)(2n+1)} =$

(A) 1

(B) $\frac{1}{2}$

(C) 0

(D) $-1$

(E) $-\frac{1}{2}$

143. A circle having its center at the origin is tangent to a straight line described by the equation $x - 2 = 0$. The area of the circle is

(A) $4\pi$

(B) $2\pi$

(C) $\pi^2/4$

(D) $\pi$

(E) $\sqrt{2}$

144. The area bounded by the sine curve between $x = 0$ and $x = 4\pi$ is

(A) 0

(B) $4\pi$

(C) $16\pi$

(D) 4

(E) 8

# THE GRADUATE RECORD EXAMINATION ENGINEERING TEST

## MODEL TEST III

## ANSWERS

| | | | | | |
|---|---|---|---|---|---|
| 1. | B | 21. | D | 41. | E |
| 2. | D | 22. | A | 42. | E |
| 3. | C | 23. | A | 43. | B |
| 4. | D | 24. | A | 44. | A |
| 5. | C | 25. | A | 45. | A |
| 6. | C | 26. | B | 46. | B |
| 7. | C | 27. | E | 47. | D |
| 8. | E | 28. | C | 48. | A |
| 9. | A | 29. | E | 49. | D |
| 10. | E | 30. | E | 50. | B |
| 11. | B | 31. | C | 51. | E |
| 12. | B | 32. | A | 52. | A |
| 13. | E | 33. | C | 53. | E |
| 14. | E | 34. | C | 54. | E |
| 15. | C | 35. | C | 55. | D |
| 16. | E | 36. | E | 56. | D |
| 17. | A | 37. | B | 57. | D |
| 18. | A | 38. | E | 58. | A |
| 19. | D | 39. | D | 59. | A |
| 20. | C | 40. | E | 60. | B |

| | | | | | |
|---|---|---|---|---|---|
| 61. | D | 89. | B | 117. | E |
| 62. | D | 90. | B | 118. | D |
| 63. | E | 91. | A | 119. | B |
| 64. | B | 92. | D | 120. | C |
| 65. | C | 93. | B | 121. | B |
| 66. | B | 94. | B | 122. | E |
| 67. | A | 95. | B | 123. | D |
| 68. | C | 96. | B | 124. | C |
| 69. | D | 97. | A | 125. | B |
| 70. | E | 98. | D | 126. | D |
| 71. | A | 99. | B | 127. | B |
| 72. | E | 100. | E | 128. | A |
| 73. | C | 101. | E | 129. | A |
| 74. | D | 102. | A | 130. | B |
| 75. | B | 103. | D | 131. | C |
| 76. | D | 104. | A | 132. | A |
| 77. | C | 105. | B | 133. | A |
| 78. | B | 106. | E | 134. | A |
| 79. | D | 107. | C | 135. | B |
| 80. | B | 108. | E | 136. | C |
| 81. | D | 109. | D | 137. | C |
| 82. | C | 110. | C | 138. | C |
| 83. | B | 111. | C | 139. | B |
| 84. | E | 112. | E | 140. | E |
| 85. | C | 113. | A | 141. | E |
| 86. | B | 114. | A | 142. | B |
| 87. | D | 115. | E | 143. | A |
| 88. | D | 116. | E | 144. | E |

# THE GRADUATE RECORD EXAMINATION ENGINEERING TEST

## MODEL TEST III

## DETAILED EXPLANATIONS OF ANSWERS

1. (B)
For an engine cycle the net heat flowing into the engine must equal the net work done by the engine since its initial and final internal energies must be the same. During the final process, system must return to its initial internal energy state. The pV diagram for the three processes is drawn below:

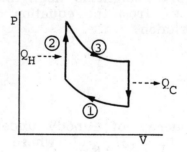

In order to return to the initial internal energy state the final process must reduce the pressure while maintaining the volume constant. The cooling of air at constant volume will complete the cycle. The first three processes explained in the question are shown on the diagram in the order they occur.

## 2. (D)

In an electric circuit the power dissipated in the resistor is given by $P = I^2 R$. I.e. the power and therefore the heat developed are proportional to the square of the current (I) in the circuit. R is the resistance of the circuit.

According to Ohm's Law, current in a circuit is proportional to the potential difference (voltage) applied to the circuit, $I = V/R$. Hence, when the potential difference is decreased the current decreases as well. Thus, the power and the heat developed in the circuit decrease.

## 3. (C)

The equation of continuity states that the mass flow rate stays constant

$$\int_{A_1} \rho_1 v_1 \cdot n \, dA = \int_{A_2} \rho_2 v_2 \cdot n \, dA$$

since a liquid is incompressible

$$A_1 v_1 = A_2 v_2.$$

Hence since the area decreases at the constriction the velocity should increase.

Bernoulli's equation for a streamline of a non-viscous fluid states that

$$p_1 + \rho g y_1 + \tfrac{1}{2} \rho v_1^2 = p_2 + \rho g y_2 + \tfrac{1}{2} \rho v_2^2$$

Since the change in height is negligible and the speed $v_2$ is greater than $v_1$ from the equation of continuity, $p_2$ is less in the constriction.

## 4. (D)

The kinetic energy of a body undergoing pure rotation is defined as: $E_k = \tfrac{1}{2} I_{AR} \omega^2$ where $I_{AR}$ is the moment of inertia about the axis of rotation and $\omega$ is the angular velocity of the body. The angular velocities and masses of the two disks are equal and $R_A = 2 R_B$,

So $\quad I_A = \tfrac{1}{2} m (2 R_B)^2 = \tfrac{1}{2} m 4 R_B^2$

$\quad I_B = \tfrac{1}{2} m (R_B)^2$

$\quad E_{KA} = \tfrac{1}{2} (\tfrac{1}{2} m R_B^2) 4 \omega^2$

$$E_{KB} = \tfrac{1}{2}(\tfrac{1}{2}mR_B^2)\omega^2$$

and $\quad E_{KA} : E_{KB} = 4:1$

## 5. (C)

The diagram shown is usually used to prove that the sum of the three angles of any triangle is equal to 180°.

Given the triangle ABC, sides AC and BC are extended, and a straight line is drawn through vertex C parallel to AB.

There are three basic principles that are used in the proof.

1. A straight angle has 180°.
2. Vertical angles are equal.
3. If two parallel lines are intersected by the same straight line, the interior angles so formed are equal.

| Statement | Reason |
|---|---|
| 1. ∠DCG + ∠GCH + ∠HCE = 180° | 1 |
| 2. ∠DCG = ∠ABC | 3 |
| 3. ∠HCE = ∠CAB | 3 |
| 4. ∠GCH = ∠ACB | 2 |
| 5. ∠ABC + ∠ACB + ∠CAB = 180° | Sums of equals are equal |

Therefore, the above steps show that the sum of the angles of a triangle is equal to 180°.

## 6. (C)

Magnesium silicate contains $Mg^{+2}$ ions and $SiO_3^{-2}$ ions. The structure of $SiO_3^{-2}$ is a chain in which one-third of the oxygen atoms are bridged from one unit to the next. This bridging produces strong Si-O bonds that are partially covalent. The $SiO_3^{-2}$ chains are bonded ionically with the $Mg^{2+}$ ion. These two types of bonding respond differently to deformation, fracture, electric fields, etc. Therefore the properties of the ceramic material are different in different directions or anisotropic.

## 7. (C)
Because titration makes use of the property of acid-base neutralization both I and II are true and the opposite of III is true.

## 8. (E)
From Coulomb's law, the electrostatic force between a pair of point charges q and q' is given by $F = \dfrac{kqq'}{r^2}$. Therefore

$$E = \dfrac{F}{q'} = \dfrac{kqq'}{r^2 q'} = \dfrac{kq}{r^2}$$

This gives the magnitude of the electric field intensity produced by a point charge q.

From above we understand that the electric field intenstiy due to a point charge varies inversely as the square of the distance from the charge. In the given figure, the distance from A to D is three times the distance from A to B. The inverse variation gives us the reciprocal of the square of 3 which is 1/9. Therefore, the field intensity at D is 1/9 E.

## 9. (A)
According to an observer outside the accelerating car, that is stationed in an inertial frame of reference, the pendulum accelerates to the right because of the unbalanced horizontal component of the tension of the rope holding the pendulum. According to an observer inside the accelerating car, the pendulum is at rest. The forces are now in equilibrium because of the so-called pseudoforce which balances the horizontal component of the tension of the rope. This force is exerted by the horizontal field from the point of view of the accelerating observer which causes a horizontal force equal to ma on any particle of mass m. This force is balanced by the horizontal component of the tension of the rope.

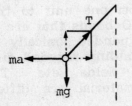
accelerating observer
(noninertial frame)
balanced forces

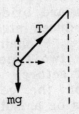
stationary observer
(inertial frame)
unbalanced forces

10. (E)

In order for 600 coulombs of charge to be delivered to the positive terminal of the 100V source, the current must flow counterclockwise. To find the current, use the definition

$$i = \frac{dq}{dt} = \frac{600 \text{ C}}{1 \text{ min} \times 60 \text{ sec}} = 10 \text{ Amperes.}$$

Now apply Kirchoff's voltage law to the circuit to solve for x.

$$x + \frac{x}{3} - 100 = 10(20)$$

$$\frac{4x}{3} = 300$$

$$\therefore \quad x = 225 \text{ V}$$

11. (B)

As explained in previous questions, a Wheatstone bridge is balanced when

$$\frac{R_{AB}}{R_{AD}} = \frac{R_{BC}}{R_{DC}} \tag{1}$$

With reference to the given circuit,

$$R_{AB} = 45\Omega, \quad R_{AD} = 400\Omega, \quad R_{DC} = 2000\Omega \text{ and}$$

$$R_{BC} = R_x$$

Substituting these values in (1) we get

$$\frac{45}{400} = \frac{R_x}{2000}$$

Hence, $$R_x = \frac{45 \times 2000}{400}$$

$$= 225\Omega.$$

12. (B)

The principle of rotational inertia is used to solve this problem.

The greater the distance between the bulk of an object's

mass and its axis of rotation, the greater is the rotational inertia. The greater the rotational inertia of an object, the tougher it is to get the object into rotational motion.

Since the ring cylinder has its mass constant rated farthest from its axis of rotation, it is tougher for it to start rolling. A ring has a greater tendency to resist a change in its motion. Hence the solid cylinder rolls faster and reaches the bottom first.

We can also consider the problem quantitatively:

At the bottom of the incline they must have equal kinetic energy which is the addition of both translational and rotational kinetic energy.

For the solid cylinder we have $\Rightarrow E = \frac{1}{2}mv^2 + \frac{1}{2}(\frac{1}{2}mr^2)\frac{v^2}{r^2}$

$$E = \frac{1}{2}mv^2 + \frac{1}{4}mv^2 = \frac{3}{4}mv^2$$

For the shell we have: $E = \frac{1}{2}mv_1^2 + \frac{1}{2}(mv^2)\frac{v_1^2}{r^2}$

$$E = \frac{1}{2}mv_1^2 + \frac{1}{2}mv_1^2 = mv_1^2$$

$$mv_1^2 = \frac{3}{4}mv^2 \Rightarrow$$

$$v_1^2 = \frac{3}{4}v^2 \Rightarrow v_1 = \frac{\sqrt{3}}{2}v \Rightarrow v > v_1$$

## 13. (E)

A liquid column of water would be unreliable since the volume of water contracts from 0°C to 4°C and expands from 4°C to 10°C. In order for a thermocouple to give rise to a voltage the junctions must be made of dissimilar metals and at different temperatures. Since the test junction is between two copper wires it would be unreliable.

## 14. (E)

The surface of water is made up of molecules that attract each other. Beneath the surface each molecule is attracted by other molecules from all directions, with the result that there is no tendency to be pulled in any particular direction. However, on the upper and lower surfaces of water, the molecular attraction from the upper and lower molecules respectively is absent. As a result the molecular attraction from other directions tends to pull the surface molecules into the liquid and the surface behaves as if it were tightened into an elastic membrane; the phenomenon is referred to as surface tension. This sort of elastic membrane prevents the water from leaking out of the container.

**15. (C)**
The rate of radiation from the surface of a body at a temperature T is given by $H = Ae\sigma T^4$. Hence, for a body at a temperature $T_1$ surrounded by walls at a temperature $T_2$, the net flow of radiation is

$$H_{net} = Ae\sigma T_1^4 - Ae\sigma T_2^4$$

Since the body is hotter than its surroundings it will emit more radiation than it will absorb. In order to maintain the temperature constant the rate of emission must equal the rate of absorption. Since the rate of emission is

$$4\pi r^2 e\sigma(T_1^4 - T_2^4)$$

the rate of energy supplied must be

$$4\pi r^2 e\sigma(T_1^4 - T_2^4)$$

**16. (E)**
A meniscus appears when there is an abrupt change of phase such as the condensation of a vapor into a liquid. In order for a path to avoid the appearance of a meniscus it must not pass through the vaporization curve at which condensation may occur. Any path that goes through the critical point of the vaporization curve from the liquid region to the vapor region will produce a gradual change from a vapor to a liquid, where there is no distinction between the liquid and vapor. At the critical temperature $T_c$ the gas phase and the liquid phase are continuous and no part can be pointed out to be called a gas or no part can be called a liquid. At this temperature a gas changes into a liquid without the appearance of a meniscus. Above this temperature a liquid does not form however great the pressure when the liquid is compressed isothermally.

**17. (A)**
When the amps are ideal, then the inverting terminal is a virtual ground and the current into it is about zero.

Hence, we take KVL around the op-amps:

$$32 = IR + 6I + V_{out}$$

When $V_{out} = 8$, $I = \dfrac{V_{out}}{4} = \dfrac{8}{4} = 2A$

Hence:
$$32 = 2R + 12 + 8$$
$$12 = 2R$$
$$R = 6\,\Omega$$

## 18. (A)
The reaction which has taken place is:

$$NaCl + AgNO_3 \rightarrow AgCl + Na^+ + NO_3^-$$

As concluded from the formula, 1M of NaCl reacts with 1M of $AgNO_3$ to form 1M of AgCl which is a white and not soluble substance.

There is $0.25 \times 0.2 = 0.05M$ of $AgNO_3$ in the solution which is added to $29.25/(23+35.5)=0.5M$ of salt. (Atomic weight of Na is 23 gr and that of chlorine is 35.5) So all of the silver nitrate in the solution will be consumed which will lead to the formation of 0.05M of AgCl.

## 19. (D)
The difference in mass is called the mass defect of the nucleus which is explained below.

When the separate nucleons are joined together to form the nucleus, work must be done to overcome the repulsive force of the positively charged protons. The source of this energy according to Einstein, is the conversion of a small portion of the mass of the nucleons into energy resulting in mass defect.

The net energy released during the synthesis of a nucleus from its component nucleons is called the binding energy of the nucleus and this accounts for the difference in mass of a nucleus and the mass of its constituent nucleons.

## 20. (C)

Properties which are dependent on the extent or size of matter are termed as extensive properties.

Volume of matter would vary directly as the amount is changed, even though the state of the matter remains unchanged. Other properties do not depend on the amount of matter in the system.

## 21. (D)
The total radiant intensity

$$\int_0^\infty R\,d\lambda = \int_0^\infty \frac{8\pi}{\lambda^4} kT \frac{c}{4} d\lambda$$

$$= \frac{8}{3} \frac{\pi}{\lambda^3} kT \frac{c}{4} \Big|_0^\infty$$

The units for the total radiant intensity are

$$\frac{1}{m^3} \left(\frac{J}{K}\right) (K) \left(\frac{m}{sec}\right) = \frac{J}{sec} \cdot \frac{1}{m^2}$$

$$= \frac{W}{m^2}$$

Therefore the units for $\sigma$ are

$$\frac{W}{K^4 \cdot m^2} \cdot \quad \text{since}$$

$$\frac{W}{m^2} = \frac{W}{m^2 \cdot K^4} (K^4)$$

## 22. (A)

This cycle is called the Otto cycle, and is one of the internal combustion engine cycles. On the diagram, the adiabatic compression process is represented by step 12, followed by isochoric heat addition, step 23 then by the adiabatic power stroke in step 34, and finally the isochoric heat expulsion in step 41. The compression ratio is very important in Otto cycle, since the thermal efficiency of the cycle depends on it.

## 23. (A)

The component of gravitational force parallel to the incline is:

$$W\sin\theta = 500 \times \frac{3}{5} = 300 \text{ lb.}$$

So: $\quad W\sin\theta - F = 300 - 200 = 100 = F_f$

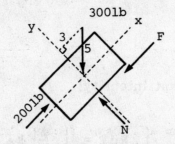

Since this component is greater than the 200 lb. force exerted on the block, the friction force exerted on the block will have upward direction. (Parallel to the incline) So we have:

$$W\sin\theta - F = 300 - 200 = F_f \quad \text{also:}$$
$$\Sigma F_y = 0$$
$$N - \frac{4}{5}(500) = 0$$
$$N = 400$$

Max Friction Force

$$F_{max} - \mu_s N = .3(400) = 120$$

Therefore, since the maximum friction force is greater than the actual friction force needed to maintain the block in equilibrium, the block must be at rest.

## 24. (A)

Entropy is a measure of randomness of the system. When liquid water changes to a gaseous state, there is an increase in the kinetic energy and therefore the disorder of water molecules. This shows an increase in entropy. Note that it is an irreversible process and for such processes, change of entropy is always positive.

## 25. (A)

Given below is the vector diagram for a coil and its associated power triangle.

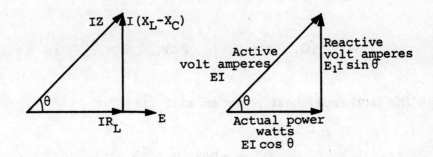

By the definition of power factor, the power factor of an a.c. circuit is equal to the cosine of the phase angle.

It is clear from the figure that when the reactive load increases, which means that the difference between $X_L = L\omega$ and $X_C = 1/C\omega$ increases, the phase angle $\theta$, increases and hence the power factor $\cos \theta$ decreases.

Also, the power is given by $P = RI_e^2$; the increase in the term $(X_L - X_C)$ leads to the increase in Z, which in turn leads to the decrease in $I_e$. $\left(\frac{V}{Z} = I_e\right)$, therefore power will decrease.

## 26. (B)

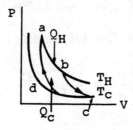

The efficiency for a Carnot heat engine is defined as

$$e = \frac{Q_H + Q_C}{Q_H} = 1 + \frac{Q_C}{Q_H}$$

The isothermal heat absorption at $T_H$ is

$$Q_H = nRT_H \ln \frac{V_b}{V_a}$$

(can be proved by using the formulas:

$$\Delta w = \Delta Q, \quad \Delta w = \int_{V_1}^{V_2} P dv, \quad P = \frac{nRT}{V} )$$

The isothermal heat rejection at $T_c$ is

$$Q_C = nRT_C \ln \frac{v_d}{v_c}$$

$$\frac{Q_C}{Q_H} = -\frac{T_C}{T_H} \frac{\ln(v_c/v_d)}{\ln(v_b/v_a)}$$

For the two adiabatic processes

$$T_H v_b^{\gamma-1} = T_C v_c^{\gamma-1}$$

$$T_H v_a^{\gamma-1} = T_C v_d^{\gamma-1}$$

$$\frac{v_b^{\gamma-1}}{v_a^{\gamma-1}} = \frac{v_c^{\gamma-1}}{v_d^{\gamma-1}} \quad \therefore \quad \frac{v_b}{v_a} = \frac{v_c}{v_d}$$

Hence,

$$\frac{\ln(v_c/v_d)}{\ln(v_b/v_a)} = 1 \quad \text{and} \quad \frac{Q_C}{Q_H} = -\frac{T_C}{T_H}$$

The efficiency of a Carnot engine

$$e = 1 - \frac{T_C}{T_H} = \frac{T_H - T_C}{T_H}$$

27. (E)
Objects are buoyant because the net force exerted by the fluid is greater than the weight of the object. However, when the weight of the object is equal to the net force exerted by the fluid, the object could remain suspended.

28. (C)
Such a setup is called a ballistic pendulum. This problem can be solved using energy conservation principle. The kinetic energy of the bullet is totally transferred to potential energy when the sphere is at its peak. But, since it is an inelastic collision, the ratio of the final energy to the initial energy is $(m_1)/(m_1 + m_2)$ where $m_1$ is the mass of the bullet and $m_2$ is the mass of the sphere. It can be verified in this way: according to conservation of linear momentum:

$$m_1 v_0 = (m_1 + m_2)v$$

in which $v_0$ is the initial velocity and $v$ is the velocity of the bullet-sphere system. The kinetic energy immediately after the collision will therefore be:

$$E = \tfrac{1}{2}(m_1 + m_2)v^2 \implies \frac{E}{E_i} = \frac{\tfrac{1}{2}(m_1 + m_2)v^2}{\tfrac{1}{2}m_1 v_0^2}$$

$$= \frac{m_1}{m_1 + m_2}$$

Therefore,

$$(m_1 + m_2)gd = \frac{m_1}{m_1 + m_2} \tfrac{1}{2}m_1 v_0^2$$

Solving for $v_0$,

$$v_0 = \frac{m_1 + m_2}{m_1}\sqrt{2gd}$$

Substituting the givens

$$v = \frac{0.050 + 1.000}{0.050}\sqrt{2(10)(0.2)} = (21)(2)$$

$$= 42 \frac{m}{sec}$$

29. (E)
An adiabatic process for an ideal gas is governed by the law

$$pV^\gamma = C$$

where $\gamma = \dfrac{C_p}{C_v}$

Since $C_p > C_v$ it cannot follow an inverse proportionality rule.

Similarly for a discharging capacitor it is governed by the law

$$v = v_0 e^{-\frac{t}{RC}}$$

and therefore cannot follow an inverse proportionality rule.

### 30. (E)
Since the bulb has a constant emitting power, the amount of energy received by both of the planes within the same period of time is constant. Therefore the ratio of the average intensities is inversely proportional to the square of the ratio of the distances form the vertex.

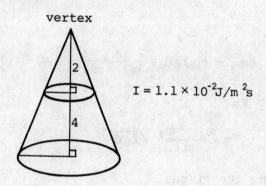

So we can write:

$$\left(\frac{2}{4+2}\right)^2 = \frac{I}{1.1 \times 10^{-2}} = \frac{1}{9}$$

$$\Rightarrow \quad I = 1.2 \times 10^{-3} \text{ J/m}^2 \cdot \text{sec}$$

### 31. (C)
Since the sled is being pulled at constant velocity, there are no unbalanced forces. We break up the tension in the rope into components parallel and perpendicular to the horizontal.

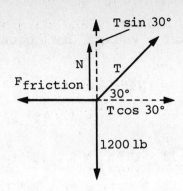

By Newton's second law $\Sigma F_x = 0$, therefore

$$F_{friction} = \mu N = T \cos 30° \quad (*)$$

$$\Sigma F_y = 0 \quad \therefore \quad N + T \sin 30 = W$$

$$N = W - T \sin 30$$

$$= 1200 - 1000(\tfrac{1}{2})$$

$$= 700$$

$$(*) \implies \mu = \frac{1000\left(\frac{\sqrt{3}}{2}\right)}{700} = \frac{5\sqrt{3}}{7}$$

### 32. (A)

When the object is hanging on the spring at equilibrium, the force exerted on the object by the spring is equal to the weight of the object. Therefore: $mg = ky_0$ where $y_0$ is the displacement of the spring from its equilibrium position. This point is the new equilibrium point. From now on, if the object is given a displacement and released, we will have:

total force exerted on the object = $k(y + y_0) - mg$

where y is the displacement from the new equilibrium point and therefore $(y + y_0)$ is the total displacement from the unstretched length of spring.

Since we have $ky_0 = mg$, then: total force on the object $= ky$. This is Hooke's law when the equilibrium point is at $y_0$ distance from the unstretched end of spring. Therefore the object will perform a simple harmonic motion about this point with maximum speed at it.

## 33. (C)

Writing a Kirchoff Voltage law equation for the loop, we obtain

$$4V + 7V - 13v_1 = i_1(1 + 8) \times 10^3$$

substitute $\quad v_1 = 1 \times 10^3 \, i_1$

$$4 + 7 - 13 \times 10^3 \, i_1 = 9 \times 10^3 \, i_1$$

$$11 = 22 \times 10^3 \, i_1$$

$$i_1 = \frac{11}{22 \times 10^3} = 0.5 \times 10^{-3} \, A = 0.5 \, mA.$$

## 34. (C)

In any reaction the nucleus is left in an excited state. When this condition is satisfied then the kinetic energy of the emitted particle is reduced. The higher the excited state of the nucleus is, the lower the kinetic energy for the emitted particle will be. If the reaction probability is to be decreased, then the number of emitted particles that are detected will decrease as well. Therefore at a lower reaction probability and with greater energy for the particles, the number of particles will be lower and so as the excited state for the nucleus.

## 35. (C)

The total clockwise moment acting on the wall (exerted by the water) is:

$$T = \int (\text{force at each point on the wall})$$
$$\cdot (\text{distance from the point O})$$

$$T = \int_0^6 \left( \rho g \left( \frac{\sqrt{3}}{2} x \right) \right) (6 - x)(2 \, dx)$$

in which (2dx) is the area of a strip of wall parallel to horizon with width dx and length 2 ft, (6 - x) is the arm of moment and $\left( \rho g \frac{\sqrt{3}}{2} x \right)$ is the pressure on the wall at a point with distance x from the top of the wall.

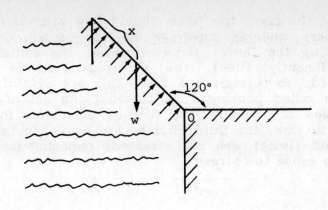

So we have:

$$T = \rho g \sqrt{3} \int_0^6 x(6-x)\,dx = \sqrt{3}\,\rho g \left[3x^2 - \frac{x^3}{3}\right]_0^6$$

$$T = \sqrt{3}\,\rho g [36] = 36\sqrt{3}\,\rho g$$

To have equilibrium, the moment of the weight of the gate must be equal to the moment of water pressure. Therefore:

$$36\sqrt{3}\,\rho g = \omega \cos 60° \left(\frac{6}{2}\right)$$

$$36\sqrt{3}\,\rho g = \frac{3\omega}{2}$$

$$\Rightarrow \quad \omega = 24\sqrt{3}\,\rho g$$

36. (E)
The periodic law says that when elements are arranged in order of increasing atomic number, there is a periodic repetition of physical and chemical properties. Thus, locating on the periodic table, where the element should be, would enable us to predict all of the given properties of the element (number of chemical bonds, chloride formula, density and melting point can all be approximated from the periodic table).

37. (B)
When the cup of juice is stationary (to a stationary observer), the juice experiences opposing forces between that of gravity and that of the cup. The result is a state of stress. Because the juice is a fluid, it cannot sustain

stress. In fact, the juice should flow when it is stressed. Containers such as cups are used to contain the fluid by preventing the flow. However, when the container or cup is sufficiently tilted, the stressed fluid is no longer contained, so it starts to flow; this act of tilting the cup is also called pouring. Notice that the act of tilting the cup does not ensure the flow of the fluid; in order for fluids to flow, the fluid must be stressed. In this problem, the fluid (juice) was not stressed; opposing forces did not exist to cause the stress.

### 38. (E)

Let W be the weight of a wet cloth. It is balanced by the tension in the rope represented by T. Writing the equilibrium equation

$$2T \sin \theta = w$$

$$T = \frac{W}{2 \sin \theta}$$

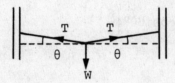

For the rope to be horizontal, $\theta$ must be equal to zero. This means that $\sin \theta$ is zero. Therefore, T must be infinite to satisfy the above equation which is an impossible task.

### 39. (D)

Due to a high temperature difference between your finger and the tray, some heat will be transferred to the tray, depending on its capacity for thermal conductivity.

Since the surface of your skin is naturally moisturized (it's not a function of the air humidity), the loss of heat will cause a thin layer of ice to form on the skin which in turn sticks to the tray.

No matter how dry the air and how cold the freezer, your finger will not stick to a wooden tray whose thermal conductivity is low. So the most important factor is the thermal conductivity of the tray.

**40. (E)**
As per Coulomb's Law, the force F exerted by a charge $q_1$ on $q_2$, separated by a distance r is given by

$$F = k \frac{q_1 q_2}{r^2}$$

where k is a constant of proportionality. Coulomb's finding hence resulted in an inverse square law of force for electrostatics. This means that when distance increases, the force decreases rapidly. This is represented well by a graph which is a hyperbola as shown by choice E.

**41. (E)**
According to the Heisenberg uncertainty principle, the limits for the uncertainties in position and momentum is given by

$$\Delta_x \Delta p_x > \frac{h}{2\pi}$$

Therefore, for a beam of light passing through a slit, the uncertainty in the position of a photon of light is determined by a. Therefore the uncertainty in the momentum of the photon is determined by the following equation.

$$\Delta p_x a > \frac{h}{2\pi}$$

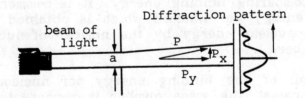

If a is decreased the momentum component $p_x$ becomes less certain since $\Delta p_x$ increases. This produces a broader diffraction pattern since the uncertainty in the position of the photons of light striking the screen has increased.

**42. (E)**
We define the amount of light passing per unit time through a unit area perpendicular to the direction in which the light is moving as the intensity of light.

As the distance from the light source increases, the light spreads over a greater area. The area (A) is proportional to the square of that distance (d).

$$A \alpha \, d^2$$

Hence the intensity of light ($\ell$) at a distance d from the source is inversely proportional to $d^2$.

$$\ell \, \alpha \, \frac{1}{d^2}$$

This inverse square relation is expressed by curve E.

## 43. (B)

A nucleus is built up of protons and neutrons. In each atom the mass of the nucleus is less than the sum of the masses of the individual protons and neutrons. The amount by which the mass of a nucleus is less than the sum of the masses of its constituent particles is known as the mass defect. The mass defect has a direct relationship to the binding energy of the nucleus. The binding energy is the energy which holds the nucleus together: The larger the mass defect, the larger the binding energy. The binding energy for any atom may be calculated by Einstein's equation $E_b = mc^2$, where $E_b$ is the binding energy, m is the loss in mass, and c the velocity of the light.

While comparing binding energy, it is common to give the binding energy per nucleon which is obtained by dividing the total nuclear energy by the number of nucleons or the mass number.

A graph of the binding energy per nucleon (in Mev), plotted against the mass number is represented by curve B.

## 44. (A)

The inductive reactance ($X_L$) is the magnitude of the reactive force opposing the flow of an AC and is measured in ohms.

It is mathematically expressed as

$$X_L = 2 \pi f L$$

where f is the frequency in cycles/second and L is the inductance in henrys. It is clear from the above relation that as the inductance is increased, the inductive reactance increases. This is represented by curve A.

### 45. (A)

The effective interest rate is the one which corresponds to the actual interest period. The effective interest rate is obtained by dividing the nominal interest rate of 8% by the number of compounded interest periods per year, which is in this case, four.

$$i = \frac{8\%}{4} = 2\%$$

### 46. (B)

In evaluating the given integral we use the formula:

$$\int u^n du = \frac{u^{n+1}}{n+1}. \text{ We obtain:}$$

$$\int_{-1}^{1} \left( x^{\frac{4}{3}} + 4x^{\frac{1}{3}} \right) dx = \frac{3}{7} x^{\frac{7}{3}} + 4 \cdot \frac{3}{4} x^{\frac{4}{3}} \Big|_{-1}^{1}$$

Now, evaluating the integral between 1 and −1, we have:

$$\frac{3}{7}(\sqrt[3]{1^7}) + 3(\sqrt[3]{1^4}) - \left[ \frac{3}{7}(\sqrt[3]{-1^7}) + 3\sqrt[3]{-1^4} \right]$$

$$= \frac{3}{7} + 3 - \left( -\frac{3}{7} + 3 \right)$$

$$= \frac{6}{7}.$$

### 47. (D)

In solving this problem we use the fact that the number of permutations P of n things taken all at a time [P(n,n)], of which $n_1$ are alike, $n_2$ others are alike, $n_3$ others are alike, etc. is

$$P = \frac{n!}{n_1! n_2! n_3! \ldots}, \text{ with } n_1 + n_2 + n_3 + \ldots = n.$$

In the given problem there are six letters (n = 6), of which two are alike (there are two N's so that $n_1 = 2$), three others are alike (there are three A's, so that $n_2 = 3$), and one is left (there is one B, so $n_3 = 1$). Notice that $n_1 + n_2 + n_3 = 2 + 3 + 1 = 6 = n$; thus,

$$P = \frac{6!}{2!3!1!} = \frac{6 \cdot 5 \cdot 4 \cdot 3!}{2 \cdot 1 \cdot 3! \cdot 1} = 60.$$

Thus, there are 60 permutations of the letters in the word BANANA.

48. (A)
We assume that the probability of parents having a blond child is $\frac{1}{4}$. In order to compute the probability that 2 of 4 children have blond hair we must make another assumption. We must assume that the event consisting of a child being blond when it is born is independent of whether any of the other children are blond. The genetic determination of each child's hair color can be considered one of four independent trials with the probability of success, observing a blond child, equal to $\frac{1}{4}$.

If X = the number of children in the family with blond hair, we are interested in finding $Pr(X = 2)$. By our assumptions X is binomially distributed with n = 4 and p = $\frac{1}{4}$.

Thus $Pr(X = 2)$ = Pr(exactly half the children are blond)

$$= \binom{4}{2}\left(\frac{1}{4}\right)^2\left(\frac{3}{4}\right)^2 = \frac{4!}{2!2!} \cdot \left(\frac{1}{4}\right)^2\left(\frac{3}{4}\right)^2$$

$$= \frac{4 \cdot 3}{2 \cdot 1} \cdot \left(\frac{1}{16}\right)\left(\frac{9}{16}\right) = \frac{27}{128}$$

49. (D)
$$\sin^2 x = \frac{1 - \cos 2x}{2}.$$

Substituting, we obtain:

$$\int_0^{\pi/4} \sin^2 x \, dx = \int_0^{\pi/4} \frac{1 - \cos 2x}{2} \, dx = \frac{1}{2}\left[\int_0^{\pi/4} dx - \int_0^{\pi/4} \cos 2x \, dx\right].$$

We can now apply the formula:

$$\int \cos u \, du = \sin u,$$

to the second integral letting $u = 2x$ and $du = 2dx$. Doing this, we have:

$$\int_0^{\pi/4} \sin^2 x \, dx = \left[\tfrac{1}{2}x - \tfrac{1}{4}\sin 2x\right]_0^{\pi/4}$$

$$= \left(\tfrac{1}{2} \cdot \tfrac{\pi}{4} - \tfrac{1}{4}\sin \tfrac{2\pi}{4}\right) - (0 - 0)$$

$$= \left(\tfrac{\pi}{8} - \tfrac{1}{4}\right)$$

$$= \frac{(\pi - 2)}{8}$$

50. (B)

$$A = \begin{bmatrix} 2 & -1 \\ 1 & 0 \\ -3 & 4 \end{bmatrix} \quad \text{and} \quad B = \begin{bmatrix} 1 & -2 & -5 \\ 3 & 4 & 0 \end{bmatrix}$$

order $3 \times 2$ \quad\quad order $2 \times 3$

$$\therefore \quad AB = \begin{bmatrix} 2 & -1 \\ 1 & 0 \\ -3 & 4 \end{bmatrix} \begin{bmatrix} 1 & -2 & -5 \\ 3 & 4 & 0 \end{bmatrix}$$

$$= \begin{bmatrix} 2(1) + (-1)(3) & 2(-2) + (-1)(4) & 2(-5) + (-1)(0) \\ 1 + 0 & 1(-2) + 0 & 1(-5) + 0 \\ -3(1) + 4(3) & (-3)(-2) + 4(4) & (-3)(-5) + 0 \end{bmatrix}$$

$$= \begin{bmatrix} -1 & -8 & -10 \\ 1 & -2 & -5 \\ 9 & 22 & 15 \end{bmatrix}$$

51. (E)

$$\frac{\tan x \cot x + \cos x \sec x}{\cos x + \sin x \tan x}$$

$$= \frac{1+1}{\cos x + \frac{\sin^2 x}{\cos x}} \qquad \left[\text{since } \tan x = \frac{\sin x}{\cos x}\right]$$

$$= \frac{2}{\frac{\cos^2 x + \sin^2 x}{\cos x}}$$

$$= \frac{2\cos x}{1} = 2\cos x$$

52. (A)
The measures of central tendency most commonly used are the mean, the median and the mode. These measures give an indication of the "middle" or "center" of the data set. A measure of central tendency attempts to identify the point that the data clusters around.

The mean, $\overline{X}$, is an average of the observations and is defined to be

$$\overline{X} = \frac{\Sigma X_i}{n} = \frac{\text{sum of observations}}{\text{number of observations}}$$

$$= \frac{19 + 23 + 27 + 19 + 23 + 28 + 27 + 28 + 29 + 27}{10}$$

$$= \frac{250}{10} = 25.$$

The median is another measure of central tendency. It is the "middle" number in a sample and is defined to be a number such that an equal number of observations lie above and below it.

To compute the median we first order the observations in the sample. This has been done below.

19  19  23  23  27  27  27  28  28  29

In the case of an even number of observations, n, the median is defined to be the average of the $\frac{n}{2}$ and $\frac{n}{2} + 1$ observations. In our example the $\frac{n}{2} = \frac{10}{2} = $ 5th and the $\frac{n}{2} + 1 = $ 6th observations are both 27.

The average is $\frac{27 + 27}{2} = 27$.

Thus the median of this sample is 27.

The mode is another measure of central tendency and is defined as the most frequently occurring observation in the sample. The mode in this sample is the observation 27 which occurs three times, more than any other observation.

53. (E)

To find $\frac{dy}{dx}$, we use the differentiation formula:

$$\frac{d}{dx} e^u = e^u \frac{du}{dx}, \text{ letting } u = \frac{x^2}{4}.$$

Then

$$\frac{du}{dx} = \frac{2x}{4} = \frac{x}{2}$$

Applying the formula we obtain:

$$\frac{dy}{dx} = e^{\frac{x^2}{4}} \cdot \frac{x}{2} = \frac{x}{2} e^{\frac{x^2}{4}}.$$

54. (E)

$\int_a^b F(x)dx$ and $\int_b^c G(x)dx$ are supposed to be considered as two different integral expressions, as $F(x)$ and $G(x)$ are two distinct functions, integrable on two different intervals; therefore the sum of $F(x)$ and $G(x)$ cannot be integrable on the same interval.

55. (D)

$$\int_a^b \int_c^d F'(x)G'(y)dxdy$$

$$= \int_c^d \int_a^b F'(x)G'(y)\,dx\,dy$$

$$= \int_c^d \left[\int_a^b F'(x)\,dx\right] G'(y)\,dy$$

$$= \int_c^d [F(b) - F(a)]G'(y)\,dy$$

$$= [F(b) - F(a)] \int_c^d G'(y)\,dy$$

$$= [F(b) - F(a)][G(d) - G(c)]$$

**56. (D)**
Using the relation

$$\int_a^b F(x)\,dx + \int_b^c F(x)\,dx = \int_a^c F(x)\,dx,$$

$$\int_0^1 F(x)\,dx + \int_2^0 F(x)\,dx + \int_1^3 F(x)\,dx$$

$$= -\int_1^0 F(x)\,dx - \int_0^2 F(x)\,dx + \int_1^3 F(x)\,dx$$

$$= -\left[\int_1^0 F(x)\,dx + \int_0^2 F(x)\,dx\right] + \int_1^3 F(x)\,dx$$

$$= -\int_1^2 F(x)\,dx + \int_1^3 F(x)\,dx$$

$$= \int_2^1 F(x)dx + \int_1^3 F(x)dx$$

$$= \int_2^3 F(x)dx$$

**57. (D)**
$F(x) = (x^2 + 1)^{\frac{1}{2}}$, and $F'(x) = \frac{1}{2}(x^2 + 1)^{-\frac{1}{2}}(2x)$,

or

$$F'(x) = \frac{2x}{2\sqrt{x^2+1}} = \frac{x}{\sqrt{x^2+1}}.$$

**58. (A)**
Let $F(x) = y = 2x^2 - 8x + 6$. To obtain the minima and maxima we find $\frac{dy}{dx}$, set it equal to 0, and solve for x. We find:

$$\frac{dy}{dx} = 4x - 8 = 0.$$

Therefore, $x = 2$ is the critical point. We now use the Second Derivative Test to determine whether $x = 2$ is a maximum or a minimum. We find: $\frac{d^2y}{dx^2} = 4$, (positive). The second derivative is positive, hence $x = 2$ is a minimum.

**59. (A)**
We can first rewrite the function as:

$$y = e^{\frac{2}{x}}$$

Now we use the formula:

$$\frac{d}{dx}e^u = e^u\frac{du}{dx},$$

letting $u = \frac{2}{x}$. Then,

$$\frac{du}{dx} = \frac{(x)(0) - (2)(1)}{x^2} = -\frac{2}{x^2}.$$

Applying the formula, we obtain:

$$\frac{dy}{dx} = e^{\frac{2}{x}} \cdot -\frac{2}{x^2}$$

$$= -\frac{2e^{\frac{2}{x}}}{x^2}.$$

**60. (B)**
Because of the analogy between infinite series and improper integrals, one might expect to find an improper integral corresponding to a power series. Indeed, a power series

$$\sum_{n=0}^{\infty} c_n x^n$$

may be written as

$$\sum_{n=0}^{\infty} f(n)x^n, \tag{1}$$

where f is a function whose value of each n is given by $f(n) = c_n$. The natural analogue of (1) is the improper integral

$$\int_0^{\infty} f(t)x^t \, dt = F(x), \tag{2}$$

and except for a minor change in notation, (2) is the Laplace transform of the function f(t). The notational change is accomplished by making the substitution

$$x = e^{-s}$$

in (2) to give

$$F(s) = \int_0^{\infty} f(t)e^{-st} \, dt, \tag{3}$$

and $F(s)$ is called the Laplace transform of $f(t)$. Letting the expression $L\{f(t)\}$ represent the operation of multiplying $f(t)$ by $e^{-st}$ and integrating from $t = 0$ to $t = \infty$ one could write

$$F(s) = L\{f(t)\} \tag{4}$$

which makes clear the interpretation of (3) as an operator, L, transforming the function $f(t)$ into the function $F(s)$.

61. (D)
$$(x + y)^2 = x^2 + y^2 + 2xy$$
$$= 2 + 2(-1)$$
$$= 2 - 2$$
$$= 0$$
$$\therefore \quad (x + y) = 0$$

Now $(x^3 + y^3) = (x + y)(x^2 + y^2 - xy)$
$$= (0)(2 + 1)$$
$$= 0$$

62. (D)
As $x \to \infty$, the function takes the form $\infty/\infty$ which is indeterminate. Using the method of dividing numerator and denominator by the largest power in x, we obtain

$$\frac{1 + 10/x^2}{6 + 2/x^2}$$

As $x \to \infty$, the fractional terms of both numerator and denominator approach zero. This means that the numerator approaches 1 and the denominator approaches 6. Therefore

$$\lim_{x \to \infty} \frac{x^2 + 10}{6x^2 + 2} = \frac{1}{6}$$

Another approach to this problem is to apply L'Hopital's rule directly.

$$\lim_{x \to \infty} \frac{x^2 + 10}{6x^2 + 2} = \frac{2x}{12x} = \frac{1}{6}$$

## 63. (E)

If A is a square matrix and it has an inverse, that inverse works both on the right and left: $AA^{-1} = A^{-1}A = I$. If A is not square, then its right and left inverses are different matrices, and we say that the inverse, which is the common value of the right and left inverses, does not exist. $AR = I$, $LA = I$, $L \neq R$.

## 64. (B)
We have

$$S_n = 1 - t + t^2 - t^3 + \ldots + (-1)^n t^n$$

$$tS_n = t - t^2 + t^3 - t^4 + \ldots + (-1)^n t^{n+1}.$$

Adding yields

$$S_n + tS_n = 1 + (-1)^n t^{n+1}$$

or

$$S_n = \frac{1 + (-1)^n t^{n+1}}{1 + t}$$

and

$$\lim_{n \to \infty} S_n = \frac{1}{1 + t}$$

for $|t| < 1$.

## 65. (C)

To answer this, we must invoke conditions that are sufficient to ensure that $f(t)$ may be expanded into a convergent trigonometric (or Fourier) series.

Firstly, a Fourier series of a periodic function, $f(t)$, is denoted by:

$$f(t) = a_o + \sum_{n=1}^{\infty} a_n \cos(n\omega_o t) + b_n \sin(n\omega_o t)$$

where $a_o$, $a_n$ and $b_n$ are the "Fourier coefficients" and $\omega_o$ is the fundamental frequency ($n\omega_o$, $x = 1, 2, \ldots$ are known as the harmonics).

Now that we see what a Fourier series looks like, we can understand the sufficient conditions that it satisfies. These conditions are known as the Dirichlet conditions.

There are four of these conditions:

1. f(t) is single-valued.
2. f(t) has a finite number of maxima and minima on the periodic interval.
3. f(t) has a finite number of discontinuities on the periodic interval.
4. $\int_{t_o}^{t_o+t} |f(t)|\, dt$ exists.

It must be stressed that these are sufficient conditions. Looking at the choices provided, (A) and (D) are eliminated. f(t) need not be positive and negative on any interval. Also, f(t) need not be smooth.

f'(t) does equal zero for at least one value of t, but this is not a sufficient condition (Dirichlet).

The area bounded by f(t) on the periodic interval is finite. This satisfies sufficient Dirichlet condition (4) and the answer is (C).

66. (B)

The constant coefficient term is the term that does not multiply a sinusoid in the Fourier series expansion. Looking at the expansion:

$$f(t) = a_o + \sum_{n=1}^{\infty} a_n \cos(n\omega_o t) + b_n \sin(n\omega_o t)$$

we see that the term we are looking for is $a_o$. The equation to find $a_o$ is:

$$a_o = \frac{1}{T} \int_{-T/2}^{T/2} f(t)\, dt$$

or, more simply stated, the average value of the function over its period T.

Looking at the function in the figure, it clearly has a period of 8. The next thing we do is a piecewise integration of f(t) over its period.

First we notice that f(t) is an even function, that is, f(t) = f(-t). Because of its symmetry we notice that

$$\int_{-T/2}^{T/2} f(t)dt = 2\int_{0}^{T/2} f(t)dt.$$

Thus, $\quad a_0 = \dfrac{2}{T}\displaystyle\int_{0}^{T/2} f(t)dt.$

Evaluating the integral:

$$\int_{0}^{T/2} f(t)dt = \int_{0}^{4} f(t)dt.$$

From the curve:

$$\int_{0}^{4} f(t)dt = \int_{0}^{2} t+4\, dt + \int_{2}^{3} 6\, dt + \int_{3}^{4} -2\, dt$$

$$= \dfrac{t^2}{2} + 4t \Big|_{0}^{2} + 6t \Big|_{2}^{3} + (-2t)\Big|_{3}^{4}$$

$$= 10 + 6 + (-2) = 14$$

Multiplying by 2: $\quad 2(14) = 28$

Dividing by $T = 8$: $\quad \dfrac{28}{8} = \dfrac{7}{2}$

Hence, $a_0 = \dfrac{7}{2}.$

67. (A)
In a Fourier Series expansion, the cosine coefficients are usually denoted by $a_n$, when f has even symmetry.

The formula for computing $a_n$ is:

$$a_n = \dfrac{4}{T}\int_{0}^{T/2} f(t)\cos(n\omega_0 t)dt$$

where T is the period, f is the function and $\omega_0 = \frac{2\pi}{T}$. We see that we must integrate over one-half of the period.

Looking at the graph of f, T = 8, so $\frac{T}{2}$ = 4. To integrate f, we must divide into intervals and integrate over them.

Hence, $\quad f = t + 4 \qquad 0 \le t \le 2$

$\qquad\qquad f = 6 \qquad\qquad 2 \le t \le 3$

$\qquad\qquad f = -2 \qquad\quad 3 \le t \le 4$

and our integral will have the form:

$$a_n = \frac{4}{T} \int_0^{T/2} f(t) \cos(n\omega_0 t)\,dt$$

$$= \frac{1}{2} \int_0^4 f(t) \cos(n\omega_0 t)\,dt$$

$$= \frac{1}{2}\left[\int_0^2 (t+4)\cos(\frac{\pi n}{4} t)\,dt + \int_2^3 6\cos(\frac{\pi n}{4} t)\,dt - \int_3^4 2\cos(\frac{\pi nt}{4})\,dt\right]$$

The first integral is evaluated by integration by parts with the result:

$$\left(\frac{4}{\pi n}\right)^2 \left[\cos\frac{\pi n}{2} + \frac{\pi n}{2}\sin\frac{\pi n}{2} - 1\right] + \frac{16}{\pi n}\sin\frac{n\pi}{2} .$$

The second integral is simply:

$$\frac{24}{n\pi}\left(\sin\frac{3\pi n}{4} - \sin\frac{\pi n}{2}\right) .$$

Finally, the third integral evaluates to:

$$\frac{8}{\pi n}\left[\sin(n\pi) + \sin(\frac{3}{4}\pi n)\right] .$$

Summing these results and dividing by 2 gives us the final result for the cosine coefficients

$$a_n = \frac{8}{(\pi n)^2}\left[\cos\frac{\pi n}{2} + \frac{\pi n}{2}\sin\frac{\pi n}{2} - 1\right] + \frac{8}{\pi n}\sin\frac{\pi n}{2}$$

$$+ \frac{12}{\pi n}\left(\sin\frac{3\pi n}{4} - \sin\frac{\pi n}{2}\right)$$

$$- \frac{4}{\pi n}\left[\sin(n\pi) - \sin(\tfrac{3}{4}\pi n)\right]$$

From this we see that $a_n$ in general is a complicated coefficient, but we can simplify it to get the following form:

$$a_n = \frac{8}{\pi n}\left[\left(\cos\frac{\pi n}{2} - 1\right)/\pi n + 2\sin(\tfrac{3}{4}\pi n)\right]$$

68. (C)
We may graphically differentiate the function $f(t)$. Remembering that the derivative is the slope of the tangent to a curve at a point, $f'(t)$ looks as follows:

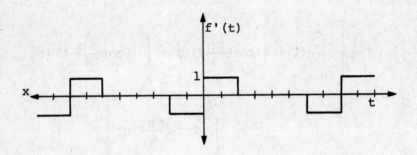

By inspection, we see that $f'(t)$ is periodic with a period of 8. It is also an odd function, that is, $f(t) = -f(-t)$.

Thus $\frac{df(t)}{dt}$ is an odd periodic function.

Note that at points where we have a cusp, the derivative is not defined, but this ignorance will not damage the basic idea.

We can also show the above result mathematically. Since we have:

$$f(t + T) = f(t) \Rightarrow f'(t + T)\cdot\frac{d(t + T)}{dt} = f'(t)$$

$$\Rightarrow f'(t) = f'(t + T)$$

Therefore the derivative of a periodic function is a periodic function with the same period.

We also have:

$$f(t) = f(-t) \Rightarrow f'(t) = f'(-t)\frac{d(-t)}{dt}$$

$$\Rightarrow -f'(t) = f'(-t).$$

Therefore the derivative of an even function is an odd function.

## 69. (D)

From the graph of f(t), by inspection, we see that f(t) is an even function. sin t is an odd function. When an even function is multiplied by an odd function, an odd function is the result. Moreover, the function g(t) will be a periodic function, because the addition, subtraction, division and multiplication of two periodic functions is a periodic function; therefore it may be represented as a Fourier Series (g(t) will be non-sinusoidal).

In an odd, non-sinusoidal periodic function the cosine coefficients, that is, the $a_n$'s are all zero.

This is because $\int_{-T/2}^{T/2} ft(t)\cos(n\omega_o t)$ calculates the area bounded on both sides of the periodic interval which are opposite in sign. Thus the integral is zero.

## 70. (E)

The rms (root mean square) value of a function is obtained through use of the following formula:

$$F_{rms} = \sqrt{\frac{1}{T}\int_{t_o}^{t_o+T} f^2(t)dt}$$

where T is the period.

We notice that f(t) is an even function. Therefore, we only have to integrate over half the periodic interval and mulitply our result by 2.

The integral becomes

$$\int_0^4 f^2(t)dt = \int_0^2 (t+4)^2 dt + \int_2^3 36 dt + \int_3^4 4 dt$$

$$= \frac{t^3}{3} + 4t^2 + 16 \Big|_0^2 + 36t \Big|_2^3 + 4t \Big|_3^4$$

Evaluating, we obtain $\frac{272}{3}$. Hence the integral over the interval $[-4,4]$ is equal to $2 \times \frac{272}{3}$. We next divide by 8 (the period) which gives us

$$\frac{2 \times 272}{24} = \frac{34 \times 2}{3}$$

Taking square root: $\sqrt{\frac{34 \times 2}{3}} = \sqrt{\frac{68}{3}}$

## 71. (A)

From the information given, we know what the initial velocity of K is and also its final velocity. We also know the distance particle K has traveled.

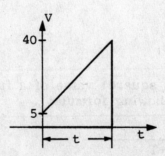

When all this information is incorporated into the solution we may construct a v-t diagram for particle K.

The area bounded by this diagram is equal to the total distance traveled by K. Hence,

$$S = \tfrac{1}{2}(t)(v_1 + v_2)$$

$$t = \frac{2S}{v_1 + v_2} = \frac{2(120)}{5 + 40} = \frac{240}{45}$$

Thus, time elapsed = $\frac{16}{3}$ s.

### 72. (E)

In the previous problem, the time interval for K traveling 120 m was found to be $\frac{16}{3}$s.

The tangential component of acceleration of a particle traveling in a circle is defined as:

$$a_t = \frac{dv}{dt}.$$

We know that the velocity changes uniformly from 5 m/s to 40 m/s, thus $a_t$ is a constant and is equal to the slope of the graph of v(t).

This slope is: $\frac{\Delta v}{\Delta t} = \frac{v_2 - v_1}{t_2 - t_1}$

Substituting values: $a_t = \frac{40 - 5}{\frac{16}{3} - 0} = \frac{35}{\frac{16}{3}} = \frac{105}{16}$

Reducing: $a_t = 6\frac{9}{16} \frac{m}{s^2}$

### 73. (C)

The normal component of acceleration of a particle undergoing circular motion is defined as:

$$a_{norm} = v\omega, \qquad (1)$$

where v is its linear velocity and ω is its angular velocity.

Since $v = r\omega$, $\omega = \frac{v}{r}$ and eq.(1) becomes:

$$a_{norm} = \frac{v^2}{r} \text{ (expressing in known quantities).}$$

At the end of the time interval, $v = 40 \frac{m}{s}$. The diameter of the circular path is 20m, so its radius is 10m.

Thus the normal acceleration is

$$a_{norm} = \frac{v^2}{r} = (40)^2/10 = 160 \, \frac{m}{s^2}$$

**74. (D)**
The first step to finding the angle turned through is to determine the velocity of K after 2 seconds have elapsed. To do this, we must go back to the v-t diagram. The slope of this diagram $\frac{\Delta v}{\Delta t}$ is the tangential acceleration, $a_t$. $a_t$ was already found to be $6\frac{9}{16} \frac{m}{s^2}$, $\Delta t = 2s$, and $v_0 = 5\frac{m}{s}$, so:

$$a_t = \frac{v_1 - v_0}{t_1 - t_0}$$

$$v_1 = a_t(t_1 - t_0) + v_0$$

$$= 6\frac{9}{16}(2) + 5 = \frac{145}{8}$$

Now that the velocity has been found, the total displacement over the first two seconds can be found by integrating v from t = 0 to t = 2 seconds.

The area bounded by v over this interval is equal to the distance:

$$S = \frac{1}{2}(\Delta t)(v_0 + v_1) = \left(\frac{1}{2}\right)2\left(5 + \frac{145}{8}\right) = \frac{185}{8}$$

To find the total angle turned through we use,

$$S = r\phi$$

where $\phi$ is the angle turned through by line $\overline{OK}$.

Thus:

$$\phi = \frac{S}{r} = \frac{185}{8(10)} = 2\frac{5}{16} \text{ radians}$$

**75. (B)**
The distance as a function of time is obtained simply by integrating v(t).

First we obtain v(t) as a function of time. Using the point-slope form of a line:

$$(v - v_0) = m(t - t_0) \quad \text{where m is the slope.}$$

$$v - 5 = \frac{\frac{35}{16}}{3} t = \frac{105}{16} t.$$

$$v = \frac{105}{16} t + 5.$$

Integrating

$$s = \int v\, dt = \int \frac{105}{16} t + 5 \; dt$$

$$= \frac{105}{32} t^2 + 5t + C.$$

C can be evaluated by:

$$s(0) = 0 = 0 + 0 + C.$$

Thus,

$$s(t) = \frac{105}{32} t^2 + 5t$$

Generally, the equation for the distance covered by the particle is: $x = \frac{1}{2}at^2 + vt$ where a is the linear acceleration, v is initial velocity and t is time.

## 76. (D)

When we look at the given distance curves we see that particle P stops at vertex D for one second at time t = 8.

This distance is plainly read off the graph as being 13.

## 77. (C)

The solution to this problem is as straightforward as the previous problem.

On the graph of the distance from point C vs. time, we see that at time t = 8, particle P resides at point D, thus the distance between points P and C is clearly equal to 12.

**78. (B)**

From the given information and curves, we conclude that particle P starts on its journey from point A. Considering the curve of the distance from C vs. time, we notice that the function starts at a distance 5 from C. Thus the distance between A and C is equal to five.

**79. (D)**

From the triangle vertices given and the distance curves we can determine the lengths of the sides of AED. The length of AE is (from the curves) 5. The length of AD is easily seen to be 13.

Since $\angle CDE = 90°$ in the triangle EDC, we can write:

$$DE^2 = EC^2 - DC^2.$$

Since EC = 16.97 (from the graph) and DC = 12, DE is computed as $\sqrt{(16.97)^2 - (12)^2} = 12$.

With sides of 5, 12, 13 we notice that the triangle has a right angle because we have: $5^2 + 12^2 = 13^2$; thus it is a right triangle.

**80. (B)**

From the given curves we find the sides of $\triangle ABC$ to be:

$$AB = d_{AB} = 3$$
$$BC = d_{BC} = 4$$
$$AC = d_{AC} = 5$$

Thus the triangle is a 3-4-5 right triangle, $\angle ABC = 90°$. Using side AB as its altitude and side BC as its base, we compute the area as follows:

$$A = \tfrac{1}{2}bh = \tfrac{1}{2}(4)(3) = 6$$

**81. (D)**

The fact is, the more points we set on the x-axis, the more the distance from the origin will be. The x-coordinate of the nth point on the x-axis (which is actually the nth point of intersection of the curve so established with the x-axis) will be equal to the following sum:

$$x_n = \sum_{m=1}^{n} \frac{1}{m}.$$

Since the limit of $x_n$ as $n \to \infty$ is $\infty$ (the series is divergent), the curve will proceed toward infinity endlessly. Therefore the answer D is correct.

(The number of points of intersection with the x-axis, or the speed at which $y \to \infty$, is unrelated to the subject.)


**82. (C)**
Segments of the curve bounded between two consecutive points of intersection with the x-axis, are quarters of circles with decreasing radii. Therefore the angle of intersection between the curve and x-axis, is always 45°, no matter how far we are from the origin and how small the radius of the 90° arc is. Therefore, although the derivative is beautifully defined over the whole positive x-axis, it does not have a limit at infinity due to the fact that f'(x) covers the whole interval [-1,1] within any interval of x bounded by the two points of intersection with x-axis.


**83. (B)**

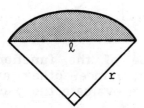

The integral is the sum of the areas under the arcs. Therefore we can find the areas under each arc and add them up in the form of an infinite series.

The shaded area shown in the figure is equal to:

shaded area = area of the quarter of circle - the area of
 the right isosceles triangle

$$\Rightarrow \quad \text{area} = \frac{\pi r^2}{4} - \frac{r^2}{2} = \left(\frac{\pi}{4} - \frac{1}{2}\right) r^2$$

$$\frac{r}{\ell} = \cos 45° = \frac{\sqrt{2}}{2} \Rightarrow r = \frac{\ell}{\sqrt{2}}$$

$$\Rightarrow \quad \text{area} = \left(\frac{\pi}{8} - \frac{1}{4}\right)\ell^2, \quad \ell_n = \frac{1}{n}$$

$$\Rightarrow \quad \text{area} = \left(\frac{\pi}{8} - \frac{1}{4}\right)\frac{1}{n^2}$$

Thus, the total area under the curve from 0 to $\infty$ is equal to the series:

$$\left(\frac{\pi}{8} - \frac{1}{4}\right) \sum_{n=1}^{\infty} \frac{(-1)^{n+1}}{n^2}$$

(the odd terms have negative signs because they are under the x-axis).

On the other hand, the above series is convergent because we have $|a_n| > |a_{n+1}|$ where $a_n$ is the nth term. They also switch their signs one by one. The above conditions are sufficient for the convergence of the series, and therefore the convergence of the integral.

84. (E)

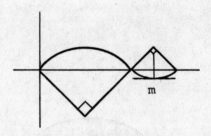

The minimum value of the function is the lowest point on the second arc, since other arcs have less radii. Therefore, we have to evaluate the y-coordinate of the point m. We have: $y = r - x$

$$x = \frac{\sqrt{2}}{2} r .$$

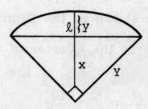

On the other hand, $r = \frac{\sqrt{2}}{2}\ell$ .

Therefore:

$$y = r - r\frac{\sqrt{2}}{2} = r\left(1 - \frac{\sqrt{2}}{2}\right) = \left(\frac{2 - \sqrt{2}}{2}\right)\frac{\sqrt{2}}{2}\ell$$

$$= \frac{(\sqrt{2} - 1)\ell}{2}.$$

So the minimum value will be:

$$\frac{-(\sqrt{2} - 1)\ell}{2} = \frac{(1 - \sqrt{2})\ell}{2}.$$

85. (C)
Since $g(x)$ is defined in this way:

$$g(x) = \int_0^x f(t)\,dt$$

we have: $\qquad g'(x) = f(x).$

Therefore $g(x)$ is a function whose derivative has a damped oscillation about zero.

On the other hand, we proved in the previous questions that $\int_0^\infty f(t)\,dt$ converges to a positive value. Therefore $g(x)$ has an asymptote parallel to the x-axis, in the upper half of the x-y plane.

Among the given curves, curve C is the only one which represents a bounded function with a certain limit and horizontal asymptote whose derivative has a damped fluctuation.

Note that since we have $g''(x) = f'(x)$, we can conclude that function g always changes its concavity, even when x is extremely large.

86. (B)
Consult the diagram: AB is the beam and C its midpoint. The weight W acts through C, since the beam is uniform and the two other forces acting on the beam are the force F exerted by the carpenter at B at right angles to AB and

a total force P exerted by the floor at A in an unspecified direction.

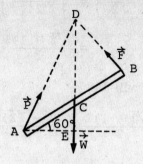

Since the beam is acted on by three forces which maintain it in equilibrium, the lines of action of the three forces must be concurrent. Thus, the direction of force P is from A to the point D at which F and W meet.

Since the board is not in motion, the second condition of equilibrium can be applied. Taking moments about point A, we have

$$\Sigma T_A = P \times 0 + F \times AB - W \times AE = 0$$

$$F = W \frac{AE}{AB} = W \frac{AC \cos 60°}{AB} = \frac{1}{2} W \cos 60° = \frac{\sqrt{1}}{4} W.$$

$$\therefore \quad F = 25 \text{ lb}.$$

87. (D)

$$\text{Speed (N)} = \frac{E_t - I_a R_a}{k\phi} \quad \text{for a motor}$$

$$= \frac{E_g}{k\phi}$$

where  $E_t$ = terminal voltage

$E_g$ = counter emf

$I_a$ = armature current

$R_a$ = armature resistance

$\phi$ = field flux

$k$ = constant.

From above

$$N \propto \frac{1}{\phi}.$$

In the series motor the $I_aR_a$ drop is very small compared to the effect of the field flux. Therefore, the speed of the series motor depends almost entirely on the field flux. An increase in load increases the armature current and therefore results in an increase in field current and field flux, causing a decrease in speed. Similarly, a decrease in load current and therefore in field current and field flux causes an increase in speed.

At no load, the flux is practically zero and since a series motor does not have a definite no-load speed, the speed rises to a dangerously high value.

88. (D)
By definition,

$$\text{percent yield} = \frac{\text{actual yield}}{\text{theoretical yield}} \times 100\%$$

Actual yield = $66g\, C_2H_4O$.

In the equation

$$C_2H_4 + \tfrac{1}{2}O_2 \rightarrow C_2H_4O$$

we see that 1 mole of $C_2H_4$ produces 1 mole $C_2H_4O$, or

$$(2 \times 12.0 + 4 \times 1.0)g\, C_2H_4$$

produces $(2 \times 12.0 + 4 \times 1.0 + 16.0)g\, C_2H_4O$,

i.e. $28g\, C_2H_4$ produces $44g\, C_2H_4O$.

$$\text{Theoretical yield} = 56g\, C_2H_4 \times \frac{44g\, C_2H_4O}{28g\, C_2H_4} = 88g\, C_2H_9O,$$

so that

$$\text{percent yield} = \frac{66g\, C_2H_4O}{88g\, C_2H_4O} \times 100\% = 75\%.$$

89. (B)
The Seebeck effect is not responsible for the difference in resistance. The Seebeck effect is an electromotive force

that results from a difference in temperature between two junctions of dissimilar metals in the same circuit.

Radiation losses and skin effects apply to any type of network, while Eddy currents and Hystensis losses are due to the presence of ferromagnetic materials in a changing magnetic field.

## 90. (B)

The continuity equation is all one needs to solve for the flow rate, $q = v_2 A_2$, because $v_2$ is the only variable.

$$v_2 A_2 = v_1 A_1,$$

$$v_2 = \frac{v_1 A_1}{A_2} = \frac{v2A}{A} = 2v,$$

so (B) is the correct answer. Bernoulli's equation could be solved as well, but would result in a much more complicated solution involving $v_2$ and $\rho$ as well:

$$\frac{v_2^2 - v_1^2}{2} = \frac{p_1 - p_2}{\rho}$$

$$v_2^2 = \frac{2}{\rho}(p_1 - p_2) + v_1^2$$

$$v_2 = \sqrt{\frac{2p_1}{\rho} + v_1^2}$$

## 91. (A)

The correct answer is (A). The pressure difference is independent of length and the coefficient for low viscosity, and depends both on flow rates and surface areas.

## 92. (D)

The equivalent circuit is shown above. Write Kirchoff's voltage law around the loop to obtain,

$$(25) + I(35) - 10 + I(10) = 0$$

∴  $I(45) = -15$

∴  $I = \frac{-15}{45} = \frac{-1}{3}$ ampere.

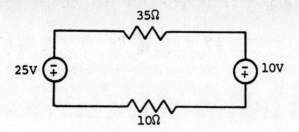

**93. (B)**

$$\frac{E_{g_1}}{E_{g_2}} = \frac{n_1}{n_2} \quad \text{or} \quad \frac{130}{E_{g_2}} = \frac{1200}{600}$$

$$\therefore \quad E_{g_2} = \frac{130 \times 600}{1200} = 65 \text{ volts}$$

**94. (B)**
An elastic object begins to resonate when it is struck with sound waves of the same frequency as its natural frequency. When successive simultaneous impulses are applied by the marching soldiers to the bridge, it starts to resonate. Due to resonance, the amplitude of the vibrations of the bridge increase and may reach a point where the bridge collapses.

**95. (B)**
The power dissipated by the resistor = $VI = \frac{V^2}{R}$

$$p = \frac{V^2}{R} = \frac{(2)^2}{5} = .8 \text{ watts}$$

This power over a period of 1 min. appears as heat energy:

$$Q = Pt = .8 \text{ watts}(1 \text{ min})\left(\frac{60 \text{ sec}}{1 \text{ min}}\right)$$

$$= 48.0 \text{ Joules}$$

$$Q = mc(T_2 - T_1) = 48.0 \text{ Joules}$$

$$\left(\frac{1000g}{1\ kg}\right)(.5\ kg)\left(1\ \frac{cal}{gC°}\right)(\Delta T) = 48.0\ \text{Joules}\left(\frac{1\ cal}{4.186J}\right)$$

$$\Delta T = \frac{0.0960}{4.186}\ C°$$

**96. (B)**

Since the man is on roller skates, the law of conservation of momentum can be used, as there is no external impulse. The man with the stone is at rest initially so

$$m_1u_1 + m_2u_2 = 0.$$

Using the conservation of momentum equation,

$$m_1u_1 + m_2u_2 = m_1v_1 + m_2v_2,$$

we get

$$0 = -(150) \times v_1 = (7.5)(20).$$

Therefore the velocity of the man

$$v_1 = -\frac{7.5 \times 20}{150} = -1\ \text{ft/sec},$$

where the negative sign indicates that the man will move backward.

**97. (A)**

The equilibrium constant expression for a balanced chemical reaction is the ratio of concentrations of products to the reactants.

The concentrations are raised to the power of moles of the substance as given in the balanced equation. So,

$$K_c = \frac{[NH_2(g)]^2}{[H_2(g)]^3[N_2(g)]}.$$

The (g) specifies that these reactants and product are in the gaseous state and can be left out of the notation.

$$K_c = \frac{[NH_2]^2}{[H_2]^3[N_2]}$$

## 98. (D)

The spring exerts an overall force of 18N, but the force exerted on each block is 9N, since the displacement of the spring on opposite sides is the same. Therefore, for the first block:

$$F_1 = m_1 a_1$$

$$18N = m_1(6 \text{ m/s}^2)$$

$$m_1 = 3 \text{ kg}$$

$$\frac{m_1}{m_2} = 3 \quad \therefore \quad m_2 = 1 \text{ kg}$$

$$F_2 = m_2 a_2$$

$$18N = m_2 a_2$$

$$a_2 = 18 \text{ m/s}^2$$

## 99. (B)

As the temperature increases, the average speed of the molecules increases, since the kinetic energy of the molecules is greater. The range of typical speeds is greater with increased temperature. Therefore, the distribution broadens but is also flattened since $\int_0^\infty N(v)dv$ (Number of molecules) must remain the same. The area under the curve of both temperatures should remain the same. The only graph that satisfies these conditions is (B).

## 100. (E)

Displacement vector, $\vec{S} = 3 \sin t \, \vec{i} + 3t^2 \, \vec{j}$

velocity vector $\vec{V} = \dfrac{d\vec{s}}{dt} = 3\cos t \, \vec{i} + 6t \, \vec{j} = v_x \vec{i} + v_y \vec{j}$

Therefore, the x-component of the velocity = $V_x = 3\cos t$.

$$V_x = 3\cos t$$

## 101. (E)

Acceleration  $\vec{a} = \dfrac{d\vec{v}}{dt} = \dfrac{d}{dt}(3\cos t\ \vec{i} + 6t\ \vec{j})$

$= -3\sin t\ \vec{i} + 6\vec{j} = a_x\vec{i} + a_y\vec{j}$

Therefore, the y-component of the acceleration $a_y = 6$.

$$a_y = 6$$

## 102. (A)

We take the bottom of the channel as the datum, and look at points on the surface. Before the jump, the elevation is 3m and the velocity is 100 m/s. The losses through the jump are defined by the parameter $h_f$, where $h_f = E_1 - E_2$.

$$E_2 = y_2 + \dfrac{V_2^2}{g2}, \quad E_1 = y_1 + \dfrac{V_1^2}{g2}$$

The jump is 3m, so $y_2 = 6$m.

The velocity $V_2$ is obtained from continuity:

$$V_2 = \dfrac{V_1 y_1}{y_2} = \dfrac{(100)(3)}{6} = 50\ \dfrac{m}{s}$$

$$E_2 = 6 + \dfrac{2500}{20} = 131.0$$

$$E_1 = 3 + \dfrac{10,000}{20} = 503.0$$

$\Delta E = 62 - 38 = 24 \quad h_f = E_1 - E_2 = 503 - 131$

$$= 372$$

To get the percent loss:

$$\%\ \text{loss} = \dfrac{372}{503} \times 100 = 74\%$$

## 103. (D)

The resultant force is the result of adding the forces, which can be done using their vector representations as shown:

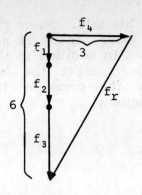

$$f_r^2 = 6^2 + 3^2$$
$$f_r = \sqrt{36 + 9}$$
$$f_r = \sqrt{45}$$

so (D) is the answer.

## 104. (A)

The polar nature of liquid water makes it an excellent solvent for an ionic solid such as NaCl. $H_2O$ is able to form ion-dipole bonds with $Na^+$ and $Cl^-$ by surrounding it with an octahedron of negative charges for $Na^+$ or positive charges for $Cl^-$. These charges come from the opposite ends of the dipole of $H_2O$. There are no dipole bonds in $C_6H_6$.

## 105. (B)

From the equations $\Delta E = q - w$, at constant volume, no PV work is done, and $\Delta E = q \neq 0$ so (A) is incorrect. If there is no energy change, $\Delta E = 0$, $w = q$ is not necessarily zero, so (C) is incorrect. At constant pressure, w is not necessarily zero because volume can change so $\Delta E = q - P\Delta V$ and (D) is incorrect.

By definition and, as has been shown by elimination of the other choices, the correct answer is (B).

## 106. (E)

When an asset is about to be sold, the accumulated depreciation and the depreciation expenses are brought up to date to reflect this sale. Since the selling price is equal to updated Book Value, there is no gain on the disposal of the fixed asset. In addition to this, there is no account for the book value because the book value reflects the difference between the price of the equipment and the accumulated depreciation.

107. (C)
Total distance S = distance covered during acceleration $S_a$
+ distance covered during deceleration $S_d$

$\therefore$ S = $S_a$ + $S_d$

Now $S_a = v_i t + \frac{1}{2} a t^2$

where $v_i$ = initial velocity = 0, since the body is initially at rest.

$S_a = \frac{1}{2} a t^2$

$= \frac{1}{2} \times 4 \times (10)^2$

$= 200$m

Also $v_f = v_i + at$

$= 0 + (4)(10)$

$= 40$ m/s

Now to find $S_d$:

$v_f^2 - v_i^2 = 2aS_d$

Here $v_f = 0$ and $(v_i)_{dec} = (v_f)_{acc} = 40$ m/s

$-(40)^2 = 2(-8) S_d$

$S_d = \frac{1600}{16} = 100$m

Thus $S = S_a + S_d = 200 + 100 = 300$m

## 108. (E)

A resistor in a monolithic integrated circuit has a resistance given by,

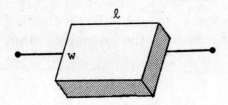

$$R = R_s \frac{L}{\omega}$$

where  $R_s$ = sheet resistance, which is a constant for a particular semiconductor with constant thickness.

$\ell$ = length of the diffused area

$\omega$ = width of the diffused area.

Hence, if a strip 1 millimeter in length and width corresponds to a resistance of R ohms, then a strip 1 millimeter in length and 3 millimeters in width will correspond to $\frac{\ell}{\omega} = \frac{1}{3}$ R ohms.

## 109. (D)

According to the Lumped Parameter analysis, the temperature as a function of time is given by:

$$\frac{T - T_f}{T_o - T_f} = e^{-t\left(\frac{hA}{\rho C_p V_o}\right)}$$

where  $T_o$ = initial temperature of the cannon ball

$T_f$ = temperature of the environment

$t$ = time

$h$ = heat transfer coefficient

$A$ = surface area

$\rho$ = density of the cannon ball

$C_p$ = specific heat

$V_o$ = volume.

According to the above equation, the "RC" time constant is equal to:

$$\frac{\rho C_p V_o}{hA}$$

Therefore, by increasing the density, the time constant would increase.

110. (C)
Because water can be considered as incompressible, the conservation of mass equation becomes: $v_1 A_1 = v_2 A_2$

$v_1 = 3$ in/sec $\qquad v_2 = ?$

$A_1 = \frac{\pi}{4} d^2 = \qquad A_2 = \frac{\pi}{4}(4)^2 = 4\pi$

$$v_2 = \frac{3 \frac{in}{sec} \pi}{4\pi} = \frac{3}{4} \frac{in}{sec}$$

111. (C)
Pressure is related to elevation, independent of surface area. (So A is wrong). The pressures are

$P_1 = p_0 + \rho gh \qquad$ ($\rho$ is density, g is gravitation constant)

$P_2 = p_1 + \rho gh = p_0 + 2\rho gh$

so $P_2$ is not twice that of $P_1$ unless $P_0 = 0$. Also, if $P_0 = 0$, $P_1$ and $P_2$ are not equal. The difference $(P_2 - P_1) = \rho gh$. So (C) is correct.

112. (E)
For any object to float in water, its weight must be balanced by the buoyant force (B).

Thus, weight of ice = B.

But weight = weight density ($\rho g$) × volume(v)

$\therefore \rho_{ice} \times g \times v_{ice} = \rho_{water} \times g \times v_{water}$

where the volume of submerged ice equals volume of water displaced.

Hence, $\dfrac{v_{water}}{v_{ice}} = \dfrac{\rho_{ice}}{\rho_{water}} = 0.92.$

So 92% of the ice is submerged and hence 8% lies above the surface of water.

## 113. (A)
Boyle's law states that the product of the volume and pressure of a gas is constant at constant temperature.

$$p_1 v_1 = p_2 v_2 \qquad (1)$$

$$v_2 = \dfrac{p_1 v_1}{p_2}$$

If the pressure is halved, from (1) we get

$$v_1 = \tfrac{1}{2} v_2$$

$$\therefore \quad v_2 = 2v_1$$

Hence, when the pressure is halved, the volume doubles.

## 114. (A)
The circuit is in a common-emitter configuration, when the emitter of the transistor is common both to the input and the output circuit. Only the circuit in choice (A) has this characteristic.

## 115. (E)
One correct representation of the Fourier's Law of Conduction is

$$\dfrac{q}{A} = -k \dfrac{\partial T}{\partial x}$$

Note that the negative sign is needed to offset the intrinsic negative value of $\dfrac{\partial T}{\partial x}$; T decreases in the direction that x increases.

**116. (E)**
The production of a laser beam is based upon the emission of a photon of light by an excited atom whose outer electrons are in a higher energy state when it is struck by another photon. This illustrates the excitation of an electron when the energy of a photon of light is equal to the difference in the allowed energy levels.

In order for the current to drop under the influence of a constantly increasing voltage, the electrons traveling through the mercury vapor must lose energy in collisions with mercury atoms. The only way the electrons can give up enough energy to cause a large drop in current is if the electron has enough energy to cause the mercury atom to make a transition to an excited state.

**117. (E)**
Precipitation hardening involves reheating the alloy to a temperature below the solvus line after quenching. The secondary phase that precipitates will exert large elastic strains on the solid solution matrix. This distortion will severely restrict dislocation movements and therefore slip. This effect will produce strain hardening.

**118. (D)**

**119. (B)**
The difference in pressure for a homogeneous liquid is

$$p_2 - p_1 = -\rho g (y_2 - y_1)$$

Two points in a fluid at the same height have the same pressure since there is no pressure gradient in the horizontal direction. This will hold for a U tube. Hence, the difference in pressure between point B and point C must be 0. Therefore the pressure force acting on the fluid element between point C and point D is due to the difference in pressures at point C and point D. The pressure at point C is the same as the pressure at point B ($p_A + \rho_1 g z_A$). Therefore the net force on the fluid is

$$(p_A + \rho_1 g(z_A) - p_a)A$$

## 120. (C)

Consider two state points 1 and 2 as initial and the final states for a system. The two paths, R and I are reversible and irreversible paths respectively. Using the inequality of Clausius we get

$$\oint \frac{\delta Q}{T} = \int_1^2 \frac{\delta Q_I}{T} + \int_2^1 \frac{\delta Q_R}{T} \leq 0$$

substituting for the reversible process we get

$$\int_1^2 \frac{\delta Q_I}{T} + S_1 - S_2 \leq 0$$

Rewriting and generalizing the above equation we get

$$\int_1^2 \frac{dQ_I}{T} \leq S_2 - S_1$$

Clearly entropy always increases for a process.

## 121. (B)

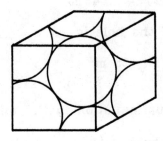

$\Rightarrow$ 1 atom centered + 8 × $\left(\frac{1}{8}\right.$ atom in each corner$\left.\right)$

= 2 atoms

## 122. (E)

Since all of the above are important deviations from the ideal, (E) is the correct answer.

123. (D)

The amount of heat change when temperature is increased or decreased is given by

$$\Delta H = mc\Delta t,$$

where $\Delta H$ is the change in heat, m is the mass of the sample, c is the specific heat of the sample, and $\Delta t$ is the change in temperature.

Converting to the CGS system,

$$m = 1 \text{ kg} = 1000 \text{g}.$$

Substituting into the equation,

$$\Delta H = mc\Delta t = (1000g)(1 \text{ cal/g°C})(100° - 10°C)$$

$$\Delta H = 90,000 \text{ cal} = 90 \text{ kcal}.$$


124. (C)

Since the flow is stated as being steady, no mass is accumulating in the tank and the discharge through the hole in the bottom of the tank is equal to the inlet rate of 1 m³/s.

The hole has a 40 cm diameter, so its area is:

$$A = \pi(.2)^2 = .04\pi \, m^2$$

The velocity out of the hole is

$$\frac{Q}{A} = \frac{1}{.04\pi} = \frac{25}{\pi} \frac{m}{s}$$

We use the momentum equation in the following form:

$$\Sigma F = \rho Q(v_{out} - v_{in})$$

To see the effect on the apparent weight of the tank we consider force components in the vertical direction. In this case, we do not consider inlet velocity since the water is flowing horizontally. Hence:

$$\Sigma F_y = \rho Q(v_{out})$$

$$= \left(1000 \frac{kg}{m^3}\right)\left(1 \frac{m^3}{s}\right)\left(\frac{25}{\pi} \frac{m}{s}\right)$$

$$= \frac{25,000}{\pi} N$$

This is approximately 8.3 kN, and since this force is positive upward, it is in a direction to reduce the apparent weight of the water tank. Thus the weight is decreased: −8.3 kN.

**125. (B)**
Usually the equivalent annual cost is determined for both investment alternatives. The alternative with the lesser cost is chosen. The comparative use value of the current equipment relative to the replacement equipment is then determined. Since the annual cost is the same for both equipment, only the present worth needs to be determined for the equal service periods. The present worth of new equipment in 11 years needs to be determined for the alternative of maintaining with the present equipment. The price of the new equipment in 11 years is $10,000 (1 + .15). The present worth of the new equipment is $\frac{\$10,000(1+.15)^{11}}{(1+.15)^{11}}$ = $10,000. Therefore the present worth of the alternative of remaining with the present equipment is $10,800. The APB Co. would save $800 by purchasing the new equipment.

**126. (D)**
Increasing the temperature will increase the thermal agitation in the molecule. This thermal agitation is the result of increased molecular vibrations. The kinked conformation can be straightened with an applied stress. At higher temperatures the molecules increase their resistance to the applied stress. Therefore, the elastic modulus increases with temperature.

**127. (B)**
The period of oscillation is related to the mass and the spring constant (or force constant). For simple harmonic motion, the period is

$$T = 2\pi\sqrt{\frac{m}{k}}$$

So, to double the period of oscillation, T, either the mass, m, would have to be increased by a factor of 4 or the spring constant, k, would have to be decreased by a factor of four. So the only answer which makes sense is (B).

## 128. (A)

Age hardening involves reheating to an intermediate temperature after quenching. The precipitation of fine particles will harden the alloy. The particles will continue to grow with time, coalescing into fewer but larger particles. Softening occurs as the precipitated particles grow. This proceeds more rapidly at elevated temperatures.

## 129. (A)

Hund's rule states that in orbitals of identical energy, electrons remain unpaired if possible (because of electron-electron repulsion). So the three p electrons of the nitrogen atom remain unpaired:

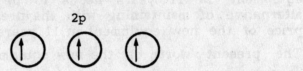

So (I) is accurate. (II) is inaccurate because the configuration does not violate the Pauli principle (that two electrons which share the same orbital must not have identical spin). Also, (III) is false since that is the ground state of the carbon atom, not the excited state. And (IV) is true, since all three atoms' configurations represent the ground state.

## 130. (B)

The definition of atomic packing factor is the ratio:

$$\frac{\text{volume of atoms per unit cell}}{\text{volume of the unit cell}}$$

Since aluminum has a 6% greater atomic packing factor, then it has a lesser inverse ratio of volume of unit cell to volume of atoms per unit cell.

## 131. (C)

Since y is a function of x, we differentiate the equation implicitly in terms of x and y. We have:

$$2x + 2y \cdot \frac{dy}{dx} = 0 \quad \text{or} \quad 2y \frac{dy}{dx} = -2x.$$

$$\frac{dy}{dx} = -\frac{x}{y}$$

## 132. (A)

(A) must be correct since if $\det(A) \neq 0$, then $A_{n \times n}$ is non-singular, but if $A_{n \times n}$ is non-singular, then no row can be expressed as a linear combination of any other, otherwise $\det(A) = 0$, so everything holds.

## 133. (A)
The sketch is:

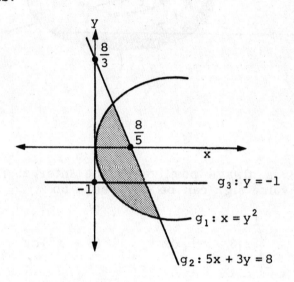

where $g_1$ and $g_2$ intersect, the coordinates must be the same, $g_1: x = y^2$ replaced in $g_2$ gives $5y^2 + 3y - 8 = 0$. The quadratic formula

$$y = \frac{-3 \pm \sqrt{9 + 160}}{10} = \frac{-3 \pm 13}{10}$$

So, $y_1 = 1$, $y_2 = -\frac{8}{5}$, $g_1$ gives $x_1 = 1$, $x_2 = \frac{64}{25}$

## 134. (A)

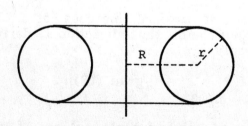

A circular torus is formed by the rotation of a circle about an axis in the plane of the circle and not cutting the circle. Let r be the radius of the revolving circle and let R be the distance of its center from the axis of rotation.

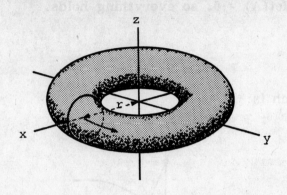

135. (B)
$3 - x^3$ is always positive in the interval $[-1,1]$, so the absolute value sign can be dropped. So

$$\int_{-1}^{1} |3 - x^3| \, dx = \int_{-1}^{1} (3 - x^3) \, dx$$

$$= \left[ 3x - \frac{x^4}{4} \right]_{-1}^{1}$$

$$= [(3(1) - \tfrac{1}{4}) - (3(-1) - \tfrac{1}{4})]$$

$$= 3 - \tfrac{1}{4} - (-3 - \tfrac{1}{4})$$

$$= 3 - \tfrac{1}{4} + 3 + \tfrac{1}{4}$$

$$= 6$$

136. (C)
The direction of maximal increase of f is the direction of the gradient and the directional derivative in that direction is simply

$$D_{\vec{A}} f(P) \bigg|_{max} = \| \mathrm{grad} \, f(P) \|.$$

Now in the case at hand, $f = x^2 + y^3$ so $\partial f/\partial x = 2x$, $\partial f/\partial y = 3y^2$ and at the point $(-1,3)$,

$$\|\text{grad } f\| = \sqrt{\left(\frac{\partial f}{\partial x}\right)^2 + \left(\frac{\partial f}{\partial y}\right)^2} = \sqrt{(2x)^2 + (3y^2)^2}$$

$$= \sqrt{4x^2 + 9y^4} = \sqrt{4(-1)^2 + 9(3)^4}$$

$$= \sqrt{733}.$$

### 137. (C)

The geometric series is $a + ar + \ldots = \sum_{n=1}^{\infty} ar^{n-1}$. The nth partial sum is

$$S_n = a + ar + ar^2 + \ldots + ar^{n-1} \qquad (1)$$

then

$$S_n = ar + ar^2 + \ldots + ar^{n-1} + ar^n \qquad (2)$$

If we subtract (2) from (1) we have

$$S_n = \frac{a(1 - r^n)}{1 - r}$$

For $r$ such that $|r| < 1$ then

$$\lim_{n \to \infty} S_n = \lim_{n \to \infty} \left( \frac{a}{1 - r} - \frac{ar^n}{1 - r} \right) = \frac{a}{1 - r}$$

### 138. (C)

$\pi$ and $e$ are two constants, so

$$\int_{\pi}^{e} \pi e \, dx = \pi e \int_{\pi}^{e} dx = \pi e x \Big|_{\pi}^{e}$$

$$= \pi e (e - \pi)$$

### 139. (B)

By Euler's formula

$$e^{i\theta} = \cos \theta + i \sin \theta$$

So
$$e^{i\pi} = \cos\pi + i\sin\pi = -1 + i0 = -1$$

**140. (E)**

We know that if $\Sigma u_k$ converges, then $\lim_{k\to\infty} u_k = 0$. We cannot say that the converse is true, because if $\lim_{k\to\infty} u_k = 0$, $\Sigma u_k$ may or may not converge, but we know that if the kth term does not approach zero, then the series diverges, so it is only a necessity that $\lim_{k\to\infty} u_k = 0$; sufficiency is established in further analysis.

**141. (E)**

This is a first order linear differential equation. Since it is in the form $y' + p(x)y = q(x)$ we can choose

$$\exp\left(\int p(x)dx\right) = \exp\left(\int \frac{4}{x} dx\right)$$
$$= \exp(4 \ln x)$$
$$= x^4$$

as the integrating factor that will make the equation a separable one.

**142. (B)**

The kth term is

$$u_k = \frac{1}{2}\left(\frac{1}{2k-1} - \frac{1}{2k+1}\right)$$

by partial fractions.

The kth partial sum is $S_k = u_1 + \ldots + u_k$

$$= \frac{1}{2}\left\{\left(1 - \frac{1}{3}\right) + \left(\frac{1}{3} - \frac{1}{5}\right) + \ldots + \left(\frac{1}{2k-1} - \frac{1}{2k+1}\right)\right\}$$

$$= \frac{1}{2}\left\{1 - \frac{1}{3} + \frac{1}{3} - \frac{1}{5} + \frac{1}{5} - \cdots + \frac{1}{2k-1} - \frac{1}{2k+1}\right\}$$

$$= \frac{1}{2}\left(1 - \frac{1}{2k+1}\right)$$

$$\lim_{k\to\infty} S_k = \lim_{k\to\infty} \frac{1}{2}\left(1 - \frac{1}{2k+1}\right) = \frac{1}{2}$$

143. (A)

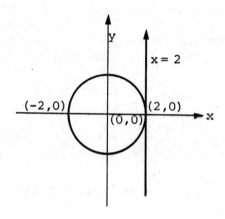

The equation of a circle is,

$$(x - h)^2 + (y - k)^2 = r^2$$

Center $(h,k) \equiv (0,0)$

Therefore we get $x^2 + y^2 = r^2$ as the equation of circle. Now the circle touches the straight line $x - 2 = 0$ means that the circle passes through point $(2,0)$.

Substituting $x = 2$, $y = 0$ in the equation of circle,

$$r^2 = (2)^2 + (0)^2 = 4$$

Therefore,

$$\text{Area of circle} = \pi r^2 = \pi(4) = 4\pi$$

144. (E)

The area of the region bounded by the sine curve and the x-axis is divided into 4 equal parts, 2 above and

2 below the x-axis. We can find the total area by multiplying the area between $x = 0$ and $x = \pi$ by four. The area is given by the integral of the function $y = \sin x$, between $x = 0$ to $x = \pi$. Hence,

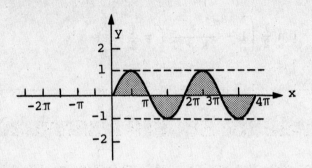

$$A_{total} = 4 \int_0^\pi \sin x \, dx$$

$$= 4[-\cos x]_0^\pi$$

$$= 4[-1(-1) - (-1)]$$

$$= 8.$$

# The Graduate Record Examination in ENGINEERING

Test 4

# GRE ENGINEERING TEST 4
## ANSWER SHEET

61. (A) (B) (C) (D) (E)
62. (A) (B) (C) (D) (E)
63. (A) (B) (C) (D) (E)
64. (A) (B) (C) (D) (E)
65. (A) (B) (C) (D) (E)
66. (A) (B) (C) (D) (E)
67. (A) (B) (C) (D) (E)
68. (A) (B) (C) (D) (E)
69. (A) (B) (C) (D) (E)
70. (A) (B) (C) (D) (E)
71. (A) (B) (C) (D) (E)
72. (A) (B) (C) (D) (E)
73. (A) (B) (C) (D) (E)
74. (A) (B) (C) (D) (E)
75. (A) (B) (C) (D) (E)
76. (A) (B) (C) (D) (E)
77. (A) (B) (C) (D) (E)
78. (A) (B) (C) (D) (E)
79. (A) (B) (C) (D) (E)
80. (A) (B) (C) (D) (E)
81. (A) (B) (C) (D) (E)
82. (A) (B) (C) (D) (E)

83. (A) (B) (C) (D) (E)
84. (A) (B) (C) (D) (E)
85. (A) (B) (C) (D) (E)
86. (A) (B) (C) (D) (E)
87. (A) (B) (C) (D) (E)
88. (A) (B) (C) (D) (E)
89. (A) (B) (C) (D) (E)
90. (A) (B) (C) (D) (E)
91. (A) (B) (C) (D) (E)
92. (A) (B) (C) (D) (E)
93. (A) (B) (C) (D) (E)
94. (A) (B) (C) (D) (E)
95. (A) (B) (C) (D) (E)
96. (A) (B) (C) (D) (E)
97. (A) (B) (C) (D) (E)
98. (A) (B) (C) (D) (E)
99. (A) (B) (C) (D) (E)
100. (A) (B) (C) (D) (E)
101. (A) (B) (C) (D) (E)
102. (A) (B) (C) (D) (E)
103. (A) (B) (C) (D) (E)
104. (A) (B) (C) (D) (E)

105. (A) (B) (C) (D) (E)
106. (A) (B) (C) (D) (E)
107. (A) (B) (C) (D) (E)
108. (A) (B) (C) (D) (E)
109. (A) (B) (C) (D) (E)
110. (A) (B) (C) (D) (E)
111. (A) (B) (C) (D) (E)
112. (A) (B) (C) (D) (E)
113. (A) (B) (C) (D) (E)
114. (A) (B) (C) (D) (E)
115. (A) (B) (C) (D) (E)
116. (A) (B) (C) (D) (E)
117. (A) (B) (C) (D) (E)
118. (A) (B) (C) (D) (E)
119. (A) (B) (C) (D) (E)
120. (A) (B) (C) (D) (E)
121. (A) (B) (C) (D) (E)
122. (A) (B) (C) (D) (E)
123. (A) (B) (C) (D) (E)
124. (A) (B) (C) (D) (E)

125. (A) (B) (C) (D) (E)
126. (A) (B) (C) (D) (E)
127. (A) (B) (C) (D) (E)
128. (A) (B) (C) (D) (E)
129. (A) (B) (C) (D) (E)
130. (A) (B) (C) (D) (E)
131. (A) (B) (C) (D) (E)
132. (A) (B) (C) (D) (E)
133. (A) (B) (C) (D) (E)
134. (A) (B) (C) (D) (E)
135. (A) (B) (C) (D) (E)
136. (A) (B) (C) (D) (E)
137. (A) (B) (C) (D) (E)
138. (A) (B) (C) (D) (E)
139. (A) (B) (C) (D) (E)
140. (A) (B) (C) (D) (E)
141. (A) (B) (C) (D) (E)
142. (A) (B) (C) (D) (E)
143. (A) (B) (C) (D) (E)
144. (A) (B) (C) (D) (E)

# THE GRADUATE RECORD EXAMINATION ENGINEERING TEST

## MODEL TEST IV

Time: 170 Minutes
      144 Questions

DIRECTIONS: Choose the best answer for each question and mark the letter of your selection on the corresponding answer sheet.

1. Which of the following is an example of a reversible process?

   (A) The isothermal expansion of an ideal gas which is absorbing heat.

   (B) The flow of gas from a high pressure compartment to a low pressure compartment.

   (C) The heating of an ideal gas at constant volume.

   (D) Heat transfer between two bodies at different temperatures which leads to thermal equilibrium at an intermediate temperature.

   (E) Both (A) and (C).

2. In the following diagram, evaluate the equivalent resistance R across the network's terminals: $\left(\text{Assume } \sum_{i=0}^{\infty} r \cong \sum_{i=1}^{\infty} r \right)$

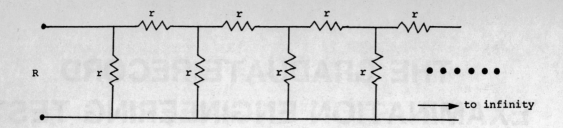

(A)  r

(B)  infinity

(C)  zero

(D)  $(\sqrt{5} - 1)\frac{r}{2}$

(E)  $\frac{\sqrt{5}\, r}{2}$

3. For an airfoil that is in motion through a fluid, the lift on the airfoil results from

   (A) the turbulent flow occurring above the airfoil.

   (B) the drag force that increases the angle of the wing relative to the flow.

   (C) the increased velocity of the fluid above the airfoil.

   (D) the reduced pressure above the airfoil.

   (E) Both (C) and (D).

4. A uniform rectangular block with height h and width a rests on an inclined surface whose coefficient of static friction is $\mu$. If we slowly increase the angle of inclination, $\theta$, what should be the relationship between a, h and $\mu$ for the block to tip over before sliding down the surface?

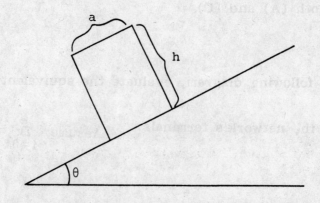

(A) μ = ah

(B) μ < $\frac{a}{h}$

(C) aμ < h

(D) μ > $\frac{a}{h}$

(E) information not sufficient

5. In the diagram, AB is the diameter of the circle, C is the center, and chord BD is 4 units long. What is the circumference of the circle?

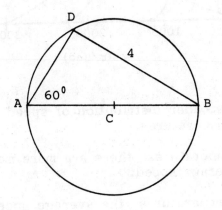

(A) 4π

(B) 4π²

(C) $\frac{4}{\sqrt{3}} \pi$

(D) $\frac{16}{\sqrt{3}} \pi$

(E) $\frac{8}{\sqrt{3}} \pi$

6. A vitreous silica such as $SiO_2$ is highly viscous at high temperatures. Which of the following will reduce the viscosity of $SiO_2$ at high temperatures?

(A) $H_2O$

(B) $Na_2O$

(C) NaCl

(D) $CaF_2$

(E) Mg

7. For the distribution curves shown, which of the following is NOT correct?

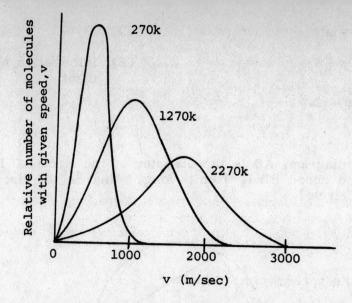

- (A) There is a broader distribution of speed among molecules at higher temperatures.
- (B) At lower temperatures there are more molecules which have the average speed.
- (C) At higher temperatures the average speed is greater.
- (D) At lower temperatures there are more molecules in motion.
- (E) Temperature and relative number of molecules for a given speed, v, are not necessarily linearly related.

8. The power p delivered in a circuit in time t is shown by the curve below:

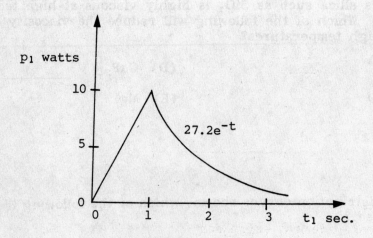

The energy absorbed from time t = 0 to 3 is

(A) $5 + 27.2e^2$ Joules

(B) $5 + 27.2\left(\dfrac{e^2-1}{e^3}\right)$ Joules

(C) $27.2e^3$ Joules

(D) $5 + 27.2\left(\dfrac{e^3-1}{e^2}\right)$ Joules

(E) $27.2\left(\dfrac{1-e}{e^4}\right)$ Joules

9. It took longer than 2 hours for a driver to cover the distance of 120 miles even though the average reading on his car's speedometer was 60 m/hr. The most logical explanation for this delay is that

(A) the speedometer is faulty.

(B) the luggage in the car impeded its progress.

(C) the tires are underinflated, thus reducing the average diameter of the tires.

(D) the headwind is too strong.

(E) the percentage of octane in the gasoline used by the driver was too low.

10. Any two-terminal linear bilateral dc network can be replaced by an equivalent circuit consisting of a voltage source and a series resistor as shown in the given diagram.

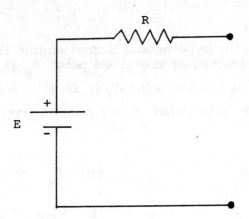

The above statement expresses

(A) the Superposition Theorem.   (D) the Maximum Power Transfer Theorem.

(B) Norton's Theorem.   (E) the Substitution Theorem.

(C) Thevenin's Theorem.

11. The Megger is an instrument used for measuring

(A) resistance below $1\Omega$.

(B) resistance between $1\Omega$ and $5\Omega$.

(C) resistance between $5\Omega$ and $50,000\Omega$.

(D) resistance above $50,000\Omega$.

(E) None of the above

12. A skater starts whirling around with both arms and one leg extended. What happens to the speed rate when the arms and the leg are drawn back?

(A) It initially increases and then remains constant.

(B) It initially decreases and then remains constant.

(C) It initially increases rapidly and then decreases slowly.

(D) It initially decreases and then increases.

(E) There is no change.

13. The efficiency for a hypothetical Carnot engine that receives heat at the temperature of the steam point $T_s$ and rejects it at the temperature of the ice point $T_i$ is 26.8%. For a thermodynamic temperature scale, what is the relative magnitude for the two points?

(A) $\dfrac{T_i}{T_s} = 0.7320$

(B) $\dfrac{T_i}{T_s} = .2680$

(C) $T_s - T_i = 100$

(D) $T_s - T_i = 180$

(E) None of the above

14. The kinetic energy for an ideal monatomic gas is dependent upon

   (A) temperature and pressure.
   (B) temperature only.
   (C) pressure only.
   (D) pressure and volume.
   (E) pressure and the molecular species.

15. One end of a bar of length L is placed in contact with steam at 100°C and the other end in contact with ice at 0°C. Which of the following represents the steady state temperature distribution along the rod which was initially at 20°C throughout?

(A)
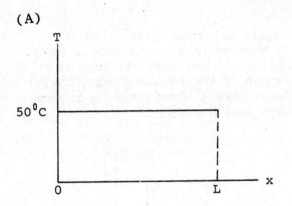

(D)
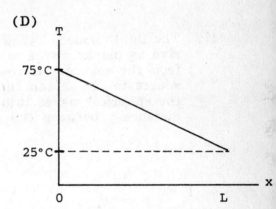

(B)
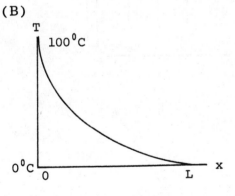

(E)
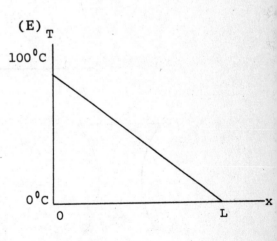

(C)
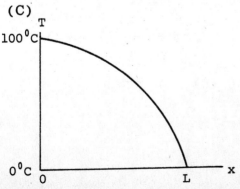

16. A tightly closed vessel containing water at room pressure and temperature, is brought to the outer space. What will happen to the water if we open the vessel?

   (A) All the water will vaporize.

   (B) Nothing will happen to the water.

   (C) Some of the water will vaporize and the rest will remain unchanged.

   (D) Small amounts of water will vaporize and the rest will freeze.

   (E) The temperature of the water will drop but no phase change will occur.

17. The light source shown emits spherical waves that arrive as planar waves at a screen which is a distance d away from the source. If the distance is not very far from the source to the screen, then which of the following will convert the spherical waves into planar leaves when situated at some distance y between the source and the screen?

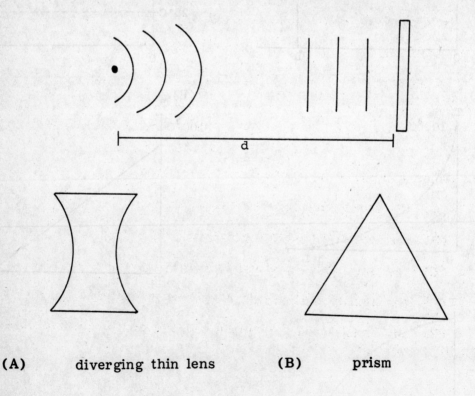

   (A)      diverging thin lens        (B)        prism

(C)   converging thin lens

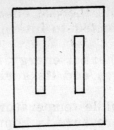
(D)   diffraction grating

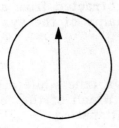
(E)   polarizer

18. 6 kg of hydrogen gas is combusted with 32 kg of oxygen gas. How much water is left over? (Atomic weights: H = 1, O = 16)

(A) 6 kg

(B) 32 kg

(C) 36 kg

(D) 38 kg

(E) None of the above

19. When a neutron is emitted from a nucleus,

(A) the atomic number decreases.

(B) the atomic number increases.

(C) the number of protons increases.

(D) the atomic number and atomic mass decrease.

(E) the atomic mass decreases but the atomic number does not change.

20. In a free expansion process,

(A) the thermal energy transferred to a gas is completely converted to internal energy.

(B) thermal energy is allowed to be transferred between the gas and its surroundings.

(C) while temperature is constant volume changes but the pressure is kept constant.

(D) the internal energy of the gas remains constant and there is no change in the gas temperature.

(E) the thermal energy added or subtracted from a system is equal to the change in the enthalpy of the system.

21. In physics, some equations are always true while others are only true under certain conditions. Which of the following equations is(are) true under all conditions?

   I. Equation of wave motion: $v = \nu \lambda$
   II. $\nabla \times \hat{E} = \frac{\partial \hat{B}}{\partial t}$
   III. Equation of kinetic energy: $E_k = \frac{1}{2}mv^2$

   (A) I only
   (B) II and III
   (C) I and II
   (D) III only
   (E) II only

22. Which of the following is true of an isentropic process for a closed system?

   (A) Change in temperature = 0
   (B) Work = 0
   (C) Change in internal energy = 0
   (D) Change in internal energy = 0 when work = 0
   (E) None of the above

23. What is the largest value of angle $\theta$ at which a gradually increasing force F applied to the 100 lb. cube would start to tip it over? (Coefficient of static friction, $\mu_s$ = 0.5 and normal force, N = 66.67 lbs.)

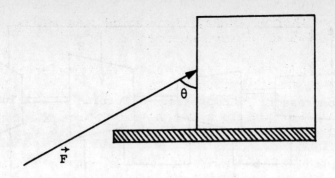

(A) 30°  (D) 45°
(B) 35°  (E) 50°
(C) 40°

24. Which of the following correctly illustrates a vapor compression refrigeration cycle?

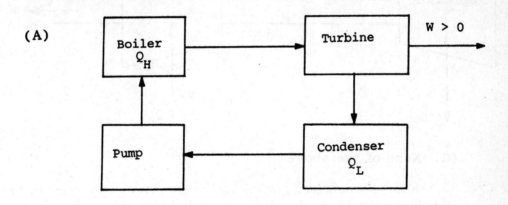

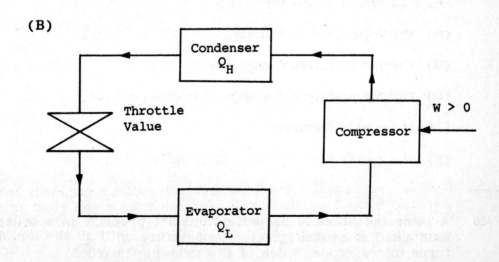

385

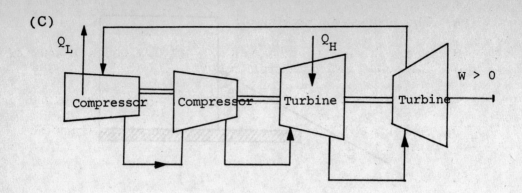

(C)

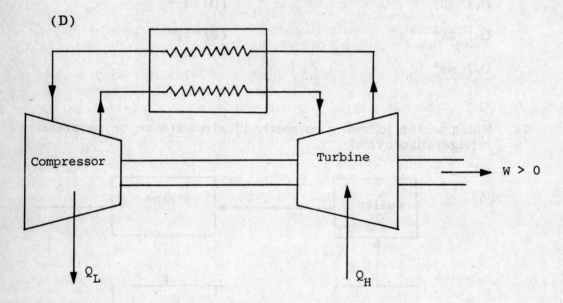

(D)

(E) None of the above

25. In a parallel L.C. circuit, at resonance and resistance nearly equal to zero,

   (A) the impedance is infinite.

   (B) the impedance is fairly low.

   (C) the circulating current is maximum.

   (D) the power factor is zero.

   (E) the power factor is less than unity.

26. A saturated liquid is heated at constant pressure in a boiler maintained at a much greater temperature until all the liquid turns into a vapor. Which of the following is true?

(A) The process is reversible.

(B) There is no net change in the entropy of the system of the boiler and saturated liquid.

(C) The temperature of the saturated liquid increases.

(D) The entropy increases for the system of the boiler and saturated liquid.

(E) Both (B) and (C).

27. If the density of air was to (hypothetically) approach that of water, what could happen to humans?

    (A) Couldn't walk because the air would weigh more.

    (B) Shrink in height because of the heavier air.

    (C) It would be easier to accelerate upward into the sky.

    (D) Nothing would change.

    (E) None of the above

28. Two satellites, of masses m and 2m, are on the same circular orbit around earth. If the velocity of the lighter satellite is $v_0$, what is the velocity of the heavier satellite?

    (A) $\frac{1}{2}v_0$     (D) $\frac{1}{4}v_0$

    (B) $v_0$     (E) $4v_0$

    (C) $2v_0$

29. The graph shown could represent the variation of

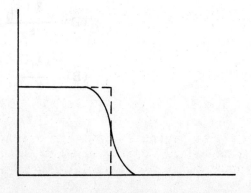

(A) the Einstein-Debye specific heat with respect to temperature.

(B) the blackbody radiation intensity with respect to wavelength.

(C) the density of states for valence electrons at T ~ 300°K with respect to energy of electron levels.

(D) the distribution of valence electrons at T ~ 300°K with respect to the energy of electron levels.

(E) the Maxwell-Boltzman speed distribtuion for molecules of a gas with respect to speed.

30. Consider a cone, with a light bulb at the vertex, cut by a plane making an angle of 30° with base. The area of the plane inside the cone has an image of area s on the base. If the area of the base is A and the average light intensity at the base is I, what is the average light intensity on the inclined plane?

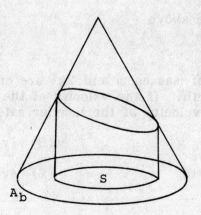

(A) $\dfrac{IA_b}{2s}$

(B) $\dfrac{I^2}{A_b s}$

(C) $\dfrac{\sqrt{3}\, IA_b}{2s}$

(D) $\dfrac{\sqrt{3}\, IA_b}{s}$

(E) $\dfrac{IA_b}{s}$

31. A 50 lb. box lies on a frictionless inclined plane that rises 3 ft. and is 5 ft. long along the incline. The force (P), parallel to the plane, which is required to prevent the box from moving downward is:

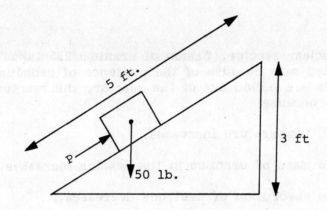

(A) 15 lb.

(B) 20 lb.

(C) 25 lb.

(D) 30 lb.

(E) 50 lb.

32. For a damped harmonic oscillator that experiences a damping force proportional to the velocity of the system, the law governing the damped harmonic motion is $x = Ae^{-bt/2m} \cos(w't + \phi)$. Which of the following remains constant?

(A) amplitude

(B) frequency

(C) maximum velocity

(D) energy

(E) None of the above

33. The power supplied by the source in the given figure is

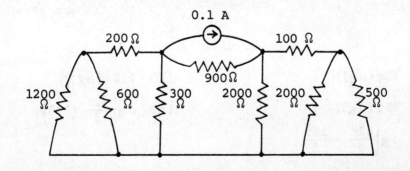

(A) 1.6 W
(B) 2.0 W
(C) 2.6 W
(D) 3.0 W
(E) 3.6 W

34. In a nuclear reactor, fission of uranium 235 takes place at a controlled rate because of the presence of cadmium rods. As the rods are pulled out of the reactor, the reaction rate increases because

   (A) the temperature increases.

   (B) the mass of uranium in the reactor increases.

   (C) the absorption of neutrons decreases.

   (D) ionization in the reactor increases.

   (E) more fissionable isotopes are produced.

35. In the arrangement shown in the figure, body W is a sphere with diameter of 4m and the width of the wall is 1m. The weight of W is (in Newtons) approximately

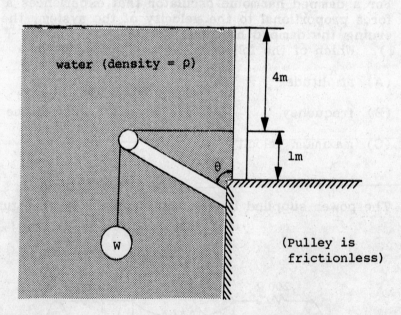

(Pulley is frictionless)

(A) $75\pi(\rho g)$ N
(B) $83\pi(\rho g)$ N
(C) $\rho g\left(\frac{32\pi}{3} + \frac{125}{6}\right)$
(D) $110\pi(\rho g)$ N
(E) $\rho g\left(\frac{32\pi}{3} + 125\right)$

36. Knowing the period in the periodic table that an element lies in and its outermost subshell's azimuthal quantum number, we can directly predict

   (A) its conductivity.

   (B) its principal quantum number.

   (C) its first ionization energy.

   (D) its relative abundance of isotopes.

   (E) its atomic radius.

37. Which of the following equations is applicable to a Newtonian fluid?

   (A) $F = ma$

   (B) $\tau = \mu \dfrac{du}{dy}$

   (C) $F \Delta t = \Delta mv$

   (D) $\tau = \mu \dfrac{d^2 u}{dy^2}$

   (E) None of the above

38. A 120 lb. boy was riding a skateboard down a 45° incline. The skateboard has a weight-measuring scale attached to the top such that when the boy is riding on the skateboard, he can also see his own weight. What will be the boy's weight, when he is coming down the above mentioned incline on his skateboard?

   (A) 0 lb.

   (B) 40 lb.

   (C) $60\sqrt{2}$ lb.

   (D) 60 lb.

   (E) insufficient information

39. Aluminum foil used for cooking food and storage sometimes has one shiny surface and one dull surface. Should the shiny side or the dull side be on the outside when the food is wrapped for baking and freezing, respectively?

   (A) shiny side, shiny side

   (B) dull side, dull side

   (C) shiny side, dull side

   (D) dull side, shiny side

   (E) it doesn't affect the process

40. A graph of a voltage source $V_{ab} = 10tu(t)$ is given below.

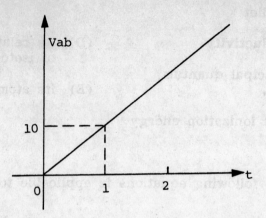

If a 2-farad capacitance is connected to $V_{ab}$, then which of the following graphs represents the current through the capacitor?

(A)
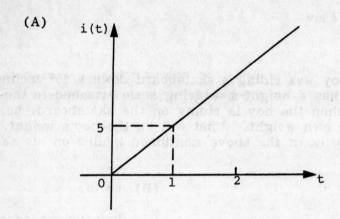

(B)
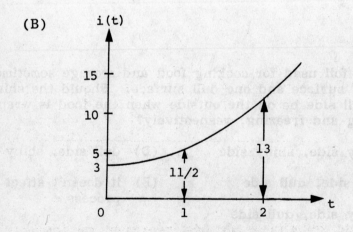

(C)

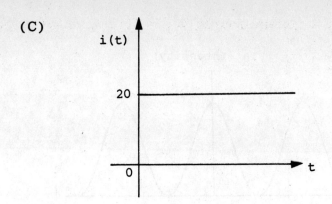

(D)

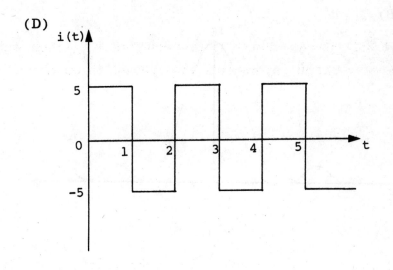

(E)

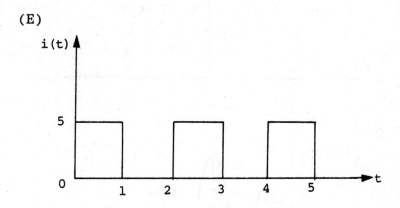

41. Which of the following would best represent the intensity pattern for an incandescent wire that is not directly behind a double slit?

(A)

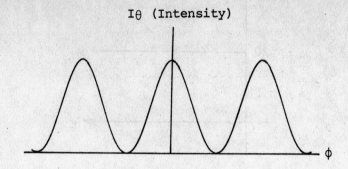

(B)

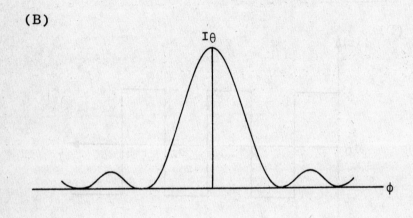

(C)

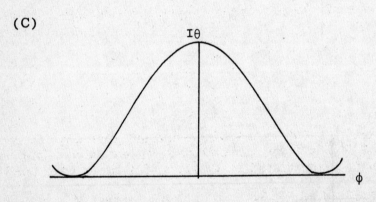

(D)

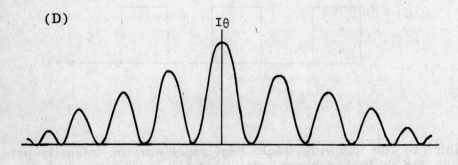

(E) None of the above

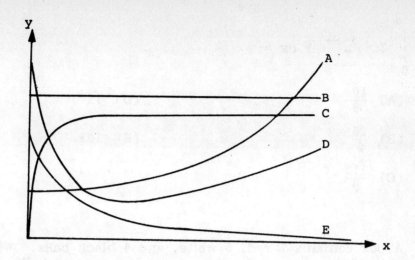

42. Which of the curves above represents the variation of electron mass (y-axis) with velocity (x-axis)?

    (A) A  (B) B  (C) C  (D) D  (E) E

43. Which of the curves above represents the variation of nuclear mass per nucleon (y-axis) with respect to increasing atomic number (x-axis) for hydrogen through uranium?

    (A) A  (B) B  (C) C  (D) D  (E) E

44. Which of the curves above represents the variation of current (y-axis) with respect to increasing external resistance (x-axis) for a given dry cell?

    (A) A  (B) B  (C) C  (D) D  (E) E

45. A man deposits $1000 in a savings account that pays interest at the rate of 6% per year, compounded semiannually. How much money will be accumulated at the end of a three year period?

    (A) $1000(1 + 0.06)^3$

    (B) $1000(1 + 0.03)^6$

    (C) $1000(1 + 0.03(1.03)^2)^6$

    (D) $1000(1 + 0.03)^3$

    (E) $1000(1 + 0.06(1.03)^2)^3$

DIRECTIONS: Choose the best answer for each question and mark the letter of your selection on the corresponding answer sheet.

NOTE: The natural logarithm of x will be denoted by lnx.

46. $\int_0^2 2x^2 \sqrt{x^3 + 1}\ dx =$

(A) $\frac{104}{9}$      (D) 12

(B) $\frac{26}{9}$      (E) 102

(C) $\frac{112}{9}$

47. A box contains 7 red, 5 white, and 4 black balls. What is the probability of your drawing at random one red ball? One black ball?

(A) $\frac{7}{9}$ red and $\frac{4}{9}$ black      (D) $\frac{7}{11}$ red and $\frac{5}{11}$ black

(B) $\frac{7}{16}$ red and $\frac{1}{4}$ black      (E) $\frac{7}{12}$ red and $\frac{4}{11}$ black

(C) $\frac{7}{12}$ red and $\frac{1}{3}$ black

48. A penny is to be tossed 3 times. What is the probability there will be 2 heads and 1 tail?

(A) $\frac{1}{2}$      (D) $\frac{1}{8}$

(B) $\frac{1}{3}$      (E) $\frac{2}{3}$

(C) $\frac{3}{8}$

49. $\int_{-\pi}^{+\pi} \sin x \cos x\ dx =$

(A) 0      (D) $\frac{1}{2}$

(B) 2      (E) $-\frac{1}{2}$

(C) -3

50. A diagonal matrix is a square matrix whose

(A) diagonal entries are all one.

(B) diagonal entries are all zero.

(C) non-diagonal entries are all zero.

(D) non-diagonal entries are all one.

(E) transpose is the matrix itself.

51. If $\pi < x < \frac{3\pi}{2}$, then

   (A) sinx > 0, cosx < 0
   (B) cosx > 0, sinx > 0
   (C) sinx < 0, cosx < 0
   (D) cosx > 0, sinx < 0
   (E) sinx = 0, cosx = -1

52. What is the modal age of the fathers of students at a Junior High school according to the chart below?

   | Age | Frequency |
   | --- | --- |
   | 35 - 39 | 5 |
   | 40 - 44 | 15 |
   | 45 - 49 | 38 |
   | 50 - 54 | 29 |
   | 55 - 59 | 13 |

   (A) 47
   (B) 44
   (C) 59
   (D) 35
   (E) 50

53. If $y = e^{2\ln(3x^2+2)}$, $\frac{dy}{dx} =$

   (A) $(3x^2 + 2)^2$
   (B) $\frac{e^{\ln(3x^2+ 2)}}{12x}$
   (C) $6xe^{2\ln(3x^2+2)}$
   (D) $12x(3x^2 + 2)$
   (E) $24x$

Questions 54 to 58 are based on the following assumption:

Let f(x) and g(x) be functions integrable over the set of all real numbers.

54. $\int f'(x)g(x)dx$ can be evaluated as,

    (A) $f'(x)g(x) - \int f''(x)g(x)dx$

    (B) $f(x)g(x) - \int f(x)g'(x)dx$

    (C) $f(x)g(x) - \int f'(x)g'(x)dx$

    (D) $f(x)g'(x) - \int f'(x)g(x)dx$

    (E) $f'(x)g'(x) - \int f'(x)g'(x)dx$

55. $\int_0^{\pi/2} x \sin x \, dx =$

    (A) 0

    (B) 1

    (C) -1

    (D) $\frac{\pi}{2}$

    (E) $\frac{-\pi}{2}$

56. $\int [F(x)G'(x)dx + G(x)F'(x)]dx =$

    (A) $G'(x)F(x) + C$

    (B) $F'(x)G(x) + C$

    (C) $G(x)F(x) + C$

    (D) $F'(x)G'(x) + C$

    (E) None of these

57. If $y = F[G(x^n)]$, $\frac{dy}{dx} =$

    (A) $nx^{n-1}F''[G(x^n)]G'(x^n)$

    (B) $nx^{n-1}F'[G(x^n)]G'(x^n)$

    (C) $(n-1)F''[G(x^n)]G'(x^n)$

    (D) $nx^n F'[G(x^n)]G(x^n)$

    (E) $nx^{n-1}F'(x^n)G'(x^n)$

58. The maxima and minima of the function $F(x) = x^4$ are

   (A) $\{1,1\}$
   (B) $\{4,0\}$
   (C) $\{4,1\}$
   (D) $\{3,3\}$
   (E) $\{0,0\}$

59. If $y = \ln \cos(e^{2x})$, $\dfrac{dy}{dx} =$

   (A) $2e^{2x} \cot(e^{2x})$
   (B) $-2e^{2x} \csc(e^{2x})$
   (C) $-2e^{2x} \tan(e^{2x})$
   (D) $-2xe^{2x} \sec(x)$
   (E) $2xe^{-2x} \tan(2x)$

60. If $i = \sqrt{-1}$, then

$$\int_{-\pi}^{\pi} (\cos x + i\sin x)(\cos x - i\sin x)\,dx =$$

   (A) 0
   (B) $\pi$
   (C) $2\pi$
   (D) $-2\pi$
   (E) None of the above

61. The 100th term of the arithmetic sequence 3, 7, 11, 15,...is

   (A) 423
   (B) 278
   (C) 399
   (D) 352
   (E) 293

62. $\lim\limits_{x \to 0^+} \dfrac{\ln x}{\frac{1}{x}} =$

   (A) 0
   (B) $\infty$
   (C) $-\infty$
   (D) $-1$
   (E) 1

63. Given

$$A(t) = \begin{bmatrix} t^2 & \cos t \\ e^t & \sin t \end{bmatrix}, \frac{dA}{dt} =$$

(A) $\begin{bmatrix} 2t & -\sin t \\ e^t & \cos t \end{bmatrix}$

(D) $\begin{bmatrix} t^2 & e^t \\ \cos t & \sin t \end{bmatrix}$

(B) $\begin{bmatrix} 2t & \sin t \\ -e^t & \cos t \end{bmatrix}$

(E) $\begin{bmatrix} t^2 & \sin t \\ e^t & \cos t \end{bmatrix}$

(C) $\begin{bmatrix} 2t & e^t \\ -\sin t & \cos t \end{bmatrix}$

64. If

$$e^x = 1 + x + \frac{x^2}{2!} + \frac{x^3}{3!} + \frac{x^4}{4!} + \ldots = \sum_{n=0}^{\infty} \frac{x^n}{n!}$$

then $\dfrac{1 - e^{-x^2}}{x^2} =$

(A) $\sum_{n=1}^{\infty} \dfrac{(-1)^{n+1} x^{2n-2}}{n!}$

(D) $\dfrac{1}{x^2}$

(B) $\sum_{n=0}^{\infty} \dfrac{(-1)^n x^{2n-2}}{n!}$

(E) an undefined function

(C) $-1$

DIRECTIONS: Choose the best answer for each question and mark the letter of your selection on the corresponding answer sheet. Assume that every function in this section has derivatives of all orders at each point of its domain unless otherwise indicated.

The following figure is to be used for questions 65 - 70,

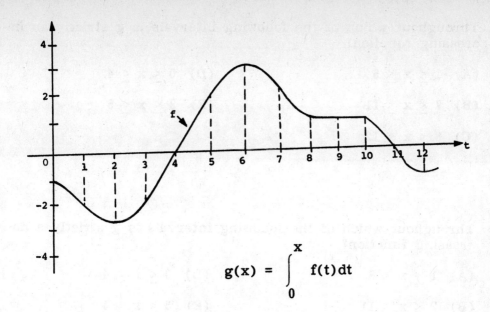

$$g(x) = \int_0^x f(t)\,dt$$

65. The value of g(0) is

    (A) -1
    (B) 4
    (C) 0
    (D) $\frac{3}{2}$
    (E) $-\frac{1}{2}$

66. g(4) is

    (A) greater than 12.
    (B) between 5 and 12.
    (C) between -5 and 5.
    (D) between -5 and -12.
    (E) less than -12.

67. g(12) is

    (A) greater than 8.
    (B) between 4 and 8.
    (C) between -4 and 4.
    (D) between -4 and -8.
    (E) less than -8.

68. g(x) is positive for all x in which of the following intervals?

    (A) $\frac{1}{2} \le x \le 4$
    (B) $3 \le x \le 6$
    (C) $4 \le x \le 11$
    (D) $5 \le x \le 9$
    (E) $9 \le x \le 12$

69. Throughout which of the following intervals is g strictly an increasing function?

(A) $1 \leq x \leq 5$

(B) $7 \leq x \leq 11$

(C) $8 \leq x \leq 12$

(D) $0 \leq x \leq 4$

(E) $3 \leq x \leq 5$

70. Throughout which of the following intervals is g strictly a decreasing function?

(A) $1 \leq x \leq 5$

(B) $7 \leq x \leq 11$

(C) $8 \leq x \leq 12$

(D) $0 \leq x \leq 4$

(E) $3 \leq x \leq 5$

Questions 71-80

Consider a circular track which lies in the x-y plane, and another one lying in the y-z plane with centers at the origin, as shown in the figure. Two particles P and Q travel on each track, respectively. At time t = 0, they are abruptly set into motion with constant velocity starting from P and Q in the directions shown. After 4 seconds, both particles have traversed their respective paths.

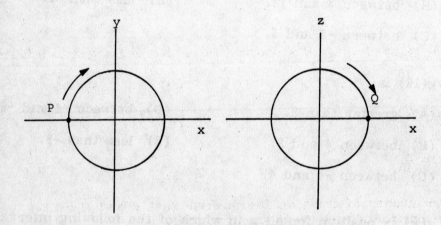

(P and Q are in their respective planes of motion.)

The following diagram is the graphical representation of distance between P and Q vs. time.

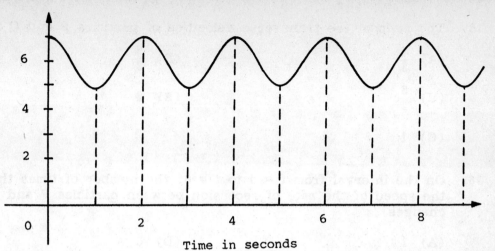

71. The magnitude of the average angular velocity of particle Q is

    (A) $\frac{\pi}{4}$ rad/s  (D) $\frac{3}{4}\pi$ rad/s

    (B) $\frac{\pi}{2}$ rad/s  (E) $2\pi$ rad/s

    (C) $\pi$ rad/s

72. The radius of the larger horizontal circle is

    (A) 4 ft.  (D) 2 ft.

    (B) 5 ft.  (E) 6 ft.

    (C) 3 ft.

73. The ratio of the areas of the two circles is (larger to smaller)

    (A) $\frac{16}{9}$  (D) 2

    (B) $\frac{22}{7}$  (E) $\frac{3}{2}$

    (C) $\frac{36}{15}$

74. The number of times on the interval that P and Q lie on the horizontal x-y plane simultaneously is

    (A) 4  (D) 11

    (B) 5  (E) 0

    (C) 8

75. The ratio of the transverse velocities of particles P and Q is

   (A) $\frac{4}{3}$         (D) $\frac{3}{2}$

   (B) $\frac{5}{2}$         (E) 2

   (C) 1

76. On the interval from t = 0 to t = 8, the number of times that the speed of the rate of recession between particles P and Q changes is

   (A) 1         (D) 5

   (B) 3         (E) 7

   (C) 4

77. If we define the function $g(x) = \int_0^x f(t)dt$, where f is the distance function defined in the figure, which of the following is true?

   I. g increases on some subintervals
   II. g is periodic
   III. g is not monotonic on the interval $0 \leq x \leq 9$

   (A) I only         (D) III only

   (B) I and II       (E) I, II and III

   (C) II and III only

78. The minimum value of g'(x) is

   (A) negative      (D) 2

   (B) 0             (E) 5

   (C) 1

79. g'(x) has an average value of

   (A) 4             (D) 10

   (B) 6             (E) 0

   (C) 8

80. If $h = \frac{df}{dx}$, then h

   (A) is periodic.
   (B) strictly decreases on all subintervals.
   (C) strictly increases on all subintervals.
   (D) is never monotonic on any subintervals.
   (E) has a positive mean value.

The following function is to be used in Questions 81-85.

$$s(t) = t^3 - \frac{15}{2}t^2 + 12t + 10$$

The function s(t) represents the position of a particle P from a fixed particle Q with respect to time.

81. When the velocity of particle P is zero, which of the following is a possible distance of P from point Q?

   (A) 0
   (B) 2
   (C) 3
   (D) 10
   (E) 28

82. Over which of the following intervals is particle P always moving backwards?

   (A) $0 < t < 5$
   (B) $-1 < t < 1$
   (C) $0 < t < 3$
   (D) $1 < t < 4$
   (E) $2 < t < 10$

83. Over which of the following intervals is P decelerating for all t?

   (A) $2 < t < 3$
   (B) $2 < t < 4$
   (C) $1 < t < 4$
   (D) $0 < t < 2$
   (E) $-1 < t < 3$

84. In the interval from t = 1 to t = 4, the average velocity of particle P is

(A) -10          (D) 40

(B) $\frac{21}{5}$   (E) -36

(C) $\frac{-27}{6}$

85. From t = 0 to t = 2, the average acceleration of particle P is

    (A) -9          (D) 7
    (B) 3           (E) -4
    (C) 5

DIRECTIONS: Choose the best answer for each question and mark the letter of your selection on the corresponding answer sheet.

86. As shown below, a 50-N tension is required to maintain the box B in equilibrium with force F. Calculate the magnitude of F given that d = 4 cm and r = 3 cm.

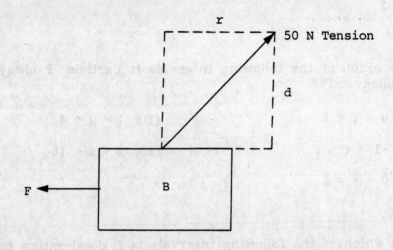

    (A) 20 N        (D) 30 N
    (B) 50 N        (E) 60 N
    (C) 40 N

87. A man falls from a bridge, but manages to grab onto a high-voltage line. The man is safe; the reason being

(A) the man's hands are dry.

(B) the man has a very high resistance.

(C) there is absolutely no current flowing through the man.

(D) the voltage between the man's hands is too low.

(E) non-existant. The man will die.

88. When 1.09g Al burns completely in oxygen, 2.06g of aluminum oxide is formed. Determine the empirical formula for the oxide. (Atomic weights: Al = 27, O = 16)

(A) $AlO_3$

(B) $AlO$

(C) $Al_2O_2$

(D) $Al_2O_3$

(E) $Al_4O_6$

89. One of the given choices below is not true for the requirements of an ideal transformer. Choose which is it.

(A) no losses

(B) no resistance

(C) leakage flux

(D) no leakage flux

(E) no magnetizing current

90. As shown below, water flows through a pipe at a rate, q, of 0.010 cubic feet per second. The average velocity, v, at 1 is 2.88 feet per second. What is the pipe's cross-sectional area, A?

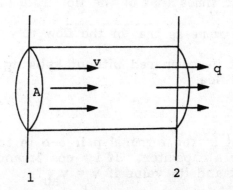

(A) 0.1 sq. in.    (D) 5 sq. in.
(B) 0.5 sq. in.    (E) 10 sq. in.
(C) 1.2 sq. in.

91. Tanks A and B contain the same low viscosity fluid, but the tank's pressure and the orifice's distance below the water level in tank A are each half that of tank B. The flow rate of the liquid through the discharge orifice of tank B:

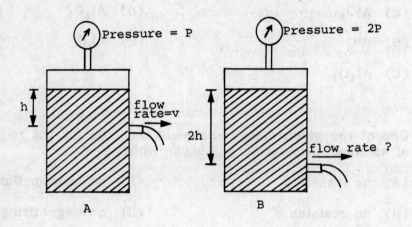

(A) is not related to the orifice's distance below the water level but related to pressure

(B) is related to both the pressure and the orifice's distance below the water level but not linearly

(C) is four times that of the flow rate of tank A

(D) is the same as that of the flow rate for tank A

(E) cannot be compared without knowing the surface area of the orifice

92. The element in the terminal pair a-b in the figure is either an inductor or a capacitor. If $i = \cos 3t$ and $v = \sin 3t$, what is the element and its value if $v = v_{ab}$?

408

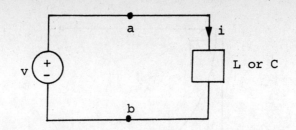

(A) A capacitor whose value is $\frac{1}{3}$F

(B) A capacitor whose value is 3F

(C) An inductor whose value is $\frac{1}{3}$H

(D) A capacitor whose value is $-\frac{1}{3}$F

(E) None of the above

93. A dc motor requires 10 kilowatts to enable it to supply its full capacity of 10 horsepower. What is its full load efficiency?

(A) 74.6 percent

(B) 1.0 percent

(C) 76.4 percent

(D) 100 percent

(E) 134 percent

94. When a mouth organ is blown harder, which one of the following changes occur in the characteristic of the sound wave?

(A) The wave frequency increases

(B) The wave frequency decreases

(C) The wave will travel faster

(D) The wave amplitude increases

(E) The wave amplitude decreases

95. A 1 kg body slides from rest down a track of radius R = 2m, but its speed at the bottom is only 4 m/s. If all the energy dissipated is in the form of heat, what amount of heat is generated? (Note: 1 cal = 4.1858 J)

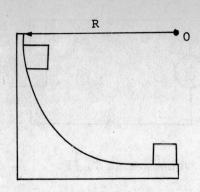

(A) $\dfrac{-11.6}{4.1858}$ cal

(B) $\dfrac{-27.6}{4.1858}$ cal

(C) $-27.6(4.1858)$ cal

(D) $-11.6(4.1858)$ cal

(E) $\dfrac{3.6}{4.1858}$ cal

96. A 60 kg man jumps from a diving board at a height of 12.5m, into a swimming pool. If it takes the water 0.5 sec. to reduce the man's velocity to zero, the average force that the water exerts on the man is

(A) $-400\sqrt{g}$ N

(B) $-450\sqrt{g}$ N

(C) $-500\sqrt{g}$ N

(D) $-550\sqrt{g}$ N

(E) $-600\sqrt{g}$ N

97. The oxidation of an element is the

(A) process which an oxidizing agent undergoes.

(B) loss of electrons.

(C) acquiring of oxygen in its gaseous state.

(D) gaining of charge due to ionic bonding.

(E) reverse process of titration.

98. To measure the acceleration of a moving cart a small body weighing 3N is fastened to one end of a massless string which is, in turn, attached to the top of the cart. If the tension in

the string is read as 5N, then what is the acceleration of the cart?

(A) $\frac{5}{3}(9.8)\frac{m}{s^2}$

(D) $\frac{3}{4}(9.8)\frac{m}{s^2}$

(B) $\frac{3}{5}(9.8)\frac{m}{s^2}$

(E) $\frac{4}{3}(9.8)\frac{m}{s^2}$

(C) $0.0\frac{m}{s^2}$

99. If the energy of the molecules in a monatomic gas sample is entirely translational kinetic $\left(\frac{1}{2}Mv^2 = \frac{3}{2}RT\right)$, then for an ideal gas containing n molecules what is the specific heat capacity at constant volume, $c_v$, for a temperature rise $\Delta T$?

(A) $c_v = \frac{2}{3}R$

(D) $c_v = \frac{3}{2}R$

(B) $c_v = \frac{1}{2}R$

(E) $c_v = \frac{3}{2}\frac{R}{n}$

(C) $c_v = \frac{2}{3}\frac{R}{n}$

## Questions 100-101

In the diagram below is a simple pendulum consisting of a small mass, m, attached to the end of a wire of length l; the other end of the wire is attached to a fine point A. When the mass is displaced slightly it oscillates with simple harmonic motion along the arc of a circle with center A. (Assume: arc OB = x, where x is measured from 0; $\angle OAB = \theta$ and $\theta$ is small)

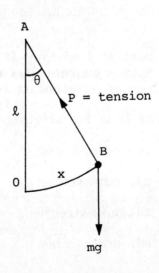

100. The acceleration along the arc OB is

   (A) $-gl(1/x)$
   (B) $(g/l)x$
   (C) $-(g/l)x$
   (D) $(g/l)x \sin\theta$
   (E) None of the above

101. Which is the correct formula for determining the acceleration due to gravity (g) using this simple pendulum? (Assume sinusoidal motion)

   (A) $\dfrac{T^2 l}{4\pi}$
   (B) $\dfrac{l}{T^2}$
   (C) $4\pi^2 \left(\dfrac{l}{T^2}\right)$
   (D) $2\pi \sin\theta \left(\dfrac{l}{T^2}\right)$
   (E) None of the above

102. A liquid flows in a circular tube at constant temperature with a Reynold's number of 1,000. In a dynamically similar situation, the same liquid flows through a tube whose radius is twice as long as the radius of the first tube. If $\nu$ is the kinematic viscosity of the fluid at that particular temperature, then the product of the velocity of fluid and the tube radius in the second tube is

   (A) 1000
   (B) $500\nu$
   (C) $1000/\nu$
   (D) $2000\nu$
   (E) $1000\nu$

103. A man can row a boat at 2 mi/hr. If he wishes to cross a river 1 mile wide with a current of 4 mi/hr, where will he land on the opposite shore if he attempts to row straight across?

   (A) Directly across from his starting point
   (B) At a point ¼ mi. downstream
   (C) At a point 4 mi. downstream
   (D) At a point 2 mi. downstream
   (E) At a point 1 mi. downstream

104. The electral conductivity for a metal such as copper decreases as the temperature increases. Which of the following is usually cited as the reason for this phenomenon when the temperature increases?

(A) The magnetization of copper increases to obstruct the path of the electrons.

(B) The band gap between the conduction band and the valence band increases.

(C) The electrons undergo Boson condensation and fall back to the ground state.

(D) The sea of electrons become more strongly bound to positive ions of the crystal.

(E) The frequency of collisions of electrons increases.

105. For a simple compressible substance the following expression can be written:

$$C_p - C_v = -T \left(\frac{\partial V}{\partial T}\right)_p^2 \left(\frac{\partial P}{\partial V}\right)_T$$

where $C_p$ and $C_v$ are specific heats at constant pressure and volume respectively, T is temperature, V is volume, and P is sure. Which of the following is true?

(A) $C_p$ minus $C_v$ is negative when $\left(\frac{\partial P}{\partial V}\right)_T$ is positive.

(B) For an ideal gas, $C_p$ minus $C_v$ is zero.

(C) $C_p$ minus $C_v$ is always positive.

(D) At constant volume, $C_p$ minus $C_v$ reduces to T.

(E) $C_p$ minus $C_v$ is linearly related to $\left(\frac{\partial V}{\partial T}\right)_p^2$.

106. The project engineering segment of an economic feasibility study will include all of the following contents except

(A) a description of the fabrication process

(B) the estimated demand for the final product

(C) raw materials and supplies

(D) the physical means of production

(E) the transportation expenses for raw materials and products

107. The diagram below shows a mass of weight w attached to a cord of length r at one end and a fixed point O at the other end. If the body is whirled in a vertical circle about point O, what is the tension in the cord when it is in the position B?

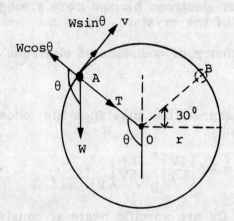

(A) $mv^2/r$

(B) $m\left(\dfrac{v^2}{r} + g\right)$

(C) $\dfrac{m}{2}\left(\dfrac{v^2}{r} - \dfrac{g}{2}\right)$

(D) $m\left(\dfrac{v^2}{r} - \dfrac{g}{2}\right)$

(E) None of the above

108. A substance, whose conduction band is empty at low temperature and therefore behaves as an insulator, but increases its conductivity with higher temperature is a(an)

(A) insulator

(B) pure semiconductor

(C) impure semiconductor

(D) metal

(E) Both (B) and (C)

109. Which of the following is a principal assumption in the development of an expression for the Boundary Layer Thickness? [Von Karmen Integral Method]

(A) Pressure does not vary in the direction perpendicular to the flow.

(B) Pressure does vary in the direction perpendicular to the flow.

(C) flow is laminar.

(D) flow is turbulent.

(E) Subsonic flow.

110. Nitrogen gas is heated in a steady-state, steady-flow process. The inlet conditions are 80 lbf/in², 100°F, and the exit conditions are 75 lbf/in², 2000°F. Assume that the changes in kinetic and potential energy are negligible. Calculate the required heat per lbm of nitrogen. $C_p$ = 3.5 Btu/(lb)(°F).

(Hint: Don't make an ideal gas assumption; make a perfect gas assumption instead)

(A) 6000 Btu/lbm

(B) 6250 Btu/lbm

(C) 6650 Btu/lbm

(D) 6850 Btu/lbm

(E) 7000 Btu/lbm

111. A U-tube is filled with two liquids (that don't mix) of different densities. Which of the following statements is <u>not</u> true?

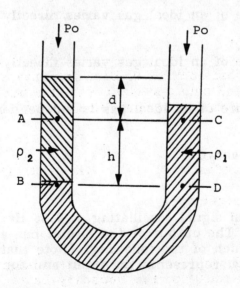

(A) The level of the liquid above A indicates that the fluid around A is less dense than the fluid around C.

(B) The pressure at B is equal to the pressure at D.

(C) The pressure difference between A and B is less than the pressure difference between C and D.

(D) The pressure difference between B and C cannot be solved without knowing $\rho_2$.

(E) The pressure difference between A and B is the same as that between A and D.

112. Given that density of ice ($\rho$ ice) = 0.92 gm/cm$^3$, and that of seawater ($\rho$ water) = 1.03 gm/cm$^3$, which of the given values is closest to the fraction of volume of an iceberg that is below the surface of seawater?

(A) 90%

(B) 80%

(C) 20%

(D) 10%

(E) 50%

113. According to Charles' Law,

(A) the temperature of an ideal gas varies directly with the pressure

(B) the volume of an ideal gas varies directly with the pressure

(C) the volume of an ideal gas varies directly with the temperature

(D) the pressure of an ideal gas is independent of the temperature

(E) All of the above

114. A 5v sinusoidal signal, oscillating at 0.25 Hz is applied to the circuit. The output of the circuit has a waveform approximate to which of the following? (Note that in the figure, the dotted signal represents the input and the solid line the output.)

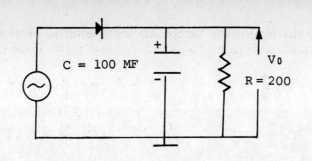

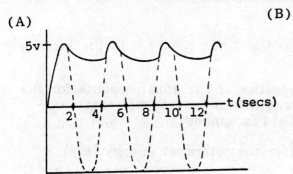

(A)

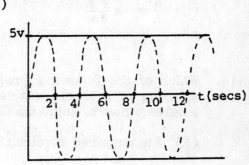

(B)

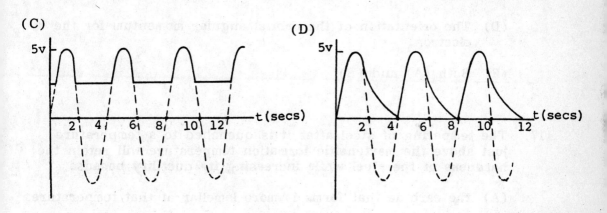

(C)　　　　　　　　　　　　(D)

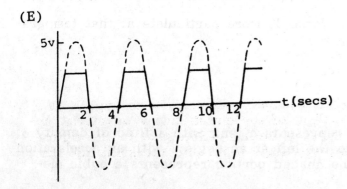

(E)

115. Which of the following correctly represents Newton's Law of Cooling?

(A) $\frac{q}{L} = -k\frac{\partial T}{\partial x}$

(B) $\frac{q}{A} = k\frac{\partial T}{\partial x}$

(C) $\frac{q}{A} = -k\frac{\partial T}{\partial x}$

(D) $\frac{q}{L} = h\Delta T$

(E) $\frac{q}{A} = h\Delta T$

116. Which of the following properties of the atom accounts for the splitting of a single wavelength into several different wavelengths when a magnetic field is applied?

   (A) An unpaired electron in the outermost energy level

   (B) The shape of the electron orbit for the smallest angular momentum quantum number

   (C) The probability density for the lowest energy level of the atom

   (D) The orientation of the orbital angular momentum for the electron

   (E) Both (A) and (B)

117. The tempering of steel after it is quenched to a temperature just above the martensitic formation temperature will retain the hardness of the steel while increasing its ductility because

   (A) the carbide that forms is more lamellar at that temperature

   (B) carbide particles do not precipitate in the structure

   (C) carbide particles diffuse from the surface

   (D) the carbide that forms is more particulate at that temperature

   (E) None of the above

118. The diagram shown represents a tank with a fluid of density $\rho$, which is deflected to the left at an angle $\theta$ with an acceleration $a_0$ to the right. The shaded portion represents a cubic ele-

ment of fluid. Note: The angle between the side face BD of the element and the surface is $\theta$. Which of the following statements is true for the tank shown above?

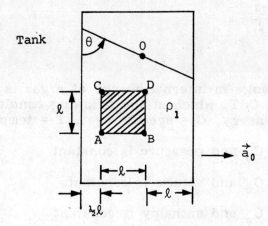

(A) The pressure at point A and the pressure at point B are the same.

(B) The pressures on the upper face and the pressure on the lower face of the fluid element are the same.

(C) The pressure on the side faces of the cubic element are the same.

(D) The pressure at point B and the pressure at point C is the same.

(E) The pressure force at B is the same at D.


119. The diagram depicts a tank with a fluid of density $\rho$. What are the forces acting on surface AC?

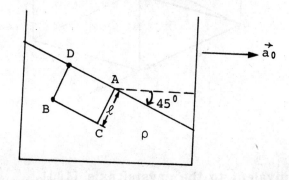

(A) 0

(B) $\rho g \dfrac{\ell^2}{2}$

(C) $\rho a_0 \dfrac{\ell^2}{2}$

(D) $\dfrac{\rho \ell^2}{2}(g^2 - a_0^2)^{\frac{1}{2}}$

(E) None of the above

120. If the change in internal energy of a gas is given by the equation du = CdT, which of the following conditions is true? (u = internal energy, C = specific heat, T = temperature)

(A) $C = C_v$ and pressure is constant

(B) $C = C_v$ and volume is constant

(C) $C = C_v$ and enthalpy is constant

(D) $C = C_p$ and pressure is constant

(E) $C = C_p$ and volume is constant

121. Using Miller indices to describe the cubic lattice represented below, the direction [111] is

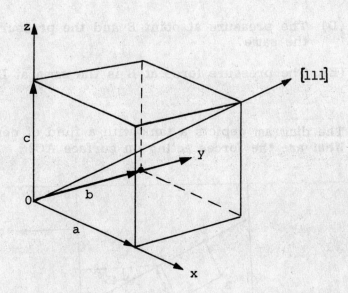

(A) equivalent to the crystal axis [100].

(B) perpendicular to the plane (111).

(C) the same as the direction $(\bar{1}\bar{1}\bar{1})$.

(D) applicable for all lattice systems, cubic or non-cubic.

(E) parallel to the plane (100).

122. The idealization of a black body is which of the following?

(A) Emits all radiation (or radiant energy) but absorbs none

(B) Absorbs and emits all radiation

(C) Absorbs all but emits none

(D) Absorbs and emits no radiation

(E) Is called black because it absorbs more radiation than it emits

123. How many Btu are needed to heat 5 lb. of water at 102°F to its boiling point of 212°F?

(A) 22

(B) 42

(C) 110

(D) 550

(E) 1060

124. Given a hollow cylindrical tube as shown in the figure below, water flows into the tube through $A_1$ with a speed of 9 m/s. What is the speed of water coming out from $A_2$ if $A_1$ is three times as large as $A_2$?

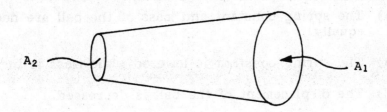

(A) 27 m/s

(B) 18 m/s

(C) 9 m/s

(D) 6 m/s

(E) 3 m/s

125. A piece of equipment has been purchased by the city for $10,000 with an anticipated salvage of $500 after 8 years. Which of the depreciation models will yield the greatest reduction in the book value of the equipment?

   (A) Straight line depreciation

   (B) Sum-of-years digits depreciation

   (C) Declining balance depreciation

   (D) Declining balance depreciation is switched to straight line

   (E) Declining balance depreciation is switched to sum-of-years digits depreciation

126. The viscoelastic modulus for a polymer will decrease with increasing time for a fixed but long-term load on the polymer. Which of the following makes this possible?

   I. The permanent displacement of the molecules caused by the stress

   II. The movement of the molecules to relax the elastic stress

   III. The increasing random orientation of the molecules with increasing time

   (A) I and II only

   (B) II and III only

   (C) I and III only

   (D) II only

   (E) I only

127. The period of oscillation of a ball on a spring will decrease when which of the following is true?

   (A) The spring constant and mass of the ball are decreased equally.

   (B) The spring constant is lowered while mass is increased.

   (C) The displacement of the ball is decreased.

   (D) The mass is increased while the spring constant is decreased.

   (E) The mass is decreased while the spring constant is increased.

128. The cooling curves below are used to construct equilibrium phase diagrams. Which of the cooling curves is applicable for a binary solid solution?

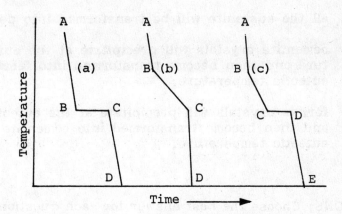

(A) only a  
(B) only b  
(C) only c  
(D) both b and c  
(E) none of the above  

129. Which of the following orbital configurations correctly represents the oxygen atom in the ground state?

(A) $1s^2 2s^2 2p^3 3s^1$

(B) $1s^2 2s^2 2p^4$

(C) $1s^2 2s^2 2p^4$

(D) $1s^2 2s^2 2p^4 3s^2$

(E) $1s^2 2s^1 2p^5$

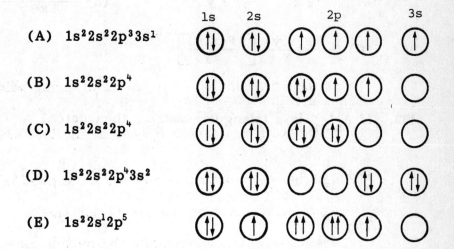

130. For a carbon-iron phase diagram, austenite steel is slowly cooled at the eutectoid composition. If the alloy is cooled below the eutectic temperature, then

423

(A) ferrite crystals will precipitate in the austenite structure.

(B) cementite crystals will precipitate in the austenite structure.

(C) all the austenite will be transformed into pearlite.

(D) cementite crystals will precipitate at the eutectic temperature and then become transformed into ferrite below the eutectic temperature.

(E) ferrite crystals will precipitate at the eutectic temperature and then become transformed into cementite below the eutectic temperature.

DIRECTIONS: Choose the best answer for each question and mark the letter of your selection on the corresponding answer sheet.

NOTE: The natural logarithm of x will be denoted by lnx.

131. Given an implicit functional relationship $F(x,y) = 0$, then $\frac{dy}{dx}$ is given by

(A) $\frac{\partial F(x,y)}{\partial x}$

(B) $\frac{\partial F(x,y)}{\partial x} + \frac{\partial F(x,y)}{\partial y}$

(C) $-\frac{\partial F(x,y)}{\partial x} \left[ \frac{\partial F(x,y)}{\partial y} \right]^{-1}$

(D) $\frac{-F_y}{F_x}$

(E) $\frac{\partial^2 F(x,y)}{\partial x \partial y}$

132. If $A_{n \times n}$ is a triangular matrix, then $\det(A) =$

(A) $\sum_{i=1}^{n} a_{ii}$

(B) $\prod_{i=1}^{n} a_{ii}$

(C) $\sum_{i=1}^{n} (-1)^n a_{ii}$

(D) $\prod_{i=1}^{n} (-1)^n a_{ii}$

(E) $\sum_{\substack{i=1 \\ j \neq i}}^{n} a_{ij}$

133. The area enclosed by the curve $\rho = a(1 - \cos\theta)^{\frac{1}{2}}$ is

   (A) $\pi a$
   (B) $\pi^2 a$
   (C) $\pi a^2$
   (D) $\pi^2 a^2$
   (E) $\pi^2 a^3$

134. The area enclosed by the ellipse $\frac{x^2}{a^2} + \frac{y^2}{b^2} = 1$ is

   (A) $\pi \frac{a}{b}$
   (B) $\pi \frac{b}{a}$
   (C) $\pi a^2 b$
   (D) $\pi(a - b)$
   (E) $\pi ab$

135. $\lim\limits_{x \to 2} \frac{2x^2 - 4x}{x - 2} =$

   (A) 0
   (B) $\infty$
   (C) 4
   (D) 2
   (E) 16

136. $\frac{d}{dx}(e^{\ln x}) =$

   (A) 1
   (B) 0
   (C) $\frac{1}{x}$
   (D) $e^x$
   (E) $\frac{x^2}{2}$

137. $\lim\limits_{x \to \infty} \frac{x^n}{e^x} =$

   (A) 0
   (B) $\infty$
   (C) $-\infty$
   (D) n
   (E) an undetermined function

138. The motion described by the equation $x(t) = A\cos(\omega t - \phi)$ is

   (A) motion with constant velocity.

   (B) motion with constant acceleration.

   (C) oscillatory motion.

   (D) motion in which the acceleration is proportional to the velocity.

   (E) motion with constant deceleration.

139. If line a is perpendicular to line b, and line a has slope $-\frac{2}{5}$, then line b has slope =

   (A) $-\frac{5}{2}$

   (B) $\frac{5}{2}$

   (C) 0

   (D) $\infty$

   (E) $\frac{2}{5}$

140. If the first term of an arithmetic series is -4 and the twelfth term is 32, the common difference is

   (A) 28

   (B) 14

   (C) -2

   (D) 16/9

   (E) 36/11

141. For the differential equation $y''(x) + \lambda^2 y(x) = 0$ with the conditions $y(0) = y^1(\pi) = 0$, the set of positive values of $\lambda$ which gives a solution of this differential equation other than $y(x) = 0$ and satisfying the given conditions is,

   (A) $\lambda = n + \frac{1}{2}$, $n = 0,1,2,3,\ldots$

   (B) $\lambda = n - \frac{1}{2}$, $n = 0,1,2,3,\ldots$

   (C) $\lambda = n$, $n = 1,2,3,\ldots$

   (D) $\lambda = n^2$, $n = 0,1,2,3\ldots$

   (E) $\lambda = n\pi$, $n = 0,1,2,3\ldots$

142. $\sum_{n=1}^{\infty} \frac{x^{n-1}}{n5^n}$ converges absolutely for

(A) all x  (D) x ∈ [-5,5]

(B) x ∈ [0,5]  (E) x ∈ (∞,5]

(C) x ∈ (0,5]

143. Given points $P_1(x_1,y_1,z_1)$, $P_2(x_2,y_2,z_2)$ and the origin O, a sufficient condition on $x_1, x_2, y_1, y_2, z_1$ and $z_2$ such that $\overline{P_1O} \perp \overline{P_2O}$ is

(A) $x_1^2 + y_1^2 + z_1^2 = 0$

(B) $x_1x_2 + y_1y_2 + z_1z_2 = 0$

(C) $x_1 + y_1 + z_1 = $ constant

(D) $(x_1-x_2)^2 = -(y_1-y_2)^2 - (z_1-z_2)^2$

(E) None of the above is sufficient

144. Which of the following implies that $f(r,\theta) = 0$ is symmetric with respect to the pole?

(1) $f(r,\theta) = 0$
(2) $f(r,\theta) = f(r,-\theta)$
(3) $f(r,\theta) = f(r,\pi-\theta)$
(4) $f(r,\theta) = f(r,\theta+n\pi)$ for all integers n
(5) $f(r,\theta) = f(-r,\theta)$

(A) (1) only

(B) (1), (2), and (5) only

(C) (1), (4), and (5) only

(D) All of them

(E) None of them

# THE GRADUATE RECORD EXAMINATION ENGINEERING TEST

## MODEL TEST IV

## ANSWERS

| | | | | | |
|---|---|---|---|---|---|
| 1. | A | 21. | C | 41. | E |
| 2. | D | 22. | D | 42. | A |
| 3. | D | 23. | D | 43. | D |
| 4. | D | 24. | B | 44. | E |
| 5. | E | 25. | A | 45. | B |
| 6. | B | 26. | D | 46. | A |
| 7. | D | 27. | C | 47. | B |
| 8. | B | 28. | B | 48. | C |
| 9. | C | 29. | D | 49. | A |
| 10. | C | 30. | C | 50. | C |
| 11. | D | 31. | D | 51. | C |
| 12. | C | 32. | B | 52. | A |
| 13. | A | 33. | E | 53. | D |
| 14. | B | 34. | C | 54. | B |
| 15. | E | 35. | C | 55. | B |
| 16. | D | 36. | B | 56. | C |
| 17. | C | 37. | B | 57. | B |
| 18. | C | 38. | C | 58. | E |
| 19. | E | 39. | D | 59. | C |
| 20. | D | 40. | C | 60. | C |

| | | | | | |
|---|---|---|---|---|---|
| 61. | C | 89. | C | 117. | D |
| 62. | A | 90. | B | 118. | D |
| 63. | A | 91. | B | 119. | D |
| 64. | A | 92. | A | 120. | B |
| 65. | C | 93. | A | 121. | B |
| 66. | D | 94. | D | 122. | B |
| 67. | C | 95. | A | 123. | D |
| 68. | E | 96. | E | 124. | A |
| 69. | B | 97. | A | 125. | D |
| 70. | D | 98. | E | 126. | A |
| 71. | B | 99. | D | 127. | D |
| 72. | A | 100. | C | 128. | B |
| 73. | A | 101. | C | 129. | B |
| 74. | B | 102. | B | 130. | C |
| 75. | A | 103. | D | 131. | C |
| 76. | C | 104. | E | 132. | B |
| 77. | A | 105. | C | 133. | C |
| 78. | E | 106. | B | 134. | E |
| 79. | B | 107. | D | 135. | C |
| 80. | A | 108. | E | 136. | A |
| 81. | B | 109. | A | 137. | A |
| 82. | D | 110. | C | 138. | C |
| 83. | D | 111. | D | 139. | B |
| 84. | C | 112. | A | 140. | E |
| 85. | A | 113. | C | 141. | A |
| 86. | D | 114. | A | 142. | D |
| 87. | D | 115. | E | 143. | B |
| 88. | D | 116. | D | 144. | C |

# THE GRADUATE RECORD EXAMINATION ENGINEERING TEST

## MODEL TEST IV

## DETAILED EXPLANATIONS OF ANSWERS

1. (A)
A process that is irreversible usually occurs spontaneously passing through a series of non-equilibrium states. In order for a process to be reversible the change must be slow and gradual so that the deviation from equilibrium will only be slight. Hence, by a certain differential change in the environment the process can be made to retrace its path. For an isothermal process, the temperature of the gas differs at all times by only a differential amount dT from the constant temperature reservoir. Therefore, the heat flow is not across a considerable temperature drop that would make it irreversible.

2. (D)
Since the resistive network is extended to infinity, we notice that the equivalent resistance across the system's terminals is the same as that of the part of the network located at the right side of dashed line AB. So we can simply replace it with the unknown final resistance of the whole system, R.

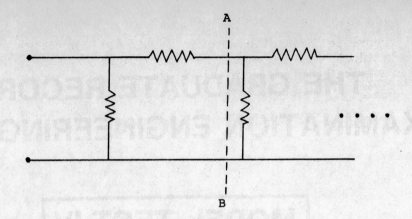

So we have:

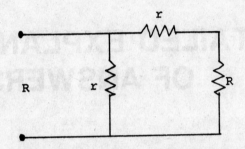

$$\Rightarrow R = \frac{r(r + R)}{2r + R} \Rightarrow 2rR + R^2 = r^2 + rR$$

$$\Rightarrow R^2 + rR - r^2 = 0 \Rightarrow R = \frac{-r \pm \sqrt{r^2 + 4r^2}}{2}$$

$$R = \frac{(\sqrt{5} - 1)r}{2} \quad \text{(chose + rather than-} \\ \text{since R must be > 0)}$$

3. (D)
The flow lines around an airfoil crowd together in the region above. Since the path of the flow is more narrow, this has the same effect as reducing the cross-sectional area of a control volume for the fluid. From the equation of continuity and Bernoulli's equation, the region above the airfoil is one of increased velocity and reduced pressure while the pressure below remains the same. This variation in pressure creates a net upward force on the airfoil.

4. **(D)**
Provided that the friction coefficient is large enough, the block will tip over when the center of mass of the block is vertically aligned with its lower corner.

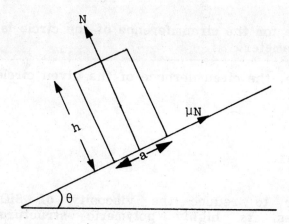

Then: $\quad N = mg \cos\theta \quad$ and $\quad mg \sin\theta = \mu N$

Therefore: $\quad \mu = \tan\theta = \dfrac{a}{h}$

The coefficient of friction must be at least $\dfrac{a}{h}$. Hence:

$$\mu > \dfrac{a}{h}$$

5. **(E)**
To solve this problem, one needs to know that any triangle inscribed in a semicircle is a right triangle. Thus, triangle ABD becomes a 30°-60°-90° triangle. In such a triangle, sides relate as follows

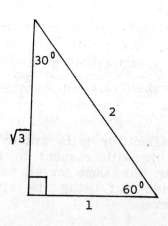

Using the property of similar triangles, diameter AB can be found

$$\frac{\sqrt{3}}{4} = \frac{2}{x} \implies \sqrt{3}\,x = 8 \implies x = \frac{8}{\sqrt{3}}$$

The formula for the circumference of the circle is $\pi d$ where d is the diameter.

Therefore, the circumference of the given circle is $\frac{8}{\sqrt{3}}\pi$.


## 6. (B)

In order to reduce the viscosity of $SiO_2$ at high temperatures, its highly polymeric structure must be disrupted. The glassy structure of the silica must be modified by disconnecting some of the oxygen bridges. A network modifier must possess bridging oxygen atoms in order to depolymerize $SiO_2$. An oxide such as $Na_2O$ will donate a non-bridging oxygen and two $Na^+$ ions.


## 7. (D)

The distribution curves show that for higher temperatures, the distribution is broader, the number of molecules at the average speed is lower (the curve's peak is lower), and the average speed is greater (the peak appears further to the right). So A, B, C are correct. Though each curve is a step of $1000°K$, the change in the number of molecules is not constant. Therefore the relationship is not linear, so E is true. The curves couldn't be drawn if the number of molecules were changed and the curves show a very small number of molecules at rest ($r = 0$).


## 8. (B)

The power, p, absorbed in a circuit is the rate at which energy is absorbed, i.e. $p = \frac{dW}{dt}$ where p is the power and W is the energy.

If an expression for p is known, W can be obtained from p by integrating with respect to time. If the expression for $p(t)$ is not the same for all values of t, the integration must be carried out using the appropriate expression for p in each interval.

For the given curve, the expression for p (in watts) is:

$$p(t) = 10t \quad \text{for} \quad 0 \leq t \leq 1$$

and $\quad p(t) = 27.2e^{-t} \quad \text{for} \quad 1 \leq t.$

Therefore the energy from t = 0 to 3 is

$$W = \int_0^3 p(t)\,dt = \int_0^1 10t\,dt + \int_1^3 27.2e^{-t}\,dt$$

$$= 5 + 27.2e^{-1} - 27.2e^{-3}$$

$$= 5 + \frac{27.2}{e} - \frac{27.2}{e^3}$$

$$= 5 + \frac{27.2}{e}\left(1 - \frac{1}{e^2}\right)$$

$$= 5 + 27.2\left(\frac{e^2 - 1}{e^3}\right) \text{ Joules}$$

9. **(C)**
When the tires are underinflated, their effective diameter is reduced. Therefore, there is a decrease in the distance covered by a single rotation of the tires. Since the speedometer mechanism measures the speed of a car only by counting revolutions of the tires, it does not know that the car is moving slower because of the reduced diameter of the tires since the number of rotations remains the same.

10. **(C)**
The statement expresses Thevenin's Theorem.

Other theorems are stated below:

Superposition Theorem: In a linear bilateral network, the current through, or voltage across any element is equal to the algebraic sum of the currents or voltages which would be produced independently by each of the sources in the network.

Norton's Theorem: Any two-terminal linear bilateral d.c. network can be replaced by an equivalent circuit consisting of a current source and a parallel resistor.

**Maximum Power Transfer Theorem:** A load receives maximum power from a linear bilateral d.c. network if its total resistance equals the Thevenin resistance of the network as seen by the load.

**Substitution Theorem:** In a linear bilateral circuit, if the voltage and current in any branch is known, then the branch can be substituted by a combination of elements that maintains the same voltage and current in that branch.

### 11. (D)

The Megger derives its name from the fact that it can be used to measure very high resistances in the range of megohm. It is basically used to measure the insulation resistance of heavy conductors in power transmission and other systems.

To measure such high resistance, a very high voltage is required. This is acquired by a hand driven d.c. generator incorporated in the megger itself. The output of the generator is a fixed voltage, typically 250, 500 or 1000 volts. The unknown resistance to be measured is connected between the terminals marked Line and Earth. Then the shaft of the d.c. generator is rotated until one sees the resistance increasing and settling at a particular point as can be made out by the movement of the pointer on the glass indicator.

### 12. (C)

The principle of angular momentum is used to solve this problem.

Angular momentum = $\Sigma \vec{r} \times m\vec{v}$, where m, v and r are mass, speed and the distance between the mass and the axis about which it rotates, respectively. When the object is a solid body rotating about an axis, we can write

$$\Sigma m\vec{r} \times \vec{v} = \Sigma mr^2\omega = \omega\Sigma mv^2 = I\omega$$

when I is the moment of inertia of the body. By the law of conservation of angular momentum for bodies in rotation, the angular momentum of the system will not change, provided that no unbalanced external torque acts on the rotating system.

When the arms and the leg are drawn in by the skater, the part of moment of inertia contributed by the arms and leg decreases because now arms and leg are closer to the axis of rotation and consequently their contribution $\Sigma mv^2$ is smaller than before. Therefore, to maintain the constancy of the angular momentum, there must be a corresponding increase in $\omega$ (angular velocity). Then the speed decreases due to the rotational friction between the skate and the ice which was neglected in the first part of our analysis.

13. (A)
The thermal efficiency for a Carnot cycle is:

$$\eta_{thermal} = 1 - \frac{Q_L}{Q_H} = 1 - \frac{T_L}{T_H}$$

Therefore

$$1 - \frac{T_i}{T_s} = 0.2680$$

The relative magnitude of the temperatures for the two points is:

$$\frac{T_i}{T_s} = 0.7320$$

14. (B)
The distribution of molecular speed for an ideal gas is a Gaussian distribution of the form $\frac{4\pi}{A'}e^{-bv^2}v^2 dv$. By integrating the kinetic energy of a gas molecule over the distribution of molecular speeds, one can find the total kinetic energy of a sample.

$$\int_0^\infty \tfrac{1}{2}mv^2(e^{-bv^2}v^2)dv = kE$$

Dividing the total kinetic energy by the number of gas molecules will give one the average kinetic energy.

$$\frac{\int_0^\infty \tfrac{1}{2}mv^2(e^{-bv^2}v^2)dv}{\int_0^\infty e^{-bv^2}v^2 dv} = \tfrac{1}{2}m\bar{v}^2 = \frac{3}{2}kT$$

Therefore the kinetic energy is dependent upon temperature.

15. (E)
Before steady state condition is reached, the heat flow is much greater near the end which provides greater temperature gradients near that end. Hence the temperature at the middle is raised. At the other end, the heat flow is out of the bar and hence the temperature gradients are in the opposite direction, lowering the temperature in that end of the bar. With increasing time, the temperature gradients are reduced and the bar attains a steady state uniform temperature distribution all along the bar. The final steady state depends only on the temperatures maintained at the ends. It is a linear function of distance from both of the bar's ends, and therefore it is a straight line passing through $(0, 100°)$ (hot end) and $(L, 0°)$ (cold end).

16. (D)
When the head of the vessel is removed, the pressure on the surface of water comes to zero. Therefore, water starts to evaporate. The change of phase from liquid to vapor needs lots of energy. Since this energy is supplied from the rest of the water (due to the absence of any other source), water will lose great amounts of energy and freeze. Therefore, except the small amount of water vaporized in the beginning, most of the water will freeze.

17. (C)
For the spherical wavefronts light rays can be drawn perpendicular to the wavefronts to indicate the direction of the light.

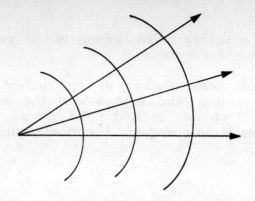

The rays indicate that the light is diverging from the source. Therefore a converging thin lens would be able to converge the light into a parallel light beam when it is placed at a distance equal to its focal point from the source.

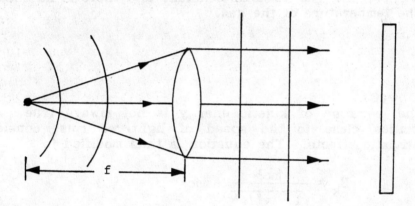

18. (C)
This reaction can be represented by the following equilibrium equation:

$$2H_2 + O_2 \underset{\longleftarrow}{\overset{Energy}{\longrightarrow}} 2H_2O$$

Since 4 kg of hydrogen reacts with 32 kg of oxygen, all of the oxygen will be consumed.

Therefore 2 moles of water will appear. Since one mole of $H_2O$ has a mass of 18 kg, two moles of $H_2O$ will have a mass of 36 kg.

**19. (E)**
The atomic number of an element is the number of protons in the nucleus of the element.

The atomic mass number of an element is the total number of protons and neutrons in the nucleus. Hence, when a neutron is emitted, the atomic mass decreases, whereas there is no change in the atomic number.

**20. (D)**
A free expansion process is one during which a system contained within adiabatic walls does no work on its surroundings. For such a process, the net heat flow into the system, Q, equals 0 and the thermal equivalent of the work done by the system on its surroundings, W, equals zero. Thus, $\Delta U$, the change in the internal energy of the system equals zero.

Therefore, in a free expansion process, the internal energy of the gas remains constant and there is no change in the temperature of the gas.

**21. (C)**
The equation of kinetic energy is not always true. At velocities close to the speed of light, we must consider relativistic effects. The equation is thus modified to

$$E_k = \frac{mc^2}{\sqrt{1 - v^2/c^2}} - mc^2$$

**22. (D)**
An isentropic process is one where the entropy remains constant. In the definition of entropy for reversible processes, $ds = 0$. Therefore we have:

$$ds = 0 = \left(\frac{dQ}{T}\right)_{rev}$$

which leads to $dQ = 0$. This means that we have no heat transfer in a reversible isentropic process. Since the change in internal energy is given by:

$$\Delta u = q - W \quad \text{where} \quad W = \int pdv.$$

Since q = 0 for an isentropic process, $\Delta u$ = W. Therefore if W = 0 (work), the change in internal energy ($\Delta u$) = 0.

23. **(D)**
Newton's laws govern this motion and must be applied in the vertical and horizontal directions separately. In order to apply Newton's laws correctly, we will draw a force diagram. The figure shows the force locations when the block just starts to tip. At this instance, the block is still in equilibrium condition. But, since it is to tip over, the normal reaction force of the flat surface is exerted on the right-most point of the cube (because the rest of the cube loses contact with the surface). Therefore, in the horizontal direction,

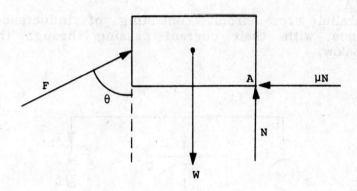

$$\Sigma F_x = f \sin \theta - \mu N = 0$$

$$f \sin \theta = 0.5 \ N.$$

**Vertically we have**

$$\Sigma F_y = f \cos \theta + N - 100 = 0$$

$$f \cos \theta = 100 - N$$

Knowing N, we can solve for $\theta$ as follows:

$$\tan \theta = \frac{f \sin \theta}{f \cos \theta} = \frac{.5 \ N}{100 - N}$$

$$\tan \theta = \frac{33.34}{33.34} = 1. \quad \theta = 45°.$$

24. (B)
Besides vapor compression, another way to achieve refrigeration is through a thermoelectric device. The basic idea of thermoelectric refrigeration is the same as that of a thermocouple, except for the fact that electricity is required for refrigeration while electricity is returned from a thermocouple. However, as of now, vapor compression refrigeration is more feasible than thermoelectric refrigeration. In vapor compression refrigeration, a refrigerant such as Freon or Ammonia is continually circulated between the evaporator, compressor and condenser. In the evaporator, heat is transferred to the refrigerant by changing its phase from liquid to vapor. Afterwards, the pressure is increased by the compressor, which enables the heat to be released to the environment through the condenser. Thus the refrigerant acts as a "heat-sponge" while it is continually circulated.

25. (A)
A parallel a.c. circuit consisting of inductance and capacitance with their current passing through them is shown below.

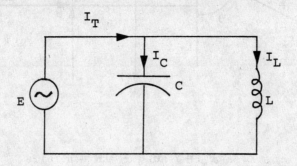

We know that $I_C$ leads E by 90° and $I_L$ lags E by 90°. Since at resonance $X_L = X_C$, the magnitude of $I_L$ equals $I_C$ but they are in the opposite direction.

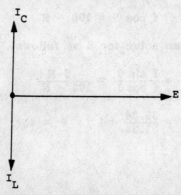

$I_L$ and $I_c$ cancel each other and so the total current, $I_T$, in the circuit at resonance is zero.

The impedance of the circuit is then

$$Z_T = \frac{E}{I_T} \quad (I_T \longrightarrow 0) = \infty$$

## 26. (D)

A quantity of heat is being transferred across a finite temperature change which makes the process irreversible. If $T_0$ is the initial temperature of the boiler which will decrease due to the transfer of heat to the liquid, and T the temperature of the saturated liquid which stays constant during vaporization because the pressure is kept constant, then the entropy change can be given as

$$\Delta S = \frac{Q}{T} - \frac{Q}{T_{0_1}}$$

where $T_{0_1}$ is a temperature between T and $T_0$. Therefore

$$\Delta S > 0 \text{ because } T_{0_1} > T \text{ and } \frac{Q}{T} > \frac{Q}{T_{0_1}}.$$

Hence, the entropy for the system of the boiler and saturated liquid increases.

## 27. (C)

Both air and water are classified as fluids: the former is compressible fluid while the latter is incompressible. Furthermore, both types of fluid exert buoyancy forces. The buoyancy force of air is usually not observed because the air's density is too small to cause a buoyancy force that is sufficiently large to overcome the gravitational force.

But, nonetheless, air does exert a buoyancy force; otherwise, hot air balloons couldn't rise. However, if the density of air would increase, then it may cause a sufficiently large force to cause objects to accelerate upward against gravity.

## 28. (B)

If a satellite is to remain in a circular orbit, the centripetal force required to keep it in its orbit is provided by the gravitational attraction between the satellite and the earth. Thus, we arrive at the following equation:

$$\frac{m_s v_s^2}{r} = G \frac{m_s m_e}{r^2}$$

where
- $m_s$ = mass of the satellite
- $v_s$ = velocity of the satellite
- $m_e$ = mass of the earth
- $r$ = distance between satellite and center of earth

Solving, we obtain

$$v_s^2 \, r = G \, m_e$$

Since G and $m_e$ are constants

$$v_s^2 \, r = \text{Constant}$$

Since the mass of the satellite has cancelled out, this shows that a light satellite and a heavy satellite, both having the same orbital speed, will occupy the same orbit. Since $M_2$ and $M_1$ are in the same orbit, they have the same speed.

## 29. (D)

The graph represents the variation in the distribtuion of valence electrons at $T \sim 300\,^{\circ}K$ with the energy of electron levels. This distribution is governed by the Fermi-Dirac distribution function

$$f_{FD}(E) = \frac{1}{e^{(E-E_F)/k^T} + 1}$$

and the density of states $D(E)dE \sim E^{\frac{1}{2}} \, dE$.

Therefore at $T \sim 300°K$ the number of electrons excited above the Fermi level is not great enough to contribute significantly to the specific heat. It is evident from the graph that the number of electrons excited above the Fermi level is less than 1% as opposed to the situation at $T \sim 0\,^{\circ}K$ where all the electrons are below the Fermi level.

## 30. (C)

The inclined plane cuts out an area given by:

$$\text{Area} = \frac{s}{\cos \theta} = \frac{s}{\cos 30°} = \frac{2s}{\sqrt{3}}$$

Let the average intensity at the incline be **x**. Then:

$$(x)\left(\frac{2s}{\sqrt{3}}\right) = (IA_b) = \text{(constant output power of bulb)}$$

Finally,

$$x = \left(\frac{\sqrt{3}}{2}\right)\left(\frac{IA_b}{s}\right)$$

## 31. (D)

From the figure it is clear that in the absence of friction we must have: $P = W_x$, while $W_y$ is balanced by the normal force N.

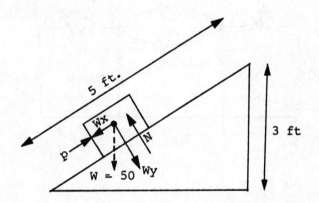

$W_x$ can be calculated by employing similar triangles.

$$\frac{W_x}{50} = \frac{3}{5}$$

$$\therefore W_x = \frac{50 \times 3}{5} = 30 \text{ lb.}$$

Thus, P should be 30 lb.

## 32. (B)

The harmonic oscillator is governed by the differential equation

$$\frac{m d^2 x}{dt^2} + \frac{b\, dx}{dt} + kx = 0$$

Therefore the frequency w is governed by

$$s^2 + \frac{b}{m} s + \frac{k}{m} = 0$$

$$\text{Img}(s) = w = \sqrt{\frac{k}{m} - \left(\frac{b}{2m}\right)^2}$$

Therefore the frequency is lowered in a damped harmonic system but is not dependent on time.

## 33. (E)

Figs. 1-4 demonstrate the process of resistor combination.

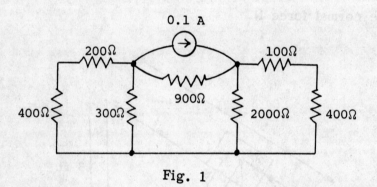

Fig. 1

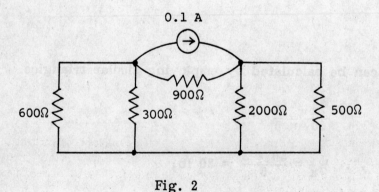

Fig. 2

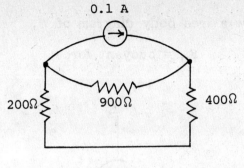

Fig. 3

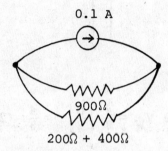

Fig. 4

Finally, we have the single loop where $v = (0.1)(360) = 36V$. It follows that the power supplied by the current source is

$$p = vi = 36(0.1) = 3.6W.$$

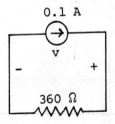

**34. (C)**

The cadmium rods in the reactor absorb neutrons that are released during fission. These absorbed neutrons cannot bombard other uranium atoms, and the reaction occurs at a slower rate.

When the rods are withdrawn, less neutrons are absorbed, and the reaction rate increases.

35. (C)

We first draw a free body diagram of W.

$F_B$ = buoyant force.

$F_B = \rho g V_W$, where $V_W = \frac{4}{3}\pi \left(\frac{d}{2}\right)^3 = \frac{4}{3}\pi (2)^3 = \frac{32}{3}\pi \text{ m}^3$

and

$F_B = \frac{32}{3}\pi \rho g \text{ N}$

$T = W - F_B = W - \frac{32}{3}\pi \rho g \text{ N}$

Next, we draw a free body diagram of the gate.

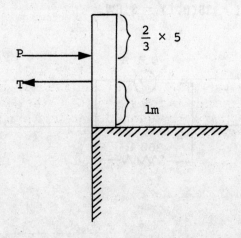

The center of pressure is at a distance equal to 2/3 of the height of the gate, measuring from the top of it (we can prove it by integration).

The total force exerted by the water on the wall is equal to the mean pressure force (at the midpoint). The pressure is considered to be spread all over the wall, which can also be proved by integration with respect to the area.
Therefore we have:

$$p = \text{pressure at midpoint} \times \text{area of the wall}$$

$$p = \rho g \left(\frac{5}{2}\right) \times (5 \times 1)$$

$$p = \rho g \frac{25}{2}$$

The moment of T and P with respect to the contact point of the gate with ground must be zero. Therefore we have:

$$p\left(\frac{1}{3} \times 5\right) = T \times (1)$$

$$\rho g \frac{25}{2} \times \frac{5}{3} = T$$

As we had before: $T = w - \frac{32}{3}\pi\rho g$

$$\Rightarrow \rho g \frac{125}{6} = w - \frac{32}{3}\pi\rho g$$

$$\Rightarrow w = \rho g \left(\frac{32}{3}\pi + \frac{125}{6}\right)$$

36. (B)
The outermost subshell is where we would place the outermost electron of its atom. By knowing its azimuthal quantum number, we know which subshell it is (i.e., s, p, d, f). Knowing this and the period of the element helps us predict the atom's principal quantum number of its outermost shell. For example:

$\quad n = \text{(period number)(for s and p)}$

$\quad n = \text{(period number)} - 1 \text{ (for d)}$

$\quad n = \text{(period number)} - 2 \text{ (for f)}$

37. (B)
For a newtonian fluid, the equation that relates the shearing stress and the velocity profile is:

$$\tau = \mu \frac{du}{dy}$$

where
- $\tau$ = shearing stress
- $\mu$ = viscosity
- $u$ = velocity
- $y$ = direction perpendicular to the velocity.

### 38. (C)

The scale enables measurement of a force which is normal to its surface.

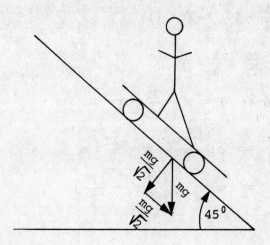

From the above free-body diagram analysis, the force normal to the scale is

$$\frac{mg}{\sqrt{2}} \quad \text{which is} \quad \frac{120}{\sqrt{2}} = 60\sqrt{2}.$$

### 39. (D)

Obviously the shiny surface of foil loses heat through radiation much faster than the dull face. When the food is to be cooked, it is customary to keep the heat inside and not let it leave the package. Since the shiny surface is more likely to lose heat through radiation, it should be kept in.

On the contrary, when the food is to be frozen, it is helpful to lose heat, so it is better to keep the shiny side out to accelerate the loss of heat through radiation.

**40. (C)**
The current through a capacitor is given as

$$i_c = C\frac{dv_c}{dt} = 2\frac{d}{dt}10t\, u(t)$$

$$i_c = 20\, u(t)\, A$$

a step function as shown in choice (C).

**41. (E)**
The light being emitted from an incandescent wire is not coherent since the atoms do not act together in the light emission process. The phase difference between diffracted beams does not remain constant with time. Therefore, there should be no interference effect from an ordinary light source. The intensity should remain approximately constant since the illumination is uniform.

**42. (A)**
The relationship for variation of mass with velocity is

$$m = \frac{m_0}{\sqrt{1-\frac{v^2}{c^2}}}$$

where $m_0$ is the mass at zero velocity and c is the velocity of light in vacuum equal to $3 \times 10^{10}$ cm/sec.

From the above relation it is clear that as the velocity of electrons increases, the mass steadily increases and this is represented by curve A.

In the above equation, if the velocity of mass equals the velocity of speed, the mass increases to infinity which is absurd. Therefore, a material body cannot have a velocity equal to the velocity of light.

**43. (D)**
A plot of nuclear masses with respect to atomic number for elements from hydrogen through uranium was shown in Test III, problem 43.

The plot of nuclear mass per nucleon, (i.e. dividing the nuclear mass by the number of nucleons in a particular nucleus) with respect to atomic number for elements from hydrogen through uranium is represented by curve D.

From the curve we observe that the masses of protons and neutrons are different when contained in different nuclei. The mass of proton is greatest for the hydrogen atom (point where curve D meets y-axis), successively decreases in atoms of increasing atomic number until the mass becomes least for iron atom (the lowest point in curve D) and then increases successfully all the way to uranium (the furthest point on the curve D with respect to x-axis).

## 44. (E)

Ohm's Law is given by $I = \frac{V}{R}$, where I is the current, V the voltage and R the resistance in the circuit. From the above relationship, it is clear that with an increase in resistance there is a decrease in the current.

So at zero resistance, the current for a dry cell equals the maximum current capacity of the dry cell, denoted by the point where curve E and the y-axis meet. As the external resistance is increased, the current decreases and approaches zero.

For a dry cell, $I = \frac{E}{r + R}$ where E is the cell emf, r the internal resistance and R the external resistance.

## 45. (B)

When the interest period is smaller than the payment period, the effective interest rate for a given interest period should be determined. Each payment for that interest period should be treated separately.

The effective interest rate is

$$i = \frac{\text{interest rate}}{\text{interest period}} = \frac{6\%}{2} = 3\%$$

The future worth of the initial payment should be determined over the number of compounding periods:

$$n = 6 \quad \text{(3 years} \times \text{2 times compounded per year)}$$

$$F = p(1 + i)^n$$

Therefore the amount that accumulates over the three years is $\$1000(1 + 0.03)^6$.

**46. (A)**
We wish to convert the given integral into a form to which we can apply the formula for

$$\int u^n du, \text{ with } u = (x^3 + 1), \, du = 3x^2 dx$$

and $n = \frac{1}{2}$.

We obtain:

$$\int_0^2 2x^2\sqrt{x^3 + 1}\, dx = \frac{2}{3} \int_0^2 (x^3 + 1)^{\frac{1}{2}} \left( \frac{3}{2} \cdot 2x^2 dx \right)$$

Applying the formula for $\int u^n du$, we obtain:

$$\frac{2}{3}\left[\frac{(x^3+1)^{3/2}}{\frac{3}{2}}\right]_0^2 = \frac{4}{9}(x^3+1)^{3/2}\Big]_0^2$$

Evaluating between 2 and 0, we have:

$$\frac{4}{9}(8+1)^{3/2} - \frac{4}{9}(0+1)^{3/2} = \frac{4}{9}(27 - 1) = \frac{104}{9}$$

**47. (B)**
There are $7 + 5 + 4 = 16$ balls in the box. The probability of drawing one red ball is

$$P(R) = \frac{\text{number of possible ways of drawing a red ball}}{\text{number of ways of drawing any ball}}$$

$P(R) = \frac{7}{16}$.

Similarly, the probability of drawing one black ball is

$$P(B) = \frac{\text{number of possible ways of drawing a black ball}}{\text{number of ways of drawing any ball}}$$

Thus,
$$P(B) = \frac{4}{16} = \frac{1}{4}$$

48. (C)
We start this problem by constructing a set of all possible outcomes:

We can have heads on all 3 tosses:         (HHH)
heads on the first 2 tosses, tails on the third:   (HHT)   (1)
heads on the first toss, tails on the next two:   (HTT)
                                           (HTH)   (2)
              ·                           (THH)   (3)
              ·                           (THT)
              ·                           (TTH)
                                           (TTT)

Hence there are eight possible outcomes (2 possibilities on first toss × 2 on second × 2 on third = 2 × 2 × 2 = 8).

We assume that these outcomes are all equally likely and assign the probability 1/8 to each. Now we look for the set of outcomes that produce 2 heads and 1 tail. We see there are 3 such outcomes out of the 8 possibilities (numbered (1), (2), (3) in our listing). Hence the probability of 2 heads and 1 tail is 3/8.

49. (A)
$$\int_{-\pi}^{+\pi} \sin x \cos x \, dx = \frac{1}{2} \int_{-\pi}^{\pi} 2 \sin x \cos x \, dx$$

$$= \frac{1}{2} \int_{-\pi}^{\pi} \sin 2x \, dx$$

$$= \frac{1}{4} [-\cos 2x]_{-\pi}^{\pi}$$

$$= \frac{1}{4} [-\cos 2\pi + \cos(-2\pi)]$$

$$= \frac{1}{4} [-1 + 1]$$

$$= 0$$

or
$$\int_{-\pi}^{\pi} \sin x \cos x \, dx = \int_{-\pi}^{\pi} \sin x \, d(\sin x)$$

$$= \frac{\sin^2 x}{2} \Big|_{-\pi}^{\pi}$$

$$= \tfrac{1}{2}(\sin^2 \pi - \sin^2(-\pi)) = \tfrac{1}{2}(0 - 0) = 0$$

50. (C)

51. (C)

For $\pi < x < \frac{3\pi}{2}$, x lies in 3rd quadrant.

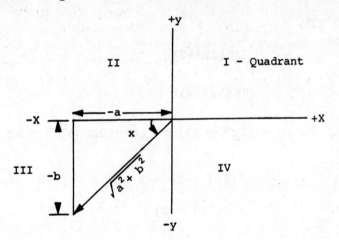

Suppose $X = -a$ and $Y = -b$ then,

$$\sin x = -\frac{b}{\sqrt{a^2 + b^2}} \quad \text{(a negative value)}$$

$$\cos x = -\frac{a}{\sqrt{a^2 + b^2}} \quad \text{(a negative value)}$$

52.  (A)

The mode is a measure of central tendency, and if all the observations are available, the mode is defined to be the most frequently occurring observation.

If the data is grouped into classes, the modal class will be the class with the most observations in it. The mode can be arbitrarily chosen to be the midpoint of the modal class.

From our data, the modal class is 45 - 49. This class has a frequency of 38 which is higher than that of any other class.

The midpoint of the modal class is

$$\frac{49 + 45}{2} = 47.$$

Thus the mode is 47.

53.  (D)
Here,
$$y = e^{2\ln(3x^2+2)}$$
$$= e^{\ln(3x^2+2)^2}$$

or $\quad y = (3x^2 + 2)^2 \quad$ (because $e^{\ln x} = x$)

$\therefore \quad \dfrac{dy}{dx} = 2(3x^2 + 2) \cdot 3(2x)$

$\qquad\quad = 12x(3x^2 + 2)$

54.  (B)
From integration by parts, if u and v are functions of x, then

$$\int u\,dv = uv - \int v\,du$$

$\therefore$ if $u = f(x)$, $v = g(x)$, then

$\quad du = f'(x)dx \quad$ and $\quad dv = g'(x)dx.$

$\therefore \displaystyle\int f(x)g'(x)dx = f(x)g(x) - \int g(x)f'(x)dx$

or interchanging the signs,

$$\int g(x)f'(x)dx = f(x)g(x) - \int f(x)g'(x)dx$$

55. **(B)**

$$\int_0^{\pi/2} x \sin x \, dx = -x \cos x \Big]_0^{\pi/2} + \int_0^{\pi/2} \cos x \, dx$$

[integration by parts]

$$= -\left[\frac{\pi}{2} \cos \frac{\pi}{2} - 0\right] + \sin x \Big]_0^{\pi/2}$$

$$= -[0] + \left(\sin \frac{\pi}{2} - 0\right)$$

$$= 1$$

56. **(C)**

$$\frac{d}{dx}[G(x)F(x)] = G'(x)F(x) + F'(x)G(x)$$

$$d[G(x)F(x)] = F(x)G'(x)dx + G(x)F'(x)dx$$

or

$$\int d[G(x)F(x)] = \int [F(x)G'(x)dx + G(x)F'(x)dx]$$

or

$$\int [F(x)G'(x)dx + G(x)F'(x)dx] = G(x)F(x) + C$$

57. **(B)**

Let $x^n = v$  $G(x^n) = u$  or  $u = G(v)$

$\therefore$ $y = F(u)$

By the chain rule

$$\frac{dy}{dx} = \frac{dy}{du} \cdot \frac{du}{dv} \cdot \frac{dv}{dx}$$

$$= F'(u) \cdot G'(v) \cdot nx^{n-1}$$

$$= F'[G(x^n)]G'(x^n)nx^{n-1}$$

$$= nx^{n-1}F'[G(x^n)]G'(x^n)$$

58. (E)

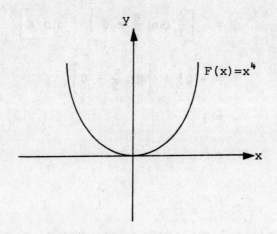

To determine maxima and minima we find $F'(x)$, set it equal to 0, and solve for $x$ to obtain the critical points. We find: $F'(x) = 4x^3 = 0$, therefore $x = 0$ is the critical value. We must now determine whether $x = 0$ is a maximum or minimum value. In this example the Second Derivative Test fails because $F''(x) = 12x^2$ and $F''(0) = 0$. We must therefore, use the First Derivative Test. We examine $F'(x)$ when $x < 0$ and when $x > 0$. We find that for $x < 0$, $F'(x)$ is negative, and for $x > 0$, $F'(x)$ is positive. Therefore there is a minimum at $(0,0)$. (See figure)

59. (C)

In this example we make use of the chain rule for differentiation. We find:

$$\frac{dy}{dx} = \frac{1}{\cos e^{2x}}(-\sin e^{2x})(e^{2x})(2) = -2e^{2x}\tan e^{2x}.$$

**60. (C)**

$$\int_{-\pi}^{+\pi} (\cos x + i\sin x)(\cos x - i\sin x)\,dx$$

$$= \int_{-\pi}^{\pi} (\cos^2 x - i^2 \sin^2 x)\,dx = \int_{-\pi}^{\pi} (\cos^2 x + \sin^2 x)\,dx \quad [i^2 = -1]$$

$$= \int_{-\pi}^{\pi} 1\,dx = x\Big]_{-\pi}^{\pi} = \pi - (-\pi) = 2\pi$$

**61. (C)**

If $a_0, a_1, a_2, a_3, \ldots, a_n$ is the arithmetic sequence, then the nth term is

$$a_n = a_0 + (n-1)d$$

where $\quad$ n = number of terms
$\qquad\quad$ d = common difference between each term.

For the sequence 3, 7, 11, 15,...

$$a_0 = 3, \quad d = 4$$

$$\therefore\ a_{100} = 3 + (100-1)(4)$$

$$= 3 + (99)(4)$$

$$= 3 + 396$$

$$\therefore\ a_{100} = 399$$

**62. (A)**

An examination of the numerator and denominator yields:

$$\lim_{x \to 0^+} \ln x = -\infty \quad \text{and} \quad \lim_{x \to 0^+} (1/x) = +\infty.$$

The ratio takes the indeterminate form $\infty/\infty$, and therefore L'Hopital's rule may be applied.

$$\lim_{x \to 0^+} \frac{\ln x}{\frac{1}{x}} = \lim_{x \to 0^+} \frac{\frac{1}{x}}{-\frac{1}{x^2}} = \lim_{x \to 0^+} (-x) = 0.$$

**63. (A)**

The elements of a matrix A are often functions of a dummy variable, say t. When it is necessary to exhibit the functional dependence of A on t, the matrix is written $A(t)$ and the entry in the $i^{th}$ row in the $j^{th}$ column is $a_{ij}(t)$. Now $A(t)$ is said to be differentiable at $t_o$ if each of the entries $a_{ij}(t)$ is differentiable at $t_o$. The derivative of A with respect to t is then

$$\frac{dA}{dt} = A'(t_o) = (a_{ij}'(t_o)).$$

Now we find the derivative of the given matrix $A(t)$:

$$a_{11} = t^2 \text{ then, } a_{11}' = \frac{da_{11}}{dt} = 2t$$

$$a_{12} = \cos t, \quad a_{12}' = -\sin t$$

$$a_{21} = e^t \quad\quad a_{21}' = e^t$$

$$a_{22} = \sin t \quad a_{22}' = \cos t.$$

Hence,

$$\frac{dA}{dt} = A'(t) = \begin{pmatrix} 2t & -\sin t \\ e^t & \cos t \end{pmatrix}$$

**64. (A)**

Starting from the fact that

$$e^a = 1 + a + \frac{a^2}{2!} + \frac{a^3}{3!} + \frac{a^4}{4!} + \cdots$$

for $-\infty < a < \infty$,

and setting $a = -x^2$ gives

$$e^{-x^2} = 1 - x^2 + \frac{x^4}{2!} - \frac{x^6}{3!} + \frac{x^8}{4!} - \cdots$$

for $-\infty < x < \infty$.

Therefore

$$1 - e^{-x^2} = x^2 - \frac{x^4}{2!} + \frac{x^6}{3!} - \frac{x^8}{4!} + \ldots$$

which means that

$$\frac{1 - e^{-x^2}}{x^2} = 1 - \frac{x^2}{2!} + \frac{x^4}{3!} - \frac{x^6}{4!} + \ldots$$

Now this series converges for all x and in addition converges uniformly for $0 \leq x \leq 1$.

Therefore

$$\frac{1 - e^{-x^2}}{x^2} = \sum_{n=1}^{\infty} \frac{(-1)^{n+1} x^{2n-2}}{n!}$$

65. (C)

Since $g(x) = \int_0^x f(t)dt$, $g(0)$ is $\int_0^0 f(t)dt$. When the limits of an integral are equal, the area that is computed is nominally zero. Thus $\int_0^0 f(t)dt = 0$.

66. (D)

$g(x) = \int_0^x f(t)dt$ defines the area bounded by f at a distance of x along the t-axis. Thus g(4) is the area bounded by f at a distance of 4 along the t-axis. Looking at the graph of f, we can see that the area in question is negative. This eliminates the first two choices.

Since $\int_0^4 f(t)dt$ is the area bounded by f, we see that this area cannot surpass -12 (in magnitude) because the minimum value of f in the interval $0 \leq t \leq 4$ is -3. Thus, the area bounded by f is contained in the rectangular area of -12.

We also note that an area of -5 is contained within the area bounded by f. Therefore, g(4) is a number between -5 and -12.

67. (C)
This problem can be solved using the very accurate Simpson's Rule approximation of the definite integral:

$$\int_a^b f(x)dx = \frac{h}{3}(f(x_0) + 4f(x_1) + 2f(x_2) + \ldots + 2f(x_{n-2}) + 4f(x_{n-1}) + f(x_n)),$$

where h is the length of each subinterval.

Hence, since g(12) is equal to $\int_0^{12} f(t)dt$:

$$g(12) = \frac{1}{3}(-1-8-6-8+0+8+6+8+2+4+2+0-1) = 2,$$

which is between -4 and 4.

68. (E)
Since g(x) is the area bounded by f a distance of x along the abscissa, g is positive only where the positive area exceeds the "negative area"; that is, when a greater area of f lies above the abscissa than below it.

For example, g is negative for all values of x between 0 and 4 because the bounded area lies completely below the abscissa.

Now, looking at the choices, the only interval given where g is positive for all values of x would be the interval from 9 to 12. This is so because g(9) and g(12) are both positive and the function g never becomes negative. At g(9) the positive area has cancelled out the negative area, and the negative area on the interval $11 \leq t \leq 12$ is cancelled by the positive area on the interval $9 \leq t \leq 11$.

**69. (B)**
When a function is increasing on an interval, its first derivative must be positive $\left(\frac{dg}{dt} > 0\right)$. Taking the derivative of g(x):

$$\frac{d}{dx} \int_0^x f(t)\,dt = f(x), \text{ by the fundamental theorem of calculus.}$$

Thus g is an increasing function when f is positive. Looking at the graph of f we see that f is always positive on the interval from 7 to 11.

**70. (D)**
When a function is decreasing on an interval, its first derivative is negative on that interval. In the previous problem, the derivative of g was found to be f. Thus, whenever $\frac{dg}{dx}$ (or f) is negative, then g is decreasing; thus, on the graph we must look for intervals where f is strictly negative. Of the choices offered, the only one that satisfies this criterion is from 0 to 4.

**71. (B)**
Particle Q makes a complete traversal of its circular path in 4 sec. Thus, its average angular velocity over this period is

$$\frac{\Delta \theta}{\Delta t} = \frac{2\pi}{4s} \text{ rad} = \frac{\pi}{2} \text{ rad/s}$$

**72. (A)**
The following equations can be set up:

(1)  R + r = 7    R = large radius
                  r = small radius

(2)  $R^2 + r^2 = 25$

The first equation is for the maximum distance between P and Q at time t = 0, 2, 4, ... sec., when the angle between the radii drawn to them is 180°.

The second equation represents the distance between the particles at time t = 1,3,5,... sec. when the angle between the radii is 90. The Pythagorean Theorem is applied as follows:

Since $r = 7 - R$, $R^2 + r^2 = R^2 + (7 - R)^2 = 25$
(Distance = 5 at t = 1, 3, 5, ...)
$\Rightarrow R^2 + 49 - 14R + R^2 = 25$
$\Rightarrow 2R^2 - 14R + 24 = 0 \qquad \Rightarrow R^2 - 7R + 12 = 0$
$\qquad\qquad\qquad\qquad\qquad\quad \Rightarrow (R - 3)(R - 4) = 0$
$\Rightarrow R = 3$ or $4$

### 73. (A)
In the previous problem, we solved for the radius of the larger circle. To get the radius of the smaller circle we use the equation:

$$R + r = 7$$

$$r = 7 - R = 7 - 4 = 3$$

Thus, the areas can now be compared:

$$\frac{A_1}{A_2} = \frac{\pi R^2}{\pi r^2} = \frac{\pi 4^2}{\pi 3^2} = \frac{16}{9}$$

### 74. (B)
P and Q reside on the horizontal plane at time t = 0. They both turn through 360° in 4 seconds. So their angular velocity, on the average, is $\frac{\pi}{2}$ rad/s. Thus, on each traversal of their respective paths, the particles lie in the x-y plane simultaneously twice. Thus in 8 seconds there are two traversals and, with the starting point, they are in the x-y plane ((2)(2) + 1 = t times.) As a matter of fact, the maxima of the curve of the function correspond to the instances at which they both locate on x-axis.

### 75. (A)
When a particle moves on a circular path, its transverse component of velocity is described by the following equation:

$$V_T = r \frac{d\theta}{dt}$$

For the larger circle: $V_{T_1} = R \frac{d\theta_1}{dt}$

For the smaller circle: $V_{T_2} = r \frac{d\theta_2}{dt}$

From the previous problem, we see that $\frac{d\theta_1}{dt} = \frac{d\theta_2}{dt}$. Thus the ratio of $\frac{V_{T_1}}{V_{T_2}}$ is just the ratio of the respective radii of the larger to the smaller circle. Hence, $\frac{R}{r} = \frac{4}{3}$.

76. (C)
When the graph changes concavity from concave up to concave down, the speed of the rate of recession changes from positive to negative. That is, the second derivative changes from positive to negative. According to the graph, the instances are those at which there is a point of inflection and f"(t) changes sign from positive to negative. There are four such points in the interval [0,8].

77. (A)
The function g is defined as the integral of the distance function whose upper limit is a running variable. That is g is the area bounded by f over the interval up to the upper limit of the integral. Examining the function f, we see that since f is always positive, the area that it bounds is always increasing on the interval. Thus, from this, we see that g passes the first criterion. At the same time, g does not pass the third criterion. Since g is increasing on the interval, it is a monotonic function.

Since g is always increasing on the interval given, then it is not periodic on that interval. Thus I is the only true statement.

78. (E)
Since g(x) is the integral of f, its derivative is just the distance function f. The minimum value that f reaches is

the minimum value of g(x). By inspection of the graph of the distance function, we can plainly see that the minimum is 5.

## 79. (B)
The average value of an arbitrary periodic function f is defined by the formula:

$$F = \frac{1}{T} \int_0^T f(t)dt, \text{ where T is the period of f.}$$

Thus, to get the mean value of g'(x) we have:

$$g'(x) = \frac{1}{T} \int_0^T g'(x)dx$$

We recognize g'(x) as being f(x). Hence:

$$f(x) = \frac{1}{T} \int_0^T f(x)dx$$

The integral need not be explicitly evaluated, since we see from the graph of f that the area above the line y = 5 is equal to the area below it. Therefore these portions will cancel each other and the average value is 5.

## 80. (A)
h is defined as being the slope of the tangent to the function f at each point in the interval. Thus we must examine the tangent lines that can be drawn to the curve. We see that the slope is positive on some intervals and negative on others. Thus h cannot be a strictly increasing or decreasing function. It is, however, monotonic on some subintervals.

The mean value of h is zero, which is not a positive number.

The function f, however, is periodic. The slope is at the same value at t as it is at t + 2. For a periodic function, we have:

$$f(x) = f(x + T)$$

where T is the period. Taking the derivative from both sides, and using the chain rule, we will have:

$$f'(x) = f'(x + T) \frac{d}{dx}(x + T) = f'(x + T)$$

Therefore the derivative of a periodic function is periodic.

81. **(B)**
The distance of P from Q is given:

$$s(t) = t^3 - \frac{15}{2}t^2 + 12t + 10$$

Since Q is fixed, we can determine the absolute velocity of P:

$$v = \frac{ds}{dt} = 3t^2 - 15t + 12$$

Setting this equal to zero:

$$3t^2 - 15t + 12 = 0,$$

$$t^2 - 5t + 4 = 0$$

$$(t - 4)(t - 1) = 0 \Rightarrow t = 4 \text{ or } 1$$

Setting t = 4 and substituting:

$$s(4) = 4^3 - \frac{15}{2}(4^2) + 12(4) + 10$$

$$= 2.$$

82. **(D)**
For particle p to be moving backwards, its velocity must be negative. We obtain the equation for velocity of p as such:

$$v = \frac{ds}{dt} = 3t^2 - 15t + 12$$

$$= 3(t - 4)(t - 1)$$

For v to be negative, (t - 4) and (t - 1) must have opposite signs.

By inspection, we see that for t > 4 and t < 1, the factors have the same sign and v > 0.

Between t = 1 and t = 4, the factors have opposite signs and v < 0.

Thus, the correct interval is 1 < t < 4.

**83. (D)**
When particle P is decelerating, its acceleration is negative. Thus, we must find the interval where

$$\frac{d^2s}{dt^2} < 0.$$

Hence: $\frac{d^2s}{dt^2} = 6t - 15.$

We solve the inequality:

$$6t - 15 < 0,$$

$$t < \frac{15}{6},$$

$$t < \frac{5}{2} = 2.5$$

Thus we look for the interval given in the choices which fits the above constraint. This interval is 0 < t < 2.

**84. (C)**
The average velocity on an interval from a to b is:

$$\frac{1}{b-a} \int_a^b v(t)dt.$$

The velocity is

$$v(t) = s'(t) = 3t^2 - 15t + 12,$$

the interval is b - a = 4 - 1 = 3.

Substituting:

$$\frac{1}{3} \int_1^4 (3t^2 - 15t + 12)dt$$

$$= \frac{1}{3}\left[ t^3 - \frac{15t^2}{2} + 12t \Big|_1^4 \right]$$

$$= \frac{1}{3}\left[ (-8) - \frac{11}{2} \right] = \frac{-27}{6}$$

**85. (A)**

The average acceleration of a particle over an interval [a,b] is:

$$\frac{1}{b-a} \int_a^b a(t)\,dt$$

$$= \frac{1}{2} \int_0^2 a(t)\,dt$$

$$= \frac{1}{2} \int_0^2 (6t - 15)\,dt = \frac{1}{2}\left[ 3t^2 - 15t \Big|_0^2 \right]$$

$$= \frac{1}{2}[-18] = -9.$$

**86. (D)**

The free-body diagram of the box is shown in the figure. It accounts for all forces acting on the box. Since only F is required, it is sufficient to consider only the x-direction. It is given that the box is in equilibrium, thus the summation of all the forces in the x-direction must be zero.

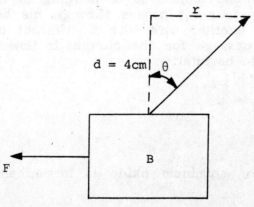

$$-F + (50N)(\sin \theta) = 0 \qquad (1)$$

From trigonometry,

$$\sin \theta = \frac{r}{\sqrt{4^2 + r^2}}$$

Substituting for $\sin \theta$ in equation (1) gives

$$F = (50N) \frac{r}{\sqrt{4^2 + r^2}}$$

But

$$r = 3 \text{cm}$$

so $F = 50 \left(\frac{3}{5}\right) = 30$

87. (D)
A person can get an electric shock if there is a current passing through his body. Currents as low as 0.05 amperes can be fatal, while a current of 0.10 amperes in the heart is usually always fatal. The amount of current passing through the body is limited by the body's resistance which can vary from 100 to 500,000 ohms according to circumstances. The resistance of a man touching two electrodes with dry fingers is about 100,000 ohms and that with both hands in salt water is 700 ohms.

In order to receive a shock, there must be a difference in electric potential between two parts of your body, since current can pass only when there is a difference of potential between two points.

For our man, who is hanging on the high voltage line, there is negligible difference of electric potential between any part of his body since he is hanging in the air. So he is safe as no current passes through his body. But if he grabs on to another wire with a different potential, it would create a passage for the current to flow through his body and it would be fatal.

88. (D)
Since 2.06g of aluminum oxide is formed and 1.09g Al

is consumed in the process then the amount of oxygen consumed is $(2.06 - 1.09)g = 0.97g$.

Converting these weights to moles:

$$1.09g\ Al \times \frac{1\ mol\ Al}{27g\ Al} = 0.04\ mol\ Al.$$

$$0.97g\ O \times \frac{1\ mol\ O}{16g\ O} = 0.06\ mol\ O.$$

Therefore the relative number of moles of atoms of these elements are $Al_{0.04}\ O_{0.06}$

Dividing the subscripts by 0.04

$$Al\ O_{1.5}$$

Multiplying subscripts by 2 yields the empirical formula. Thus, the empirical formula for aluminum oxide is $Al_2O_3$.

89. (C)
An ideal transformer is defined as one which has:
(a) zero winding resistance
(b) no losses
(c) no leakage flux. All the flux produced by a current in either winding links the other winding.
(d) reluctance of the magnetic core equal to zero.

The transformer operates on the principle of mutual induction. A transformer does not operate without induction.

90. (B)
Average velocity is defined as: $v \equiv \frac{q}{A}$

Therefore, $q = vA$

or $A = \frac{q}{v}$

$$A = (0.010\ ft^3/sec)(144\ in^2/ft^2)\left(\frac{1}{2.88}\ ft/s\right)$$

$$A = \frac{1.44}{2.88} = .5\ in^2$$

which is (B).

91. (B)
From Bernoulli's equation

$$P + \rho gh + \tfrac{1}{2}\rho v^2 = \text{constant}$$

so
$$\tfrac{1}{2}\rho v^2 \; \alpha \; P + \rho gh$$

$$v^2 \; \alpha \; P + h$$

(flow rate is proportional to pressure and height)

so doubling both increases velocity but not linearly.

92. (A)
We know that the dual relations

$$v = L\frac{di}{dt} \quad (1)$$

$$i = C\frac{dv}{dt} \quad (2)$$

define the parameters for the unknown element in (a) $v = v_{ab}$. We notice that $\frac{dv}{dt} = 3\cos 3t$ but $\frac{di}{dt} = -3\sin 3t$.
Since both sides of equation (2) remain positive, the unknown element in part (a) must be a capacitor.

Since $i = C\frac{dv}{dt} = 3c[\cos 3t]$, but given $i = \cos 3t$, the capacitor has a value of 1/3 F.

93. (A)
Efficiency = $\frac{\text{output}}{\text{input}}$

1 horsepower = 746 watts.

Efficiency = $\frac{10 \times 746}{10{,}000}$ = .746 or 74.6%

94. (D)
When the mouth organ is blown harder, the amplitude

of the sound wave increases. This is analogous to pushing a swing. When the swing is given a harder push, it rises higher (i.e. its amplitude increases).

95. (A)
If the work done by friction, $W_f$, is converted to heat, the energy equation would be:

$$\tfrac{1}{2}mv_1^2 + mgh_1 + W_F = \tfrac{1}{2}mv_2^2 + mgh_2$$

$$W_F = mg(h_2 - h_1) + \tfrac{1}{2}mv_2^2 - \tfrac{1}{2}mv_1^2$$

$$W_f = 1.0 \text{ kg}(9.8 \text{ m/s}^2)(0.0\text{m} - 2.0\text{m}) + \tfrac{1}{2}(1 \text{ kg})(4 \text{ m/s})$$

$$W_f = -19.6\text{J} + 8.0\text{J}$$

$$W_f = -11.6\text{J}$$

Assuming that all the frictional work is converted to heat,

$$Q = W_f = -11.6\text{J}\left(\frac{\frac{1}{4.1858}\text{ cal}}{1.0\text{J}}\right) = \frac{-11.6}{4.1858}\text{ cal}$$

96. (E)
We have to first find the velocity of the man on striking the water. In falling from a height h, the initial potential energy of the weight, mgh, is converted into kinetic energy, $\tfrac{1}{2}mv_0^2$.

So $\tfrac{1}{2}mv_0^2 = mgh$.

Hence $v_0 = \sqrt{2gh}$

$$= \sqrt{2 \times g \times 12.5}$$

$$= 5\sqrt{g} \text{ m/sec}$$

Therefore the man's momentum on striking the water is

$$p_1 = mv_0 = (60)(5\sqrt{g})$$

$$= 300\sqrt{g} \text{ kg m/sec}.$$

The final momentum being zero, the average force exerted is:

$$\vec{F} = \frac{\Delta p}{\Delta t} = \frac{p_2 - p_1}{\Delta t}$$

$$= \frac{0 - 300\sqrt{g}}{-5}$$

$$= -600\sqrt{g} \text{ N},$$

where the negative sign indicates that the retarding force of the water was opposite to the downward velocity of the man.

### 97. (A)

Many chemical reactions involve transfer of electric charge from one atom to another. Because this process is so common and important, words which describe these changes are in common use in the field of chemistry. Oxidation refers to the loss of electrons (so the correct answer is A) while the term 'reduction' refers to the gaining of electrons. The element which supplies the electrons is called the reducing agent, while the element which accepts the electron and permits oxidation to take place is the oxidizing agent.

### 98. (E)

The force diagram for the small body is shown below:

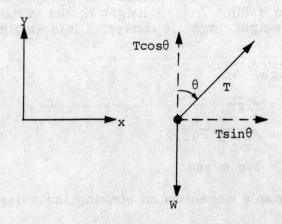

$$m = \frac{3N}{9.8 \text{ m/s}^2}$$

$$\Sigma F_y = 0$$

$$T\cos\theta - mg = 0$$
$$T\cos\theta = 3N$$
$$\therefore \quad T\sin\theta = 4N$$
$$\Sigma F_x = ma$$
$$T\sin\theta = ma$$
$$a = \frac{4}{3}(9.8)\frac{m}{s^2}$$

Therefore the acceleration of the cart is the acceleration of the small body in the horizontal direction.

## 99. (D)

Since the average translational kinetic energy per molecule is 3/2 RT, then the total kinetic energy for N molecules is 3/2 nRT. The internal energy which is dependent on the energy of the molecules is proportional to the temperature.

$$\therefore \quad \Delta U = \frac{3}{2} nR\Delta T$$

From a macroscopic point of view, the change in the internal energy of the system is dependent on the work done on the system and the heat added to the system.

$$\Delta U = Q - W$$

For a constant volume process $p\Delta V = 0$.

$$\therefore \quad W = 0$$
$$\Delta U = nC_v \Delta T$$
$$\frac{3}{2} nR\Delta T = nC_v \Delta T$$
$$\therefore \quad C_v = \frac{3}{2} R$$

## 100. (C)

The force, $mg\sin\theta$, directed along the tangent at B is responsible for the acceleration along the arc OB. The tension P has no component in this direction.

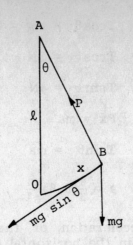

Therefore $-mg \sin \theta = ma$ where $a$ is the acceleration along the arc OB.

Note, the reason for the minus sign is the force $mg\sin\theta$ is directed toward O while the displacement $x$ is measured along the arc from O.

Since $\theta$ is small, $\sin\theta \approx \theta$ so that $-mg\theta = ma$

$$a = -g\theta$$

Since $\ell\theta = x$

$$a = -g\left(\frac{x}{\ell}\right)$$

101. (C)
Since the motion is simple harmonic

$$a = -\omega^2 x$$

However, from question (100)

$$a = \frac{-g}{\ell} x$$

so that

$$-\omega^2 x = -\frac{g}{\ell} x$$

and

$$g = \ell\omega^2 \qquad (1)$$

By definition

$$T = \frac{2\pi}{\omega}$$

476

Therefore

$$T^2 = \frac{4\pi^2}{\omega^2}$$

$$\omega^2 = 4\pi^2 \left[\frac{1}{T^2}\right]$$

Substituting for $\omega^2$ in equation (1)

$$g = 4\pi^2 \left[\frac{\ell}{T^2}\right]$$

102. (B)
It is stated in the problem that the flow in both tubes are dynamically similar. Thus, the Reynold's number for the two flows is 1,000.

$$R_{e_1} = \frac{V_1 D_1}{\nu} = R_{e_2} = \frac{V_2 D_2}{\nu}$$

($\nu$ is constant because the fluids and temperatures are the same in both situations.)

$$R_e = 1,000$$

so
$$\frac{V_2 D_2}{\nu} = 1000$$

$$V_2 D_2 = 1000\nu$$

$$V_2 (2R_2) = 1000\nu$$

$$V_2 R_2 = 500\nu$$

103. (D)
This problem is solved using the equation $s = vt$. If there were no current, the man would be able to reach the opposite side of the river in a time, t, given by:

$$t = \frac{s}{v} = \frac{1 \text{ mile}}{2 \text{ miles hr}^{-1}} = \tfrac{1}{2} \text{ hour}$$

In this time t, the current of velocity 4 miles $hr^{-1}$ displaces the boat in a perpendicular direction downstream a distance s, given by:

$$s = vt = (4 \text{ miles hour}^{-1})(\tfrac{1}{2} \text{ hour}) = 2 \text{ miles}$$

**104. (E)**
With an increase in temperature the acceleration of the electrons by the applied electric field is reduced by collisions to a drift velocity. This reduces the mean free path of the electrons, the distance traveled between collisions.

**105. (C)**
Since $\left(\frac{\partial V}{\partial T}\right)_P^2$ and T must always be positive and $\left(\frac{\partial P}{\partial V}\right)_T$ is always negative, $C_p$ minus $C_v$ is always positive. So (C) is the correct answer.

**106. (B)**
The project engineering segment of an economic feasibility study contains all the economic and technical aspects of the production process for the project. The report will comment on the effect of technology, size, start-up, and major equiment. The estimated demand for the final product belongs in the market study.

**107. (D)**
Consider the body at position A. The normal component of the acceleration is provided by the net force in the normal direction, so that

$$F_N = T - W\cos\theta = m\frac{v^2}{r} \quad (1)$$

and

$$F_T = W\sin\theta \quad (2)$$

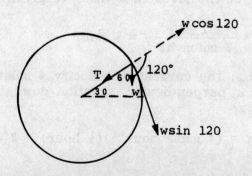

From equation (1)

$$T = \frac{mv^2}{r} + W\cos\theta = \frac{mv^2}{r} + mg\cos\theta$$

Thus
$$T = m\left(\frac{v^2}{r} + g\cos\theta\right)$$

At position B, $\theta = 120°$, so that $T = m(\frac{v^2}{r} + g\cos(120°))$

$$T = m\left(\frac{v^2}{r} - \frac{g}{2}\right)$$

108. (E)

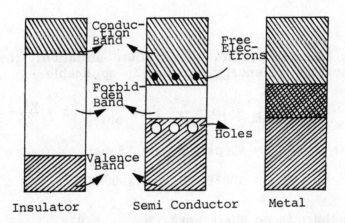

The forbidden band of a semiconductor is considerably less than that of an insulator. Hence, the valence band remains full, the conduction band empty, and these materials are insulators at low temperatures. As the temperature is increased, some of these valence electrons acquire thermal energy greater than the value associated with the forbidden band, and hence, move into the conduction band. Hence, the low temperature insulator has now become slightly conducting and this conductivity increases rapidly with rising temperature. This type of conduction is possible in pure semiconductors. It is also possible to have conduction from energy states introduced in between the forbidden band by doping the semiconductor with impurities. This type of conduction is possible in doped or impure semiconductors.

## 109. (A)

It was necessary to assume uniform pressure in the direction perpendicular to the flow because the model of the flow dictated it; the model is that the fluid is composed of infinite plates sliding on top of one another.

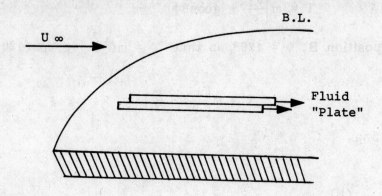

## 110. (C)

For a steady-state steady-flow situation, the following conservation of energy equation is applicable:

$$q + W_{shaft} = \left[h + \frac{v^2}{2} + gz\right]_{out} - \left[h + \frac{v^2}{2} + gz\right]_{in}$$

Given:  $P_{in} = 80$ psi, $P_{out} = 75$ psi

$T_{in} = 100°F$, $T_{out} = 2000°F$

Since there is no shaft work, $W_{shaft} = 0$.

Also, since there are negligible changes in the kinetic and potential energy between the inlet and the exit,

$$\left[\frac{v^2}{2} + gz\right]_{out} \approx \left[\frac{v^2}{2} + gz\right]_{in}$$

Therefore, the energy equation becomes:

$$q = h_{out} - h_{in}$$

$$q = \Delta h$$

For an ideal gas $= \Delta h = \int C_p dT$ where $C_p = f(T)$.

For a perfect gas $= \Delta h = C_p \Delta T$ where $C_p = $ constant.

At this point it is worthwhile to note the difference between an ideal gas assumption and a perfect gas assumption.

Recall from the equation of state ($P = \rho RT$), that any two states define a third state perfectly. Since $h = U + Pv$, $h$ is also a state function. Therefore it too can be defined by any two state variables such as $T$ and $v$.

Therefore, $h = f(T,P)$.

Therefore,
$$dh = \left(\frac{\partial h}{\partial T}\right)_P dT + \left(\frac{\partial h}{\partial P}\right)_T dP$$

defining, $C_p \equiv \left(\frac{\partial h}{\partial T}\right)_P$

Therefore,
$$dh = C_p dT + \left(\frac{\partial h}{\partial P}\right)_T dp$$

A similar development can be done for $dU$.

The equation is
$$dU = C_v dT + \left(\frac{\partial u}{\partial v}\right)_T dv$$

It has been confirmed experimentally by Joule that $\left(\frac{\partial u}{\partial v}\right)_T$ is approximately zero at low pressure.

It also happens that the ideal equation of state also holds for low pressure of gas. Therefore, it can be said that

$$\left(\frac{\partial u}{\partial v}\right)_T = 0$$

for ideal gas assumption. It can thus be said that $du = C_v dT$ for ideal gas assumption.

From the above equation, $du = f(T)$. Therefore, it can be said that under ideal gas assumption, the internal energy is solely a function of temperature. Since $h = U + Pv$ and since $Pv = RT$, $h = U + RT$. Therefore in an ideal gas assumption, $h$ is also solely a function of $T$. Earlier we had just derived that $dh = C_p dT + \left(\frac{\partial h}{\partial P}\right)_T dp$. Now since we have shown $h$ is a function of $T$ only, $\left(\frac{\partial h}{\partial P}\right)_T = 0$.

Therefore $dh = C_p dT$. To sum it up, under an ideal gas assumption both $dh$ and $du$ are functions of $T$ only.

Likewise $C_p$ and $C_v$ are functions of T. In thermodynamics literature, one can easily find charts of $C_p$ and $C_v$ as functions of T for nitrogen. In which case, one would find $C_p$ at $T_1$, $C_p$ at $T_2$ and find $\overline{C}_p = \dfrac{C_{p_{T_1}} + C_{p_{T_2}}}{2}$. Using $\overline{C}_p$, one can then find $\Delta h = \overline{C}_p \Delta T$. Note that it is necessary to use $\overline{C}_p$ because $C_p$ is a function of temperature and because one usually cannot find the function that relates $C_p$ and T. The same things hold true for $C_v$ and $\Delta u$. However in a perfect gas assumption, $C_p$ and $C_v$ are considered to be constant. In essence, a perfect gas assumption is even more "ideal" than an ideal gas assumption.

In any case, the solution to the problem is $q = \Delta h = C_p \Delta T$ since

$$C_p = 3.5 \frac{\text{Btu}}{(\text{lb})(°F)}$$

$$q = \frac{3.5 \text{ Btu}}{(\text{lb})(°F)} (2000°F - 100°F)$$

$$q = (3.5)(1900) \frac{\text{Btu}}{\text{lb}} = 6650 \frac{\text{Btu}}{\text{lbm}}$$

111. (D)
Since the level is higher on the left, the liquid of $\rho_2$ is less dense than the liquid of $\rho_1$. The pressure at equal heights in the tube are equal for equal densities, but it is not necessary to know $\rho_2$ in order to solve $P_C - P_B = \rho_1 g(h)$.

112. (A)
At equilibrium, the weight of the iceberg is balanced by the buoyant force (B) of the displaced water.

Thus, weight of iceberg = B

But weight = specific weight ($\rho g$) × volume (v).

$\therefore \quad \rho_{ice} \times g \times V_{ice} = \rho_{water} \times g \times V_{water}$

where the volume of the submerged iceberg equals the volume of the water displaced.

Hence $\dfrac{V_{water}}{V_{ice}} = \dfrac{\rho_{ice}}{\rho_{water}} = \dfrac{0.92}{1.03} = 89\%$

Thus nearly 90% of the iceberg is below the seawater.

113. (C)
Charles' Law for the effect of temperature upon the volume of confined gases at constant pressure is stated as follows:

The volume of a given mass of gas is directly proportional to the temperature on the Kelvin scale when the pressure is held constant or $V \propto T$.

114. (A)
During the positive cycle of the input, the diode is forward biased and hence the signal appears at the output as it is. Also during this cycle the capacitor is charged up to the peak value of the input. During the negative part of the input the diode becomes reverse biased and the input signal is blocked. But during this cycle the capacitor gets discharged exponentially with a time constant RC = 100mF × 200 Ω = 20 sec. Since the input signal has a period of $\dfrac{1}{.25 \text{Hz}}$ = 4 sec, the capacitor does not get discharged fully. Note that 20 sec > 4 sec. Thus the figure in (A) is the more appropriate one. The figure in (D) shows that during the negative half of the cycle the capacitor has discharged considerably, implying RC ≈ 2 sec. But actually RC = 20 sec. Hence Figure (D) is ignored.

115. (E)
The correct representation of Newton's Law of Cooling is:

$$\dfrac{q}{A} = h \Delta T$$

where     q = rate of heat transfer
              A = surface area
              h = heat transfer coefficient
             $\Delta T$ = temperature difference

116. (D)

A magnetic moment, μ, is produced by the orbiting electron which interacts with the applied magnetic field, β. The energy v = -μ·β associated with this interaction splits the energy level and produces both a greater and lesser energy state depending upon the alignment of the magnetic moments. Therefore, an atom can emit more photons when a magnetic field is present due to the splitting of the energy levels.

117. (D)

In tempered martensite the carbide that forms is made up of extremely fine particles that are uniformly distributed. These fine particles will greatly interfere with dislocation movements.

118. (D)

For the case of uniform linear acceleration the pressure gradient $\nabla P = \bar{g} - \bar{a}_0$ as shown.

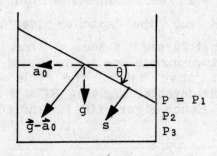

$$\frac{d\vec{P}}{ds} = \rho(\vec{g} - \vec{a}_0)$$

Therefore the lines of constant pressure are normal to the pressure gradient. Since the fluid element is at an angle θ from the surface the pressure is the same at B and C.

119. (D)

For the case where there is uniform linear acceleration, the pressure gradient is: $\nabla p = g - a_0$.

$$\therefore \frac{d\vec{p}}{ds} = \rho(\vec{g} - \vec{a}_0)$$

Therefore:
$$d\vec{p} = \rho(\vec{g} - \vec{a}_0)(ds)$$
$$|d\vec{p}| = dp = \rho(\sqrt{g^2 + a_0^2})ds$$

Also,
$$p = \rho(\sqrt{g^2 + a_0^2})s$$

Since
$$F = \int p\,dA = \int p\,ds = \int \rho(\sqrt{g^2 + a_0^2})s\,ds$$

Finally,
$$F = \rho(\sqrt{g^2 + a_0^2})\left(\frac{\ell^2}{2}\right)$$

### 120. (B)

Consider $u = f(T, v)$. Differentiating we get

$$du = \left(\frac{\partial u}{\partial T}\right)_V dT + \left(\frac{\partial u}{\partial v}\right)_T dv$$

Since,

$$\left(\frac{\partial u}{\partial T}\right)_V \equiv C_v$$

Therefore,

$$du = C_v dT + \left(\frac{\partial u}{\partial v}\right)_T dV$$

At constant V, $du = C_v dT$.

(For ideal gasses, internal energy is a function of temperature only and is independent of volume.)

### 121. (B)

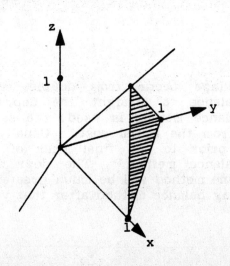

485

The plane (1 1 1) is shown above; it can be seen that the direction [1 1 1] is perpendicular to it. This is true in general for Miller indices. But Miller indices apply only to cubic lattice structures.

122. (B)
The definition of a black body is that it absorbs and emits all of its radiant energy.

123. (D)
Btu is the abbreviation for the British thermal unit. One Btu is the amount of heat needed to raise the temperature of one pound of water by one degree Fahrenheit.

Thus to raise 5 lb. of water from 102°F to 212°F, we need $5 \times (212 - 102)$ Btu of heat, which is 550 Btu.

124. (A)
To solve this problem, we must apply the continuity equation in the form $A_1 V_1 = A_2 V_2$ (p = const).

We are told that $A_1 = 3A_2$ and that $V_1 = 9$ m/s. Hence:

$$V_2 = \frac{A_1}{A_2} V_1$$

$$V_2 = \frac{3A_2(9)}{A_2} = 27 \text{ m/s}$$

125. (D)
All switchings occur from double-declining balance or declining balance to straight line depreciation. When a declining balance method is used, the salvage value is not subtracted from the initial cost. Usually the salvage value is reached prior to the final year of depreciation for a declining balance method. The depreciation charges for a straight-line method will be much greater than the charges for a declining balance method after that year.

126. (A)
With increasing time, creep, a slow strain, will lead to excessive distortions of the material. A slow displacement of the molecules will take place as a result of the stress. When the stress is removed, the molecules will rekink among their newly obtained neighbors, retaining a permanent displacement. Stress relaxation is the decay of stress by creep but with the displacement remaining constant. This stress relaxation is brought about by the movement of the molecules to maintain the displacement constant as the viscous stress increases with time.

127. (D)
For simple harmonic motion $T = 2\pi\sqrt{\frac{m}{k}}$ where m = mass, k = spring constant and T is independent of displacement. Only D makes sense.

128. (B)
For a binary system consisting of two metals forming a solid solution, the number of degrees of freedom is governed by $F = c - p + 1$; c = the number of components and p = the number of phases. During the freezing period, there are two components and two phases present and thus one degree of freedom.

$$F = c - p + 1 = 2 - 2 + 1 = 1$$

Therefore, the temperature does not remain constant but drops along line BC until the whole mass has completely solidified.

129. (B)
Oxygen, which has eight electrons, will fill up the orbitals as shown in (B). Since (A) is not in the ground state, (A) is not correct and (C), (D) and (E) are impossible because of violations of either Hund's rule (electrons will remain unpaired in an orbital if possible) or the Pauli exclusion principle (electrons with the same spin can't share an orbital).

130. (C)

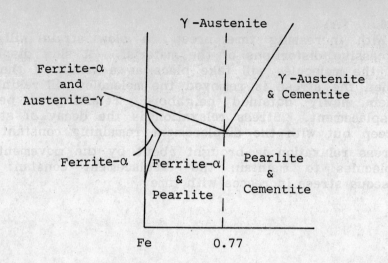

According to a portion of the carbon iron phase diagram, cooling at the eutectic composition occurs at .77% weight carbon. Ferrite or cementite will not precipitate in the austenite structure prior to reaching the eutectic temperature. Therefore all the austenite will be completely transformed into pearlite.

**131. (C)**
The implicit relationship $F(x,y) = 0$ has total derivative

$$\frac{\partial F(x,y)}{\partial x} dx + \frac{\partial F(x,y)}{\partial y} dy = 0$$

Solving for $\frac{dy}{dx}$ gives

$$\frac{dy}{dx} = \frac{-\partial F(x,y)}{\partial x}\left(\frac{\partial F(x,y)}{\partial y}\right)^{-1} = -\frac{F_x}{F_y}$$

**132. (B)**
From linear algebra, for $A_{n \times n}$ a triangular matrix

$$\det(A) = \prod_{i=1}^{n} a_{ii},$$

the product of the diagonal entries of A.

## 133. (C)

The total area of this curve covers each of the four quadrants. Therefore θ goes from 0 to $2\pi$, which are the limits of the integral that gives the required area. The formula for area in polar coordinates is:

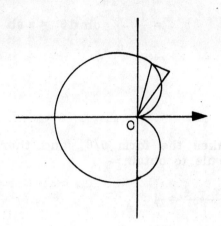

$$A = \int_{\alpha}^{\beta} \tfrac{1}{2}\rho^2 d\theta.$$

$$A = \int_{0}^{2\pi} \tfrac{1}{2}(a(1 - \cos\theta)^{\frac{1}{2}})^2 d\theta$$

$$= \frac{a^2}{2} \int_{0}^{2\pi} (1 - \cos\theta) d\theta = \frac{a^2}{2}[\theta - \sin\theta]_{0}^{2\pi}$$

$$= \frac{a^2}{2}(2\pi - 0 - 0 + 0) = \frac{a^2 \cdot 2\pi}{2} = \pi a^2.$$

## 134. (E)

We can make use of a result of Green's theorem which says that the area bounded by a simple closed curve C is

$$\tfrac{1}{2} \oint_{C} xdy - ydx$$

then introduce the parameter θ and let $x = a(\cos\theta)$, $y = b(\sin\theta)$, so

$$\text{Area} = \frac{1}{2} \oint_C x\,dy - y\,dx = \frac{1}{2}\int_0^{2\pi} ab(\cos^2\theta + \sin^2\theta)\,d\theta$$

$$= \frac{1}{2}\int_0^{2\pi} ab\,d\theta = \pi ab$$

**135. (C)**
The function takes the form $0/0$, and therefore we can apply L'Hopital's rule to obtain:

$$\lim_{x\to 2} \frac{4x-4}{1} = 4$$

**136. (A)**
$e^{\text{Ln}(x)} = x$ since $e^x$ and $\text{Ln}(x)$ are inverse functions.

So $\quad \frac{d}{dx}(e^{\text{Ln}(x)}) = \frac{d}{dx}(x) = 1$

**137. (A)**
The function takes the form $\infty/\infty$ which is indeterminate, but allows the application of L'Hopital's rule. Accordingly,

$$\lim_{x\to\infty} \frac{x^n}{e^x} = \lim_{x\to\infty} \frac{nx^{n-1}}{e^x}$$

It is seen that $\infty/\infty$ continues to be obtained after successive use of the rule. We note, however, that the $n^{\text{th}}$ derivative tends to $n!$ for the numerator. Therefore, since n is assumed finite,

$$\lim_{x\to\infty} \frac{n!}{e^x} = 0.$$

**138. (C)**
$x(t) = A\cos(\omega t - \phi)$ is oscillatory motion as any one

who might have studied differential equations can recall, it is the equation of motion of a Simple Harmonic Oscillator.

### 139. (B)
From basic analytical geometry, if two lines are perpendicular then their slopes are negative reciprocals, so the slope of line b must be $\frac{5}{2}$.

### 140. (E)
Since the first and last terms and the number of terms are known, the formula for the nth term, or last term of the series

$$\ell = a_1 + (n - 1)d$$

where   $a_1$ = first term of the series
        n = number of terms
        d = common difference
        $\ell$ = nth term, or last term

can be solved for d.

$\ell = a_1 + (n - 1)d$ with $a_1 = -4$, $n = 12$, $\ell = 32$ gives

$$32 = -4 + (12 - 1)d$$

$$36 = 11\,d$$

$$\frac{36}{11} = d$$

### 141. (A)
$y''(x) + \lambda^2 y(x) = 0$

or   $(D^2 + \lambda^2)y(x) = 0$   {where $D = \frac{d}{dx}$}

$\therefore$   $D^2 + \lambda^2 = 0 \Rightarrow D^2 = -\lambda^2$

$\therefore$   $D = \pm \lambda i$

$\therefore$   $y(x) = A\cos \lambda x + B\sin \lambda x$

$y(0) = 0 = A\cos\lambda(0) + B\sin\lambda(0) \Rightarrow 0 = A + 0$
$$\Rightarrow A = 0$$

$\therefore \quad y(x) = B\sin\lambda x$

Now $\quad y'(x) = B\lambda\cos\lambda x$

$\quad\quad y'(\pi) = B\lambda\cos\lambda\pi = 0$

Since, for a non-trivial solution $B \neq 0$

$$\cos\lambda\pi = 0 = \cos(n + \tfrac{1}{2})\pi$$

where $n = 0, 1, 2, 3, \ldots$ because we want $\lambda$ positive.

$\therefore \quad \lambda = (n + \tfrac{1}{2})$

Check:

We have $y(x) = B\sin(n + \tfrac{1}{2})x$

At $x = 0$, $\quad y(0) = B\sin(n + \tfrac{1}{2})(0) = 0$

(First condition is satisfied)

Now $\quad y'(x) = B(n + \tfrac{1}{2})\cos(n + \tfrac{1}{2})x$

$\therefore \quad y'(\pi) = B(n + \tfrac{1}{2})\cos(n + \tfrac{1}{2})(\pi) = 0$

(Second condition is satisfied)


142. (D)
From the ratio test,

$$\lim_{n\to\infty} \left|\frac{a_{n+1}}{a_n}\right| = \lim_{n\to\infty} \left|\frac{x^n}{(n+1)5^{n+1}} \cdot \frac{n5^n}{x^{n-1}}\right|$$

$$= \lim_{n\to\infty} \left|\frac{nx}{5(n+1)}\right| = \frac{|x|}{5} \lim_{n\to\infty} \frac{n}{n+1}$$

$$= \frac{|x|}{5}$$

Absolute convergence occurs if $\frac{|x|}{5} < 1$; that is, $x \in (-5, 5)$.

We cannot say anything for $x = -5$ or $x = 5$ without further analysis:

$$\sum_{n=1}^{\infty} \frac{5^{n-1}}{n5^n} = \frac{1}{5}\sum_{n=1}^{\infty}\frac{1}{n}, \text{ not convergent.}$$

For x = -5

$$\sum_{n=1}^{\infty} \frac{x^{n-1}}{n5^n} = \sum_{n=1}^{\infty} \frac{(-5)^{n-1}}{n5^n}$$

$$= \sum \frac{(-1)^{n-1} 5^{n-1}}{n5^n} = \sum \frac{(-1)^{n-1}}{n5}$$

$$= \frac{1}{5} \sum \frac{(-1)^{n-1}}{n}$$

and by alternating series test this converges, so that

$$\sum_{n=1}^{\infty} \frac{x^{n-1}}{n5^n} \text{ converges for } x \in [-5, 5)$$

143. (B)
If line segments $\overline{P_1 O}$ and $\overline{P_2 O}$ are perpendicular to each other, then the angle between them is 90° or 270°. But the cosine of 90° and 270° equals 0 and these are the only values (below 360°) for which cosine equals 0. Thus, a sufficient condition for two lines to be perpendicular is for the cosine of the angle formed between them to be 0. We know that

$$\cos P_1 O P_2 = \frac{x_1 x_2 + y_1 y_2 + z_1 z_2}{d_1 d_2}$$

which is equal to zero only when the numerator ($x_1 x_2 + y_1 y_2 + z_1 z_2$) equals zero.

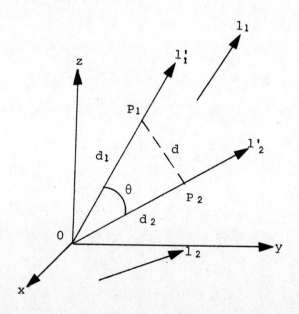

Therefore, given two segments with common endpoint at the origin O and the other endpoints being $P_1(x_1,y_1,z_1)$ and $P_2(x_2,y_2,z_2)$, a sufficient condition for line segments $\overline{P_1O}$ and $\overline{P_2O}$ to be perpendicular is:

$$x_1x_2 + y_1y_2 + z_1z_2 = 0.$$

144. (C)
If (1), then $f(r, \theta) = 0$ for the entire plane. The plane is symmetric with respect to the pole.

If (2), then $f(r, \theta) = 0$ must be symmetric with respect to the x-axis, but not the pole. $f(r, \theta) = r - \cos\theta$ is a counter-example.

If (3), then $f(r, \theta) = 0$ must be symmetric with respect to the y-axis, but not the pole. $f(r, \theta) = r - \sin\theta$ is a counter-example.

If (4) then if $f(r, \theta) = 0$, $f(r, \theta + \pi) = 0$ and $f(r, \theta) = 0$ is therefore symmetric with respect to the pole.

If (5), then if $f(r, \theta) = 0$, $f(-r, \theta) = 0 = f(r, \theta + \pi)$ and $f(r, \theta) = 0$ is therefore symmetric with respect to the pole.

# The Graduate Record Examination in
# ENGINEERING

## Test 5

# GRE ENGINEERING TEST 5
## ANSWER SHEET

| | | | | | | | | | | | |
|---|---|---|---|---|---|---|---|---|---|---|---|
| 1. | Ⓐ | Ⓑ | Ⓒ | Ⓓ | Ⓔ | 11. | Ⓐ | Ⓑ | Ⓒ | Ⓓ | Ⓔ |
| 2. | Ⓐ | Ⓑ | Ⓒ | Ⓓ | Ⓔ | 12. | Ⓐ | Ⓑ | Ⓒ | Ⓓ | Ⓔ |
| 3. | Ⓐ | Ⓑ | Ⓒ | Ⓓ | Ⓔ | 13. | Ⓐ | Ⓑ | Ⓒ | Ⓓ | Ⓔ |
| 4. | Ⓐ | Ⓑ | Ⓒ | Ⓓ | Ⓔ | 14. | Ⓐ | Ⓑ | Ⓒ | Ⓓ | Ⓔ |
| 5. | Ⓐ | Ⓑ | Ⓒ | Ⓓ | Ⓔ | 15. | Ⓐ | Ⓑ | Ⓒ | Ⓓ | Ⓔ |
| 6. | Ⓐ | Ⓑ | Ⓒ | Ⓓ | Ⓔ | 16. | Ⓐ | Ⓑ | Ⓒ | Ⓓ | Ⓔ |
| 7. | Ⓐ | Ⓑ | Ⓒ | Ⓓ | Ⓔ | 17. | Ⓐ | Ⓑ | Ⓒ | Ⓓ | Ⓔ |
| 8. | Ⓐ | Ⓑ | Ⓒ | Ⓓ | Ⓔ | 18. | Ⓐ | Ⓑ | Ⓒ | Ⓓ | Ⓔ |
| 9. | Ⓐ | Ⓑ | Ⓒ | Ⓓ | Ⓔ | 19. | Ⓐ | Ⓑ | Ⓒ | Ⓓ | Ⓔ |
| 10. | Ⓐ | Ⓑ | Ⓒ | Ⓓ | Ⓔ | 20. | Ⓐ | Ⓑ | Ⓒ | Ⓓ | Ⓔ |
| 21. | Ⓐ | Ⓑ | Ⓒ | Ⓓ | Ⓔ | 31. | Ⓐ | Ⓑ | Ⓒ | Ⓓ | Ⓔ |
| 22. | Ⓐ | Ⓑ | Ⓒ | Ⓓ | Ⓔ | 32. | Ⓐ | Ⓑ | Ⓒ | Ⓓ | Ⓔ |
| 23. | Ⓐ | Ⓑ | Ⓒ | Ⓓ | Ⓔ | 33. | Ⓐ | Ⓑ | Ⓒ | Ⓓ | Ⓔ |
| 24. | Ⓐ | Ⓑ | Ⓒ | Ⓓ | Ⓔ | 34. | Ⓐ | Ⓑ | Ⓒ | Ⓓ | Ⓔ |
| 25. | Ⓐ | Ⓑ | Ⓒ | Ⓓ | Ⓔ | 35. | Ⓐ | Ⓑ | Ⓒ | Ⓓ | Ⓔ |
| 26. | Ⓐ | Ⓑ | Ⓒ | Ⓓ | Ⓔ | 36. | Ⓐ | Ⓑ | Ⓒ | Ⓓ | Ⓔ |
| 27. | Ⓐ | Ⓑ | Ⓒ | Ⓓ | Ⓔ | 37. | Ⓐ | Ⓑ | Ⓒ | Ⓓ | Ⓔ |
| 28. | Ⓐ | Ⓑ | Ⓒ | Ⓓ | Ⓔ | 38. | Ⓐ | Ⓑ | Ⓒ | Ⓓ | Ⓔ |
| 29. | Ⓐ | Ⓑ | Ⓒ | Ⓓ | Ⓔ | 39. | Ⓐ | Ⓑ | Ⓒ | Ⓓ | Ⓔ |
| 30. | Ⓐ | Ⓑ | Ⓒ | Ⓓ | Ⓔ | 40. | Ⓐ | Ⓑ | Ⓒ | Ⓓ | Ⓔ |
| 41. | Ⓐ | Ⓑ | Ⓒ | Ⓓ | Ⓔ | 51. | Ⓐ | Ⓑ | Ⓒ | Ⓓ | Ⓔ |
| 42. | Ⓐ | Ⓑ | Ⓒ | Ⓓ | Ⓔ | 52. | Ⓐ | Ⓑ | Ⓒ | Ⓓ | Ⓔ |
| 43. | Ⓐ | Ⓑ | Ⓒ | Ⓓ | Ⓔ | 53. | Ⓐ | Ⓑ | Ⓒ | Ⓓ | Ⓔ |
| 44. | Ⓐ | Ⓑ | Ⓒ | Ⓓ | Ⓔ | 54. | Ⓐ | Ⓑ | Ⓒ | Ⓓ | Ⓔ |
| 45. | Ⓐ | Ⓑ | Ⓒ | Ⓓ | Ⓔ | 55. | Ⓐ | Ⓑ | Ⓒ | Ⓓ | Ⓔ |
| 46. | Ⓐ | Ⓑ | Ⓒ | Ⓓ | Ⓔ | 56. | Ⓐ | Ⓑ | Ⓒ | Ⓓ | Ⓔ |
| 47. | Ⓐ | Ⓑ | Ⓒ | Ⓓ | Ⓔ | 57. | Ⓐ | Ⓑ | Ⓒ | Ⓓ | Ⓔ |
| 48. | Ⓐ | Ⓑ | Ⓒ | Ⓓ | Ⓔ | 58. | Ⓐ | Ⓑ | Ⓒ | Ⓓ | Ⓔ |
| 49. | Ⓐ | Ⓑ | Ⓒ | Ⓓ | Ⓔ | 59. | Ⓐ | Ⓑ | Ⓒ | Ⓓ | Ⓔ |
| 50. | Ⓐ | Ⓑ | Ⓒ | Ⓓ | Ⓔ | 60. | Ⓐ | Ⓑ | Ⓒ | Ⓓ | Ⓔ |

61. A B C D E
62. A B C D E
63. A B C D E
64. A B C D E
65. A B C D E
66. A B C D E
67. A B C D E
68. A B C D E
69. A B C D E
70. A B C D E
71. A B C D E

72. A B C D E
73. A B C D E
74. A B C D E
75. A B C D E
76. A B C D E
77. A B C D E
78. A B C D E
79. A B C D E
80. A B C D E
81. A B C D E
82. A B C D E

83. A B C D E
84. A B C D E
85. A B C D E
86. A B C D E
87. A B C D E
88. A B C D E
89. A B C D E
90. A B C D E
91. A B C D E
92. A B C D E
93. A B C D E

94. A B C D E
95. A B C D E
96. A B C D E
97. A B C D E
98. A B C D E
99. A B C D E
100. A B C D E
101. A B C D E
102. A B C D E
103. A B C D E
104. A B C D E

105. A B C D E
106. A B C D E
107. A B C D E
108. A B C D E
109. A B C D E
110. A B C D E
111. A B C D E
112. A B C D E
113. A B C D E
114. A B C D E

115. A B C D E
116. A B C D E
117. A B C D E
118. A B C D E
119. A B C D E
120. A B C D E
121. A B C D E
122. A B C D E
123. A B C D E
124. A B C D E

125. A B C D E
126. A B C D E
127. A B C D E
128. A B C D E
129. A B C D E
130. A B C D E
131. A B C D E
132. A B C D E
133. A B C D E
134. A B C D E

135. A B C D E
136. A B C D E
137. A B C D E
138. A B C D E
139. A B C D E
140. A B C D E
141. A B C D E
142. A B C D E
143. A B C D E
144. A B C D E

# THE GRADUATE RECORD EXAMINATION ENGINEERING TEST

## MODEL TEST V

Time: 170 Minutes
144 Questions

DIRECTIONS: *Choose the best answer for each question and mark the letter of your selection on the corresponding answer sheet.*

1. Water entering the boiler of a steam engine is heated to its boiling point at 100°C and is vaporized. The process of vaporization can be described as

    (A) adiabatic
    (B) isothermal
    (C) isochoric (constant volume)
    (D) isobaric (constant pressure)
    (E) Both (B) and (D)

2. An undesirable side effect of motional inductance of large conducting specimens is

    (A) skin effect
    (B) hysterisis
    (C) eddy currents
    (D) dielectric loss
    (E) electromagnetic radiation

3. A liquid flows out of an orifice of an enclosed tank filled to a depth h that contains air above the liquid at a pressure p. If the speed of the fluid flowing out is to be reduced

(A) the tank should be open to the atmosphere

(B) the cross-sectional area of the orifice is increased

(C) the air pressure above the liquid should be increased

(D) the amount of liquid in the tank should be increased

(E) Both (B) and (D)

4. Suppose a uniform cube of edge b is at rest at the top of a cylindrical surface where there is enough friction to prevent slipping. What should be the relationship between b and the radius of the cylindrical surface r to have stable equilibrium at the top of the cylinder?

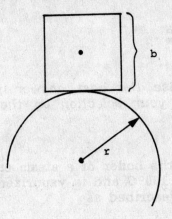

(A) $r > b$

(B) $r > \frac{b}{2}$

(C) $r^2 > \frac{b}{4}$

(D) information not sufficient

(E) It always has stable equilibrium because there is sufficient friction to prevent slipping.

5. In the diagram below, what is the value of sec x?

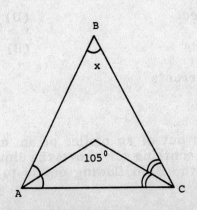

(A) $\dfrac{2}{\sqrt{2}}$

(B) $\dfrac{3}{\sqrt{3}}$

(C) 1

(D) $\dfrac{2}{\sqrt{3}}$

(E) 2

6. All of the following comparisons between metals and metal alloys and ceramics are true except

   (A) ceramic compounds crystallize more slowly than metallic compounds
   (B) ceramic compounds have higher melting temperatures than metallic compounds
   (C) metallic compounds are better conductors of current than ceramic compounds at higher temperatures
   (D) ceramic compounds are more viscous than metallic compounds below their respective melting temperatures
   (E) ceramic compounds are able to resist greater tensile stresses than metallic compounds at room temperatures

7. The kinetic theory of gases theoretically explains the behavior of an ideal gas and provides a framework for understanding how real gases differ from this ideal behavior. This theory assumes which of the following?

   I. A gas is made of molecules of negligible volume in constant motion which collide elastically.
   II. Each of the point molecules possesses a constant kinetic energy.
   III. The size of the container affects the amount of motion of the molecules.
   IV. The speed of each molecule varies according to the different collisions it might have.

   (A) I only
   (B) II only
   (C) I and IV only
   (D) II and III only
   (E) IV and II only

8. Which of the following curves represents the function θ = ln(r) in a system of polar coordinates?

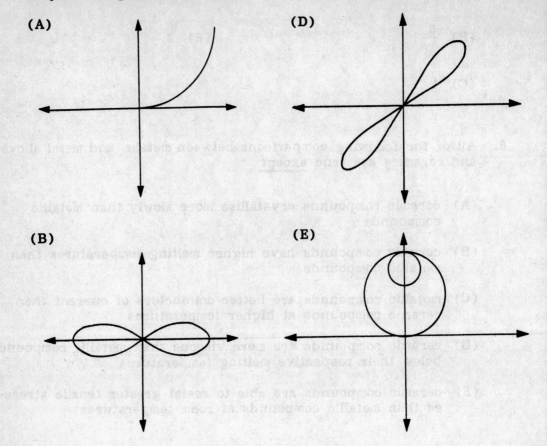

(A)  (D)  (B)  (E)  (C)

9. A human cannon angled 15° with the horizontal, shoots a stuntman with an initial velocity of 20 m/sec. How far from the cannon should the net be positioned so that the stuntman lands safely in it?

Note: g = 10 m/sec²

(A)  5m

(B)  10m

(C)  15m

(D)  20m

(E)  25m

10. The open circuit terminal voltage of a car battery is x volts. When the car is started, the motor takes a current of z amperes with a resulting battery terminal voltage equal to (x - y) volts. The Thevenin resistance for the battery is equal to

(A)  $\frac{xy}{z} \Omega$

(B)  $\frac{x - y}{z} \Omega$

(C)  $\frac{x}{z} \Omega$

(D)  $\frac{y}{z} \Omega$

(E)  $\frac{x + y}{z} \Omega$

11. The number of branches and nodes in the given circuit are

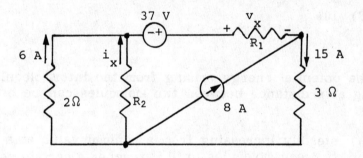

(A)  6 and 6 respectively

(B)  7 and 4 respectively

(C)  6 and 4 respectively

(D)  5 and 3 respectively

(E)  5 and 4 respectively

12. A circus acrobat leaves a swing with his arms and legs extended. As he immediately pulls in his arms and legs he starts spinning at a rate faster than when he left the swing. Which of the following makes this possible?

(A) Since he increases his moment of inertia when he pulls in his arms and legs, there must be a corresponding increase in his spin rate.

(B) The torque about his center of mass caused by the force of gravity forces him to spin faster.

(C) Since he decreases his moment of inertia when he pulls in his arms and legs, there must be an increase in his spin rate to maintain a constant angular momentum.

(D) The combined effect of increasing his moment of inertia and a gravitational torque acting on his center of mass causes him to spin faster.

(E) In order to maintain a constant rotational kinetic energy, the square of the spin rate must change inversely with the inertia.

13. A cube of some material of side 3m, is held at the normal freezing point of water. The temperature is then increased to the normal boiling point. If the coefficient of expansion is $6 \times 10^{-4}$ $^0C^{-1}$, then the cube volume is changed by: (assume material withstands thermal stresses)

(A) 4%

(B) 6%

(C) 10%

(D) 7%

(E) 12%

14. The potential energy arising from the intermolecular force acting at a distance between two molecules can be best described as:

(A) steadily increasing from a minimal value at a close distance between molecules until it reaches a certain separation distance after which it remains constant

(B) steadily decreasing from a maximum value at a close distance till it becomes nearly zero at a great distance

(C) steadily decreasing when the molecular force is repulsive at a close distance and then increasing when the distance is greater and the force is attractive

(D) having a minimum value inside a potential well from which the molecule can be excited

(E) Both (C) and (D)

15. An electrical conductor starts dissipating electrical energy at the rate of 50 watts (170.65 Btu/hr) when a switch is turned on.

It is cooled by a liquid flowing over the surface and carrying away energy at the rate of 25 Btu/hr. The temperature of the conductor

(A) remains constant

(B) decreases

(C) increases until it attains a steady state

(D) decreases until it attains a steady state

(E) increases

16. Heat is added to a saturated liquid of a pure substance at constant pressure until the substance reaches its critical temperature. The substance passes through

(A) the saturated vapor state only

(B) the saturated vapor state and then through the superheated vapor state

(C) the compressed liquid state and then through the saturated vapor state

(D) the compressed liquid state and then through the superheated vapor state

(E) the compressed liquid state only

17. A metallic chain of length L and mass M is vertically hung above a surface with one end in contact with it. The chain is then released to fall freely. If x is the distance covered by the end of the chain, how much force (exerted by the surface) will the chain experience at any instance during the process?

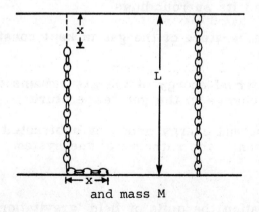
and mass M

(A) $N = Mg - M\ddot{x}$

(B) $N = 3Mg$

(C)  N = Mg - 2Mẍ

(D)  $N = \frac{3M}{L} gx$

(E)  N = Mg

18. A solution containing 14.5 g sodium chloride is mixed with a solution of sulfuric acid. How many liters of gas will be released at standard temperature and pressure if all of the salt is consumed? (Atomic weights: Na = 23, Cl = 35, S = 32)

(A) 4.8 liters

(B) 0.25 liter

(C) 22.4 liters

(D) 5.6 liters

(E) information not sufficient

19. Chemical behavior exhibited by the atoms is mainly due to

(A) neutrons

(B) protons

(C) electrons

(D) neutrons and electrons

(E) neutrons and protons

20. In an adiabatic process,

(A) the thermal energy transferred to a gas is completely converted to internal energy

(B) no thermal energy is allowed to be transferred between the gas and its surroundings

(C) the temperature of the gas is kept constant throughout the process

(D) the internal energy of the gas remains constant and there is no change in the gas temperature

(E) the thermal energy added or subtracted from a system does not change the enthalpy of the system

21. In mks notation the units of field 'gravitational intensity' are

(A) Joule/sec$^2$

(B) N/kg  
(C) N/m²  
(D) kg/m²  
(E) kg/(m·sec)²

22. 20°C is equal to which of the following?

    (A) 273°K
    (B) 43.11°F
    (C) 68°F
    (D) 72°F
    (E) None of the above

23. In the figure, a block has been placed on an inclined plane and the slope angle θ of the plane has been adjusted until the block slides down the plane at constant speed, once it has been set in motion. What is the value of angle θ?

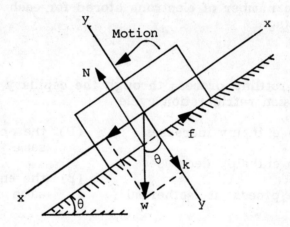

    (A) $\tan^{-1}\mu_k$
    (B) $\sin^{-1}\mu_k$
    (C) $\cos^{-1}\mu_k$
    (D) $\sin^{-1}\frac{N}{W}\mu_k$
    (E) $\cos^{-1}\frac{N}{W}\mu_k$

24. Which of the following correctly identifies the usage of a heat pump?

    (A) transforms work into heat as in an electric heater
    (B) transforms heat into work

(C) pumps heat from an area of higher temperature to one of lower temperature

(D) transforms heat into work or work into heat, depending on the application

(E) None of the above

25. The capacitance of a capacitor increases when which of the following is decreased?

   (A) distance between the plates

   (B) the area of the plates

   (C) the number of plates

   (D) the dielectric constant

   (E) the number of electrons stored for each volt of applied voltage

26. For a throttling process through the capillary tube in a vapor compression refrigeration cycle

   (A) the enthalpy increases

   (B) the enthalpy decreases

   (C) the process is isothermal

   (D) the entropy remains the same

   (E) the enthalpy remains the same

27. A person looks into a common mirror and sees himself nearly perfectly. This is possible because

   (A) the mirror is a perfect plane

   (B) the silicon structure of the glass reflects light

   (C) the glass is doped with a highly reflective material

   (D) when the flat mirror was produced, the molecular bonds were stressed in such a way that light reflects

   (E) None of the above

28. What must the minimum length of a wall mirror be so that a person of height h can see himself from head to toe?

(A) $\dfrac{h}{4}$

(B) $\dfrac{h}{2}$

(C) h

(D) 2h

(E) None of the above

29. Which of the following graphs represents the motion of an object moving with a linearly increasing acceleration against time?

(A)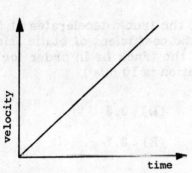

(D)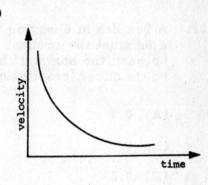

(B)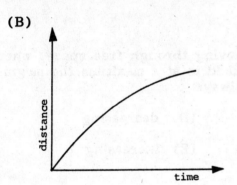

(E)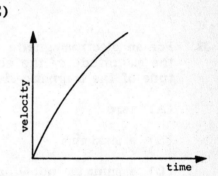

(C)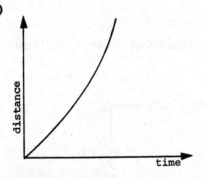

30. If the amount of solar energy striking the outer edge of the earth's atmosphere is approximately $1.4 \times 10^3$ J/sec·m² and the amount of the earth's surface area exposed to sunlight is $1.6 \times 10^{13}$ m²

how much energy does earth lose through radiation within each 24 hours? The average earth temperature increase within the same period is $10^{-6}°C$, the average specific heat of earth is $100 J/kg °C$ and earth's mass is $6 \times 10^{24}$ kg.

(A) $10^{18}$ J

(B) $2 \times 10^{10}$ J

(C) $13.35 \times 10^{20}$ J

(D) $58.7 \times 10^{20}$ J

(E) $9 \times 10^{18}$ J

31. A box lies in a moving truck. If the truck decelerates at 5 m/sec$^2$, what must the smallest value of the coefficient of static friction ($\mu_s$) between the box and the floor of the truck be in order for the box not to slide? (gravitation acceleration is 10 m/s²)

(A) 0.3

(B) 0.4

(C) 0.5

(D) 0.6

(E) 0.7

32. For an electromagnetic wave moving through free space, when the magnitude of the electric field is at a maximum, the magnitude of the magnetic field is always

(A) zero

(B) a maximum

(C) a minimum not equal to zero

(D) decreasing

(E) increasing

33. In the circuit shown below, the value of source voltage ($v_s$) is

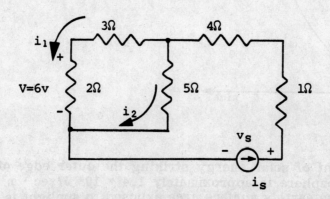

(A) 45V  (D) 20V
(B) 30V  (E) 15V
(C) 25V

34. The tremendous amount of energy required to activate a fusion reaction of two isotopes is mainly due to

   (A) electrostatic repulsion between nuclei

   (B) high binding energy of the isotopes

   (C) the high velocities necessary to collide isotopes

   (D) the low initial temperature of the reactants

   (E) the stability of the isotopes

35. The arrangement shown below is the cross-section of a tank with two separate fluids. What is the force required to keep gate AB closed? (Neglect atmospheric pressure and weight of the gate).

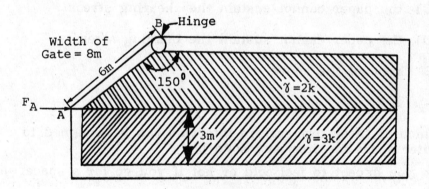

(A) 72K  (D) 192K
(B) 144K (E) 60K
(C) 30K

36. If an element is in Group O, then which of the following is true?
   I. Its outermost subshell must be a filled p subshell.
   II. It will react violently with hydrogen.
   III. Its outermost electrons are paired.

(A) I only  (D) III only
(B) I and II only  (E) I, II, and III
(C) II and III only

37. Which of the following correctly identifies the expression of Reynold's Number for a flat plate?

   (A) $Re = \dfrac{\mu V}{\nu}$  (D) $Re = \dfrac{VL}{\nu}$

   (B) $Re = \dfrac{\rho L}{\nu}$  (E) None of the above

   (C) $Re = \dfrac{\rho L}{\mu}$

38. A pair of scissors are capable of cutting paper because

   (A) the scissors are capable of causing high normal stress
   (B) the paper cannot sustain the normal stress
   (C) the paper cannot sustain the shearing stress
   (D) the paper cannot sustain the bending stress
   (E) None of the above

39. Which of these temperature intervals can be assigned to outer space far from any solid body?
    Do you expect to feel cold or hot if you go for a space walk?

   (A) 2°-4°K; feel cold
   (B) 273°K; feel the same way as if we were on earth at the same temperature
   (C) 0°K; feels extremely cold
   (D) Near, but not 0°K; feel normal
   (E) Temperature cannot be assigned to outer space

40. Which of the curves given below shows the speed-load characteristics for a d.c. series motor?

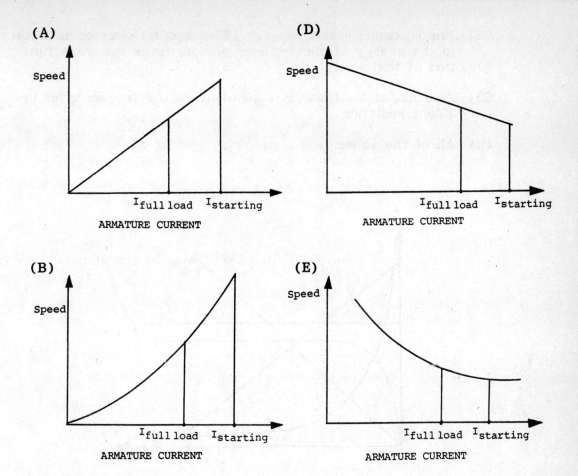

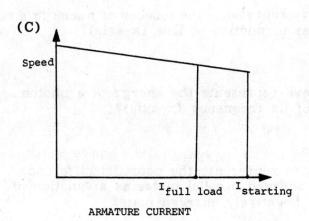

41. Which of the following statements concerning the photoelectric effect is untrue?

   (A) The number of emitted photoelectrons increases with the intensity of the incident radiation.

   (B) The intensity of the radiation determines whether or not the photoelectric effect occurs.

(C) The maximum kinetic energy of an emitted electron is equal to the energy of the incident photon minus the work function of the metal.

(D) Emission of electrons is dependent on the frequency of incident radiation.

(E) All of the above

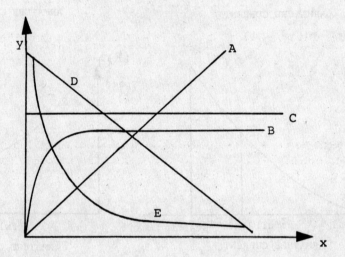

42. Which of the curves above represents the number of nuclei in a radioactive sample (y-axis) as a function of time (x-axis)?

43. Which of the curves above represents the energy of a photon (y-axis) as a function of its frequency (x-axis)?

44. Which of the curves above represents the potential difference between two oppositely charged parallel plates as a function of the distance r from the negatively charged plate?

45. A company issued a note for $6,000 at an interest rate of 6% per year, compounded bimonthly. The interest is paid out every six months. If the note was issued 40 days prior to the end of the year, the amount that is debited to the interest expense is

(A) $30

(B) $200

(C) $20

(D) $400

(E) $40

DIRECTIONS: *Choose the best answer for each question and mark the letter of your selection on the corresponding answer sheet.*

Note: *The natural logarithm of x will be denoted by ln x.*

46. Integration of $\int \frac{2x}{x^2 + 1} dx$ yields

    (A) $\frac{x^2 + 1}{2} + c$

    (B) $\ln(x^2 - 1) + c$

    (C) $\frac{x^2 - 1}{2} + c$

    (D) $\ln(x^2 + 1) + c$

    (E) $4x^2 + c$

47. There is a box containing 5 white balls, 4 black balls, and 7 red balls. If two balls are drawn one at a time from the box and neither is replaced, find the probability that both balls will be white.

    (A) $\frac{1}{12}$

    (B) $\frac{1}{5}$

    (C) $\frac{1}{7}$

    (D) $\frac{5}{16}$

    (E) $\frac{1}{16}$

48. If a card is drawn from a deck of playing cards, what is the probability that it will be a jack or a ten?

    (A) $\frac{1}{10}$

    (B) $\frac{1}{5}$

    (C) $\frac{5}{13}$

    (D) $\frac{1}{2}$

    (E) $\frac{2}{13}$

49. $\int_{\pi/4}^{\pi/6} \cos 2x \, dx =$

(A) $\dfrac{\sqrt{3}+1}{4}$

(B) $\dfrac{\sqrt{3}+2}{4}$

(C) $\dfrac{\sqrt{3}-2}{4}$

(D) $\dfrac{\sqrt{3}+2}{6}$

(E) $\dfrac{\sqrt{3}+\sqrt{2}}{4}$

50. The inverse of $\begin{bmatrix} 1 & 1 \\ 0 & 1 \end{bmatrix}$ is

(A) $\begin{bmatrix} 1 & -1 \\ 0 & 1 \end{bmatrix}$

(B) $\begin{bmatrix} -1 & 1 \\ 0 & -1 \end{bmatrix}$

(C) $\begin{bmatrix} 1 & -1 \end{bmatrix}$

(D) $\begin{bmatrix} 1 & 0 \\ 0 & 1 \end{bmatrix}$

(E) Non-existing

51. For $0 < x < \pi$, $x \neq \dfrac{\pi}{2}$,

(csc x)(cot x)(sin x)(sec x)(tan x)(cos x) =

(A) 1

(B) 0

(C) $\dfrac{\pi}{2}$

(D) $\pi$

(E) x

52. A couple has six children whose ages are 6, 8, 10, 12, 14, and 16. The variance in ages is

(A) 2

(B) 12

(C) $\dfrac{35}{3}$

(D) $\dfrac{25}{4}$

(E) $\dfrac{14}{3}$

53. If $y = e^{1/x^2}$, $D_x y =$

(A) $2xe^{1/x^2}$

516

(B) $-\dfrac{2e^{1/x^2}}{x^3}$     (D) $\dfrac{e^{1/x^2}}{2x^3}$

(C) $\dfrac{2x^3}{e^{1/x^2}}$     (E) $-\dfrac{e^{1/x^2}}{2x}$

Questions 54 to 58 are based on the following assumption:

Let F(x) and G(x) be functions over the set of all real numbers.

54. If f'(x) = F(x) and g'(x) = G(x) then

$$\int_a^b F(x)dx + \int_a^b G(x)dx =$$

(A) [f(b) - g(b)] + [f(a) - g(a)]

(B) [f(b) + g(b)] - [f(a) + g(a)]

(C) [f(b) - f(a)] - [g(b) - g(a)]

(D) [f(b) - f(a)] - [g(a) + g(b)]

(E) [f(b) + g(b)] - [g(a) - f(a)]

55. $\int_0^\infty e^{-x}dx =$

(A) 0     (D) -1

(B) 1     (E) ∞

(C) ½

56. $\int F'[G(x)]G'(x)dx =$

(A) $\int [F'(x) + G'(x)]dx$     (D) F[G(x)] + C

(B) F(x) + G(x) + C     (E) None of the above

(C) F(x)G'(x) + F'(x)G(x) + C

57. If $F(x) = \dfrac{ax^n - 2}{b}$, $F'(x)$ is

    (A) $\dfrac{1}{ab}(nx^{n-1})$            (D) $(2ab)nx^{n-1}$

    (B) $\dfrac{1}{b}(nax^{n-1})$            (E) None of the above

    (C) $\dfrac{nax^{n-1}}{2b}$

58. Let $F(x)$ and $G(x)$ be two functions representing perpendicular straight lines such that $F(x) = ax + b$. Then the slope of line $G(x)$ is

    (A) $-\dfrac{b}{a}$            (D) $\dfrac{1}{a}$

    (B) $-\dfrac{1}{a}$            (E) $-\dfrac{1}{b}$

    (C) $a$

59. Find the derivative of the expression:

$$y = xe^{\tan x}$$

    (A) $e^{\tan x}(x \sin^2 x - 1)$            (D) $e^{\tan x}(x \sec^2 x - 1)$

    (B) $e^{\sin x}(x \cos^2 x + 1)$            (E) $e^{\tan x}(x \sec^2 x + 1)$

    (C) $e^{\cos x}(x \tan^2 x + 1)$

60. $\int (x^3 - 7)^8 3x^2 \, dx =$

    (A) $\dfrac{(x^3 - 7)^9}{9} + C$            (D) $2x(x^3 + 7)^8 + C$

    (B) $3x \dfrac{(x^3 - 7)^8}{8} + C$            (E) $6x(x^3 - 7)^8 + C$

    (C) $\dfrac{x^2(x^3 - 7)^9}{8} + C$

61. $\exp\left(\int_1^e \frac{dx}{x}\right) =$

(A) e

(B) (e − 1)

(C) 1

(D) 0

(E) ∞

62. $\lim_{x \to 0} \frac{\sin nx}{x} =$

(A) ∞

(B) 0

(C) $\frac{1}{n}$

(D) 1

(E) n

63. Let

$$A = \begin{bmatrix} 2 & 3 & 7 \\ 4 & m & \sqrt{3} \\ 1 & 5 & a \end{bmatrix} \quad B = \begin{bmatrix} \alpha & \beta & \delta \\ \sqrt{5} & 3 & 1 \\ p & q & 4 \end{bmatrix}$$

A + B =

(A) $\begin{bmatrix} 2+\alpha & 3+\delta & 7+\beta \\ 4+\sqrt{5} & m+1 & \sqrt{3}+3 \\ q+1 & p+5 & a+4 \end{bmatrix}$

(B) $\begin{bmatrix} 2+\alpha & 3+\beta & 7+\delta \\ 4+\sqrt{5} & m+3 & \sqrt{3}+1 \\ 1+p & 5+q & a+4 \end{bmatrix}$

(C) $\begin{bmatrix} 2+\delta & 10+\beta & 3+\alpha \\ 7 & 4+\sqrt{5} & 1+\sqrt{3} \\ 1+p & 5+q & 4+a \end{bmatrix}$

(D) $\begin{bmatrix} 2+\alpha & 7+\delta & 3+\beta \\ 4+\sqrt{5} & m+3 & \sqrt{3}+1 \\ 4+a & 1+q & 5+p \end{bmatrix}$

(E) $\begin{bmatrix} 2+\alpha & 3+\beta & 7+\delta \\ 4+\sqrt{3} & m+\sqrt{5} & \sqrt{3}+1 \\ 1+q & 5+a & p+4 \end{bmatrix}$

64. $\lim_{n \to \infty} \sum_{k=0}^{n} \frac{(-1)^k \cdot \pi^{2k}}{(2k)!} =$

(A) ∞

(B) 0

(C) −1

(D) 1

(E) π

DIRECTIONS: *Choose the best answer for each question and mark the letter of your selection on the corresponding answer sheet. Assume that every function in this section has derivatives of all orders at each point of its domain unless otherwise indicated.*

Questions 65-70 concern the function, g, the graph of which is not shown. The graph below is that of the function f.

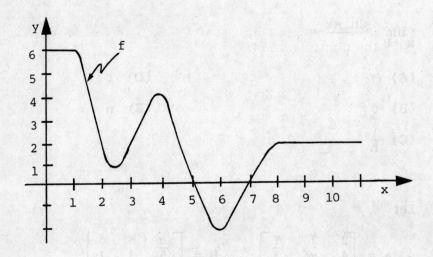

$$g(x) = \int_0^x f(t)\,dt \quad \text{for all } x, \ 0 \leq x \leq 10$$

where f is the function shown in the graph above.

65. The function, g, is a decreasing function throughout which of the following intervals?

   (A) from 0 to 1

   (B) from 1 to 2

   (C) from 2 to 3

   (D) from 3 to 4

   (E) from 5 to 7

66. A local maximum for g(t) occurs at which value of t?

   (A) 1

   (B) 4

(C) 5

(D) 6

(E) 8

67. For which interval of x is g(x) most nearly constant?

   (A) from 0 to 1
   (D) from 5 to 7

   (B) in the neighborhoods of the points 5 and 7
   (E) from 6 to 8

   (C) from 3 to 4

68. The value of g at x = 3 is between

   (A) 0 and 1
   (D) 10 and 14

   (B) 2 and 4
   (E) 15 and 20

   (C) 6 and 8

69. The value g(10) is between

   (A) 0 and 5
   (D) 22 and 25

   (B) 5 and 15
   (E) 25 and 30

   (C) 18 and 22

70. Which of the following has the largest value?

   (A) g(1) - g(0)
   (D) g(8) - g(5)

   (B) g(5) - g(3)
   (E) g(10) - g(9)

   (C) g(6) - g(5)

Questions 71-80 are based on the following information.

   A circle is rolling around the inner side of another circle at a constant rate without slipping (in a two-dimensional plane).

$P_1$ denotes a constant point on the circumference of the smaller circle which has radius $r_1$. Similarly, $P_2$ denotes a constant point on the larger circle's circumference of radius $r_2$ (obviously $r_2 > r_1$).

The graph below shows the straight-line distance in inches between $P_1$ and $P_2$ at all times t for $0 \leq t \leq 11$ seconds.

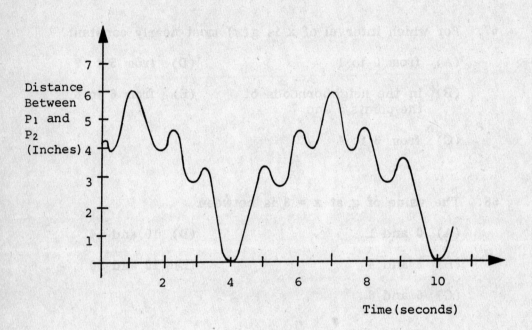

71. As the smaller circle rotates and completes one revolution around the larger circle, the number of times that $P_1$ touches the larger circle's circumference is which of the following?

   (A) 2         (D) 8

   (B) 4         (E) 10

   (C) 6

72. The diameter of the larger circle is

   (A) 3 inches        (D) 6 inches

   (B) 4 inches        (E) 8 inches

   (C) 4½ inches

73. What is the area of the triangle defined by the following

three vertices : $P_2$, the center of the larger circle, and the point on the larger circle where $P_1$ has completed one rotation away from $P_2$?

(A) 0.75 sq. in.

(B) 1.0 sq. in.

(C) 2 sq. in.

(D) $\frac{9}{4}\sqrt{3}$ sq. in.

(E) $2\pi$ sq. in.

74. The circumference of the smaller circle is

(A) 1 in.

(B) $\pi$ in.

(C) $2\pi$ in.

(D) 6 in.

(E) $\frac{3}{2}\pi$ in.

75. What is the number of times between $t = 0$ and $t = 11$ seconds that the rate of change in distance between $P_1$ and $P_2$ per second changes from negative to positive?

(A) 2

(B) 3

(C) 6

(D) 8

(E) 9

Questions 76-80 involve the following additional information.

f is the infinitely differentiable function whose graph appears before question 71. F is defined by

$$F(x) = \int_0^x f(t)dt \quad \text{for each } x, \quad 0 \le x \le 11$$

76. The number of t in the interval $4 \le t \le 10$ for which $f'(t) = 0$ is

(A) one

(B) two

(C) five

(D) six

(E) eleven

77. On the interval from 1 to 4, F is a function which can be described as

    (A) increasing on some subintervals and decreasing on others

    (B) periodic

    (C) strictly decreasing

    (D) strictly increasing

    (E) non-monotonic

78. Which is the best approximation to F(8) - F(6) ?

    (A) 0          (D) 10

    (B) 2          (E) 20

    (C) 6

79. The number t, 0 < t < 4, for which f"(t) = 0 is

    (A) 4          (D) 10

    (B) 7          (E) 12

    (C) 9

80. The best approximate value for the average of the <u>derivative</u> of F is

    (A) 0          (D) 8

    (B) 3          (E) 11

    (C) 6

Questions 81-85 relate to the following situation.

   A faucet which releases liquid at a constant rate is used to fill four rectangular containers which differ in height and base areas. The time it takes to fill all four containers, stopping for one second between each, is 17 seconds. The height of the liquid is observed from the top of the second container. The level of the liquid with respect to the top of this container is shown graphically below.

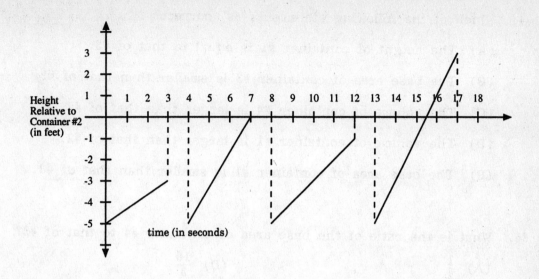

81. The height of container #1 is

    (A) 1 foot          (D) 4 feet
    (B) 2 feet          (E) 5 feet
    (C) 3 feet

82. The height of container #2 is

    (A) 4 feet
    (B) less than the height of container #1
    (C) greater than the height of container #3
    (D) one-half of the height of container #4
    (E) inconclusive from the graph

83. Which of the following statements is correct?

    (A) Container #1 is the same height as container #2.
    (B) Container #1 has the same width as container #2.
    (C) The difference in height of containers #4 and #1 is greater than the difference between the height of containers #2 and #3.
    (D) The time it takes to fill container #4 is greater than the time it takes to fill container #3.
    (E) The time it takes to fill container #1 is equal to the time it takes to fill container #3.

84. Which of the following statements is correct?

   (A) The height of container #1 is equal to that of #3.

   (B) The base area of container #3 is smaller than that of #2.

   (C) The volume of container #4 is larger than that of #2.

   (D) The volume of container #1 is larger than that of #2.

   (E) The base area of container #1 is smaller than that of #3.

85. What is the ratio of the base area of container #4 to that of #2?

   (A) $\frac{1}{2}$

   (B) $\frac{5}{6}$

   (C) $\frac{6}{5}$

   (D) $\frac{10}{3}$

   (E) 1

DIRECTIONS: *Choose the best answer for each question and mark the letter of your selection on the corresponding answer sheet.*

86. The magnitude of the resultant of forces $F_1$ and $F_2$ acting at point S in the given figure is approximately

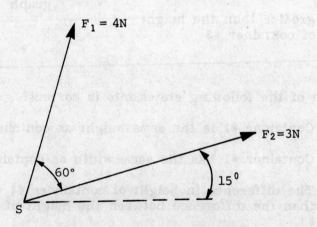

   (A) 5N

   (B) 6N

   (C) 7N

   (D) 8N

   (E) 9N

87. In order to continuously induce EMF in a secondary winding, the primary of a transformer

   (A) must be connected to an ac source

   (B) must be connected to a dc source

   (C) must be shunted with a resistor to prevent excessive losses in the coils

   (D) must be connected in series with a varying resistor

   (E) must be connected in series with a capacitor

88. The decomposition of an oxide of potassium produces 0.5 mol $O_2$ and 29.1g of K. What is the empirical formula for the oxide? (Assume atomic weights: K = 39.0, O = 16.0)

   (A) KO
   (B) $KO_3$
   (C) $K_2O_2$
   (D) $K_4O_5$
   (E) $K_2O$

89. Which one of the theories given below is related to the theory of the thermocouple?

   (A) Piezoelectric effect
   (B) Skin effect
   (C) Seebeck effect
   (D) Faraday's Law
   (E) Hall effect

90. The average velocity of water through an inlet pipe is 288 feet per second. If the cross-sectional area of the pipe is 5 square inches, then the rate at which the water will leave the faucet is

   (A) 0.5 $ft^3$/sec
   (B) 10 $ft^3$/sec
   (C) 50 $ft^3$/sec
   (D) 100 $ft^3$/sec
   (E) 288 $ft^3$/sec

91. Air flows through a nozzle from point 1 to point 2 as represented in the figure. P is the pressure, v is the velocity, and A is the cross-sectional area of the nozzle.

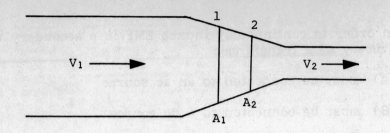

It is appropriate to say that:

(A) $P_2$ is greater than $P_1$ and $v_2$ is greater than $v_1$

(B) $P_1$ is greater than $P_2$ and $v_1$ is less than $v_2$

(C) no force acts on the nozzle because there is no change in momentum between points 1 and 2

(D) the kinetic energy remains conserved between points 1 and 2

(E) none of the choices apply; insufficient information

92.

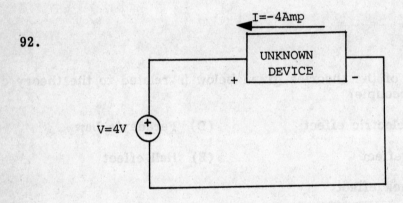

For the circuit shown, the device contained in the box has voltage-current relationship given by which of the following figures?

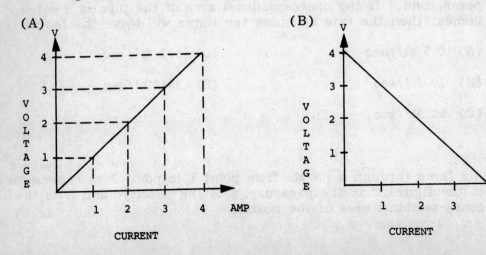

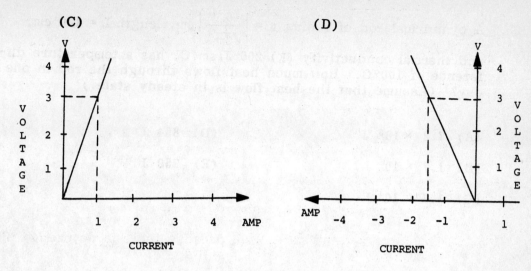

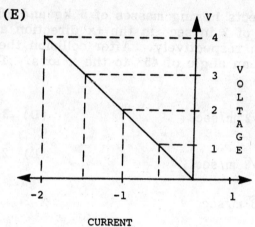

93. A certain shunt motor generates a torque of 200 lb-ft. and has an armature current of 100 amp. What is the armature current when the field is reduced by half and the torque developed is 250 lb-ft.?

(A) 150 amp

(B) 13.33 amp

(C) 75 amp

(D) 250 amp

(E) 100 amp

94. To double the frequency of a stretched string, the tension in the string must be:

(A) reduced by half

(B) reduced by a quarter

(C) increased by a quarter

(D) increased to double its value

(E) increased four times it value

95. A cylindrical rod of radius $r = \left(\dfrac{10^{-1}}{\sqrt{2\pi}}\right)$ cm, length L = 100 cm, and thermal conductivity (K) 200 J/sm°C, has a temperature difference of 100°C. How much heat flows through the rod in one day? (Assume that the heat flow is in steady state.)

(A) $7.2 \times 10^3$ J

(B) $1.7 \times 10^6$ J

(C) $1.7 \times 10^{10}$ J

(D) 864 J

(E) 250 J

96. Two objects having masses of 5 kg and 10 kg, have initial velocities of 2 in/sec in the +x direction and 1 in/sec in the +y direction respectively. After collision they stick together and move at an angle of 45° to the x-axis. Their combined final velocity is

(A) $\dfrac{2}{3}\sqrt{2}$ m/sec

(B) $\dfrac{2}{3}\sqrt{3}$ m/sec

(C) $2\sqrt{3}$ m/sec

(D) $3\sqrt{2}$ m/sec

(E) $2\sqrt{2}$ m/sec

97. The reduction of an element is the

(A) process which a reducing agent undergoes

(B) gaining of electrons

(C) loss of neutrons

(D) loss of charge due to ionic bonding

(E) reverse process of titration

98. In the diagram shown below, the direction in which an extra force must act in order to maintain equilibrium is

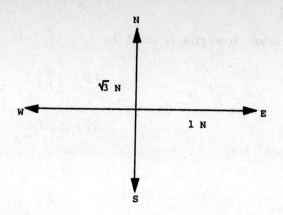

(A) E 60° S  
(B) W 30° N  
(C) W 60° S  
(D) E 30° N  
(E) E 60° N  

99. An atom of mass m and velocity v strikes the wall of a container at a right angle and rebounds with the same speed. The change in magnitude of momentum is

(A) $\dfrac{mv^2}{2}$  
(B) $2mv^2$  
(C) $2mv$  
(D) $mv$  
(E) $mv^2$  

**Questions 100-101**

Consider the U-tube shown below containing a liquid. When the liquid on one side is depressed by blowing air gently down that side, the levels of the liquid will oscillate with simple harmonic motion about their respective undisturbed positions O and P. Ignore damping due to friction.

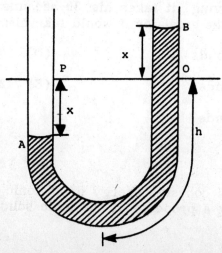

100. The acceleration towards O or P is

   (A) $\left(\dfrac{1}{g}\right) hx$

   (B) $\left(-\dfrac{g}{h}\right) x$

   (C) $g\left(\dfrac{h}{x}\right)$

   (D) $\left(\dfrac{g}{h}\right) x$

   (E) None of the above

101. The period of the oscillation is

   (A) $2\pi \left(\dfrac{h}{g}\right)^{\frac{1}{2}}$

   (B) $\left(\dfrac{h}{\pi g}\right)^{\frac{1}{2}}$

   (C) $2\pi \sqrt{gh}$

   (D) $\left(\dfrac{1}{2\pi}\right)\left(\dfrac{h}{g}\right)^{\frac{1}{2}}$

   (E) None of the above

102. A stationary propeller of diameter 4 m is mounted in a circular pipe and has incoming water of 2 m/s. The velocity of water downstream from the propeller is increased by 4 m/s relative to the velocity upstream. The power transferred by the propeller motion is (in KW): (Assume adiabatic and inviscid process and that the density of water is 1000 kg/m$^3$)

   (A) 32

   (B) 27

   (C) 18

   (D) 402

   (E) 128

103. A man walks up a stalled escalator in 90 seconds. When the escalator is moving, it takes him 30 seconds to walk up. If he were to stand on the escalator it would take him

   (A) $\sqrt{60}$ seconds

   (B) 30 seconds

   (C) 45 seconds

   (D) 60 seconds

   (E) 125 seconds

104. Based on the cohesive energy of an ionic solid, which of the following is not a property of the ionic solid?

(A) It forms relatively stable and hard crystals.

(B) It is a poor electric conductor.

(C) It has a high vaporization temperature.

(D) It has a high melting point.

(E) None of the above

105. From the definition of enthalpy and the equation of state of an ideal gas, h = u + PV = u + RT, where u is internal energy, V is volume, P is pressure, R is the gas constant, T is temperature and h is enthalpy. Which of the following is not true?

    (A) Enthalpy of an ideal gas is a function of temperature only, since R is constant and u depends only on temperature.

    (B) On a pressure-volume diagram, the lines of constant temperature are also lines of constant internal energy.

    (C) The constant-volume specific heat, $C_v$, is zero, since enthalpy depends on temperature.

    (D) The constant-volume and constant-pressure specific heats, $C_v$ and $C_p$ are both functions of temperature only.

    (E) None of the above

106. The minimum attractive rate of return is different for various types of projects. All of the following factors will affect the minimum attractive rate of return for a new project except:

    (A) the amount of capital that is available in the retained earnings

    (B) the interest paid on capital, which is obtained through debt financing

    (C) the rates used by the lending institutions or other firms

    (D) the amount of risk associated with the project

    (E) None

107. A driver takes an automobile trip. He travels at the rate of 45 miles/hr. for the first 90 miles, 60 miles/hr. for the next 1.5 hours and 25 miles/hr. for the next 12.5 miles. His average speed for the entire trip is

(A) 25 mph

(B) 45 mph

(C) 48.1 mph

(D) between 25 and 40 mph

(E) between 40 and 45 mph

108. A certain photoconductive cell is known to have a critical wavelength of 0.5 micrometer. Then the minimum energy required to move an electron from the valence band to the conduction band is

   (A) 4.96 electron volt

   (B) 0.31 electron volt

   (C) 0.62 electron volt

   (D) 2.48 electron volt

   (E) 1.0 electron volt

109. Natural convection differs from forced convection in that

   (A) the two velocity profiles are different

   (B) one does not obey Newton's Law of Cooling

   (C) one does not obey Fourier's Law of Conduction

   (D) in forced convection, the fluid flow is caused by temperature induced density changes

   (E) in natural convection, the fluid flow can be caused by pressure differences perpendicular to the gravitational force

110. Which of the following graphs disproves the ideal gas assumption?

(A)
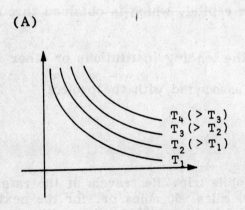

(B)
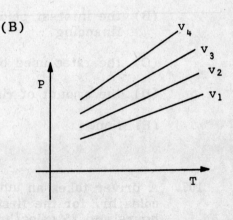

(C)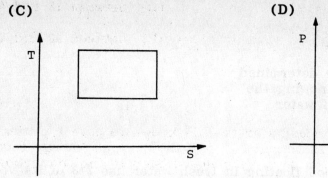

(D)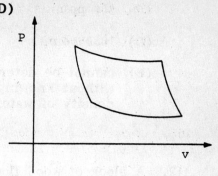

(E)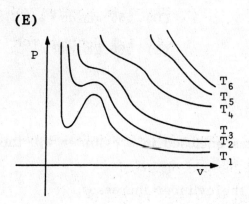

111. For the hydraulic press shown below, the small piston has a cross-sectional area of 2 square inches, while the large piston has a cross-sectional area of 18 square inches. If a force of 120 pounds is applied to the small piston, what is the force on the large piston neglecting friction and assuming constant density of water through the press?

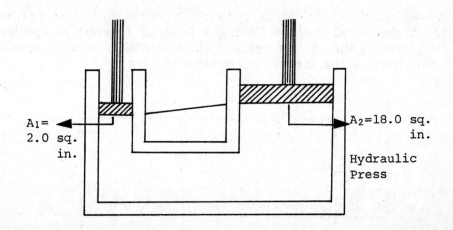

(A) 13 pounds             (B) 120 pounds

(C) 200 pounds

(D) 1080 pounds

(E) cannot be determined without knowing the density of water

112. A block of wood floating in fresh water has 2/3 of its volume above the surface of the water. The density of the block is

(A) .22 gm/cm³

(B) .33 gm/cm³

(C) .44 gm/cm³

(D) .55 gm/cm³

(E) .66 gm/cm³

113. The amount of an ideal gas enclosed in a cylinder will increase if

(A) the pressure inside the cylinder increases

(B) the volume of the cylinder increases

(C) the temperature of the cylinder increases

(D) the pressure is kept constant

(E) None of the above

114. A sinusoidal voltage having a peak of 10 volts is applied to the input of the circuit below, which contains an ideal diode. The output of the circuit is represented by

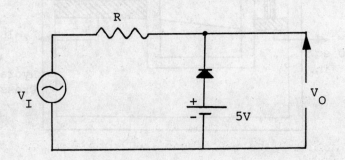

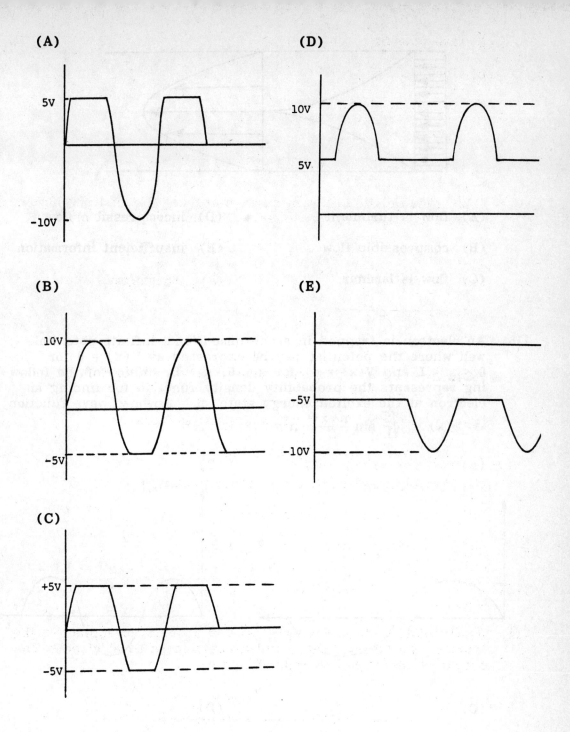

115. What can definitely be said about the tube-flow in the diagram below?

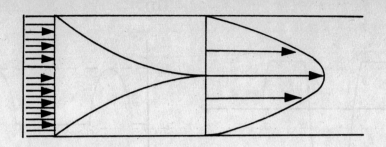

(A) flow is turbulent

(B) compressible flow

(C) flow is laminar

(D) incompressible flow

(E) insufficient information

116. An electron is trapped in a one-dimensional infinite potential well where the potential may be expressed as $V(x) = 0$ for $0 \leq x \leq L$ and $V(x) = \infty$ for $x < 0$, $x > L$. Which of the following represents the probability density function for finding an electron at the excited energy state, $n = 2$, whose wave function is $\psi(x) = \sqrt{\frac{2}{L}} \sin \frac{n\pi x}{L}$, $n = 1, 2, 3, \ldots$?

(A)

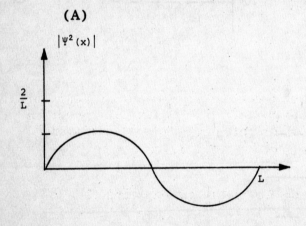

(B)

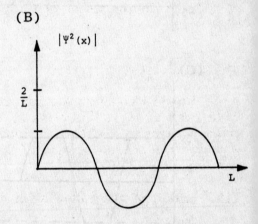

(C)

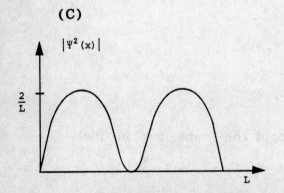

(D)

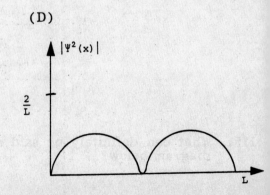

(E)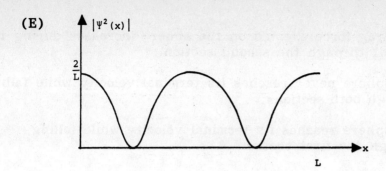

117. Forging iron at a temperature below the recrystallization temperature will increase the hardness of iron because

   (A) internal stresses do not appear in the iron

   (B) crystal defects such as dislocations are formed

   (C) surface shrinkage from the forging operation puts pressure on the interior of the iron

   (D) the increase in the density will cause pores and cavities in the metal to disappear

   (E) the decrease in the density will cause pores and cavities in the metal to disappear


Questions 118-119

In an experiment, a sphere of density $\rho_1$ and radius r is dropped in a tank of oil of viscosity $\mu_1$ and density $\rho_2$. The time of descent for the sphere through the first section of height d is recorded as $t_1$ and through the second section of the same height as $t_2$, $0 < t_2 - t_1 \ll 1$.

118. Which of the following is true for the below experiment?

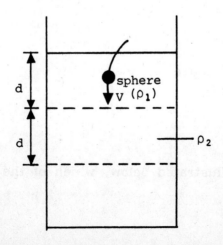

(A) The drag force exerted on the sphere increases during the descent through the second section.

(B) The sphere never reaches its terminal velocity while falling through both sections.

(C) The sphere reaches its terminal velocity while falling through the first section.

(D) The drag force exerted on the sphere decreases during the descent through the second section.

(E) None of the above

119. The drag force exerted on the sphere during its descent through the second section is

(A) $(\rho_1 - \rho_2)g\frac{4}{3}\pi r^3$

(B) $\rho_1 g \frac{4}{3}\pi r^3$

(C) $\frac{4}{3}\pi r^3 \rho_1 g - 6\pi\mu_1 r \left(\frac{d}{t_2}\right)$

(D) $\frac{4}{3}\pi r^3 \rho_1 g + 6\pi\mu_1 r \left(\frac{d}{t_2}\right)$

(E) 0

120. Which of the following is an example of a reversible process?

(A) A block sliding down an incline

(B) Heating a resistor by the passage of a current

(C) Carnot cycle

(D) Rankine cycle

(E) None of the above

121. For the planes illustrated below, which of the following is correct?

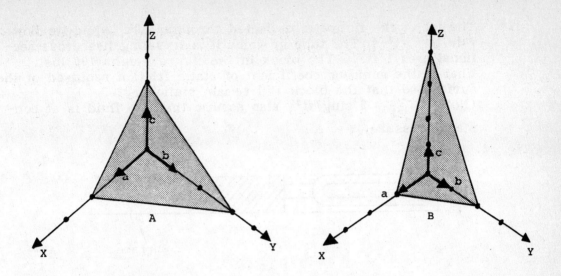

(A) Miller indices for A and B respectively are (1 2 1) and (0 1 3).

(B) Direction vector [2 3 2] is parallel to the plane A.

(C) Plane A accurately represents only body-centered crystal lattice structures.

(D) Planes A and B are parallel for face-centered lattice structures.

(E) Miller indices for A and B respectively are (2 3 2) and (1 2 4).

122. In a project to build solar energy collectors, a material which warms up considerably in sunlight is to be used. If $\alpha$ is the absorptance of the surface in sunlight and $\varepsilon$ is the emittance of the surface, then the material to be used will have

(A) a large $(\alpha\varepsilon)$

(B) a large $(\alpha/\varepsilon)$

(C) a small $(\alpha\varepsilon)$

(D) a small $(\alpha/\varepsilon)$

(E) $(\alpha\varepsilon) = 1$

123. When 60g of metal X at 80°C is dipped into 30g of water at 10°C the final temperature is 15°C. Therefore, the specific heat of metal X is close to

(A) 0.02 cal/g-°C

(B) 0.03 cal/g-°C

(C) 0.04 cal/g-°C

(D) 0.05 cal/g-°C

(E) 0.06 cal/g-°C

124. The jet in the figure is deflected through 180°, while its flow rate is 4 ft³/s. The tube in which it is traveling has cross-sectional area 1 ft². The block in the figure weighs 100 lbs. What is the minimum coefficient of static friction required of the surface so that the block will remain stationary?
(Note: $\rho_{jet}$ = 2 slug/ft³; also assume that the fluid is at constant pressure.)

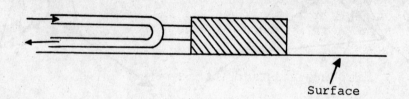

(A) 0.42

(B) 0.64

(C) 0.72

(D) 0.76

(E) 0.93

125. The ABC Corporation wants to determine its tax bill. The corporation has recently sold a piece of equipment for $6,000 which was initially purchased for $5,000. Its current book value is $3,000 with an annual depreciation of $1,000. The net taxable income of the company prior to the sale was $11,000. If the tax rate on income is 50% and the tax rate on capital gains is 25%, then the tax bill is

(A) $7,250

(B) $6,750

(C) $6,250

(D) $5,250

(E) $5,750

126. The viscoelastic behavior of an amorphous polymer will change with increasing temperature from its rigid structure at low temperatures. The polymer's behavior may proceed through the following stages:

    I. leathery

   II. rubbery

  III. viscous

What is the consecutive order of the stages it will pass through?

(A) I, III only

(B) II, III only

(C) II, I, III

(D) II, I only

(E) I, II, III

127. A hollow sphere filled with water is used as a pendulum by tying a long thread to it and hanging it from a point. If water starts to leak out from the bottom of the sphere then the period of the pendulum

(A) first increases and then decreases

(B) first decreases and then increases

(C) increases

(D) decreases

(E) remains unaffected

128. Which of the following cooling paths will have the following microstructure at the eutectic temperature?

eutectic

(A) path a

(B) path e

(C) path b

(D) path c

(E) path d

129. Which of the following orbital configurations is/are impossible?

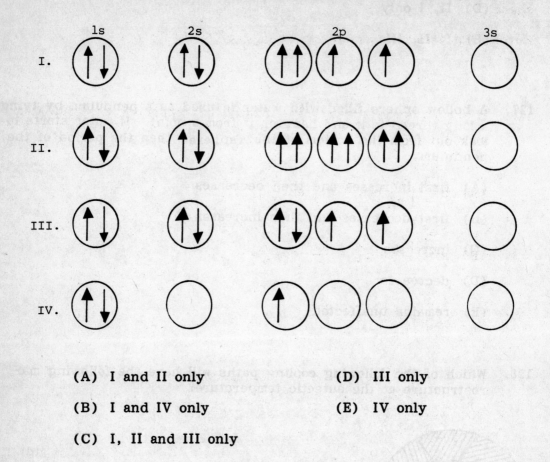

(A) I and II only

(B) I and IV only

(C) I, II and III only

(D) III only

(E) IV only


130. The microstructure composition of pearlite for an $Fe_3C$ diagram consists of

(A) carbon dissolved in alpha iron having a body-centered cubic structure

(B) carbon dissolved in gamma iron having a face-centered cubic structure

(C) a mixture of body-centered alpha iron and face-centered gamma iron

(D) carbon dissolved in body-centered alpha iron and an $Fe_3C$ compound of higher carbon content

(E) carbon dissolved in face-centered gamma iron and an $Fe_3C$ compound of higher carbon content

DIRECTIONS: Choose the best answer for each question and mark the letter of your selection on the corresponding answer sheet.

Note: The natural logarithm of $x$ will be denoted by $\ln x$.

131. If the equation $e^{\frac{x^2}{y^3}} \cdot \cos \frac{x}{y} = 1$ defines $y$ implicitly as a function of $x$, then $\frac{dy}{dx} =$

(A) $e^{\frac{x^2}{y^3}} \left( \frac{\partial x}{y^3} \cos \frac{x}{y} - \frac{1}{y} \sin \frac{x}{y} \right)$

(B) $e^{\frac{x^2}{y^3}} \left( \frac{x}{y^2} \sin \frac{x}{y} - \frac{3x^2}{y^4} \cos \frac{x}{y} \right)$

(C) $\dfrac{\left( \frac{\partial x}{y^2} \cos \frac{x}{y} - \sin \frac{x}{y} \right)}{\left( \frac{1}{y} \sin \frac{x}{y} - \frac{3}{y^3} \cos \frac{x}{y} \right)}$

(D) $\dfrac{\left( \frac{1}{y} \sin \frac{x}{y} - \frac{3}{y^3} \cos \frac{x}{y} \right)}{\left( \frac{\partial x}{y^2} \cos \frac{x}{y} - \sin \frac{x}{y} \right)}$

(E) $\dfrac{\left( \sin \frac{x}{y} - \frac{\partial x}{y^2} \cos \frac{x}{y} \right)}{\left( \frac{x}{y} \sin \frac{x}{y} - \frac{3x^2}{y^3} \cos \frac{x}{y} \right)}$

132. Which of the following is a Hermitian matrix?

(A) $\begin{bmatrix} 3 & i & 5+2i \\ -i & 2 & 6 \\ 5-2i & 6 & 1 \end{bmatrix}$

(B) $\begin{bmatrix} i & -2 & 1 \\ 2 & 2i & 3 \\ 1 & 3 & 0 \end{bmatrix}$

(C) $\begin{bmatrix} 5 & 3-2i & 6 \\ 3+2i & i & 0 \\ 8i & 0 & 6 \end{bmatrix}$

(D) $\begin{bmatrix} 0 & 3+i & 0 \\ 3+i & 0 & 2 \\ 0 & 2 & 0 \end{bmatrix}$

(E) $\begin{bmatrix} 1 & 2 & 4i \\ 2 & 1 & 6i \\ 4i & 6i & 1 \end{bmatrix}$

133. The curves shown in the sketch below are defined by the equations,

$$2(y-1)^2 = x$$
$$(y-1)^2 = x - 1$$

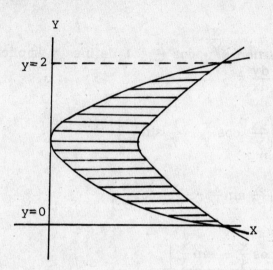

The area bounded by both curves is

(A) 2

(B) $\frac{4}{3}$

(C) $\frac{1}{4}$

(D) $\frac{2}{3}$

(E) $\frac{8}{3}$

134. In 3-dimensions, the vector $\vec{C}$ such that $\vec{C} = \vec{a} \times \vec{b}$ is

(A) in the plane of $\vec{a}$ and $\vec{b}$

(B) in a plane parallel to that containing $\vec{a}$ and $\vec{b}$

(C) in a plane perpendicular to that containing $\vec{a}$ and $\vec{b}$

(D) parallel to $\vec{a}$

(E) a linear combination of $\vec{a}$ and $\vec{b}$

135. $\int_C \vec{F} \cdot \vec{dc}$ where $\vec{F}(x,y) = (x^2, xy)$, $C: x=y^2$ between $(1,-1)$ and $(1,1)$ has the value

(A) $\frac{1}{2}$                     (D) $\frac{1}{6}$

(B) 6                                  (E) $-\frac{1}{2}$

(C) 0

136. If $y = x^x$, then $\frac{dy}{dx} =$

   (A) $xx^{x-1}$                      (D) $x^x \ln x$

   (B) $x \ln x$                       (E) $x(1 + \ln x)$

   (C) $x^x(1 + \ln x)$

137. $\int_0^\pi \left( \sum_{k=1}^\infty \frac{\sin kx}{k^3} \right) dx =$

   (A) 0                               (D) $\sum_{k=1}^\infty \frac{2}{(2k-1)^3}$

   (B) 1

   (C) $\sum_{k=1}^\infty \frac{1}{(2k-1)^3}$    (E) $\sum_{k=1}^\infty \frac{2}{(2k-1)^4}$

138. $\int_{3\pi}^{7\pi/2} \cos x \, dx =$

   (A) 1                               (D) $-1$

   (B) $\sqrt{2} + 1$                  (E) $\sqrt{2}$

   (C) 0

139. If the slope of line a is $\frac{3}{8}$ and the slope of line b is $\frac{3}{8}$, then the lines a and b are

   (A) parallel to the x-axis          (D) perpendicular to each other

   (B) parallel to each other but not to the x-axis or y-axis    (E) intersecting lines

   (C) parallel to the y-axis

140. If $\sum_{n=0}^{\infty} (a_0 + a_1 + \ldots + a_n)x^n = \sum_{n=0}^{\infty} \frac{a_n x^n}{1-x}$ and $a_n = \frac{1}{n!}$, then $\sum_{m=0}^{\infty} \left[ \sum_{n=0}^{\infty} a_n \right] x^m$, for $|x| < 1$, is:

(A) $e^x$

(B) 1

(C) $\frac{e}{1-x}$

(D) $\frac{e^x}{1-x}$

(E) $\infty$

141. The equation

$$U_t - a^2 U_{xx} = 0$$

represents a(n)

(A) elliptic P.D.E.

(B) hyperbolic P.D.E.

(C) parabolic P.D.E.

(D) exact D.E.

(E) nonhomogeneous D.E.

142. $\sum_{n=0}^{\infty} \frac{(-1)^n}{2^{2n}} =$

(A) $\infty$

(B) $\frac{3}{2}$

(C) $\frac{3}{4}$

(D) $\frac{4}{5}$

(E) 0

143. Which of the following is the equation of the plane passing through the point $(4, -1, 1)$ and parallel to the plane $4x - 2y + 3z - 5 = 0$?

(A) $4x - 2y + 3z = -1$

(B) $4x - y + z = 0$

(C) $4x - 2y + 3z = 21$

(D) $2y - 3z = 18$

(E) Not necessarily any of the above

144. A point moves on the parabola $6y = x^2$ in such a way that when $x = 6$ the abscissa is increasing at the rate of 2 ft. per second. At what rate is the ordinate increasing at that instant?

   (A) 3 ft/sec

   (B) 1 ft/sec

   (C) 4 ft/sec

   (D) 2 ft/sec

   (E) 6 ft/sec

# THE GRADUATE RECORD EXAMINATION ENGINEERING TEST

## MODEL TEST V

# ANSWERS

| | | | | | |
|---|---|---|---|---|---|
| 1. | E | 21. | B | 41. | B |
| 2. | C | 22. | C | 42. | E |
| 3. | A | 23. | A | 43. | A |
| 4. | B | 24. | E | 44. | D |
| 5. | D | 25. | A | 45. | E |
| 6. | E | 26. | E | 46. | D |
| 7. | A | 27. | E | 47. | A |
| 8. | C | 28. | B | 48. | E |
| 9. | D | 29. | C | 49. | C |
| 10. | D | 30. | C | 50. | A |
| 11. | D | 31. | C | 51. | A |
| 12. | C | 32. | B | 52. | C |
| 13. | B | 33. | A | 53. | B |
| 14. | E | 34. | A | 54. | B |
| 15. | C | 35. | D | 55. | B |
| 16. | B | 36. | D | 56. | D |
| 17. | D | 37. | D | 57. | B |
| 18. | D | 38. | C | 58. | B |
| 19. | C | 39. | D | 59. | E |
| 20. | B | 40. | E | 60. | A |

| | | | | | |
|---|---|---|---|---|---|
| 61. | A | 89. | C | 117. | B |
| 62. | E | 90. | B | 118. | C |
| 63. | B | 91. | E | 119. | A |
| 64. | C | 92. | A | 120. | C |
| 65. | E | 93. | D | 121. | E |
| 66. | C | 94. | E | 122. | B |
| 67. | B | 95. | D | 123. | C |
| 68. | D | 96. | A | 124. | B |
| 69. | C | 97. | B | 125. | B |
| 70. | A | 98. | C | 126. | E |
| 71. | C | 99. | C | 127. | A |
| 72. | D | 100. | B | 128. | D |
| 73. | D | 101. | A | 129. | A |
| 74. | B | 102. | D | 130. | D |
| 75. | E | 103. | C | 131. | E |
| 76. | E | 104. | E | 132. | A |
| 77. | D | 105. | C | 133. | B |
| 78. | D | 106. | E | 134. | C |
| 79. | B | 107. | C | 135. | C |
| 80. | B | 108. | D | 136. | C |
| 81. | B | 109. | A | 137. | E |
| 82. | C | 110. | E | 138. | D |
| 83. | C | 111. | D | 139. | B |
| 84. | C | 112. | B | 140. | D |
| 85. | B | 113. | E | 141. | C |
| 86. | B | 114. | D | 142. | D |
| 87. | A | 115. | C | 143. | C |
| 88. | A | 116. | C | 144. | C |

# THE GRADUATE RECORD EXAMINATION ENGINEERING TEST

## MODEL TEST V

## DETAILED EXPLANATIONS OF ANSWERS

1. **(E)**
Throughout the vaporization of water, heat is added but the temperature does not change. Since there is no degree of freedom in the vaporization process, the pressure must also remain the same. The volume of the water expands when it undergoes a phase change from a liquid to a vapor. Hence, this process can be described as both isothermal and isobaric.

2. **(C)**
When a conductor is moved toward the north pole of a bar magnet, it will experience a change of magnetic flux as a result of which there will be an induced emf. This emf will give rise to an induced current in the conductor, in a direction (determined by Lenz's law) that will oppose the movement of the conductor. Such currents are called eddy currents.

In general, eddy currents occur in conductors that experience a change of flux. This flux change can be due either to the mechanical motion in a magnetic field as in the armature of a motor, or to a changing current in the wires wound around the soft iron core of a transformer. In both cases, eddy currents cause $i^2R$ losses due to heating. These can be minimized by laminating the iron core, i.e. building it up with thin sheets covered by a thin coating of insulating varnish. This reduces the eddy currents.

3. (A)

Applying Bernoulli's equation to points at the orifice of the tank and at the surface of the fluid and taking the bottom as our reference point (1), we get

$$p_1 + \frac{1}{2}\rho v_1^2 + \rho g h_1 = p_T + \frac{1}{2}\rho v_2^2 + \rho g h_2$$

since $h_1 = 0$(datam)

at the orifice $v_1^2 = v_2^2 + 2\dfrac{p_T - p_1}{\rho} + 2gh_2$

where $p_r$ = Tank pressure.

From the continuity equation

$$v_1 = \frac{A_2}{A_1} v_2$$

Since $A_1 \gg A_2$, $v_2$ is small and can be neglected leaving

$$v_1^2 = 2\frac{p_T - p_1}{\rho} + 2gh_2$$

By opening the top of the tank to the atmosphere. Bernoulli's equation reduces to classic Torricelli result

$$v_1^2 = 2gh$$

Therefore the speed of the fluid flowing out is reduced when the tank pressure equals the orifice pressure.

4. (B)

In order to have stable equilibrium at the top of the cylinder, a slight displacement from the initial state should increase the gravitational potential energy. In such a case, the cube will be forced to its initial state to reach the minimum level of potential energy.

Since the potential energy is a function of height, we only have to make the height of the centroid of the cube larger than its initial value upon a slight disturbance from the equilibrium state.

Measuring the heights from the center of the cylinder (only for convenience), we can write: height after a slight displacement = OC' + C'D + D'M > initial height = $r + \frac{b}{2}$ but we have:

$$OC' = r \cos\theta$$

C'D = CO $\sin\theta$ = $r\theta\sin\theta$ (the length of CD is equal to that of arc AC, because there is no slipping)

and
$$D'M = \frac{b}{2} \cos\theta$$

$$\Rightarrow r\cos\theta + r\theta\sin\theta + \frac{b}{2}\cos\theta > r + \frac{b}{2}$$

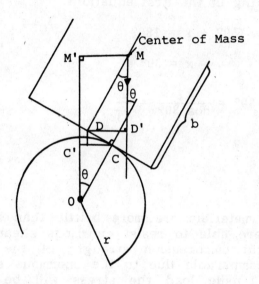

In a neighborhood near zero, we have:

$$1 - 2\frac{\sin^2\theta}{2} = \cos\theta = 1 - \frac{\theta^2}{2} \text{ and } \theta = \sin\theta$$

Hence:

$$r\left(1 - \frac{\theta^2}{2}\right) + r\theta^2 + \frac{b}{2}\left(1 - \frac{\theta^2}{2}\right) > r + \frac{b}{2}$$

$$\Rightarrow r + \frac{b}{2} + r\theta^2 - \frac{r\theta^2}{2} - \frac{b\theta^2}{4} > r + \frac{b}{2}$$

$$\left(\frac{r}{2} - \frac{b}{4}\right)\theta^2 > 0$$

$$\Rightarrow \frac{r}{2} > \frac{b}{4}$$

$$r > \frac{b}{2}$$

5. (D)
To solve the problem, we set up the following system of equations:

$$\angle A + \angle C + x = 180°$$

$$\tfrac{1}{2}\angle A + \tfrac{1}{2}\angle C + 105° = 180°$$

From the second equation:

$$\tfrac{1}{2}(\angle A + \angle C) = 75°$$

$$\angle A + \angle C = 150°$$

Substituting in the first equation:

$$150° + x = 180°$$

$$x = 30°$$

$$\sec 30° = \frac{1}{\cos 30°} = \frac{2}{\sqrt{3}}$$

6. (E)
Ceramic materials are more brittle than metals. Ceramic materials are able to resist enormous shear stresses. The tension and compression strength of the ceramic material are not comparable due to the enormous shear resistance. Under a tensile load the stress will be concentrated at cracks and flaws in the ceramic material. A more ductile material will be able to lower the stress concentration through plastic deformation. However, if plastic deformation cannot occur due to high shear resistance, the stress concentration will continue to increase as the load increases. The crack will extend and a brittle fracture will take place.

7. (A)
The kinetic molecular theory of gases makes assumptions about the simplest way molecules interact - they exert no other forces on each other except those due to elastic collisions. It assumes that molecules will be traveling at constant speed which is conserved after the collision with the other molecules. Therefore, the speed of molecules remains constant before and after the collision. (Answer IV is wrong.) The total amount of motion of the gas is equal to the total kinetic energy, which is in turn a function of temperature. Therefore the size of the container does not affect it.

8. (C)

In polar coordinates, each point is specified by:

1) The angle between the x-axis and the line joining the point to the origin (measured counterclockwise).
2) The distance between the point and the origin.

Hence $\quad r = f(\theta)$

$$e^\theta = e^{\ln r}$$

$$r = e^\theta$$

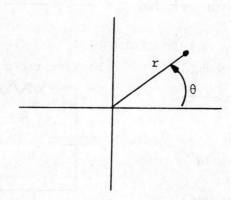

Therefore the locus of points whose polar coordinates are related by the function $r = e^\theta$ has a shape similar to the curve given in answer (C), because as $\theta$ increases, the distance between the point and the origin increases too. The more turns made around the origin, the more the distance from the origin will be.

9. (D)

The formula for the range of a projectile is given by

$$R = \frac{V_0^2}{g} \sin 2\theta_0$$

where $V_0$ is the initial velocity of the projectile and $\theta_0$ is the angle with the horizontal of the initial velocity vector. In this case

$$R = \frac{(20 \text{ m/sec})^2}{10 \text{ m/sec}^2} \sin 2(15°) = \frac{400 \text{ m}^2/\text{sec}^2}{10 \text{ m/sec}^3} \sin 30°$$

$$= (40\text{m})(\tfrac{1}{2}) = 20\text{m}$$

Note then that the maximum range can be obtained if $\theta_0$ is equal to 45° since this makes $\sin 2\theta_0 = 1$, its maximum value.

## 10. (D)

To find the Thevenin resistance, we first have to find the Thevenin voltage. For the battery circuit the Thevenin voltage is the open circuit voltage which is given to be x volts. The voltage drop across the Thevenin resistor, since it is connected in series to the Thevenin voltage source, is equal to the voltage drop when the battery supplies z amperes to the motor for starting. Hence the voltage drop across the Thevenin resistor equals the open circuit voltage minus the terminal voltage which is given to be (x - y) volts, i.e. voltage drops x - (x - y) volts.

Hence Thevenin resistance = $\dfrac{x - (x - y)}{z} \Omega$ .

CIRCUIT DIAGRAMS

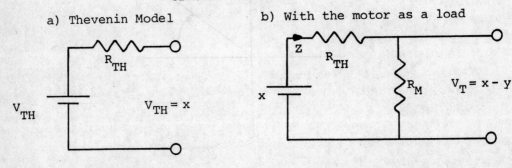

a) Thevenin Model     b) With the motor as a load

## 11. (D)

The given circuit can be presented in a simpler form as follows:

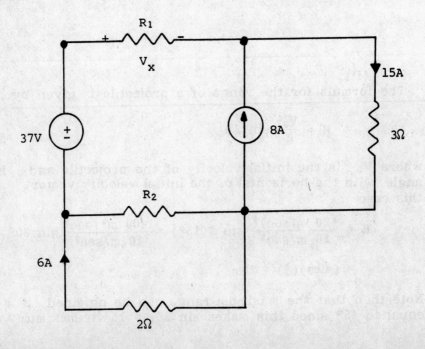

Now let us define node and branch. A branch is that part of a network which consists of a single component or a group of components connected in series. A point connecting two or more branches is said to be a node of a network.

To determine the nodes and branches of a circuit, simplify a circuit such that it uses a minimum number of nodes without the circuit elements as follows:

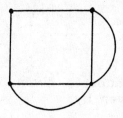

From the above figure we get three nodes as shown by dots. Note that the upper left vertex is not a node since it connects two components in series.

Now inserting the circuit elements we get the following:

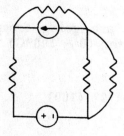

From the above figure it is clear that there are five branches.

It is also possible to draw the above two figures from the original circuit diagram as follows:

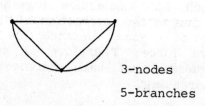

3-nodes
5-branches

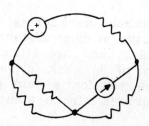

**12. (C)**
As soon as the acrobat leaves the swing, his angular momentum is $I\omega$. Since there is no external torque acting on the acrobat, angular momentum must be conserved. (Note: Gravity exerts no external torque since it acts on the center of mass of the acrobat.) Once he pulls in his arm and legs, the acrobat is decreasing his inertia, $I$, and therefore his angular velocity, $\omega$, must be increased.

**13. (B)**
The volume of the cube is given: $s^3 = 3^3 = 27 m^3$. The change in volume is given by:

$$\Delta V = K V_0 \Delta T$$

where $\Delta T$ is the temperature change, $V_0$ is the initial volume and $K$ is the expansion coefficient. Hence

$$\Delta V = (.0006)(27) \Delta T$$

$\Delta T = 100$ since we are going from the freezing point to the boiling point of water ($0 - 100°C$).

$$\Delta V = (.0006)(27)(100)$$
$$= 1.62 m^3$$

So $\quad \frac{\Delta V}{V_0} = .06 = 6\%$

**14. (E)**
Intermolecular forces which hold molecules together are repulsive at close distances due to the interaction of charged particles, and are attractive at a great distance due to the action of a weak attractive force. Therefore an inverse square relation should predominate at small distances until the force is attractive and increasing.

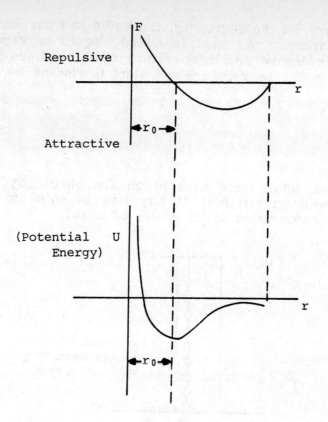

Since potential energy is related by $F = \frac{-du}{dr}$ it has a minimum when the force = 0. This function is best described by (E).

### 15. (C)

The difference between the energy generated by the conductor and the energy transferred to the fluid is the energy stored in the conductor. The increase in stored energy with time will result in an increase in the conductor temperature due to unsteady-state conditions. All of the energy has not yet been conducted to the surface of the conductor to be convected to the cooling fluid. When it does, the rate at which electrical energy is being dissipated is the same as the rate at which energy is convected to the fluid. Therefore the steady state will be reached.

### 16. (B)

Heating a saturated liquid at constant pressure will cause it to vaporize without an increase in temperature. Once all the liquid is vaporized further addition of heat will cause the temperature to rise.

Therefore, in the beginning, the liquid and gas states exist in equilibrium. As heat is added, liquid is vaporized at constant pressure and temperature until we have only gas. Since then the temperature will start to rise as we add more heat.

17. (D)
The net total force exerted on the chain (by both the surface and gravitation) at any time is equal to its mass times the acceleration of its center of mass.

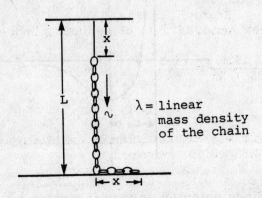

$\lambda$ = linear mass density of the chain

To find the equation of motion of the center of mass, according to the figure, we can write (all the distances are evaluated with respect to the hanging point):

$$x_{cm} = \frac{(x\lambda)L + (L-x)\lambda(x + \frac{L-x}{2})}{L\lambda}$$

$$= x + \frac{L^2 - x^2}{2L}$$

$$\Rightarrow \dot{x}_{cm} = \dot{x} - \frac{x\dot{x}}{L}$$

$$\ddot{x}_{cm} = \ddot{x} - \frac{x\ddot{x} + \dot{x}^2}{L} \Rightarrow M\ddot{x}_{cm} = Mg - N = M\left(\ddot{x} - \frac{x\ddot{x} + \dot{x}^2}{L}\right)$$

N: the normal force of the surface

But $\ddot{x} = g$ since the chain is falling freely and also we have:

$$\dot{x}^2 = 2gx$$

(equation of motion with constant acceleration)

So we have

$$N = \frac{M}{L}(xg + 2gx) = \frac{3M}{L}gx$$

**18. (D)**
The following equation represents the reaction

$$NaCl + H_2SO_4 \rightarrow NaHSO_4 + HCl \uparrow$$

HCl is removed in the form of a gas and causes the reaction to proceed.

On the other hand, 14.5 g of sodium chloride is:

$$\frac{14.5}{23+35} = \frac{14.5}{58}$$

$$= 0.25 \text{ mol}$$

Since one mole of sodium chloride releases one mole of HCl, 0.25 mole of HCl will be released.

At the given condition of standard pressure and temperature, 1 mole of each gas will have 22.4 liters of volume. Therefore $22.4 \times 0.25 = 5.6$ liters of HCL will be released from the reaction.

**19. (C)**
An atom consists of a positively charged nucleus, surrounded by one or more negatively charged electrons. The nucleus contains protons and neutrons.

When two atoms come close enough to combine chemically, that is to form a chemical bond, each atom "sees" only the outermost electrons which surround the nucleus of the other atom. Hence, these outer electrons play a very important part in the chemical behavior of atoms. Neutrons have little effect on chemical behavior while protons are significant since they determine how many electrons surround the nucleus in a neutral atom.

**20. (B)**
An adiabatic process is one that allows no thermal energy to be transferred between the gas and its surroundings.

The work done by the expanding gas is at the expense of the internal energy, so the temperature decreases. Alternatively, if the gas is compressed adiabatically, the temperature rises, because

$$\Delta Q = 0 \text{ and } \Delta\omega < 0, \text{ therefore we can write}$$

$$\Delta u = \Delta Q - \Delta\omega > 0, \text{ which means an increase in temperature.}$$

**21. (B)**
By definition, the 'gravitational intensity' is the force per unit mass in a gravitational field.

Thus, gravitational intensity has units of

$$\frac{N}{kg} \text{ or } \frac{kg\ m}{sec^2} \times \frac{1}{kg} = \frac{m}{sec^2}$$

Note that these are also the units of acceleration.

**22. (C)**
The relationship between the Fahrenheit and Celsius temperature scales is given by:

$$T_F = \frac{9}{5} T_C + 32°F$$

where:
$T_F$ = temperature in Fahrenheit
$T_C$ = temperature in Celsius

Hence we have

$$T_F = \frac{9}{5}(20) + 32 = 36 + 32 = 68°F$$

**23. (A)**
The forces on the block are its weight w and the normal and frictional components of the force exerted by the plane. The angle $\theta$ of the inclined plane is adjusted until the block slides down the plane. Since motion exists, the friction force is $f_k = \mu_k N$. Take axes perpendicular and parallel to the surface of the plane. Then, applying Newton's

Second Law to the x and y components of the block's motion, we obtain

$$\Sigma F_x = \mu_k N - w \sin \theta = 0$$

$$\Sigma F_y = N - w \cos \theta = 0$$

where $\Sigma F_x$ and $\Sigma F_y$ are the x and y components of the net force on the block. Both of these equations are equal to zero because the block accelerates neither parallel nor perpendicular to the plane. Hence

$$\mu_k N = w \sin \theta,$$

$$N = w \cos \theta.$$

Dividing the former by the latter, we get

$$\mu_k = \tan \theta, \text{ so } \theta = \tan^{-1} \mu_k$$

It follows that a block, regardless of its weight, slides down an inclined plane with constant speed if the tangent of the slope angle of the plane equals the coefficient of kinetic friction. Measurement of this angle then provides a simple experimental method of determining the coefficient of kinetic friction.

24. (E)
Another name for heat pump is refrigerator. It is used to transfer heat from areas of lower temperature, against the temperature gradient, to areas of higher temperature, by inputting work.

25. (A)
The capacitance of a capacitor is determined by the amount of electric charge that may be stored in it for each volt of applied voltage.

i.e. $C = \dfrac{Q \text{ (coulombs)}}{V \text{ (volts)}}$ Farads.

This eliminates choice (E).

Also, capacitance is determined by:
1) The material used as a dielectric.

2) The area of the plates.

3) The distance between the plates.

These factors are related in the mathematical formula as,

$$C = \frac{kA\varepsilon_0}{d}$$

where    $k$ = dielectric constant,

$A$ = area of one side of one plate in square inches

$d$ = the distance between the plates in inches.

From the above formula, it is obvious that the capacitance increases as the distance between the plates decreases.

26. (E)
In any throttling, the fluid flows through a constriction from a high pressure region to a low pressure region without any heat transfer occurring.

The fluid changes volume in going from one pressure region to another. There is no significant change in the kinetic energy of the fluid. Since

$$h_i + \frac{V_i^2}{2g_c} = h_e + \frac{V_e^2}{2g_c}$$ , we have that

$h_i = h_e$. The enthalpy remains the same.

27. (E)
A common mirror reflects light because the back of the glass is coated with a layer of metal. This metallic backing reflects light.

28. (B)
Suppose that the person stands at a distance, $x$, from the wall and that her eyes (E) are a distance, $y$, from the top of her head (H). To look at her toes (F) she looks at point A, which is the point of reflection of a light ray from her foot. "A" must be at a height halfway between her eyes and feet so that the angle of incidence equals the angle of reflection. Similarly, to look at the top of her head she

looks at point B. If OP = h and BP = y/2, then the length of the mirror is AB = PO - BP - OA = h - y/2 - 1/2(h - y) = h/2. Thus the minimum length of the mirror is h/2, and this does not depend on the distance x that the person is standing away from the mirror.

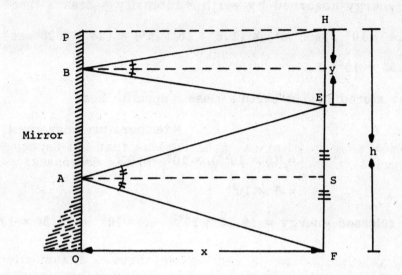

### 29. (C)

An acceleration that varies linearly with time can be expressed as

$$a = \frac{dv}{dt} = a_0 + kt$$

There is no curve with acceleration shown versus time.

$$v = v_0 + a_0 t + \frac{kt^2}{2}$$

Curves (a), (d) and (e) are drawn with velocity against time. None of them satisfies the above relations.
However, we know that

$$s = s_0 + v_0 t + \frac{a_0 t^2}{2} + \frac{kt^3}{6}$$

and only (C) satisfies this relation.

### 30. (C)

Most of the solar energy obtained from the sun through the day hours is sent back to space through radiation from the surface of the earth. This released energy is equal to the difference between the total energy absorbed from the sun and the stored heat which has raised the earth's temperature.

So we have:

total energy absorbed by earth = intensity × area × time

$= (1.4 \times 10^3 \text{ J/sec m}^2) \times (1.6 \times 10^{13} \text{ m}^2) \times (24 \times 3600 \text{ sec})$

$= 19.35 \times 10^{20}$ J

The stored heat = earth's mass × specific heat
$\hspace{4cm}$ × temperature increment

$\hspace{2cm} = (6 \times 10^{24}) \times 10^2 \times 10^{-6}$

$\hspace{2cm} = 6 \times 10^{20}$

$\Rightarrow$ released energy $= 19.35 \times 10^{20} - 6 \times 10^{20} = 13.35 \times 10^{20}$ J

### 31. (C)

When the box is just at the point of slipping, the frictional force ($f_c$) should be equal to the stopping force on the box caused by the deceleration of the truck.

The stopping force $F = ma = m \times 5 \text{m/sec}^2$ where m is the mass of the box. So $f_c$ = stopping force = $m \times 5 \text{m/sec}^2$.

We know that the maximum value of the frictional force for a constant coefficient of friction is:

$$F_{f_{max}} = \mu N$$

where N is the normal force exerted by the ground.

Since we are looking for the smallest value of the friction coefficient to prevent the block from sliding, we must use the maximum value of friction force to do the job. Therefore we must have: $f_c = \mu_s N$ or $\mu_s = \dfrac{f_c}{N}$

Hence

$$\mu_s = \frac{m \times 5}{m \times g}$$

$$= \frac{5}{10} = 0.500$$

## 32. (B)
From Faraday's Law

$$\oint E \, dl = \frac{-d\phi}{dt}$$

$$\therefore \quad \oint \nabla \times E \, ds = -\oint \frac{\partial B}{\partial t} \cdot ds$$

$$\nabla \times E = -\frac{\partial B}{\partial t}$$

if $E = E_0 \sin(\omega t - kx)$ $\quad \frac{\partial E}{\partial x} = -\frac{\partial B}{\partial t}$

$$\therefore \quad KE = \omega B \Rightarrow E = cB$$

(The E-field and B-field have to be in phase.)

Therefore when $E = E_{max}$, we must have $B = B_{max}$.


## 33. (A)
Knowing the voltage across the 2-$\Omega$ resistor makes it possible to find the current through the branch containing the 2-$\Omega$ and the 3-$\Omega$ resistors.

$$i_1 = \frac{V}{R} = \frac{6}{2} = 3A$$

$i_1$ must be equal to $i_2$ because they both flow through 5-$\Omega$ resistances. The series resistor is equivalent to

$$R = 3 + 2 = 5\Omega.$$

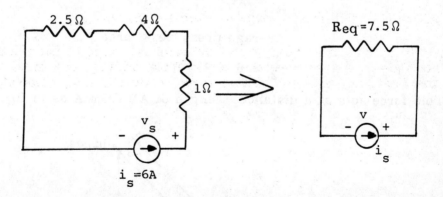

Since $\quad i_s = i_1 + i_2$ we have, $\quad i_s = 6A$.

To find $v_s$, calculate the equivalent resistance across the terminals of the source.

$$R_{eq} = 7.5 \Omega$$

It follows that

$$v_s = R_{eq} i_s = 7.5(6) = 45V.$$

### 34. (A)
Isotopes of the same element have positively charged nuclei which makes it very difficult for them to be brought together, because of the electrostatic repulsion between them. In order to bring these nuclei together, a tremendous amount of energy (high temperature) is required.

### 35. (D)
Pressure at A = $\gamma 6 \cos 60° = 6K$

Total pressure force, P

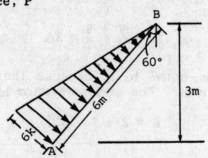

= average pressure × area

= $\frac{6K}{2} \times 6 \times 8 = 144K$

This force acts at a distance $\frac{1}{3}$ length of AB from A as in fig.

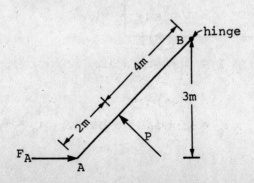

Taking moment about B

$$F_A \times 3 = P \times 4$$

$$F_A = 144K \times \frac{4}{3} = 192K$$

## 36. (D)

All elements in Group O have outermost subshells that are filled. Hence all electrons in them are paired. For example:

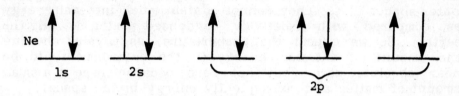

We see that all electrons in the 2p subshell are paired. Therefore III must be true.

II is not true because the outermost shells are filled, which makes the Group O element very non-reactive. (This is why they are called the inert gases.) I is not true because helium is in Group O and its outermost subshell is a filled s subshell, not a filled p.

## 37. (D)

For flows over a flat plate, the Reynold's Number is given by:

$$Re = \frac{VL}{\nu} = \frac{\rho VL}{\mu}$$

where:  Re = Reynold's Number
V = Velocity of flow
$\rho$ = Density
L = distance (from the leading edge) at which Re is computed
$\mu$ = viscosity
$\nu$ = kinematic viscosity

38. (C)

If one has ever observed a pair of scissors closely, one would discover that neither edge is sharp in any sense. Upon further observation, one would also find that the two edges do not close onto each other, but merely pass each other. Such a situation, where two equal and opposite forces act upon an object but whose lines of action are displaced, is that of a shearing action; the paper is cut (torn) because it cannot sustain the shearing stress.


39. (D)

Temperature assignment is a way to show how much internal energy exists in the environment (or system). There are very small amounts of matter scattered throughout outer space. They cannot contain much internal energy as compared with relatively condensed materials on the earth. So we expect the temperature, as a value showing the amount of internal energy in the environment, to be near absolute zero (but not zero because there is a small amount of matter and consequently energy in the space).

On the other hand you cannot lose or gain much heat through convection due to the unavailability of matter in the boundary environment, so you won't feel cold! (Loss of energy by means of radiation is not considerable.)


40. (E)

The speed of a d.c. series motor varies inversely with the field flux. Any change in load causes a change in armature current and so in the field flux. The IR drop of the armature also varies with the change in load, but its effect is very small as compared to that of the field flux. Thus the speed of a d.c. series motor is very much dependent on the field flux and a stronger flux results in a lower speed and vice versa. Hence, the speed varies from very high speed at light load to a low speed at full load. Such a characteristic of speed-load is shown by choice (E).

Choices (A) and (B) show the torque-load characteristic of shunt and series d.c. motors respectively. Choices (C) and (D) show the speed-load characteristic of shunt and compound d.c. motors respectively.


41. (B)

The photoelectric effect is described by a collision

process. Photons strike electrons at the surface of the metal. The photon's energy must be at least as large as the work function of the metal.

Since the photon's energy is proportional to the frequency of the incident radiation, it is frequency and not intensity that determines whether or not the photoelectric effect occurs.

### 42. (E)

It is found experimentally that the number of nuclei of a radioactive sample which decay per unit time is proportional to the number of nondecayed nuclei present.

Hence, the rate at which nuclei decays is given by

$$\frac{dn}{dt} = -\lambda n$$

where n is the number of nuclei still not decayed at time t and $\lambda$ is a proportionality constant called the decay constant.

The above equation can be integrated to give

$$\ln n = -\lambda t + C$$

At time $t = 0$, $C = \ln n_0$, where $n_0$ is the number of nuclei at time $t_0$. Substituting the value of C and taking inverse logs we get

$$n = n_0 e^{-\lambda t}$$

A plot of $n/n_0$ against t gives us the curve (E).

### 43. (A)

The energy of a photon (E) is given by $E = hf_1$ where h is Planck's constant and f is the wave frequency of the photon. Since h is a constant, E is proportional to f. Therefore an increase in f increases the energy. Such a proportional relationship is represented by the straight line (A).

### 44. (D)

The equation for a potential difference between the two plates is as follows:

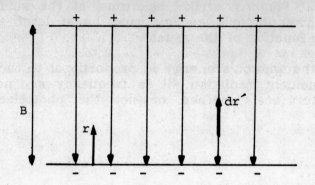

$$V_B - V_r = -\int_r^B \vec{E} \cdot \vec{dr'}$$

$$V_B - V_r = \int_r^B E\, dr'$$

$$V_B - V_r = E(B - r)$$

Therefore curve D most closely resembles the potential difference.

**45. (E)**
The effective interest rate for an annual interest rate of 6% that is compounded biannually is 1%. Therefore the interest accrued every 2 months on the issued note is $60. The interest that is accumulated in 40 days is $\frac{2}{3}(\$60) = \$40$.

**46. (D)**
Use the rule: $\int \frac{du}{u} = \ln|u| + C$, letting $u = x^2 + 1$ making $du = 2x\,dx$.

$$\int \frac{2x\,dx}{x^2 + 1} = \ln(x^2 + 1) + C.$$

We may omit the absolute value sign here, since a number, x, when squared, is always positive.

47. (A)

This problem involves dependent events. Two or more events are said to be dependent if the occurrence of one event has an effect upon the occurrence or non-occurrence of the other. If you are drawing objects without replacement, the second draw is dependent on the occurrence of the first draw. We apply the following theorem for this type of problem. If the probability of the occurrence of one event is p and the probability of the occurrence of a second event is q, then the probability that both events will happen in the order stated is pq.

To find the probability that both balls will be white, we express it symbolically.

p (both balls will be white) =

p (first ball will be white and the second ball will be white) =

p (first ball will be white) · p (second ball will be white) =

$$= \left( \frac{\text{number of ways to choose a white ball}}{\text{number of ways to choose a ball}} \right) \cdot \left( \frac{\text{number of ways to choose a second white ball after removal of the first white ball}}{\text{number of ways to choose a ball after removal of the first ball}} \right)$$

$$= \frac{5}{16} \cdot \frac{4}{15} = \frac{1}{4} \cdot \frac{1}{3} = \frac{1}{12}$$

48. (E)

If A and B are independent events, then the probability that A or B occurs, but not both at the same time, $P(A \cup B) = P(A) + P(B)$. Here the symbol "U" stands for "or".

In this particular example, we only select one card at a time. Thus, we either choose a jack "or" a ten. P (a jack or a ten) = P (a jack) + P(a ten).

$$P(\text{a jack}) = \frac{\text{number of ways to select a jack}}{\text{number of ways to choose a card}} = \frac{4}{52} = \frac{1}{13}.$$

$$P(\text{a ten}) = \frac{\text{number of ways to choose a ten}}{\text{number of ways to choose a card}} = \frac{4}{52} = \frac{1}{13}.$$

$$P(\text{a jack or a ten}) = P(\text{a jack}) + P(\text{a ten}) = \frac{1}{13} + \frac{1}{13} = \frac{2}{13}.$$

**49. (C)**
Consider the formula: $\int \cos u \, du = \sin u + C$. Let $u = 2x$, then $du = 2dx$. To obtain $\int \cos u \, du$, we multiply $dx$ by 2 under the integral sign, and by $\frac{1}{2}$ outside the sign. We obtain:

$$\frac{1}{2} \int \cos 2x \cdot 2dx.$$

We apply the formula: $\int \cos u \, du = \sin u + C$. Then,

$$\int_{\pi/4}^{\pi/6} \cos 2x \, dx = \frac{1}{2} \sin 2x \Big]_{\pi/4}^{\pi/6}$$

$$= \int_{\pi/4}^{\pi/6} \cos 2x \, dx = \frac{1}{2} \left[ \sin \frac{2\pi}{6} - \sin \frac{2\pi}{4} \right]$$

$$= \frac{1}{2} \left[ \sin \frac{\pi}{3} - \sin \frac{\pi}{2} \right]$$

$$= \frac{1}{2} \left( \frac{\sqrt{3}}{2} - 1 \right) = \frac{\sqrt{3} - 2}{4}$$

**50. (A)**

$$\begin{bmatrix} 1 & 1 \\ 0 & 1 \end{bmatrix} = \begin{bmatrix} a_{11} & a_{12} \\ a_{21} & a_{22} \end{bmatrix}$$

$C_{11}$ = cofactor of $a_{11} = (-1)^{1+1} a_{22} = (-1)^2 (1) = 1$

$C_{12}$ = cofactor of $a_{12} = (-1)^{1+2} a_{21}$

$\qquad = (-1)^3 (0)$

$\qquad = 0$

$C_{21}$ = cofactor of $a_{21} = (-1)^{2+1} a_{12}$

$\qquad = (-1)^3 (1)$

$\qquad = -1$

$C_{22}$ = cofactor of $a_{22} = (-1)^{2+2} a_{11}$

$\qquad = (-1)^4 (1)$

$\qquad = 1$

$$[C] = \text{cofactor matrix} = \begin{bmatrix} C_{11} & C_{12} \\ C_{21} & C_{22} \end{bmatrix} = \begin{bmatrix} 1 & 0 \\ -1 & 1 \end{bmatrix}$$

Transpose of $[C] = \begin{bmatrix} 1 & -1 \\ 0 & 1 \end{bmatrix}$

$\det[A] = 1(1) - 1(0) = 1 - 0 = 1$

$$[A]^{-1} = \frac{\text{Adj}[A]}{|A|} = \frac{[C]^T}{|A|} = \begin{bmatrix} 1 & -1 \\ 0 & 1 \end{bmatrix}$$

51. **(A)**
Rearranging, we have

(csc x)(sin x)(cot x)(tan x)(sec x)(cos x)

$= \left(\dfrac{1}{\sin x}\right) \cdot (\sin x) \cdot \left(\dfrac{1}{\tan x}\right) (\tan x) \left(\dfrac{1}{\cos x}\right) (\cos x)$

$= 1$

52. **(C)**
The variance in ages is a measure of the spread or dispersion of ages about the sample mean.

To compute the variance we first calculate the sample mean.

$\overline{X} = \dfrac{\Sigma X_i}{n} = \dfrac{\text{sum of observations}}{\text{number of observations}}$

$= \dfrac{6 + 8 + 10 + 12 + 14 + 16}{6} = \dfrac{66}{6} = 11.$

The variance is defined to be

$$s^2 = \dfrac{\sum\limits_{i=1}^{n}(X_i - \overline{X})^2}{n}$$

$= \dfrac{(6-11)^2+(8-11)^2+(10-11)^2+(12-11)^2+(14-11)^2+(16-11)^2}{6}$

$= \dfrac{25 + 9 + 1 + 1 + 9 + 25}{6} = \dfrac{70}{6} = \dfrac{35}{3}$

53. (B)

To find $D_x y = \frac{dy}{dx}$, we use the differentiation formula:

$$\frac{d}{dx} e^u = e^u \frac{du}{dx}, \text{ with } u = \frac{1}{x^2}. \text{ We obtain:}$$

$$D_x y = e^{\frac{1}{x^2}} \cdot \left(-\frac{2x}{x^4}\right)$$

$$= e^{\frac{1}{x^2}} \left(-\frac{2}{x^3}\right) = -\frac{2e^{\frac{1}{x^2}}}{x^3}$$

54. (B)
$F(x)dx = f'(x)$

$$\int_a^b F(x)dx = \int_a^b f'(x) = f(x) \Big]_a^b$$

$$= f(b) - f(a)$$

Similarly for $\int_a^b G(x)dx = g(b) - g(a)$

$$\therefore \int_a^b F(x)dx + \int_a^b G(x)dx$$

$$= f(b) - f(a) + g(b) - g(a)$$

$$= f(b) + g(b) - f(a) - g(a)$$

$$= [f(b) + g(b)] - [f(a) + g(a)]$$

55. (B)

$$\int_0^\infty e^{-x} dx = -e^{-x} \Big]_0^\infty$$

$$= -\left[e^{-(\infty)} - e^{-(0)}\right]$$

$$= -\left[\frac{1}{e^{\infty}} - 1\right]$$

$$= -\left[\frac{1}{\infty} - 1\right]$$

$$= -[0 - 1]$$

$$= 1$$

**56. (D)**
Let $G(x) = u$, $G'(x)dx = du$

$$\therefore \int F'(u)du = F(u) + C$$

$$= F[G(x)] + C$$

**57. (B)**
Let $y = F(x)$, then

$$y = \frac{a}{b}x^n - \frac{2}{b}$$

Applying the theorem for $d(u^n)$,

$$\frac{dy}{dx} = \frac{a}{b} \cdot n \cdot x^{n-1} - 0,$$

or

$$\frac{dy}{dx} = \frac{na}{b} \cdot x^{n-1}.$$

The same result may be obtained by writing

$$y = \frac{1}{b}(ax^n - 2),$$

and solving for $\frac{dy}{dx} = \frac{1}{b}(nax^{n-1})$, where $F'(x) = \frac{dy}{dx}$.

58. **(B)**
For two perpendicular straight lines,

$$F'(x)G'(x) = -1$$

$$F'(x) = a$$

$$\therefore \quad G'(x) = -\frac{1}{a}$$

Therefore slope of $G(x)$ is $-\frac{1}{a}$.


59. **(E)**
In this example, we make use of the product rule, $d(uv) = vdu + udv$. We find:

$$\frac{dy}{dx} = x\left(e^{\tan x}\right)\sec^2 x + e^{\tan x}$$

$$= e^{\tan x}(x\sec^2 x + 1).$$


60. **(A)**
We use the formula: $\int u^n du = \frac{u^{n+1}}{n+1} + C$, letting $u=(x^3-7)$, $du = 3x^2 dx$, and $n = 8$.

Therefore,

$$\int (x^3 - 7)^8\, 3x^2 dx = \frac{(x^3 - 7)^9}{9} + C.$$


61. **(A)**

Let $\exp\left(\int_1^e \frac{dx}{x}\right) = I$

Now $\int_1^e \frac{dx}{x} = \ln x \Big|_1^e$

$$= \ln e - \ln 1$$
$$= 1 - 0 = 1 \qquad (\ln e = 1; \ln 1 = 0)$$
$$\therefore \quad I = e^1 = e$$

**62. (E)**
Let the numerator be
$$f(x) = \sin nx,$$
and let the denominator be
$$F(x) = x.$$

Then $f(0) = 0$, $F(0) = 0$. Therefore, the function takes the form 0/0 and we apply L'Hospital's rule.

$$\lim_{x \to 0} \frac{f(x)}{F(x)} = \lim_{x \to 0} \frac{f'(x)}{F'(x)} = \lim_{x \to 0} \frac{n \cos nx}{1} = n.$$

**Alternate Method:**

$$\lim_{x \to 0} \frac{\sin nx}{x}$$
$$= \lim_{x \to 0} \frac{n \sin nx}{nx}$$
$$= n \left[ \lim_{x \to 0} \frac{\sin nx}{nx} \right] \qquad \left[ \text{since } \lim_{\theta \to 0} \frac{\sin \theta}{\theta} = 1 \right]$$
$$= n \times 1 = n$$

**63. (B)**
Using the definition of matrix addition, add the (i,j)th element of A to the (i,j)th element of B. Thus,

$$A + B = \begin{bmatrix} 2 & 3 & 7 \\ 4 & m & \sqrt{3} \\ 1 & 5 & a \end{bmatrix} + \begin{bmatrix} \alpha & \beta & \delta \\ \sqrt{5} & 3 & 1 \\ p & q & 4 \end{bmatrix} = \begin{bmatrix} 2+\alpha & 3+\beta & 7+\delta \\ 4+\sqrt{5} & m+3 & \sqrt{3}+1 \\ 1+p & 5+q & a+4 \end{bmatrix}$$

**64. (C)**
For all real x, we have

$$\cos x = 1 - \frac{x^2}{2!} + \frac{x^4}{4!} - \ldots = \lim_{n\to\infty} \sum_{k=0}^{n} \frac{(-1)^k x^{2k}}{(2k)!}$$

$$\therefore \quad \lim_{n\to\infty} \sum_{k=0}^{n} \frac{(-1)^k \pi^{2k}}{(2k)!} = \cos \pi = -1$$

**65. (E)**
The function g, being the integral of f over a specified interval, is the area bounded by the curve (of the function f). As long as the graph is positive, g is increasing in that interval. Only after 5, when the function f takes on negative values, does the area of g begin to decrease. Hence in the interval from x = 5 to x = 7 where f is negative the function g(x) is decreasing. So the correct answer is (E).

**66. (C)**
Critical points (maxima and minima) for g occur when the function f crosses the x-axis (y = ϕ). These are points at which g changes either from decreasing to increasing values (minima) or increasing to decreasing values (maxima). Since the two places on f where the curve crosses the x-axis are at 5 and 7, we must check whether these are local maxima or minima on g. At x = 7, the function f is changing from negative values (which implies decreasing values of g) to positive values (which implies increasing values of g). This would then be a local minimum (decreasing to increasing values) of g. At x = 5, the opposite is the case, where the curve of g has a local maximum; therefore the correct answer is (C).

**67. (B)**
Since g is the integral of f, values of g may be interpreted as the area under the curve f. For the interval from x = 0 to 1, the value of f(x) is always positive, so g(x) steadily increases. As f(x) decreases (but remains positive) in the interval from x = 1 to 2, g increases but less quickly since the derivative is a positive decreasing function (at that interval). Also from 3 to 4 , g increases

rapidly and from 5 to 7 it changes direction (decreases). Since the rate of change of g is equal to f, it is at a minimum where |f| is at a minimum, i.e. in the neighborhoods where |f| is close to zero. In fact when f is small, the area under the curve of |f| is small and therefore a small value is added to g. Therefore at points 5 and 7, where f = g' = 0, we have least change.

Hence, in the neighborhood of points 5 and 7, the change in g is at a minimum.

68. (D)
Again, the value of g at x = 3 may be interpreted as the area under the curve from x = 0 to x = 3. The approximated area is shown in the figure. Each block represents one unit of area.

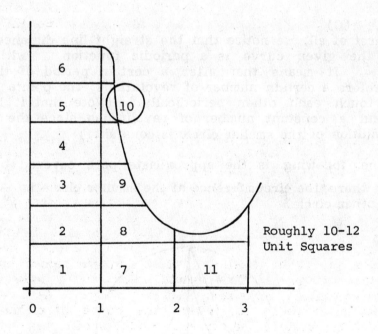

So g is between 10 and 14. The answer is (D).

69. (C)
Again the value of g at x = 10 is the area under the curve over the interval [0,10]. Here the part of the curve that goes below the x-axis indicates negative values of f, thus decreasing the area. When this negative area is taken into account (roughly 3 units), the area under the curve from 0 to 10 is $17 - 3 + 5 \approx 19$ or 20 square units. So the answer is (C).

## 70. (A)

Again the concept of interpreting g as the area under the curve must be used to arrive at the proper conclusion.

The answers for each part are rough estimates:

$$g(1) - g(0) \approx 6 - 0 = 6$$
$$g(5) - g(3) \approx 16 - 12 = 4$$
$$g(6) - g(5) \approx 14 - 15 = -1$$
$$g(8) - g(5) \approx 13 - 15 = -2$$
$$g(10) - g(9) = 17 - 15 = 2$$

So the largest value is 6, or the difference $g(1) - g(0)$.

## 71. (C)

First of all, we notice that the straight line distance shown by the given curve is a periodic function with period $T = 6$. It means that after a certain period of time and therefore a certain number of revolutions, the points $P_1$ and $P_2$ touch each other periodically. (Note that it happens within a constant number of revolutions since the rate of revolution of the smaller circle is constant.)

The following is the approximate path covered by point $P_1$ where the circumference of the smaller circle is $\frac{1}{6}$ that of the other circle.

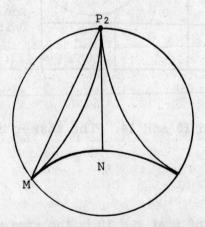

We notice that along the path $P_2M$, the distance between $P_1$ and $P_2$ increases from zero to a local maximum value equal to $P_2M$. After that, the distance $P_1P_2$ starts to decrease until it reaches its local minimum at point N, which is located on the intersection of the radius passing through $P_1$ and the path of motion.

Therefore, it can be purported that in the general case, when the circumference of the smaller circle is $\frac{1}{n}$ times that of the other one (and therefore point $P_1$ touches the inner side of the other circle n times during one revolution), the points of contact of $P_1$ with the inner side are the local maxima of the curve (except at the time when $P_1$ touches $P_2$ which is the absolute minimum of the curve).

Therefore, for our special case, since we have 5 local maxima in each period, it suggests that excluding the time when $P_1$ touches $P_2$, there are five other times where $P_1$ touches the inner side of the other circle. Therefore, $P_1$ touches the inner side at six different points in each revolution.

**72. (D)**

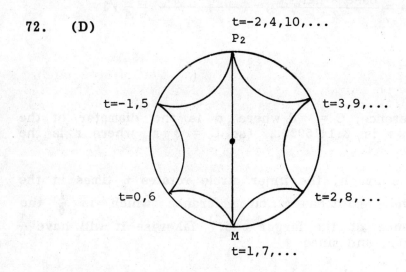

According to the previous problem, since the point $P_1$ touches the inner side of the other circle at 6 points, and the segments of the circumference of the large circle bounded by two of them are equal to the circumference of the smaller circle and are therefore equal, we can conclude that at $t = 1, 7, 13, \ldots$ point $P_1$ touches the inner side exactly the opposite side of point $P_2$. Therefore points M and $P_2$ are located at the same diameter. Since point M is the third point of contact with the inner side, the distance $P_2M$ is the y-coordinate of the curve at $t = 1, 7, 13, \ldots$, and is therefore equal to 6.

**73. (D)**
Since the points of contact are at equal distances, the triangle $P_2ON$ is an equilateral triangle. Therefore we have

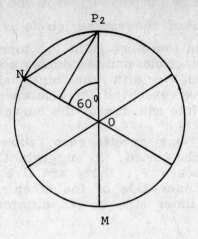

$$\text{Area} = \frac{\text{base} \times \text{height}}{2} = \frac{3 \times \left(3 \times \frac{\sqrt{3}}{2}\right)}{2} = \frac{9}{4}\sqrt{3}$$

**74. (B)**
Circumference, $C = \pi d$ where $d$ is the diameter of the circle and $\pi$ is 3.141592... (so $C = 2\pi r$ where $r$ is the radius).

From the graph, the inner circle rotates 6 times in the larger one, so it has a circumference which is $\frac{1}{6}$ the circumference of the larger one. Likewise it will have $\frac{1}{6}$ the diameter, and since

$$d_s = \frac{1}{6}d_2 = \frac{1}{6}(6 \text{ in.}) = 1 \text{ in.},$$

then the circumference of the smaller circle is

$$C = \pi d = \pi(1) = \pi,$$

so (B) is the correct answer.

**75. (E)**
The places on the graph where the rate of change in distance goes from negative to positive are the local minima of the curve. Inspecting the graph reveals that this occurs 9 times. So the answer is (E).

**76. (E)**
The derivative, f'(t), is the rate of change of t with respect to the distance, the vertical axis of the graph. So the derivative can be thought of as the slope of the curve. Where the slope is zero, that is, where a line drawn tangent to the curve is horizontal, the derivative is zero. The sketch below shows two such extrema points - a local minimum and a local maximum. With successive tangent lines drawn to the curves, the tangent line which is horizontal crosses the curve at a value for which the derivative of the function is zero.

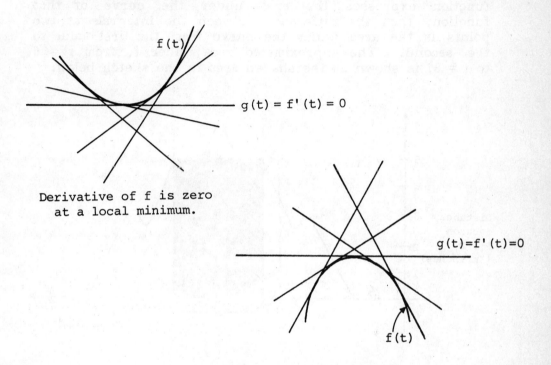

Derivative of f is zero at a local minimum.

Derivative of f is zero at a local maximum.

A quick scan of the graph, then, provides the answer. Between t = 4 and t = 10 there are eleven minimum and maximum or extrema points. So the number of times for which f'(t) = 0 is eleven.

**77. (D)**
Since F is the integral of the graphed function, f, F can be expressed as the area under the curve of f for the given interval. From t = 1 to 4, the area under the curve is increasing, so F is said to be strictly increasing. It is also monotonically increasing because it is always increasing in that interval. For the same reason it's not periodic in

that interval. Several of the answers do apply correctly to the function f, but the question was about F. Therefore (D) is the correct answer.

78. (D)
This problem requires finding an approximate answer by finding the integral of two places on the curve and calculating their difference. Since the integral of the function expresses the area under the curve of that function, then the difference between the integrals at two points is the area under the curve from the first value to the second. The approximated area under f, from t = 6 to t = 8, is shown as the shaded area in the sketch below:

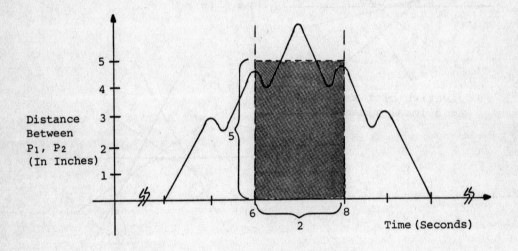

$5 \times 2 = 10$ units

So the answer is (D).

79. (B)
In problems that require information about the extreme points, the second derivative is either positive or negative, and this tells whether the extreme point is a maximum or minimum. But in this problem what is asked for are the places on the curve where the second derivative is zero. Since the first derivative is the slope of the curve, the second derivative is the rate of change of the slope. Where the first derivative is constant, the second derivative is zero. On the graph this occurs midway between each minimum and maximum where the concavity of the function

changes; that is where the derivative of the curve changes the status from increasing to decreasing or vice versa. Therefore the number is seven, so the correct answer is (B).

80. (B)
Since F is the integral of f, its derivative is simply the value of f. So the average value of the derivative of F is the average value of the function f. Since the function f nearly travels equally between 0 and 6, the average, which is defined by the value $\frac{1}{b-a} \int_a^b f(t)dt$, is roughly 3.

81. (B)
The height of the first container is measured from -5 to -3 which is 2 feet. It is the first container filled and takes 3 seconds to be filled.

82. (C)
The height of container #2, the second container filled, is measured from -5 to 0 or 5 feet. So (A),(B) and (E) are incorrect. Container #4 is 8 feet, so (D) is incorrect. Since container #3 is only 4 feet, answer (C) is the correct response.

83. (C)
Since container #1 is 2 feet high and container #2 is 5 feet high, (A) is incorrect. Since they both fill up in the same amount of time (3 seconds), we can assume that they have different widths, so (B) is incorrect. The times to fill containers 1, 3 and 4 are 3 seconds, 4 seconds and 4 seconds, respectively, so (D) and (E) are incorrect. The difference in height of containers #4 and #1 is 8 - 2 = 6, which is greater than the difference between #2 and #3 or 5 - 4 = 1. So (C) is the correct answer.

84. (C)
(A) is not correct because the height of #1 is 2 feet while

that of #3 is 4 feet. (B) and (E) are not correct because of the following: For each container we have: $\frac{hA}{t} = cte$ where h is the height of water surface at time t, A is the basic area which is constant, t is time and cte is the rate at which the faucet releases water. Therefore we may write: $h = \frac{(cte) \, t}{A}$; obviously $\frac{cte}{A}$ is the slope of the straight lines shown on the graph. Therefore, the more the slope of the containers, the less is the base area. Since the slope of #3 is larger than that of #1, its base area is smaller; the same statement can be made about #2 and #3.

On the other hand, since the rate of releasing liquid is the same, the volume of each container directly depends on the time it takes to be filled. Therefore the volume of #1 is the same as that of #2 ((D) is wrong), and finally the right answer is (C), which states that the volume of #4 is larger than #2.

85. (B)
As we mentioned in the previous problem, the ratios of the base areas of the containers are equal to the inverse of the ratios of the corresponding slopes (because the rate at which containers are being filled are the same). Hence, for the case of #4 and #2, we have:

$$\frac{\text{base area of \#4}}{\text{base area of \#2}} = \frac{\text{slope of \#2}}{\text{slope of \#4}} = \frac{\frac{5}{3}}{\frac{8}{4}} = \frac{5}{6}$$

Thus answer (B) is correct.

86. (B)
The resultant of the two forces $F_1$ and $F_2$ is sketched as shown in the figure, using the law of cosines. Its magnitude is determined using the relation

$$R^2 = F_1^2 + F_2^2 - 2F_1F_2 \cos \theta.$$

From inspection, angle θ = 180° - 60° = 120°. Therefore

$$R^2 = (4)^2 + (3)^2 - 2(4)(3)\cos 120°$$

$$= 25 - (2 \times 4 \times 3 \times -\tfrac{1}{2})$$

$$= 25 + 12 = 37.$$

∴ R = 6 N approximately.

### 87. (A)
A transformer is never connected to a source of steady direct current for its safe operation.

The transformer operates on the principle of mutual induction. Since induction takes place in the secondary of the transformer whenever the current in the primary changes, if the primary is supplied with alternating current, there will always be an efm induced in the secondary.

If the primary is connected to a source of steady direct current, no induction takes place in the secondary. Since the primary of a transformer is made of thick wire to reduce its resistance to a negligible amount, it would draw a constant voltage source. This current may be large enough to burn out the transformer primary winding.

### 88. (A)
To convert the quantities to moles,

$$0.5 \text{ mol } O_2 = 1 \text{ mol oxygen atoms}$$

$$39.1 \text{ g K} \times \frac{1 \text{ mol K}}{39.0 \text{ g K}} = 1.01 \text{ mol K}$$

So the empirical formula is KO.

### 89. (C)
"An electromotive force results from a difference of temperature between two junctions of dissimilar metals in the same circuit."

This effect was first noted by Thomas Johann Seebeck.

The difference of electric potential which results from

the difference in junction temperatures is dependent upon the metals used, the temperatures of the junction, and the intimacy of contact between the two metals.

Such a combination of dissimilar metals, when used to produce an emf, is called a thermocouple, and has many uses in the measurement of temperature.

90. (B)
Volumetric flow rate is the product of the velocity and the cross-sectional area, so

$$q = vA$$

$$q = (288 \text{ ft/sec}) \left[ \frac{1 \text{ ft}^2}{144 \text{ in}^2} \right] (5 \text{ in}^2)$$

$$q = 10 \text{ ft}^3/\text{sec}$$

So (B) is the correct response.

91. (E)
The relation between $P_1$ and $P_2$ depends upon the inlet velocity. If the inlet velocity is supersonic, $P_2$ could be greater than $P_1$. Conversely, if the inlet velocity is subsonic, $P_1$ could be greater than $P_2$.

92. (A)
Although little is given about the linearity of the device, one can still deduce an answer. In this case, we know that there is a current of 4 amps at 4 volts; the only graph that has this point is graph (A).

93. (D)
Using the following relation:

$$T = k_m I_a \Phi$$

where   $T$ = Motor torque (lb-ft)
        $k_m$ = Motor constant

$I_a$ = Armature current (amp)

$\Phi$ = Flux per pole.

We can establish the armature current relation in both cases as follows,

$$\frac{I_2}{I_1} = \left(\frac{T_2}{T_1}\right)\left(\frac{\Phi_1}{\Phi_2}\right)$$

$$\therefore \quad I_2 = (I_1)\left(\frac{T_2}{T_1}\right)\left(\frac{\Phi_1}{\Phi_2}\right) = 100 \times \left(\frac{250}{200}\right)\left(\frac{\Phi_1}{0.5\Phi_1}\right)$$

$$= 250 \text{ Amp}$$

94. (E)
To solve this problem, it is important to know the laws of vibrating strings which are given below.

1. The frequency varies inversley with the square root of the mass per unit length.

2. The frequency varies directly with the square root of the tension.

3. The frequency varies inversely with the length of the string.

From the second statement, it is clear that to double the frequency of a vibrating string, the tension in the string has to be increased to four times its original value.

95. (D)
(1) First compute cross-sectional area.

$$A = \pi r^2 = \pi \left(\frac{10^{-2}}{2\pi}\right) = \frac{10^{-2}}{2} = 0.005 \text{ cm}^2$$

(2) Compute heat flow rate, q.

$$q = kA\frac{\Delta T}{\Delta L} = \left(\frac{200 \text{ J}}{\text{m}°\text{C sec}}\right)\left(0.005 \text{ cm}^2 \cdot \frac{1 \text{ m}^2}{10,000 \text{ cm}^2}\right)\left(\frac{100°\text{C}}{1\text{m}}\right)$$

$$q = 0.01 \frac{\text{J}}{\text{sec}}$$

(3) In one day:

$$Q = qT$$

$$= 0.01 \frac{J}{sec} (24 \text{ hr}) \times \frac{60 \text{ min}}{1 \text{ hr}} \times \frac{60 \text{ sec}}{1 \text{ min}}$$

$$Q = 864 \text{ Joules.}$$

## 96. (A)

The total x and y components of linear momentum must be conserved after the collision. The mass of the body resulting after the collision is

$$m = m_1 + m_2$$

and the velocity $\vec{v}$ is inclined at an angle of 45° to the x-axis. We know that the total momentum vector is unchanged, and we can write down the x and y components of momentum.

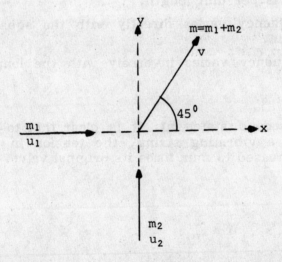

|  | INITIAL MOMENTUM | FINAL MOMENTUM |
|---|---|---|
| x component | $m_1 u_1$ | $(m_1 + m_2) v \cos \theta$ |
| y component | $m_2 u_2$ | $(m_1 + m_2) v \sin \theta$ |

Thus $\quad m_1 u_1 = (m_1 + m_2) v \cos \theta$

Hence

$$v = \frac{m_1 u_1}{(m_1 + m_2)} \times \frac{1}{\cos \theta}$$

$$= \frac{5 \times 2}{(5 + 10)} \times \frac{1}{\cos 45°}$$

$$= \frac{10}{15} \times \sqrt{2}$$

$$= \frac{2}{3} \sqrt{2} \quad \text{m/sec.}$$

## 97. (B)

Many chemical reactions involve transfer of electronic charge from one atom to another. Because this process is so common and important, words which describe these changes are in common use in the field of chemistry. Reduction refers to the gaining of electrons (and thus the reduction of positive charge) so (B) is the correct answer. Oxidation, the complementary process, is the loss of electrons. The element which supplies the electrons is called the reducing agent, while the element which accepts the electron and permits oxidation is the oxidizing agent.

## 98. (C)

From the diagram below the resultant

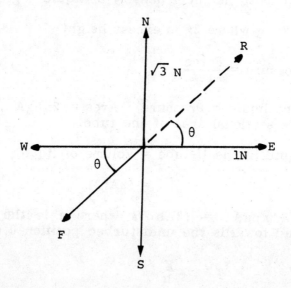

$$R = \sqrt{(\sqrt{3})^2 + (1)^2} \text{ N}$$

$$R = 2\text{N}$$

$$\theta = \tan^{-1}(\sqrt{3}) = 60°$$

For equilibrium, we must have a force acting along the line of action of the resultant R, but in the opposite direction.

Thus

$$F = 2\text{N in the direction W } 60° \text{ S.}$$

## 99. (C)

The collision of the atom with the wall is perfectly elastic since it rebounds with the same speed. The momentum of the atom when striking the wall is mv. The atom rebounds with momentum $-mv$. The change in momentum is

$$\Delta MV = mv_2 - mv_1 = -mv - (mv) = -2mv$$

$$|\Delta MV| = |-2mv| = 2mv$$

## 100. (B)

In the position shown, the excess pressure on the whole liquid is

$$= \text{excess height} \cdot \text{density of liquid} \cdot g$$

$$= 2x\rho g \text{ where } 2x \text{ is excess height}$$

Since Pressure = $\dfrac{\text{Force}}{\text{Area}}$

Force on the liquid = Pressure · Area = $2x\rho g A$, where A is the cross-sectional area of the tube.

Mass of liquid in the U-tube = Volume of liquid · density

$$= (2Ah)\rho$$

Therefore $-2x\rho gA = (2Ah\rho)a$ where $a$ is the acceleration of the liquid towards the undisturbed position 0 or P. Thus

$$a = -x\frac{g}{h}$$

$$a = \frac{-g}{h}x$$

101. (A)

From question 100, $a = \dfrac{-g}{h} x$.

Since the motion is simple harmonic, $a = -\omega^2 x$. Thus

$$\dfrac{-g}{h} x = -\omega^2 x$$

so that

$$\dfrac{g}{h} = \omega^2 \qquad (1)$$

By definition,

$$T = \dfrac{2\pi}{\omega}$$

$$T^2 = \dfrac{4\pi^2}{\omega^2}$$

$$\omega^2 = \dfrac{4\pi^2}{T^2}$$

Substituting for $\omega^2$ in equation (1)

$$\dfrac{g}{h} = \dfrac{4\pi^2}{T^2}$$

$$T^2 = 4\pi^2 \dfrac{h}{g}$$

$$T = 2\pi \sqrt{\dfrac{h}{g}}$$

102. (D)
Since the change in velocity is 4 m/s, the downstream velocity is:

$$V_2 = V_1 + 4 = 6 \text{ m/s}.$$

From the general energy equation:

$$W_s = \tfrac{1}{2}(V_2^2 - V_1^2) + \text{(rest of the equation)}$$

Therefore,

$$W_s = \tfrac{1}{2}(36 - 4) = 16 \text{ m}^2/\text{s}^2$$

Power = $(W_s) \times$ (mass flow rate)

The mass flow rate is equal to $\rho V A$.

Therefore:

$$\dot{m} = \rho_1 V_1 A_1 = \left(1000 \frac{kg}{m^3}\right)\left(2 \frac{m}{s}\right)\left(\frac{\pi}{4} 16 m^2\right)$$

$$\dot{m} = 25,120 \frac{kg}{sec}$$

Power = 401,920 W = 402 kW

## 103. (C)

It takes 90 seconds for the man to move up the stalled escalator, or he moves $\frac{1}{90}$ th of the total distance in one second. When the escalator and the man are moving, it takes them 30 seconds, or the escalator and the man together move $\frac{1}{30}$ th of the total distance in one second. If the man were to stand on the escalator, he would move $\frac{1}{30} - \frac{1}{90}$ of the total distance in one second, or he moves

$$\frac{1}{30} - \frac{1}{90} = \frac{3}{90} - \frac{1}{90} = \frac{2}{90} = \frac{1}{45} \text{ th}$$

the total distance in one second. So the full distance would take 45 seconds.

## 104. (E)

The cohesive energy, which is the energy needed to take the solid apart into positive and negative ions, is relatively high for ionic solids. Therefore, they should be poor electrical conductors since there are no free electrons available. They should be hard and stable with high vaporization temperatures and high melting points.

## 105. (C)

(A) is correct since the equation shows that enthalpy is a function of temperature only since R is constant and u is related to temperature only. In fact the definition of constant volume specific heat is

$$C_v = \left(\frac{\partial u}{\partial T}\right)_v$$

and for constant pressure is

$$C_p = \left(\frac{\partial h}{\partial T}\right)_p$$

so both are functions of temperature only, and (D) is correct. (B) is correct as well since internal energy and enthalpy are related to temperature only for these conditions. So if there is a temperature difference there must be a value for $C_p$ or $C_r$ to have and enthalpy change.

106. (E)
The capital cost is the single most influential factor on the minimum rate of return. It is the factor affecting the minimum attractive rate of return the most. The cost of capital will depend on the interest paid on bonds in the debt financing, or the lending rate at a bank. The need for capital will depend on the amount of retained earnings the company has. The greater the risk involved in the project, the higher the cost of capital will be. Therefore all of the factors will affect the minimum attractive rate of return.

107. (C)
The distance covered and time taken by the driver in each stage:

first stage is 90 miles and 2 hours

second stage is $60 \times 1.5 = 90$ miles and 1.5 hours

third stage is $25 \times \frac{1}{2} = 12.5$ miles and 0.5 hour

Hence the total distance covered is $90 + 90 + 12.5 = 192.5$ miles and the total time taken is $2 + 1.5 + 0.5 = 4$ hours.

Hence the average speed is

$192.5/4 = 48.1$ mph.

108. (D)
The minimum energy of a photon required to excite an electron from the valence band to the conduction band is the forbidden energy gap of that semiconductor material. The relationship between the wavelength of the photon and the energy gap is given by,

$$\lambda_c = \frac{1.24}{E_g}$$

$E_g$ = energy gap = minimum energy required to move an electron from the valence band to the the conduction band, in electron volts

$\lambda_c$ = Critical wavelength of the photon (micrometers)

1.24 = constant that arises from the Planck's constant

Substituting the given data,

$$E_g = \frac{1.24}{.5} = 2.48 \text{ electron volt}$$

## 109. (A)
The two velocity profiles are:

Natural Convection        Forced Convection

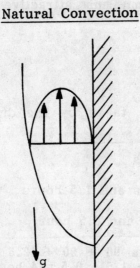 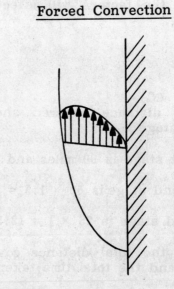

## 110. (E)
According to the ideal gas assumption, $Pv = RT$. For constant T, Pv plot should resemble that of a hyperbola only.

## 111. (D)
Pascal's law states that pressure applied to an enclosed fluid is transmitted throughout the fluid in all directions without loss. In the hydraulic press, this means that the

pressure applied to the smaller piston is transmitted unchanged to the larger piston.

$$P_1 = P_2$$

Since $\quad F = PA$

$$\frac{F_1}{A_1} = \frac{F_2}{A_2}$$

So, since the second piston has a larger area, it experiences a greater force.

$$F_2 = \left(\frac{A_2}{A_1}\right) F_1$$

$$F_2 = \left(\frac{18.0}{2.0}\right) (120 \text{ lbs.})$$

$$F_2 = 1080 \text{ lbs.}$$

112. (B)
For the block to float, its weight must be balanced by the buoyant force.

Thus, weight of block = B

But weight = weight density ($\rho g$) × volume (v).

$$\therefore \quad \rho_{block} \times g \times v_{block} = \rho_{water} \times g \times v_{water}$$

where the volume of submerged block equals volume of water displaced.

Volume of submerged block = $1 - \frac{2}{3} = \frac{1}{3}$.

Hence, $\quad \dfrac{v_{water}}{v_{block}} = \dfrac{\rho_{block}}{\rho_{water}} = \dfrac{1}{3} = .33$

So density of block = .33 × density of fresh water gm/cm$^3$

$$= .33 \times 1$$

$$= .33$$

113. (E)
The amount of gas enclosed in a cylinder does not change.

As per Charles' Law, the volume varies with temperature for an ideal gas at constant pressure.

This question should not be confused with Charles' Law. If the amount of gas would increase, it would mean that mass is created by varying temperature, pressure and/or volume, which is not true.

114. (D)
Since the diode is an ideal one, it acts as a short circuit when the input voltage is less than 5v. As soon as the input reaches 5v, the diode becomes reverse biased and no current passes through the diode. Hence, the output voltage follows the input voltage, when it is greater than 5v. When the input is less than 5v, the output voltage is simply 5 volt DC supply biasing the diode. Hence the answer is choice (D).

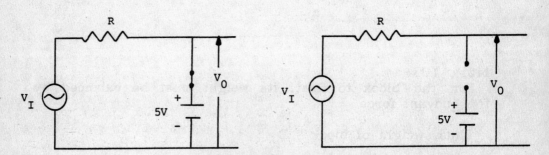

a) Equivalent circuit, when the diode is on, i.e. $v_I < 5v$.

b) Equivalent circuit, when the diode is off, i.e. $v_I \geq 5v$.

115. (C)
When the flow is turbulent, the velocity profile is not that of a parabola. In fact, it looks like the one below:

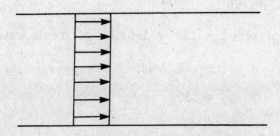

However, when the flow is laminar, regardless of whether or not it is compressible, the velocity profile is parabolic.

**116. (C)**
The probability density for finding the electron for an electron whose wave function is $\psi(x)$ is:

$$p(x)dx = |\psi(x)|^2 \, dx$$

The probability of finding the particle between 0 and L is

$$\int_0^L |\psi(x)|^2 \, dx$$

The probability density for finding the electron must satisfy the condition of a zero probability for finding the electron at $x = 0$ and $x = L$ since the potential is infinite. Since graph (a) represents the wave function $\psi(x)$, the probability density $|\psi(x)|^2$ must be $> 0$ over the range $0 \leq x \leq L$. Since graph (d) represents $|\psi(x)|$, $|\psi(x)|^2$ must be greater than graph (d). Choice (C) is the only one that satisfies all these conditions.

**117. (B)**
The forging operation distorts the equiaxed microstructure of the metals and causes the formation of crystal defects such as dislocations. Elongated coarse grains are also produced in the direction of formation. These effects of the forging operation create iron that is brittle.

**118. (C)**

**119. (A)**
Since $t_2 \approx t_1$ the velocity of the sphere $\frac{d}{t_1} \approx \frac{d}{t_2}$. If the sphere accelerates throughout most of the first section then $t_1 \ll t_2$. Therefore the terminal velocity of the sphere was reached during the descent through the first section.

When the sphere reaches its terminal velocity the sum of the forces acting on the sphere is 0. The force due to gravity is $\rho_1 g \frac{4}{3}\pi r^3$ and the buoyant force is $\rho_2 g \frac{4}{3}\pi r^3$. Therefore the drag force acting on the sphere is

$$(\rho_1 - \rho_2)g\frac{4}{3}\pi r^3.$$

120. (C)
A reversible process is defined as a process in which the system passes from initial state to the final state by departing slightly from an equilibrium state at all times.

The Carnot cycle, by definition, passes from one state to the next by slightly deviating from the point of equilibrium, and thus is reversible.


121. (E)
Since Miller indices apply to all cubic lattice structures, (C) is incorrect. Since the planes are not parallel, (D) is incorrect. The direction vector [232] is perpendicular to (not parallel to) the plane (232), so (B) is incorrect.

The Miller indices are the distances in each direction to a plane. For the planes A and B these are, respectively, (232) and (124).


122. (B)
Since the surface is to absorb radiant solar energy, $\alpha$ should be large, and to emit very little, it should have a small $\varepsilon$ or else a large $\frac{1}{\varepsilon}$, so (B) is the correct answer.


123. (C)
Heat lost by x = Heat gained by water.

mass × specific heat × temperature change (of x) = mass

specific heat × temperature change (of water)

60 × specific heat × (80 - 15) = 30 × 1 × (15 - 10)

∴ specific heat of x = $\frac{30 \times 15}{60 \times 65} \simeq$ .04 cal/gC°


124. (B)
The jet impinging on the vane, which is attached to the block, exerts a force to the right on the block. The friction of the surface opposes this force. The horizontal force is calculated by the momentum principle:

$$F_x = \rho Q(V_2 - V_1)$$

Since the area is constant, the velocities are equal and opposite (Q = const.).

Thus $F_x = \rho A(2V^2)$

$$V = \frac{Q}{A} = \frac{4}{1} = 4 \text{ ft/s}$$

and $F_x = 2(1)(2)(4)^2 = 64$ lbs.

The friction force for static friction is $F \leq \mu N$ where $\mu$ is the friction coefficient and N is the normal force.

We use the "=" sign when the block is about to move. Hence: N = W = 100 lbs.

$$F_x = 64 \text{ lbs.}$$

$$F_x = \mu N$$

$$\mu = \frac{64}{100} = .64$$

125. (B)
Long term gains are aggregated separately and taxed at a different rate than the income.

Tax rate (Selling price - original value) = tax on long term gain

Tax on long term gain = .25($1000) = $250

The net taxable income = net income - depreciation

$$= \$11,000 - \$1,000$$

$$= \$10,000$$

The ordinary income on the sale is

(Selling price - book value) $6000 - $3,000 = $3,000

Tax on net income = .5($13,000) = $6,500

Therefore the total tax bill is $6,750.

**126. (E)**

The polymer goes from a leathery stage, to a rubbery and then a viscous stage. In the leathery stage the polymer can be deformed readily but it cannot regain its shape quickly if the stress is removed. In the rubbery stage the polymer can regain its original shape quickly. In the viscous stage the polymer deforms extensively by viscous flow.

**127. (A)**

The period will first increase and then decrease because the center of mass keeps moving downward as the water flows out and then moves back to the center of the sphere.

To understand this consider the sphere. Its center of mass is at its geometrical center when it's empty or full of water. But when it is not completely full the center of mass shifts.

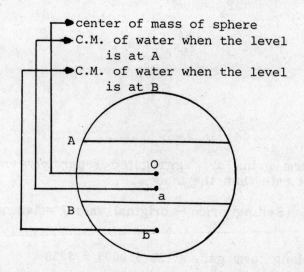

When the water level is at A the center of mass is at point a and the center of mass of the pendulum is between O and a. This is true for any other level, hence the center of mass keeps moving to a lower height and the effective length of the pendulum keeps increasing. When the sphere is empty it again shifts to O or the length of the pendulum decreases and the period increases.

**128. (D)**

Cooling path c will have the microstructure shown at the

eutectic point. Above the eutectic temperature the microstructure will be all liquid. The cooling path along the eutectic composition will not pass through the liquid line and thus no crystals of α or β will form. At the eutectic temperature all of the liquid will crystallize as a lamellar structure made up of fine crystals of α and β.

129. (A)
Configurations represented in I and II are impossible because of violation of the Pauli exclusion principle. This principle excludes electrons with the same spin (↑ or ↓) from sharing the same orbital. The configurations in III and IV are both possible, though IV represents an excited state. So the correct answer is (A).

130. (D)
Pearlite structure is an intimate mixture of ferrite and cementite. Ferrite is gamma iron with face centered cubic structure capable of dissolving up to 0.025%. Cementite is a compound of composition $Fe_3C$ with a carbon content of 6.69%.

131. (E)
Recall for $F(x,y) = 0$ an implicit function of x and y,

$$\frac{dy}{dx} = \frac{-F_x}{F_y}$$

$$F(x,y) = e^{\frac{x^2}{y^3}} \cos \frac{x}{y} - 1 = 0$$

$$F_x = -e^{\frac{x^2}{y^3}} \left[\sin \frac{x}{y}\right]\left(\frac{1}{y}\right) + \left[\cos \frac{x}{y}\right]\left(\frac{2x}{y^3}\right) e^{\frac{x^2}{y^3}}$$

$$= \frac{1}{y} e^{\frac{x^2}{y^3}} \left(\frac{2x}{y^2} \cos \frac{x}{y} - \sin \frac{x}{y}\right)$$

$$F_y = e^{\frac{x^2}{y^3}} \left(-\sin \frac{x}{y}\right)\left(-\frac{x}{y^2}\right) + \left[\cos \frac{x}{y}\right] \frac{(-3x^2)}{y^4} e^{\frac{x^2}{y^3}}$$

$$= \frac{1}{y} e^{\frac{x^2}{y^3}} \left( \frac{x}{y} \sin \frac{x}{y} - \frac{3x^2}{y^3} \cos \frac{x}{y} \right)$$

$$\frac{dy}{dx} = -\frac{F_x}{F_y} = \frac{\left( \sin \frac{x}{y} - \frac{2x}{y^2} \cos \frac{x}{y} \right)}{\left( \frac{x}{y} \sin \frac{x}{y} - \frac{3x^2}{y^3} \cos \frac{x}{y} \right)}$$

**132. (A)**
A matrix [A] is Hermitian if $\bar{A}^T = A$

Check:

Let $A = \begin{bmatrix} 3 & i & 5+2i \\ -i & 2 & 6 \\ 5-2i & 6 & 1 \end{bmatrix}$

$\bar{A} = \begin{bmatrix} 3 & -i & 5-2i \\ i & 2 & 6 \\ 5+2i & 6 & 1 \end{bmatrix}$

$\bar{A}^T = \begin{bmatrix} 3 & i & 5+2i \\ -i & 2 & 6 \\ 5-2i & 6 & 1 \end{bmatrix}$

Thus $\bar{A}^T = A$.

**133. (B)**
From the sketch of the curves given in the problem, it is clear that the limits of the integral which give us the required area are the points of intersection of the two curves, which can be found by setting the functions equal to one another and solving for y.

$$A = \int_0^2 [(y-1)^2 + 1 - 2(y-1)^2] dy$$

$$= \int_0^2 [1 - (y-1)^2] dy$$

$$= \left[ y - \frac{1}{3}(y-1)^3 \right]_0^2$$

$$= \left(2 - \frac{1}{3}\right) - \left(\frac{1}{3}\right)$$

$$= \frac{4}{3} \text{ sq. units.}$$

### 134. (C)

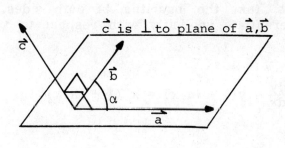

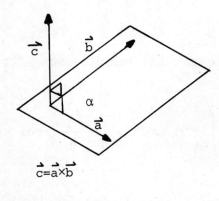   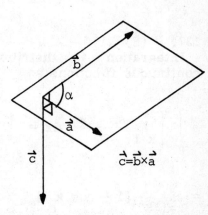

As illustrated, $\vec{C}$ is always in a plane $\perp$ to the plane of $\vec{a}$ and $\vec{b}$.

### 135. (C)

Since $\int \vec{F} \cdot d\vec{C}$ is being evaluated, first compute $\vec{F} \cdot d\vec{C} = (x^2, xy) \cdot (dx, dy) = x^2 dx + xy\,dy$. To parameterize let $y = y$; $x = y^2$ so that $dy = dy$; $dx = 2y\,dy$ and

$$\int_C \vec{F} \cdot d\vec{C} = \int_C x^2 dx + xy dy = \int_{-1}^{1} y^4 \cdot 2y dy + y^3 dy$$

$$= \frac{2y^6}{6} + \frac{y^4}{4} \Big|_{-1}^{1} = \left(\frac{1}{3} + \frac{1}{4}\right) - \left(\frac{1}{3} + \frac{1}{4}\right) = 0.$$

### 136. (C)

For $y = x^x$ take the logarithm of both sides, $\ln y = x \ln x$. Differentiate implicitly with respect to $x$, solve for $\frac{dy}{dx}$ to give

$$\frac{dy}{dx} = y(1 + \ln x) = x^x(1 + \ln x)$$

### 137. (E)

Integration is distributive over an infinite sum for continuous functions so

$$\int_0^\pi \left( \sum_{k=1}^\infty \frac{\sin kx}{k^3} \right) dx = \sum_{k=1}^\infty \frac{1}{k^3} \int_0^\pi \sin kx \, dx = \sum_{k=1}^\infty \frac{1}{k^4} [-\cos kx]_0^\pi$$

$$= \sum_{k=1}^\infty \frac{1}{k^4} [1 - \cos k\pi]$$

Now $1 - \cos k\pi = \begin{cases} 0 \text{ for } k = 2, 4, \ldots, 2n \\ 2 \text{ for } k = 1, 3, \ldots, n-1 \\ \text{where n is in the set of natural numbers} \end{cases}$

Hence

$$\int_0^\pi \left( \sum_{k=1}^\infty \frac{\sin kx}{k^3} \right) dx = \sum_{k=1}^\infty \frac{2}{(2k-1)^4}$$

138. (D)

$$\int_{3\pi}^{7\pi/2} \cos\theta\, d\theta = \sin\theta \Big]_{3\pi}^{7\pi/2}$$

since $\sin 3\pi = \sin \pi = 0$

and $\sin \dfrac{7\pi}{2} = \sin\left(3\pi + \dfrac{\pi}{2}\right) = -\sin \dfrac{\pi}{2} = -1$

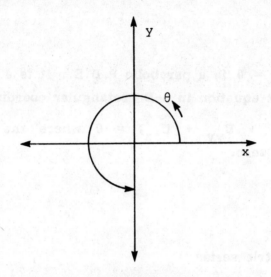

Since $7\pi/2$ means that the angle $\theta$ has already completed one and one-half rotations, i.e. $3\pi = 2\pi + \pi$ and now lies on the y-axis where the value of sine is $-1$.

$$\therefore \int_{3\pi}^{7\pi/2} \cos\theta\, d\theta = \left[\sin\left(\dfrac{7\pi}{2}\right) - \sin 3\pi\right]$$

$$= [-1 - 0] = -1$$

139. (B)
Recall that two lines are parallel if and only if they have equal slopes.

140. (D)

$$\sum_{m=0}^{\infty} \sum_{n=0}^{\infty} a_n x^m = \sum_{m=0}^{\infty} \left(\dfrac{1}{0!} + \dfrac{1}{1!} + \dfrac{1}{2!} + \ldots + \dfrac{1}{n!}\right) x^m$$

$$= \sum_{m=0}^{\infty} \frac{x^m}{n!} \frac{1}{1-x}$$

$$= \frac{1}{1-x} \sum_{m=0}^{\infty} \frac{x^m}{n!}$$

$$= \frac{e^x}{1-x} \qquad \left[\text{since } \sum_{n=0}^{\infty} \frac{x^n}{n!} = e^x\right]$$

**141. (C)**
$U_t - a^2 U_{xx} = 0$ is a parabolic P.D.E.. It is a particular form of the heat equation in 3-D rectangular coordinates

$U_t - a^2(U_{xx} + U_{yy} + U_{zz}) = 0$ where the y and z dependence are zero.

**142. (D)**
By the geometric series

$$\sum_{n=1}^{\infty} ar^{n-1} = a + ar + ar^2 + \ldots$$

which converges to $\frac{a}{(1-r)}$ if $|r| < 1$ (it diverges for $r \geq 1$), we have, with $r = -x^2$ and $a = 1$, that

$$\frac{1}{1+x^2} = 1 - x^2 + x^4 - x^6 + \ldots, \quad -1 < x < 1,$$

where we can write

$$1 - x^2 + x^4 - x^6 + \ldots = \sum_{n=0}^{\infty} (-1)^n x^{2n}$$

$$\therefore \quad \sum_{n=0}^{\infty} (-1)^n x^{2n} = \frac{1}{1+x^2}, \quad -1 < x < 1$$

Thus $\quad \sum_{n=0}^{\infty} \frac{(-1)^n}{2^{2n}} = \sum_{n=0}^{\infty} (-1)^n (\tfrac{1}{2})^{2n}$

$$= \frac{1}{1 + (\frac{1}{2})^2}$$

$$= \frac{4}{5}$$

143. (C)
The general form of the equation of a plane is

$$ax + by + cz + d = 0$$

Hence, to answer this question we must determine a, b, c, and d.

To find a, b, and c we must draw on an analogy from the two-dimensional case.

The equation of a line is given by

$$ex + fy + k = 0.$$

By the definition of slope, any line parallel to this line is of the form

$$ex + fy + \ell = 0.$$

Any two parallel lines have identical coefficients preceding the variables in their equations.

While the slope concept is inapplicable to planes, the above rule of thumb can be extended to planes. Hence,

$ax + by + cz + d = 0$ is parallel to $ax + by + cz + e = 0$.

We know (4,-1,1) is a point on the plane. This will allow us to find d, by substituting these coordinates into the following equation:

$$4x - 2y + 3z + d = 0.$$

It follows that, $4(4) - 2(-1) + 3(1) + d = 0$

$$16 + 2 + 3 + d = 0$$

Hence, $\qquad d = -21.$

Therefore, the equation of the required plane is

$$4x - 2y + 3z - 21 = 0.$$

144. (C)

Since
$$6y = x^2,$$
$$6\frac{dy}{dt} = 2x\frac{dx}{dt}, \text{ or}$$
$$\frac{dy}{dt} = \frac{x}{3} \cdot \frac{dx}{dt}.$$

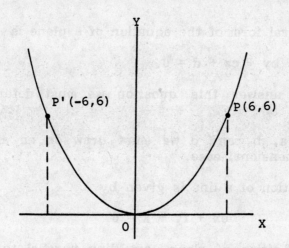

This means that, at any point on the parabola, the rate of change of ordinate = $\left(\frac{x}{3}\right)$ times the rate of change of abscissa.

When $x = 6$, $\frac{dx}{dt} = 2$ ft. per second.

Thus, substitution gives:
$$\frac{dy}{dt} = \frac{6}{3} \cdot 2$$
$$= 4 \text{ ft/sec.}$$

# The Graduate Record Examination in ENGINEERING

## Appendix

# VARIABLES

| | | |
|---|---|---|
| $a$ | = | acceleration |
| $a_t$ | = | tangential acceleration |
| $a_r$ | = | radial acceleration |
| $d$ | = | distance |
| $e$ | = | coefficient of restitution |
| $f$ | = | frequency |
| $F$ | = | force |
| $g$ | = | gravity = 32.2 ft/sec$^2$ or 9.81 m/sec$^2$ |
| $h$ | = | height |
| $I$ | = | mass inertia |
| $k$ | = | spring constant, radius of gyration |
| $KE$ | = | kinetic energy |
| $m$ | = | mass |
| $M$ | = | moment |
| $PE$ | = | potential energy |
| $r$ | = | radius |
| $s$ | = | position |
| $t$ | = | time |
| $T$ | = | tension, torsion, period |
| $v$ | = | velocity |
| $w$ | = | weight |
| $x$ | = | horizontal position |
| $y$ | = | vertical position |
| $\alpha$ | = | angular acceleration |
| $\omega$ | = | angular velocity |
| $\theta$ | = | angle |
| $\mu$ | = | coefficient of friction |

# EQUATIONS

## Kinematics

### Linear Particle Motion

Constant velocity

$$s = s_o + vt$$

Constant acceleration

$$v = v_o + at$$

$$s = s_o + v_o t + \left(\frac{1}{2}\right)at^2$$

$$v^2 = v_o^2 + 2a(s - s_o)$$

**Projectile Motion**

$$x = x_o + v_x t$$
$$v_y = v_{yo} - gt$$
$$y = y_o + v_{yo}t - \left(\frac{1}{2}\right)gt^2$$
$$v_y^2 = v_{yo}^2 - 2g(y - y_o)$$

**Rotational Motion**

Constant rotational velocity
$$\theta = \theta_o + \omega t$$
Constant angular acceleration
$$\omega = \omega_o + \alpha t$$
$$\theta = \theta_o + \omega_o t + \left(\frac{1}{2}\right)\alpha t^2$$
$$\omega^2 = \omega_o^2 + 2\alpha(\theta - \theta_o)$$
Tangential velocity
$$v_t = r\omega$$
Tangential acceleration
$$a_t = r\alpha$$
Radial acceleration
$$a_r = r\omega^2 = \frac{v_t^2}{r}$$
Polar coordinates
$$a_r = \frac{d^2r}{dt^2} - r\left(\frac{d\theta}{dt}\right)^2 = \frac{d^2r}{dt^2} - r\omega^2$$
$$a_\theta = r\left(\frac{d^2\theta}{dt^2}\right) + 2\left(\frac{dr}{dt}\right)\left(\frac{d\theta}{dt}\right) = r\alpha + 2\left(\frac{dr}{dt}\right)\omega$$
$$v_r = \frac{dr}{dt}$$
$$v_\theta = r\left(\frac{d\theta}{dt}\right) = r\omega$$

## Relative and Related Motion

Acceleration

$a_A = a_B + a_{A/B}$

Velocity

$v_A = v_B + v_{A/B}$

Position

$x_A = x_B + x_{A/B}$

## Kinetics

$w = mg$

$F = ma$

$F_c = ma_n = \dfrac{mv_t^2}{r}$

$F_f = \mu N$

## Kinetic Energy

$KE = \left(\dfrac{1}{2}\right)mv^2$

Work of a force = $\int F ds$

$KE_1 + \text{Work}_{1-2} = KE_2$

## Potential Energy

Spring $PE = \left(\dfrac{1}{2}\right)kx^2$

Weight $PE = wy$

$KE_1 + PE_1 = KE_2 + PE_2$

## Power

Linear power $P = Fv$

Torsional or rotational power $P = T\omega$

## Impulse-Momentum

$mv_1 + \int F dt = mv_2$

## Impact

$m_A v_{A1} + m_B v_{B1} = m_A v_{A2} + m_B v_{B2}$

$e = \dfrac{v_{B2} - v_{A2}}{v_{A1} - v_{B1}}$

Perfectly plastic impact ($e = 0$)

$m_A v_{A1} + m_B v_{B1} = (m_A + m_B)v'$

One mass is infinite

$$v_2 = ev_1$$

**Inertia**

Beam  $I_A = \left(\frac{1}{12}\right)ml^2 + m\left(\frac{1}{2}\right)^2 = \left(\frac{1}{3}\right)ml^2$

Plate  $I_A = \left(\frac{1}{12}\right)m(a^2+b^2) + m\left[\left(\frac{a}{b}\right)^2 + \left(\frac{b}{2}\right)^2\right] = \left(\frac{1}{3}\right)m(a^2+b^2)$

Wheel  $I_A = mk^2 + mr^2$

**Two-Dimensional Rigid Body Motion**

$$F_x = ma_x$$
$$F_y = ma_y$$
$$M_A = I_A \alpha = I_{cg}\alpha + m(a)d$$

**Rolling Resistance**

$$F_r = \frac{mga}{r}$$

**Energy Methods for Rigid Body Motion**

$$KE_1 + \text{Work}_{1-2} = KE_2$$
$$\text{Work} = \int F ds + \int M d\theta$$

**Mechanical Vibration**

Differential equation

$$\frac{md^2x}{dt^2} + kx = 0$$

Position

$$x = x_m \sin\left[\sqrt{\frac{k}{m}}\,t + \theta\right]$$

Velocity

$$v = \frac{dx}{dt} = x_m \sqrt{\frac{k}{m}} \cos\left[\sqrt{\frac{k}{m}}\,t + \theta\right]$$

Acceleration

$$a = \frac{d^2x}{dt^2} = -x_m\left(\frac{k}{m}\right)\sin\left[\sqrt{\frac{k}{m}}\,t + \theta\right]$$

Maximum values

$$x = x_m, v = x_m\sqrt{\frac{k}{m}}, a = -x_m\left(\frac{k}{m}\right)$$

Period

$$T = \frac{2\pi}{\left(\sqrt{\frac{k}{m}}\right)}$$

Frequency

$$f = \frac{1}{T} = \frac{\sqrt{\frac{k}{m}}}{2\pi}$$

Springs in parallel

$$k = k_1 + k_2$$

Springs in series

$$\frac{1}{k} = \frac{1}{k_1} + \frac{1}{k_2}$$

## AREA UNDER NORMAL CURVE

$$\frac{1}{\sqrt{2\pi}} \int_0^z e^{-\frac{z^2}{2}} dz$$

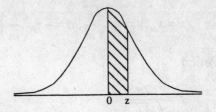

| Z | 0 | 1 | 2 | 3 | 4 | 5 | 6 | 7 | 8 | 9 |
|---|---|---|---|---|---|---|---|---|---|---|
| 0.0 | .0000 | .0040 | .0080 | .0120 | .0160 | .0199 | 0239 | .0279 | .0319 | .0359 |
| 0.1 | .0398 | .0438 | .0478 | .0517 | .0557 | .0596 | .0636 | .0675 | .0714 | .0754 |
| 0.2 | .0793 | .0832 | .0871 | .0910 | .0948 | .0987 | .1026 | .1064 | .1103 | .1141 |
| 0.3 | .1179 | .1217 | .1255 | .1293 | .1331 | .1368 | .1406 | .1443 | .1480 | .1517 |
| 0.4 | .1554 | .1591 | .1628 | .1664 | .1700 | .1736 | .1772 | .1808 | .1844 | .1879 |
| 0.5 | .1915 | .1950 | .1985 | .2019 | .2054 | .2088 | .2123 | .2157 | .2190 | .2224 |
| 0.6 | .2258 | .2291 | .2324 | .2357 | .2389 | .2422 | .2454 | .2486 | .2518 | .2549 |
| 0.7 | .2580 | .2612 | .2642 | .2673 | .2704 | .2734 | .2764 | .2794 | .2823 | .2852 |
| 0.8 | .2881 | .2910 | .2939 | .2967 | .2996 | .3023 | .3051 | .3078 | .3106 | .3133 |
| 0.9 | .3159 | .3186 | .3212 | .3238 | .3264 | .3289 | .3315 | .3340 | .3365 | .3389 |
| 1.0 | .3413 | .3438 | .3461 | .3485 | .3508 | .3531 | .3554 | .3577 | .3599 | .3621 |
| 1.1 | .3643 | .3665 | .3686 | .3708 | .3729 | .3749 | .3770 | .3790 | .3810 | .3830 |
| 1.2 | .3849 | .3869 | .3888 | .3907 | .3925 | .3944 | .3962 | .3980 | .3997 | .4015 |
| 1.3 | .4032 | .4049 | .4066 | .4082 | .4099 | .4115 | .4131 | .4147 | .4162 | .4177 |
| 1.4 | .4192 | .4207 | .4222 | .4236 | .4251 | .4265 | .4279 | .4292 | .4306 | .4319 |
| 1.5 | .4332 | .4345 | .4357 | .4370 | .4382 | .4394 | .4406 | .4418 | .4429 | .4441 |
| 1.6 | .4452 | .4463 | .4474 | .4484 | .4495 | .4505 | .4515 | .4525 | .4535 | .4545 |
| 1.7 | .4554 | .4564 | .4573 | .4582 | .4591 | .4599 | .4608 | .4616 | .4625 | .4633 |
| 1.8 | .4641 | .4649 | .4656 | .4664 | .4671 | .4678 | .4686 | .4693 | .4699 | .4706 |
| 1.9 | .4713 | .4719 | .4726 | .4732 | .4738 | .4744 | .4750 | .4756 | .4761 | .4767 |
| 2.0 | .4772 | .4778 | .4783 | .4788 | .4793 | .4798 | .4803 | .4808 | .4812 | .4817 |
| 2.1 | .4821 | .4826 | .4830 | .4834 | .4838 | .4842 | .4846 | .4850 | .4854 | .4857 |
| 2.2 | .4861 | .4864 | .4868 | .4871 | .4875 | .4878 | .4881 | .4884 | .4887 | .4890 |
| 2.3 | .4893 | .4896 | .4898 | .4901 | .4904 | .4906 | .4909 | .4911 | .4913 | .4916 |
| 2.4 | .4918 | .4920 | .4922 | .4925 | .4927 | .4929 | .4931 | .4932 | .4934 | .4936 |
| 2.5 | .4938 | .4940 | .4941 | .4943 | .4945 | .4946 | .4948 | .4949 | .4951 | .4952 |
| 2.6 | .4953 | .4955 | .4956 | .4957 | .4959 | .4960 | .4961 | .4962 | .4963 | .4964 |
| 2.7 | .4965 | .4966 | .4967 | .4968 | .4969 | .4970 | .4971 | .4972 | .4973 | .4974 |
| 2.8 | .4974 | .4975 | .4976 | .4977 | .4977 | .4978 | .4979 | .4979 | .4980 | .4981 |
| 2.9 | .4981 | .4982 | .4982 | .4983 | .4984 | .4984 | .4985 | .4985 | .4986 | .4986 |
| 3.0 | .4987 | .4987 | .4987 | .4988 | .4988 | .4989 | .4989 | .4989 | .4990 | .4990 |
| 3.1 | .4990 | .4991 | .4991 | .4991 | .4992 | .4992 | .4992 | .4992 | .4993 | .4993 |
| 3.2 | .4993 | .4993 | .4994 | .4994 | .4994 | .4994 | .4994 | .4995 | .4995 | .4995 |
| 3.3 | .4995 | .4995 | .4995 | .4996 | .4996 | .4996 | .4996 | .4996 | .4996 | .4997 |
| 3.4 | .4997 | .4997 | .4997 | .4997 | .4997 | .4997 | .4997 | .4997 | .4997 | .4998 |
| 3.5 | .4998 | .4998 | .4998 | .4998 | .4998 | .4998 | .4998 | .4998 | .4998 | .4998 |
| 3.6 | .4998 | .4998 | .4999 | .4999 | .4999 | .4999 | .4999 | .4999 | .4999 | .4999 |
| 3.7 | .4999 | .4999 | .4999 | .4999 | .4999 | .4999 | .4999 | .4999 | .4999 | .4999 |
| 3.8 | .4999 | .4999 | .4999 | .4999 | .4999 | .4999 | .4999 | .4999 | .4999 | .4999 |
| 3.9 | .5000 | .5000 | .5000 | .5000 | .5000 | .5000 | .5000 | .5000 | .5000 | .5000 |

# POWER SERIES FOR ELEMENTARY FUNCTIONS

$\dfrac{1}{x} = 1 - (x-1) + (x-1)^2 - (x-1)^3 + (x-1)^4 - \ldots + (-1)^n (x-1)^n + \ldots,\qquad 0 < x < 2$

$\dfrac{1}{1+x} = 1 - x + x^2 - x^3 + x^4 - x^5 + \ldots + (-1)^n x^n + \ldots,\qquad -1 < x < 1$

$\ln x = (x-1) - \dfrac{(x-1)^2}{2} + \dfrac{(x-1)^3}{3} - \dfrac{(x-1)^4}{4} + \ldots + \dfrac{(-1)^{n-1}(x-1)^n}{n} + \ldots,\qquad 0 < x \leq 2$

$e^x = 1 + x + \dfrac{x^2}{2!} + \dfrac{x^3}{3!} + \dfrac{x^4}{4!} + \dfrac{x^5}{5!} + \ldots + \dfrac{x^n}{n!} + \ldots,\qquad -\infty < x < \infty$

$\sin x = x - \dfrac{x^3}{3!} + \dfrac{x^5}{5!} - \dfrac{x^7}{7!} + \dfrac{x^9}{9!} - \ldots + \dfrac{(-1)^n x^{2n+1}}{(2n+1)!} + \ldots,\qquad -\infty < x < \infty$

$\cos x = x - \dfrac{x^2}{2!} + \dfrac{x^4}{4!} - \dfrac{x^6}{6!} + \dfrac{x^8}{8!} - \ldots + \dfrac{(-1)^n x^{2n}}{(2n)!} + \ldots,\qquad -\infty < x < \infty$

$\arctan x = x - \dfrac{x^3}{3} + \dfrac{x^5}{5} - \dfrac{x^7}{7} + \dfrac{x^9}{9} - \ldots + \dfrac{(-1)^n x^{2n+1}}{2n+1} + \ldots,\qquad -1 \leq x \leq 1$

$\arcsin x = x + \dfrac{x^3}{2 \cdot 3} + \dfrac{1 \cdot 3 x^5}{2 \cdot 4 \cdot 5} + \dfrac{1 \cdot 3 \cdot 5 x^7}{2 \cdot 4 \cdot 6 \cdot 7} + \ldots + \dfrac{(2n)! x^{2n+1}}{(2^n n!)^2 (2n+1)} + \ldots,\qquad -1 \leq x \leq 1$

$(1+x)^k = 1 + kx + \dfrac{k(k-1)x^2}{2!} + \dfrac{k(k-1)(k-2)x^3}{3!} + \dfrac{k(k-1)(k-2)(k-3)x^4}{4!} + \ldots,$
$\qquad\qquad -1 < x < 1$

$(1+x)^{-k} = 1 - kx + \dfrac{k(k+1)x^2}{2!} - \dfrac{k(k+1)(k+2)x^3}{3!} + \dfrac{k(k+1)(k+2)(k+3)x^4}{4!} - \ldots,$
$\qquad\qquad -1 < x < 1$

# TABLE OF MORE COMMON LAPLACE TRANSFORMS

| $f(t) = L^{-1}\{F(s)\}$ | $F(s) = L\{f(t)\}$ |
|---|---|
| $1$ | $\dfrac{1}{s}$ |
| $t$ | $\dfrac{1}{s^2}$ |
| $\dfrac{t^{n-1}}{(n-1)!}; n=1,2,\ldots$ | $\dfrac{1}{s^n}$ |
| $e^{at}$ | $\dfrac{1}{s-a}$ |
| $t\,e^{at}$ | $\dfrac{1}{(s-a)^2}$ |
| $\dfrac{t^{n-1}e^{-at}}{(n-1)!}$ | $\dfrac{1}{(s+a)^n}; n=1,2,\ldots$ |
| $\dfrac{e^{-at}-e^{-bt}}{b-a}; a\neq b$ | $\dfrac{1}{(s+a)(s+b)}$ |
| $\dfrac{a\,e^{-at}-b\,e^{-bt}}{a-b}; a\neq b$ | $\dfrac{s}{(s+a)(s+b)}$ |
| $\sin st$ | $\dfrac{a}{s^2+a^2}$ |
| $\cos at$ | $\dfrac{s}{s^2+a^2}$ |
| $\sinh at$ | $\dfrac{a}{s^2-a^2}$ |

| $f(t) = L^{-1}\{F(s)\}$ | $F(s) = L\{f(t)\}$ |
|---|---|
| $\cosh at$ | $\dfrac{s}{s^2 - a^2}$ |
| $\dfrac{1}{a^2}(1 - \cos at)$ | $\dfrac{1}{s(s^2 + a^2)}$ |
| $\dfrac{1}{a^3}(at - \sin at)$ | $\dfrac{1}{s(s^2 + a^2)}$ |
| $\dfrac{t}{2a}\sin at$ | $\dfrac{s}{(s^2 + a^2)^2}$ |
| $\dfrac{1}{b}e^{-at}\sin bt$ | $\dfrac{1}{(s+a)^2 + b^2}$ |
| $e^{-at}\cos bt$ | $\dfrac{s+a}{(s+a)^2 + b^2}$ |
| $h_1(t - a)$ | $\dfrac{1}{s}e^{-as}$ |
| $h_1(t) - h_1(t - a)$ | $\dfrac{1 - e^{-as}}{s}$ |
| $\dfrac{1}{t}\sin kt$ | $\arctan\dfrac{k}{s}$ |

# REA's Problem Solvers

The "PROBLEM SOLVERS" are comprehensive supplemental textbooks designed to save time in finding solutions to problems. Each "PROBLEM SOLVER" is the first of its kind ever produced in its field. It is the product of a massive effort to illustrate almost any imaginable problem in exceptional depth, detail, and clarity. Each problem is worked out in detail with a step-by-step solution, and the problems are arranged in order of complexity from elementary to advanced. Each book is fully indexed for locating problems rapidly.

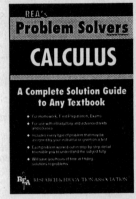

ACCOUNTING
ADVANCED CALCULUS
ALGEBRA & TRIGONOMETRY
AUTOMATIC CONTROL
   SYSTEMS/ROBOTICS
BIOLOGY
BUSINESS, ACCOUNTING, & FINANCE
CALCULUS
CHEMISTRY
COMPLEX VARIABLES
COMPUTER SCIENCE
DIFFERENTIAL EQUATIONS
ECONOMICS
ELECTRICAL MACHINES
ELECTRIC CIRCUITS
ELECTROMAGNETICS
ELECTRONIC COMMUNICATIONS
ELECTRONICS
FINITE & DISCRETE MATH
FLUID MECHANICS/DYNAMICS
GENETICS
GEOMETRY
HEAT TRANSFER
LINEAR ALGEBRA
MACHINE DESIGN
MATHEMATICS for ENGINEERS
MECHANICS
NUMERICAL ANALYSIS
OPERATIONS RESEARCH
OPTICS
ORGANIC CHEMISTRY
PHYSICAL CHEMISTRY
PHYSICS
PRE-CALCULUS
PROBABILITY
PSYCHOLOGY
STATISTICS
STRENGTH OF MATERIALS &
   MECHANICS OF SOLIDS
TECHNICAL DESIGN GRAPHICS
THERMODYNAMICS
TOPOLOGY
TRANSPORT PHENOMENA
VECTOR ANALYSIS

*If you would like more information about any of these books, complete the coupon below and return it to us or visit your local bookstore.*

---

**RESEARCH & EDUCATION ASSOCIATION**
61 Ethel Road W. • Piscataway, New Jersey 08854
Phone: (908) 819-8880

**Please send me more information about your Problem Solver Books**

Name _____

Address _____

City _____ State _____ Zip _____

# REA's Test Preps
## The Best in Test Preparation

- REA "Test Preps" are **far more** comprehensive than any other test preparation series
- Each book contains up to **eight** full-length practice exams based on the most recent exams
- **Every** type of question likely to be given on the exams is included
- Answers are accompanied by **full** and **detailed** explanations

*REA has published over 60 Test Preparation volumes in several series. They include:*

**Advanced Placement Exams (APs)**
Biology
Calculus AB & Calculus BC
Chemistry
Computer Science
English Language & Composition
English Literature & Composition
European History
Government & Politics
Physics
Psychology
Statistics
Spanish Language
United States History

**College-Level Examination Program (CLEP)**
American History I
Analysis & Interpretation of Literature
College Algebra
Freshman College Composition
General Examinations
General Examinations Review
Human Growth and Development
Introductory Sociology
Principles of Marketing
Spanish

**SAT II: Subject Tests**
American History
Biology
Chemistry
English Language Proficiency Test
French

**SAT II: Subject Tests (continued)**
German
Literature
Mathematics Level IC, IIC
Physics
Spanish
Writing

**Graduate Record Exams (GREs)**
Biology
Chemistry
Computer Science
Economics
Engineering
General
History
Literature in English
Mathematics
Physics
Political Science
Psychology
Sociology

**ACT** - American College Testing Assessment

**ASVAB** - Armed Services Vocational Aptitude Battery

**CBEST** - California Basic Educational Skills Test

**CDL** - Commercial Driver's License Exam

**CLAST** - College Level Academic Skills Test

**ELM** - Entry Level Mathematics

**ExCET** - Exam for Certification of Educators in Texas

**FE (EIT)** - Fundamentals of Engineering Exam

**FE Review** - Fundamentals of Engineering Review

**GED** - High School Equivalency Diploma Exam (US & Canadian editions)

**GMAT** - Graduate Management Admission Test

**LSAT** - Law School Admission Test

**MAT** - Miller Analogies Test

**MCAT** - Medical College Admission Test

**MSAT** - Multiple Subjects Assessment for Teachers

**NJ HSPT**- New Jersey High School Proficiency Test

**PPST** - Pre-Professional Skills Tests

**PRAXIS II/NTE** - Core Battery

**PSAT** - Preliminary Scholastic Assessment Test

**SAT I** - Reasoning Test

**SAT I** - Quick Study & Review

**TASP** - Texas Academic Skills Program

**TOEFL** - Test of English as a Foreign Language

---

**RESEARCH & EDUCATION ASSOCIATION**
61 Ethel Road W. • Piscataway, New Jersey 08854
Phone: (908) 819-8880

**Please send me more information about your Test Prep Books**

Name _____

Address _____

City _____ State _____ Zip _____

# "The ESSENTIALS" of Math & Science

Each book in the ESSENTIALS series offers all essential information of the field it covers. It summarizes what every textbook in the particular field must include, and is designed to help students in preparing for exams and doing homework. The ESSENTIALS are excellent supplements to any class text.

The ESSENTIALS are complete and concise with quick access to needed information. They serve as a handy reference source at all times. The ESSENTIALS are prepared with REA's customary concern for high professional quality and student needs.

## Available in the following titles:

Advanced Calculus I & II
Algebra & Trigonometry I & II
Anatomy & Physiology
Anthropology
Astronomy
Automatic Control Systems / Robotics I & II
Biology I & II
Boolean Algebra
Calculus I, II & III
Chemistry
Complex Variables I & II
Data Structures I & II
Differential Equations I & II
Electric Circuits I & II
Electromagnetics I & II
Electronics I & II
Electronic Communications I & II
Finite & Discrete Math
Fluid Mechanics / Dynamics I & II
Fourier Analysis
Geometry I & II
Group Theory I & II
Heat Transfer I & II
LaPlace Transforms
Linear Algebra
Math for Engineers I & II
Mechanics I, II & III
Microbiology
Modern Algebra
Numerical Analysis I & II
Organic Chemistry I & II
Physical Chemistry I & II
Physics I & II
Pre-Calculus
Probability
Psychology I & II
Real Variables
Set Theory
Statistics I & II
Strength of Materials & Mechanics of Solids I & II
Thermodynamics I & II
Topology
Transport Phenomena I & II
Vector Analysis

*If you would like more information about any of these books, complete the coupon below and return it to us or go to your local bookstore.*

---

RESEARCH & EDUCATION ASSOCIATION
61 Ethel Road W. • Piscataway, New Jersey 08854
Phone: (908) 819-8880

**Please send me more information about your Math & Science Essentials Books**

Name _____

Address _____

City _____ State _____ Zip _____